Pearson Education Test Prep Series for

AP® BIOLOGY

Fred W. Holtzclaw
Theresa Knapp Holtzclaw

To accompany:

PEARSON'S CAMPBELL BIOLOGY PROGRAMS

Boston Columbus Indianapolis New York San Francisco Upper Saddle River Amsterdam
Cape Town Dubai London Madrid Milan Munich Paris Montreal Toronto Delhi
Mexico City São Paulo Sydney Hong Kong Seoul Singapore Taipei Tokyo

Vice President/Editor-in-Chief: *Beth Wilbur*
Senior Acquisitions Editor: *Josh Frost*
Senior Editorial Manager: *Ginnie Simione Jutson*
Editorial Project Editor: *Katie Cook*
Executive Marketing Manager: *Lauren Harp*
Managing Editor, Production: *Michael Early*
Production Project Manager: *Jane Brundage*
Photo Editor: *Donna Kalal*
Image & Text Permissions Coordinators: *Alison Bruckner, Tom Wilcox*
Manufacturing Buyer: *Jeffrey Sargent*
Production Management and Composition: *S4Carlisle Publishing Services*
Cover Design Production: *Seventeenth Street Studios*
Text and Cover Printer: *Manufactured in the United States by LSC Communications*
Cover Photo Credits: (Left): *Martin Turner / Getty Images;* **(Right):** *Chris Hellier / Photo Researchers, Inc.*

19 2019

PearsonSchool.com/Advanced

ISBN 10: 0-13-345814-8
ISBN 13: 978-0-13-345814-5

Brief Contents

Photo credits:

Figure 2.10: Biophoto/Science Source
Figure 4.1: CNRI/Science Source
Figure 4.6a: Getty Images
Figure 4.6b: Anthony Loveday/Pearson Science
Figure 5.11: Charles Duguet/Science Source
Figure 5.15: Repligen Corporation

About Your Pearson AP Guide

Pearson Education is the leading education solution provider worldwide. With operations on every continent, we make it our business to understand the changing needs of students at every level, from kindergarten to college. We think that makes us especially qualified to offer this series of Test Prep Workbooks for AP, tied to some of our best-selling programs.

Our reasoning is that as you study for your course, you're preparing along the way for the AP exam. If you can connect the material in the book directly to the exam, it makes the material that much more relevant and enables you to focus your time most efficiently. And that's a good thing!

The AP exam is an important milestone in your education. A high score means you're in a better position for college acceptance and possibly puts you a step ahead with college credits. Our goal is to provide you with the tools you need to succeed.

Good luck!

Revisions to This Edition

Part I: *Introduction to the AP Biology Examination*

- The introduction aligns with the new Curriculum Framework (CF) launched in the 2012–2013 academic year.
- The new CF outline is thoroughly explained.
- The seven science practices are introduced.
- The description of the exam and the testing hints are revised to reflect the changes in the course, beginning with the 2013 exam.

Part II: *A Review of Topics with Sample Questions*

- **You Must Know** boxes are edited to reflect the new CF's change in emphasis.
- **What's Important to Know?** boxes are used for material that is no longer explicitly in the CF but may be used for illustrative examples and explanations.
- New boxes titled **Connect with the Curriculum Framework** target specific information about the new CF.
- Although information about the science practices is woven throughout the book, special boxes titled **Science Practices: Can You . . . ?** give students the opportunity to zoom in, make connections, and apply the science practices.
- Many new test questions that reflect the new exam are added to each topic, including questions that require interpretation of data and application of knowledge.
- Grid-in questions are included for each topic as well as revised free-response questions that align with released sample exams.
- Over 330 sample questions with explanations are included.

Part III: *The Laboratory*

- This section is heavily revised to reflect the new investigations released by the College Board. All 13 investigations are discussed, with hints, explanations, and sample questions. This section includes a strong link to inquiry and science practices.
- The CF expects students to be able to apply mathematics to a variety of topics. The formula sheet, provided at the time of the exam, is included in this edition, along with a number of tutorials and problems that take students through sample mathematical applications.
- More than 40 practice questions over lab topics with answers and explanations are included.

Part IV: *Sample Test*

- The sample test has been completely rewritten. Most items are new or significantly revised to accurately reflect the types of questions students may encounter at exam time.
- Grid-in questions are included in the sample test. Students will be expected to have a calculator for the new exam and provide numerical responses with a grid-in system.
- Free-response questions will be of varying lengths. The sample test follows this new format.

Part V: *Answers and Explanations*

- The format of this section remains similar to past editions, but some explanations stress connections to the curriculum the student must make, along with a description of pertinent knowledge. Many explanations describe the relevant science practice and its required application to correctly answer.

Part I

Introduction to the AP Biology Examination

This section gives an overview of the Advanced Placement* Program and the AP Biology Examination. Part I introduces the types of questions you will encounter on the exam, explains the procedures used to grade the exam, and provides helpful test-taking strategies. A correlation chart shows where in *Campbell Biology*, Tenth Edition, by Reece et al. and in *Campbell Biology in Focus* by Urry et al., you will find key information that commonly appears on the AP Biology Examination. Review Part I carefully before trying the sample test items in Part II and Part III.

The Advanced Placement Program*

You are probably reading this book because you are a student in an Advanced Placement (AP) Biology class, and you want to have help preparing for the exam. Let us be your guides. Fred and Theresa Holtzclaw have taught AP Biology and worked with the College Board in developing exams and served as graders of the exams for more than 25 years. We know the course well, have been involved with the changes that have been made in it, and have successfully prepared hundreds of our own students for the exam. If you study hard, our hints will give you an edge and you will be poised for success.

This book will help you in several important ways. The first part of this book introduces you to the AP Biology course and the AP Biology Exam. You'll learn helpful details about the different question formats—multiple-choice, grid-ins, and free-response—that you'll encounter on the exam. In addition, you'll find many test-taking strategies that will help you prepare for the exam. A correlation chart at the end of Part I shows how to use your textbook, *Campbell Biology*, Tenth Edition, by Reece et al. or *Campbell Biology in Focus* by Urry et al., to find the information you'll need to know to score well on the AP Biology Exam. By the way, this chart is also useful in helping to identify any extraneous material that won't be tested. Part II of this book contains a review of everything that you learned in your textbook that could be on the AP Biology test. Many questions will be posed along the way so that you can get used to being tested on the concepts in the way that the College Board will test you. The more familiar you are with the AP Biology Exam ahead of time, the more comfortable you'll be on testing day. In Part III, inquiry labs have been reviewed, including sample test questions, to check your understanding. In Part IV, there is a practice test for you to try on your own. Part V is where you will find answers and explanations to all the questions asked in this book.

The AP Program is sponsored by the College Board, a nonprofit organization that oversees college admissions examinations. The AP Program offers 34 different exams in college-level courses to qualified high school students. Over 3,000 colleges and universities around the world grant credit or placement to students who have performed well on AP exams. Each college or university decides whether or not to grant college credit for an AP course, and each bases this decision on what it considers satisfactory grades on AP exams. Depending on what college you attend and what area of study you pursue, your decision to take the AP Biology Exam could save you tuition money. You can contact schools directly to find out their guidelines for accepting AP credits, or use College Board's online feature, "AP Credit Policy Info" (www.collegeboard.com).

Why Take an AP Course?

You may be taking an AP course simply because you like challenging yourself and you are thirsty for knowledge. Another reason may be that you know that colleges look favorably on applicants who have AP courses on their secondary school transcripts. College admissions officers may see your willingness to take these courses as evidence of your work ethic and commitment to your education. Because AP course work is more difficult than average high school work, many admissions officers evaluate AP grades on a higher academic level. For example, if you receive a B in an AP class, it might carry the same weight as an A in a regular-level high school class.

Your AP Biology course prepares you for many of the skills you will need to succeed in college. For example, your teacher may assign a major research paper or require you to perform several challenging laboratory exercises using proper scientific protocol. AP Biology teachers routinely give substantial reading assignments, and students learn how to take detailed lecture notes and participate vigorously in class discussions. Students gain experience with science practices through inquiry and class activities designed to apply these skills. The AP Biology course will challenge you to gather and consider information in new—and sometimes unfamiliar—ways. You can feel good knowing that your ability to use these methods and skills will give you a leg up as you enter college.

Taking an AP Examination

The AP Biology Exam is given annually in May and the exams are scored in June. In July the results will be available online for you, your high school, and any colleges or universities you indicated on your answer sheet. Your AP teacher or school guidance counselor can give you information on how to register for an AP exam. Remember, the deadline for registration and payment of exam fees is usually in March, two months before the actual exam date in May. The cost of the exam is subject to change and can differ depending on the number of exams taken. However, in 2014 a single exam costs $89. For students who can show financial need, the College Board will reduce the price by $26, and your school might also waive its regular rebate of $8, so the lowest possible total price is usually $55. Some states pay the exam fee for the student. If you feel you may qualify for reduced rates, ask your school administrators for more information.

To learn more about all things related to AP courses and exams, visit this very helpful website: https://apstudent.collegeboard.org/home. You will always find the latest information here.

AP Biology: Course Goals

Beginning with the 2012–2013 school year, the AP Biology course was revised, and the May examination reflects a change in emphasis. Although you may have friends who took the AP exam in earlier years, their course and yours are not the same. We will help you be prepared for the current format.

The new course and exam focuses on enduring, conceptual understandings and the content that supports them. Factual recall is less important, and there is more emphasis on being able to use science practices. Knowing content is not enough! You must work on gaining skills all year long. The goal is to help you develop the science skills necessary to be prepared for the study of advanced topics in college science courses.

The increased emphasis on science practices will help you develop advanced inquiry and reasoning skills. You should be able to design plans for collecting data, analyze data, make predictions, apply mathematical routines, and connect concepts across domains (levels of organization). Throughout this guide, we will make references to these science practices. Let's be clear: You will still need to master content! And then you will apply your newfound knowledge.

In the revised AP Biology course, content, inquiry, and reasoning are equally important. The exam will be based on a series of Learning Objectives (LOs), and each LO combines content and science practices. Here is an abbreviated list of the Science Practices from pages 97–102 of the College Board's publication *Course and Exam Description 2012.*

Science Practices for AP Biology

Source: AP Biology—Course and Exam Description. © 2012. The College Board. www.collegeboard.org. Reproduced with permission.

Science Practice 1: The student can use representations and models to communicate scientific phenomena and solve scientific problems.
Science Practice 2: The student can use mathematics appropriately.
Science Practice 3: The student can engage in scientific questioning to extend thinking or to guide investigations within the context of the AP course.
Science Practice 4: The student can plan and implement data collection strategies appropriate to a particular scientific question.
Science Practice 5: The student can perform data analysis and evaluation of evidence.
Science Practice 6: The student can work with scientific explanations and theories.
Science Practice 7: The student is able to connect and relate knowledge across various scales, concepts, and representations in and across domains.

Overview of the Course

The key concepts and related content of the revised AP Biology course and exam are organized around four underlying principles called **Big Ideas**. These encompass the core scientific principles, theories, and processes governing living organisms and biological systems.

Each Big Idea has several **Enduring Understandings**, which are the fundamental concepts you should retain from your course. Each Enduring Understanding has statements of **Essential Knowledge** that you should know. All of the details in the outline are required elements of the course and may be included in the AP Biology Exam. Finally, your teacher may elect to help you learn about biology this year by selecting from many possible **Illustrative Examples**. This is what will make each course unique and allow your teacher to teach a rich course where the teacher may select unique examples of his or her own choosing. For this reason, this guide may cover content that your teacher has not have selected. You will need to pick and choose through these areas, realizing that there are many ways a teacher may elect to teach this course.

We took the Curriculum Framework (all 91 pages) and summarized it in an outline that correlates with the concepts in Campbell Biology, Tenth Edition and Campbell Biology in Focus. These outlines are found at the front of Campbell Biology, Tenth Edition (p. vi), and also Campbell Biology in Focus (p. ix). Expanded outlines that include Illustrative Examples are available at the Pearson website (www.PearsonSchool.com/AdvancedCorrelations).

It is important for you to know that this outline gives you an idea of the topics that will be covered on the AP exam in May, but the new course will also emphasize Science Practices linked to this content. Facts will not be sufficient! You will need to work with the Science Practices all year long to do well on the exam.

CONNECT WITH THE CURRICULUM FRAMEWORK

Go to page (21–26) to see a brief summary of the Big Ideas and included topics as well as details on the Science Practices. These are all things you should know and be able to do!

Understanding the AP Biology Examination

The AP Biology Exam takes 3 hours and includes both a 90-minute multiple-choice section and a 90-minute free-response (essay) section that begins with a 10-minute reading period. The multiple-choice section will be one-half of your exam grade, and the free-response section will account for the other half. Both sections include questions that assess students' understanding of the Big Ideas, Enduring Understandings, and Essential Knowledge statements. The exam probably looks like many other tests you've taken. At the core of the examination are questions designed to measure your knowledge and understanding of modern biology. You should be prepared to recall basic facts and concepts, to apply scientific facts and concepts to particular problems, to synthesize facts and concepts, and to demonstrate reasoning and analytical skills by organizing written answers to broad questions. You will hear us repeat this throughout the book . . . you must be able to *apply* your knowledge. We will give you many chances to practice this in this book.

The AP Biology Exam is very challenging. When you sit down to take the test, you are expected not only to be fluent in the areas of biology that you find fascinating (the ones that probably inspired you to take a special interest in the subject originally), but also to have an intimate knowledge of topics you may find less interesting. Whatever those topics might be—DNA replication, the dizzying details of gene expression or the immune system—you need to be comfortable with and knowledgeable about all of the AP Biology topics.

Section I, Part A: Multiple-Choice Questions

The first part of the exam consists of 63 standard multiple-choice questions and 6 questions that involve mathematical calculation for which you will grid in a response. You will have 90 minutes to complete Section I. Each of the multiple-choice questions will be directly paired with a Learning Objective from the Curriculum Framework. The questions require both an understanding of important concepts and biological processes and the ability to apply information that is given to you in the question. Because of this, the stem of the questions may be longer than many multiple-choice questions you have seen before. Since the questions require a deep conceptual understanding, rather than the recall of facts, the number of questions was reduced from 100 items in the years prior to 2013 to 63 multiple-choice and 6 grid-in questions today. This portion of the exam is followed by a 5–10 minute break—the only official break during the examination. The following chart summarizes this information.

Section	Question Type	Number of Questions	Timing
I	Part A: Multiple Choice	63	90 minutes
	Part B: Grid-In	6	
II	Long Free Response	2	80 minutes + 10-minute reading period
	Short Free Response	6	

Source: AP Biology—Course and Exam Description. © 2012. The College Board. www.collegeboard.org. Reproduced with permission.

Here is a sample item from page 152 of the College Board's *Course and Exam Description 2012.*

A human kidney filters about 200 liters of blood each day. Approximately two liters of liquid and nutrient waste are excreted as urine. The remaining fluid and dissolved substances are reabsorbed and continue to circulate throughout the body. Antidiuretic hormone (ADH) is secreted in response to reduced plasma volume. ADH targets the collecting ducts in the kidney, stimulating the insertion of aquaporins into their plasma membranes and an increased reabsorption of water.

If ADH secretion is inhibited, which of the following would initially result?

(A) The number of aquaporins would increase in response to the inhibition of ADH.
(B) The person would decrease oral water intake to compensate for the inhibition of ADH.
(C) Blood filtration would increase to compensate for the lack of aquaporins.
(D) The person would produce greater amounts of dilute urine.

Essential Knowledge	3.D.3: Signal transduction pathways link signal reception with cellular response.
Science Practice	1.5: The student can re-express key elements of natural phenomena across multiple representations in the domain.
Learning Objective	3.36: The student is able to describe a model that expresses the key elements of signal transduction pathways by which a signal is converted to a cellular response.

Your course may not have included a study of kidney function, but all the information you need to answer this item is given in the stem. You are told what ADH does and how it is controlled and given the role of aquaporins. It is expected that you can use this information to select the correct answer. If you are certain you know the answer, fill in the corresponding oval on the answer sheet. However, what if you're not certain? The next step is to see if you can eliminate one or more of the choices.

> ***TIP FROM THE READERS***
> In the past, there was a penalty for guessing, but that has been eliminated. *Answer every question!*

Let's look again at the question. The correct answer is D. Choice *A* suggests a response when ADH is inhibited. You must know that this means there is less ADH. The stem tells you that ADH "stimulates the insertion of aquaporins" so you will eliminate this choice. Continue to methodically analyze the other choices, and try to eliminate another choice based on your understanding of the information that is given in the stem. Confidently mark your answer sheet and continue. Remember, answer *every* question!

Lab-Based or Experimental Questions

Another type of question you will see in Section I is the lab-based or experimental question. These questions either present you with a set of data in graph (or other) form, or they describe an experiment and ask you to make predictions, select appropriate data, form hypotheses, analyze data mathematically, and perform other science practices. These questions often occur in groups that use the same data set.

It is known that plant cells require oxygen in order to obtain ATP. Those who work with plants have long known that it is possible to quickly kill a plant by overwatering. The graph below shows the results of a study of the effect of soil air spaces on plant growth.

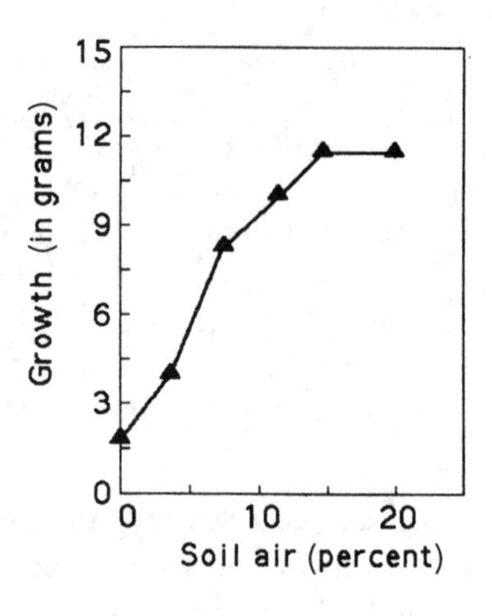

1. The data from the above graph show that the plant
   (A) grows fastest when the soil is 5–10% air.
   (B) grows fastest when the soil is 15–20% air.
   (C) grows at the same rate regardless of the soil air percentage.
   (D) grows most slowly when the soil is 5–10% air.

The correct choice is *A*. The graph shows the line with the greatest slope (the highest degree of change over the shortest amount of time) between the percentages 5 and 10. During this time, the plant grows by about $9 - 5 = 4$ grams. Just to be sure, check the amount this plant grows when the soil is 15–20% air. At the start, when the air was 15% air, the plant weighed 12 grams. At the end, when the soil is 20% air, the plant weight is the same—12 grams. Virtually no growth occurred during this time.

Clearly this question requires you to be able to interpret a graph, but at this point in your biology education you should be quite capable of doing that. In order to brush up on the various ways that graphs present information, you might review Appendix B of the *AP Biology Investigation Manual.*

2. It is seen from the graph that plant growth is negatively affected by decreased soil oxygen. Which of the following statements best justifies the reason for this effect?
   (A) Plant root cells require oxygen because no photosynthesis occurs underground.
   (B) Plant stem and leaf cells are able to do photosynthesis at a higher rate when there is more oxygen in the soil.
   (C) Water in the soil eliminates oxygen from the soil spaces, so the root cells are unable to produce ATP.
   (D) The rate of cellular respiration is decreased under conditions of high soil moisture.

The correct choice is *C*. This question requires you to recall the reactants and products of photosynthesis and cellular respiration. *All* plant cells do cellular respiration, consuming O_2 in order to produce ATP. Choice *C* justifies (explains) why plant growth is low when there is low soil O_2. Choice *D* restates the information from the graph. Although it is a true statement, it does not justify the reason for this correlation.

Section I, Part B: Grid-In Questions

These questions will require you to calculate an answer for the question and enter it in a grid in that section on your answer sheet, as shown on the next page. Be sure to practice gridding responses correctly prior to the exam. Let's take a look at some sample grid-in responses.

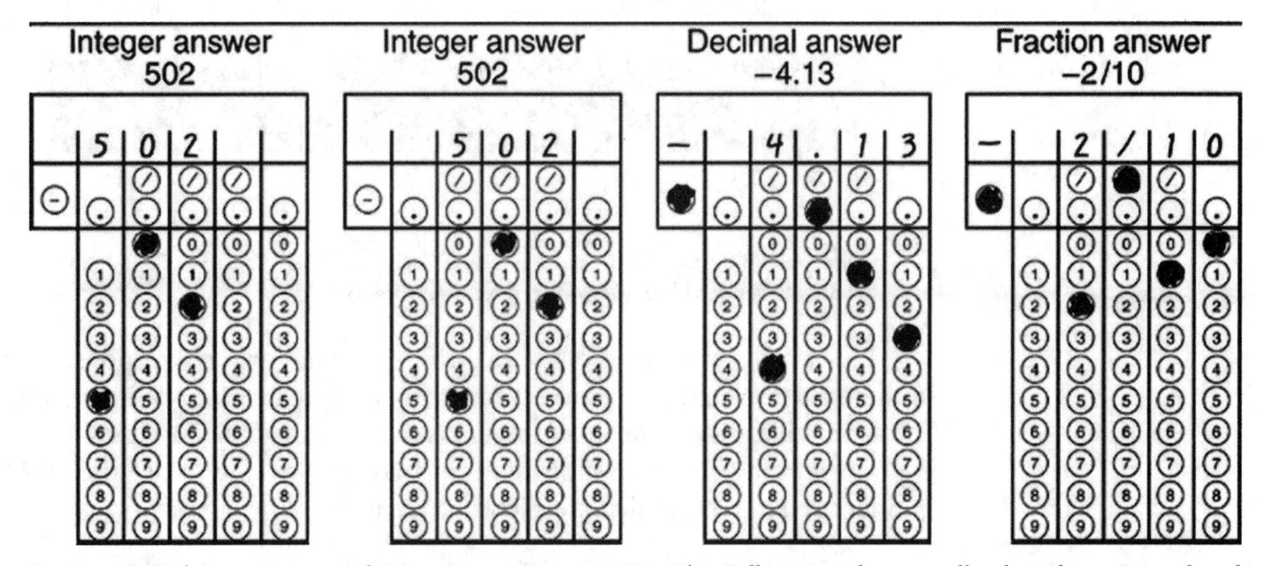

Source: AP Biology—Course and Exam Description. © 2012. The College Board. www.collegeboard.org. Reproduced with permission.

Notice that you can bubble your response in any of the boxes, such as is shown in the samples for 502. Start your answer in any column, space permitting. Unused columns should be left blank. You can see that you are able to indicate negative numbers, fractions, and the location of decimal points. Be sure to follow instructions for the calculation you are to make, and express your answer with the precision that is specified, such as "round to the nearest tenth." If the answer results in a negative number, remember to bubble the minus (–) sign. The final sample grid-in shows you how to express a fractional answer. Fractions only need to be reduced enough to fit in the grid, so 5/10 is acceptable. One detail here is that you should not use mixed numbers such as "3 ¾" because it would be scored as "3 3/4."

You will be able to use a four-function calculator during the exam as well as a formula sheet (Appendix B of the *Investigation Manual*), because the emphasis here is on your ability to apply mathematical techniques. Now, try your hand at this sample problem.

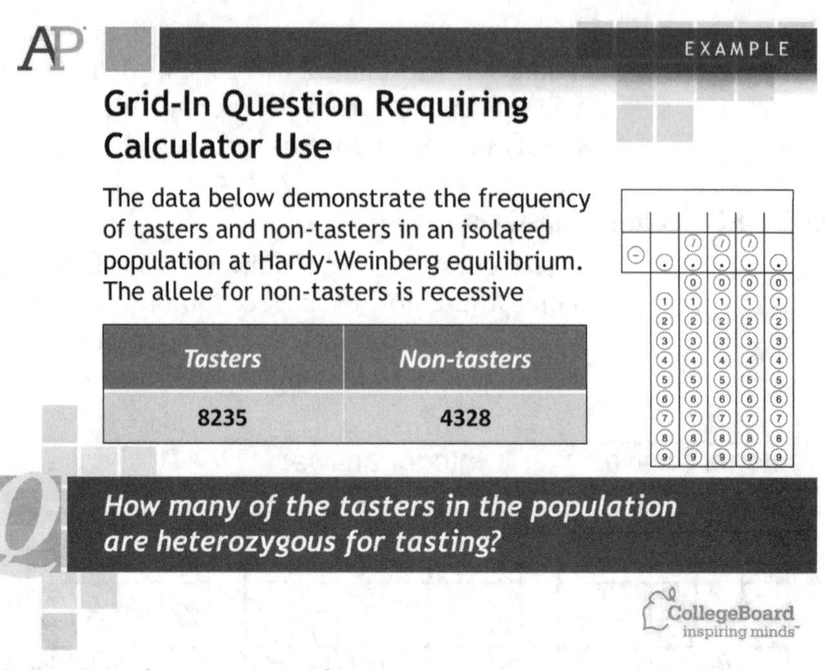

Source: AP Biology—Course and Exam Description. © 2012. The College Board. www.collegeboard.org. Reproduced with permission.

To set your mind at ease, the electronic scoring for these responses is set so there may be a range of correct scores to allow for variations in rounding. The acceptable answer for this problem is in the range of 6030–6156, depending on rounding. This question requires that you be able to use the Hardy-Weinberg equation to calculate allelic frequencies.

Section II: Free-Response Questions

At the beginning of Section II of the AP Biology Exam, you will be given a 10-minute period where you may read the questions and plan and outline your responses before you may begin writing. You should use this time wisely to read and then reread each question. Underline key words or phrases such as "Select *two*" or "*Explain* and *predict*" or "Using an example from *each* domain." Use key words to prepare an outline of your response to each of the 10-point questions if time allows. If a graph will be required, decide what type is appropriate, and what will go on each axis. During the planning time, you can jot notes on some blank pages, but the instructions are very clear that anything written here will not be graded. Do NOT begin writing on the lined pages until the proctor tells you to do so.

Section II of the AP Biology Exam has eight free-response questions. There are two 10-point questions, three questions worth 4 points each, and three questions worth 3 points each. The long questions will consist of several sections or related tasks and will be evaluated on a 10-point scale. You should allow approximately 20 minutes for a long free-response question.

The 3-point and 4-point questions call for briefer responses. The instructions may say something like "In a sentence or two. . . ." Heed these parameters to use your time wisely. The short free-response questions should be completed in about 6 minutes or so. For *all* free-response questions, you must write your answers out, and you cannot use an outline form. Let's look at an example. What follows will give you an idea about how these questions may appear in your exam booklet.

BIOLOGY

Section II

Time—10 minutes to plan, 80 minutes to write

Directions: Answer all questions.

At a Glance

Total Time
1 hour, 30 minutes
Number of Questions
8
Percent of Total Score
50%
Writing Instrument
Pen with black or dark blue ink
Electronic Device
Four-function calculator (with square root)

Reading Period

Time
10 minutes. Use this time to read the questions and plan your answers.

Writing Period

Time
1 hour, 20 minutes
Suggested Time
Approximately 22 minutes per long question, and 6 minutes per short question.
Weight
Approximate weights
Questions 1 and 2: 25% each
Questions 3–5: 10% each
Questions 6–8: 7% each

IMPORTANT Identification Information

PLEASE PRINT WITH PEN:

1. First two letters of your last name ☐☐

   First letter of your first name ☐

2. Date of birth

   ☐☐ ☐ ☐☐☐☐
   Month Day Year

3. Six-digit school code

   ☐☐☐☐☐☐

4. Unless I check the box below, I grant the College Board the unlimited right to use, reproduce, and publish my free-response materials, both written and oral, for educational research and instructional purposes. My name and the name of my school will not be used in any way in connection with my free-response materials. I understand that I am free to mark "No" with no effect on my score or its reporting.

   No, I do not grant the College Board these rights. ☐

Instructions

The questions for Section II are printed in this booklet. You may use the unlined pages to organize your answers and for scratch work, but you must write your answers on the labeled pages provided for each question.

The proctor will announce the beginning and end of the reading period. You are advised to spend the 10-minute period reading all the questions, and to use the unlined pages to sketch graphs, make notes, and plan your answers. The focus of the reading period should be the organization of questions 1 and 2. Do not begin writing on the lined pages until the proctor tells you to do so.

Each answer should be written in paragraph form; an outline or bulleted list alone is not acceptable. Do not spend time restating the questions or providing more than the number of examples called for. For instance, if a question calls for two examples, you can earn credit only for the first two examples that you provide. Labeled diagrams may be used to supplement discussion, but unless specifically called for by the question, a diagram alone will not receive credit. Write clearly and legibly. Begin each answer on a new page. Do not skip lines. Cross out any errors you make; crossed-out work will not be scored.

Manage your time carefully. You may proceed freely from one question to the next. You may review your responses if you finish before the end of the exam is announced.

Here is a sample question for Section II.

1. Water comprises roughly 70% of the human body; cells are roughly 70–95% water, and water covers about three-quarters of the Earth's surface.

   (a) **Describe** the major physical properties of water that make it unique from other liquids.

   (b) **Explain** the properties of water that enable it to travel up through the roots and stems of plants to reach the leaves.

(c) **Explain** why the temperature of the oceans can remain relatively stable and support vast quantities of both plant and animal life, when air temperature fluctuates so significantly throughout the year.

Like many free-response questions on the AP Biology Exam, this sample is broken into three distinct parts. Each contains a clear directive. In fact, they are printed in boldface to help you focus on exactly how you should answer the question. First you will need to explain the uniqueness of water by describing its major physical properties. (In your response to this first part of the question, you might wish to include a labeled diagram of the structure of water, complete with electrons and bonds.) Then you must explain the properties of water that allow it to travel from root to leaf. Finally, you should explain the reason(s) why ocean water temperature remains stable and supports plant and animal life—even in the face of great air temperature variations. Of course, limiting your answer by addressing exactly what the question asks will make writing the essay easier for you and earn you a higher score. Always take the time to determine precisely what is being asked before you begin to formulate a concrete thesis and focus on writing your relevant supporting paragraphs.

Grading Procedures for the AP Biology Examination

The raw scores of the AP Biology Examination are converted to the following 5-point scale:

5—Extremely Well Qualified
4—Well Qualified
3—Qualified
2—Possibly Qualified
1—No Recommendation

Some colleges give undergraduate course credit to students who achieve scores of 3 or better on AP exams. Other colleges require students to achieve scores of 4 or 5. You may check the policy for individual colleges on the College Board website (www.collegeboard.com). Here you can also get information about student performance in prior years. Below is a breakdown of how the grading of the AP Biology Exam works.

Section I: Multiple-Choice Questions

The multiple-choice section of the exam is worth 50% of your total score. The raw score of Section I is determined by crediting one point for each correctly answered question. Therefore, it is important to answer each question. *There is no penalty for guessing!*

Section II: Free-Response Questions

Section II counts for 50% of your examination grade. The free-response section is scored by several hundred faculty consultants, including high school teachers and college instructors from all over the country who work in a central location to grade the essays. This period of scoring exams is called the "Reading." To ensure that scoring of all exams is consistent, grading rubrics, or standards, are developed and then faculty consultants are trained in their application. Because of this intense training, group discussion, and supervision, your essay should receive the same score regardless of who reads it. Ongoing internal checks during the Reading ensure this. Each of your essays will be evaluated by a faculty consultant trained to score that single response.

Your answers to the free-response questions must be presented in essay form. Outlines or unlabeled and unexplained diagrams are not given credit. Your performance on any single essay is evaluated independently of the other essays. Do not assume that information provided in one question will be considered during the grading of another essay. You should repeat information from question to question if it is necessary to illustrate your point.

Test-Taking Strategies for the AP Biology Examination

Here are a few tips for preparing in the weeks leading up to the examination.

- The earlier you start studying for the AP Biology Exam, the better. Some students use this AP Biology prep book along with their textbook throughout the course, taking notes in the margin to supplement their teacher's lectures. You should definitely begin serious preparation for the test at least one month in advance.
- Each topic in Part II is correlated to *Campbell Biology*, Tenth Edition, by Reece et al. and *Campbell Focus on Biology* by Urry et al. For each topic, review your lecture notes, study the figures in your text that explain key concepts, and then make your way through the corresponding section of Part II of this book. If possible, retake your own unit test on the topic, and also answer the questions in this guide for each unit. This will help you identify areas that will require further study. Do not try to reread your text; use it as a tool for those topics that need further study. Pace yourself!

AP Review: Lab Essays

The College Board has published a manual with 13 recommended inquiry labs. It is expected that you will complete two labs for each Big Idea during the course of the school year. Your teacher may elect to use some of the labs from the College Board's *AP Biology Investigative Labs*, or others of his or her choosing. All the laboratories will emphasize the science practices, and you will become more proficient in working and thinking like a scientist only by engaging in this type of work. These labs will often run over multiple days or even weeks and will include components where you learn new techniques and then use them to answer your own experimental questions.

Here is a list of the labs in the College Board's *AP Biology Investigative Labs.* An asterisk is placed by those labs that use techniques or cover concepts many teachers are familiar with from the "old" *AP Biology Laboratory Manual.*

Big Idea 1: Evolution

Investigation 1: Artificial Selection
Investigation 2: Mathematical Modeling: Hardy-Weinberg
Investigation 3: Comparing DNA Sequences to Understand Evolutionary Relationships with BLAST

Big Idea 2: Cellular Processes: Energy and Communication

Investigation 4: Diffusion and Osmosis*
Investigation 5: Photosynthesis*
Investigation 6: Cellular Respiration*

Big Idea 3: Genetics and Information Transfer

Investigation 7: Cell Division: Mitosis and Meiosis*
Investigation 8: Biotechnology: Bacterial Transformation*
Investigation 9: Biotechnology: Restriction Enzyme Analysis of DNA*

Big Idea 4: Interactions

Investigation 10: Energy Dynamics
Investigation 11: Transpiration*
Investigation 12: Fruit Fly Behavior*
Investigation 13: Enzyme Activity*

It is likely that at least one essay on the exam will be based on an AP investigation, although it may not be one that you have done. Not to fear! Because the questions are based on objectives for the course, your teacher will have prepared you to work with data from a variety of sources, so you will simply need to apply your expertise in scientific thinking acquired over the course to a novel situation. If you have done a similar experiment, this may be helpful—but be sure you read carefully and do not anticipate what is being asked.

You may be asked to "design an experiment to determine . . ." If you performed a lab in your AP class that would answer this question, it is fine to describe this lab. There are a number of items that would generally be included in your design, so consider including each of these:

- **State a hypothesis** as an "**If** . . . (conditions), **then** . . . (results)" statement. Your hypothesis must be testable.
- **Identify the variable factor** for the experiment (e.g., temperature).
- **Identify a control**. You must explain the control for the experiment.
- **Hold all other variables constant**. Explain how you would do this.

- **Manipulate the variable** (for example, one group at 10°C, one at 20°C, and one at 30°C).
- **Measure the results** (for example, cm grown, mass increase in grams).
- **Discuss results expected** as related to hypothesis.
- **Replication or verification**. The experiment must be repeated or large sample sizes must be used.

If appropriate, you could also consider using statistical analysis of data and review of the literature.

Graphing Data

It is likely that you will be asked to graph data. You will need to consider the type of graph that is appropriate for your data. Bar graphs are used when data points are discrete, that is, not related to each other, such as the number of girls in AP Biology vs. the number of boys in AP Biology. Line graphs are used when the data are continuous, such as the change in an individual's height at each birthday. Consider if there is a data point at 0 on the graph. Be sure to extend your line to 0 if there is, but do not take the line to 0 if there is no measurement for that data point. Also,

- Label the graph with a descriptive title.
- Label the *x* and *y* axes. Include units (such as "Time in minutes"). Be sure you know which variable is independent and which is dependent.
- Keep all measurement units constant. Each division on the graph must be a unit equal to all the others.
- If you are asked to draw a line to predict what would be shown if some change occurred, be sure that you include a legend for the second line.

Sample Tests

When you are ready to check your preparation, take the sample exam in Part III of this book. Keep track of your time, and try to simulate test conditions. Circle the items you get wrong or could not answer, and keep a list of the subject matter of those questions. The Answers and Explanations section will help you address any errors you have made, so be sure to spend time reviewing the helpful information there. Analyze the items missed to look for patterns—are you having a hard time with the grid-ins or the process of photosynthesis specifically? Spend the next week or so studying the topics in which you are weak. Spend the final days before the test looking through this guide, your class notes, and your textbook to fill in any remaining gaps.

The Day of the Exam

If you have followed this suggested study plan, you should feel well prepared by test day. Plan your schedule so that you get two very good nights of uninterrupted sleep before exam day. The night before the exam, relax, think positive thoughts, and focus on getting a good night's rest. Like some endurance races we have run, success will go to the strong, not necessarily the swift. Not sure about the analogy, but the idea is that the AP Biology Exam is a grueling mental exercise, so you need to be rested and well nourished. A little more cramming will *not* help! Below is a brief list of basic tips and strategies to think about before you arrive at the exam site.

1. **Arrive early!** It's a good idea to arrive at the exam site 30 minutes before the start time. On the day of the exam, make sure that you eat a good, nutritious meal. These tips may sound corny or obvious, but your body must be in peak form in order for your brain to perform well. Remember, you are going to need ATP to fuel brain cells at peak efficiency for more than 3 hours.
2. **Bring a photo ID.** (It's essential if you are taking the exam at a school other than your own.) Carrying a driver's license or a student ID card will allow you to prove your identity.
3. **Bring at least two sharpened #2 pencils** for the multiple-choice section. Also, bring a clean pencil eraser with you. Many pencils today have cheap erasers that smudge. Invest in a good eraser. The machine that scores Section I of the exam recognizes only marks made by a #2 pencil. Poorly erased responses are often scored incorrectly.
4. **Bring two black ballpoint pens** for the free-response portion of the test. Felt-tip pens run and pencils and inks of other colors are harder to read.
5. **Bring a watch** with you to the exam. Although many testing rooms have clocks, having your own watch makes it easy to keep close track of your own pace. Watches with calculators or alarms are not permitted in the exam room.
6. **Bring a four-function calculator with square root** (or your school may arrange to have them available). Our own students are so accustomed to their graphing calculators that we pull these out the month before the exam and have them get used to them. The small buttons and different positions for functions may take some practice, and you don't want to be figuring this out during your precious exam minutes. Practice early!

Forbidden Items and Rules of Conduct!

Some items that are forbidden from the testing room are books, notes, laptops, beepers, cameras, and portable listening or recording devices. If you bring a cellular phone with you, be prepared to turn it off and to give it to the test proctor until you are finished with your exam. For a complete list of what not to bring, see the College Board website.

Educational Testing Service prohibits the objects listed above in the interest of fairness to all test-takers. Similarly, the test administrators are very clear and very serious about what types of conduct are not allowed during the examination. Below is a list of actions to avoid at all costs because each can result in your immediate dismissal from the exam room.

- Do not consult any outside materials during the 3 hours of the exam period. Remember, the break is technically part of the exam—you are not free to review any materials at that time either.
- Do not speak during the exam. If you have a question for the test proctor, raise your hand to get the proctor's attention.

- When you are told to stop working on a section of the exam, you must stop immediately. (*Important hint:* The grid-ins are included in Part I of the exam, so finish them before your break!)
- Do not open your exam booklet before the test begins.
- Never tear a page out of your test booklet or try to remove the exam from the test room.
- Do not behave disruptively—even if you're distressed about a difficult test question or because you've run out of time. Stay calm and make no unnecessary noise.

Section I: Strategies for Multiple-Choice Questions

Obviously, having a firm grasp of biology is the key to doing well on the AP Biology Examination. In addition, being well informed about the exam itself increases your chances of achieving a high score. Below is a list of strategies that you can use to increase your comfort, your confidence, and your chances of excelling on the multiple-choice section of the exam.

- Become as familiar as possible with the format of Section I. The more comfortable you are with the multiple-choice format and with the kinds of questions you'll encounter, the easier the exam will be. Remember, Part II and Part III of this book provide you with invaluable practice on the kinds of multiple-choice questions you will encounter on the AP Biology Exam.
- Every question you answer correctly is a point, so pacing is important. If you have done all the suggested practice tests, you should have a good sense of how to pace yourself. You will have 90 minutes to answer 63 multiple-choice and 6 grid-in questions (about 78 seconds per question). Keep track of time!
- Some of the questions will require calculations or may have complicated data sets. If you encounter a question that will require extra time, leave it blank and make a note. Your goal should be to reach the end of the test, picking up all the points from questions you can answer easily.
- Our students tell us that pacing is very important! They say to never spend too long on an item and then not have time to finish. They report this as their most common error.
- The test is organized with three types of questions: standard multiple-choice, lab sets, and grid-ins. Lab sets are generally the most tedious. When a data table or graph is presented, proceed directly to the related questions. Determine what information is needed to answer the questions, and then return to the data table or graph and seek the information. Sometimes, although the data appear daunting, the questions are actually very easy. Practice doing this!
- Make a light mark in your test booklet next to any questions you can't answer. Return to these questions after you reach the end of Section I. Sometimes questions that appear later in the test will refresh your memory on a particular topic, and you will be able to answer one or more of those earlier questions.
- Always read the entire question carefully, and underline key words or ideas. You might wish to circle words such as NOT or EXCEPT in multiple-choice questions.

- Read each and every one of the answer choices carefully before you make your final selection. When we give our students practice exams, the mistake that troubles them most is when the question is missed because they did not read or mark carefully. Don't let this happen to you!
- Use the process of elimination to help you arrive at the correct answer. Even if you are quite sure of an answer, cross out the letters of incorrect choices in your test booklet as you eliminate them. This cuts down on the incorrect choices and allows you to narrow the remaining choices even further.
- Become completely familiar with the instructions for the multiple-choice questions before you take the exam. By knowing the instructions cold, you'll save yourself the time of reading them carefully on exam day.
- If you finish early, you should go back over as many items as possible to catch any careless errors. This is not the time to second-guess your responses but to be sure you marked the answer you intended and did not misread a question or choice.

Section II: Strategies for Free-Response Questions

Below is a list of strategies that you can use to increase your chances of excelling on the free-response section of the exam.

- You will have a 10-minute period to review the essay questions and make notes before you begin to write your graded responses. During this time, you should organize your thoughts and outline your essays in the space provided. Since the exam now covers eight different questions, our students suggest we advise you to focus primarily on the first two 10-point questions. They tell us that it is almost impossible to work through eight different questions during the 10-minute pre-read time, so they suggest being well-prepared for the longer essays. After the preparation time, you should record your answer on the pages provided for each question in the response book. You have about 20 minutes to spend on each 10-point essay and about 6 minutes for each of the short questions.
- *Read the question; then read the question again!* Be sure you answer the question that is asked and that you address each part of the question. As you read a question, underline any directive words (usually the first word in an essay) that indicate how you should answer and focus the material in your essay. Some of the most frequently used directives on the AP Biology Exam are listed below, along with descriptions of what you need to do in your writing to answer the question.
    - *Analyze* (show relationships between events; explain)
    - *Compare* (discuss similarities and differences between groups; be sure to address *each* group!)
    - *Contrast* (discuss points of difference or divergence between two or more things)
    - *Describe* (give a detailed account)
    - *Design* (create an experiment and convey its ideas)
    - *Predict* (tell what you expect to happen when conditions change; should be scientifically plausible)
    - *Justify* (give a scientific basis for why a response is reasonable)

- *Explain* **This is an important term that has very specific requirements!** Here is what you need to include:

    1. Make a *claim* (an assertion or conclusion)
    2. Provide *evidence* (scientific data that support the claim)
    3. Give a *justification* (your reasoning that links the evidence to the claim)

  In the grading of the AP Biology Exam, we can tell you that it is *essential* to include all three components of a scientific explanation.

- Reread the question as many times as necessary to make sure that you will cover each aspect of the topic. Free-response questions frequently have several parts, so you will need to take this into account as you outline your ideas.
- Write an essay! As the exam states clearly in the directions to free-response questions, a diagram or graph by itself is never an acceptable way to answer a free-response question. However, you should think about whether you could use a labeled diagram or graph to develop your written answer in some useful way.
- The essay you craft for this exam is not the same type of essay you should write for an English course. Yes, it should be well-organized; however, introductory sentences and conclusions are absolutely not necessary. Readers are interested in what you know and how well you express your knowledge. Spend your time packing the essay with the biological information you have worked so hard to learn.
- If the question has several parts, answer the parts in the sequence given. We suggest that you skip a line between sections (a, b, c, etc.) so it is very clear that you have addressed each part of the question. Use a letter or some other indication for each part so that the faculty consultant does not overlook a section of your response.
- If you are asked to perform a calculation, be sure to show the steps used to arrive at your answer. You have heard this before: Show your work!
- If you cannot remember a specific term, describe the structure or process.
- Define any scientific term you use that is directly related to your response. For example, if you discuss hydrogen bonding and how it relates to properties of water, be sure to explain what hydrogen bonds are, and then describe or define adhesion, cohesion, and so on.
- Your handwriting can affect your results. Although faculty consultants make every attempt to read each essay, sometimes it is impossible to decipher messy handwriting. When your handwriting is poor, the reader may lose concentration or patience and miss an important word or phrase.
- Don't leave any part of any essay blank. Every essay point is worth approximately 1.5 times as much as each multiple-choice point.
- If time allows, proofread your essays. Don't worry about crossing out material—readers understand that your responses are first drafts and that you are writing down ideas under the pressure of time. A single line through the material you wish to omit is sufficient. However, anything that you cross out will not be graded!

A Sample Free-Response Question and How to Approach It

The success of your free-response essays will depend a great deal on how clearly and extensively you answer the questions posed. Of course, the structure of your essays will depend entirely on your knowledge of the subjects at hand. Take a look at an example of a free-response question below.

1. In cancer, the cell's reproductive machinery experiences a loss of control that makes cancer cells reproduce continually, and eventually form a tumor.
   (A) **Describe** three DNA-related cellular events that could lead to the loss of cell division control that contributes to cancer.
   (B) **Describe two** ways in which tumors can disrupt normal function of the body.
   (C) **Predict** two cell processes that could be targeted to find a cure for cancer and include a biological explanation for each.

In order to answer this question, you should isolate exactly what it is that you must answer. You may want to underline the relevant information in the question to remind yourself of your focus. *Circle* the number of examples you are to use in any response because this is all the reader will give you credit for doing. Go with your best choices only! To answer part (a), you'll need to identify three appropriate DNA-related cellular events. Here are three likely choices:

1. A mutation or change in the original DNA sequence
2. Errors in DNA replication that go undetected by the cell's proofreading devices
3. A translocation error

Under each of the three events, you should list any and all details you remember about those events to use in your description. When you flesh out these details, you'll need to clearly connect them to the concept of the loss of cell division control leading to cancer.

To answer part (b), list the two best reasons you can think of as to why tumors are harmful to the body. Do not bother with more! Remember, only your first two responses will be graded. These might include cancerous cells' ability to metastasize; tumors' ability to occur almost anywhere in the body; their tendency to block the flow of blood when they grow near blood vessels; or disruption of the natural function of an organ in the body. Discuss how each of your choices can alter the natural function of any organ in the body and how that will endanger homeostasis.

To answer part (c), you'll need to consider first what causes cancer. Then you must think creatively in order to predict reasonable approaches to dealing with each specific cause.

AP Biology Concepts at a Glance

Source: AP Biology—Investigative Labs: An Inquiry-Based Approach. © 2012. The College Board. www.collegeboard.org. Reproduced with permission.

BIG IDEA 1: The process of evolution drives the diversity and unity of life.	
Enduring understanding 1.A: Change in the genetic makeup of a population over time is evolution.	**Essential knowledge 1.A.1:** Natural selection is a major mechanism of evolution.
	Essential knowledge 1.A.2: Natural selection acts on phenotypic variations in populations.
	Essential knowledge 1.A.3: Evolutionary change is also driven by random processes.
	Essential knowledge 1.A.4: Biological evolution is supported by scientific evidence from many disciplines, including mathematics.
Enduring understanding 1.B: Organisms are linked by lines of descent from common ancestry.	**Essential knowledge 1.B.1:** Organisms share many conserved core processes and features that evolved and are widely distributed among organisms today.
	Essential knowledge 1.B.2: Phylogenetic trees and cladograms are graphical representations (models) of evolutionary history that can be tested.
Enduring understanding 1.C: Life continues to evolve within a changing environment.	**Essential knowledge 1.C.1:** Speciation and extinction have occurred throughout the Earth's history.
	Essential knowledge 1.C.2: Speciation may occur when two populations become reproductively isolated from each other.
	Essential knowledge 1.C.3: Populations of organisms continue to evolve.
Enduring understanding 1.D: The origin of living systems is explained by natural processes.	**Essential knowledge 1.D.1:** There are several hypotheses about the natural origin of life on Earth, each with supporting scientific evidence.
	Essential knowledge 1.D.2: Scientific evidence from many different disciplines supports models of the origin of life.
BIG IDEA 2: Biological systems utilize free energy and molecular building blocks to grow, to reproduce, and to maintain dynamic homeostasis.	
Enduring understanding 2.A: Growth, reproduction, and maintenance of the organization of living systems require free energy and matter.	**Essential knowledge 2.A.1:** All living systems require constant input of free energy.

	Essential knowledge 2.A.2: Organisms capture and store free energy for use in biological processes.
	Essential knowledge 2.A.3: Organisms must exchange matter with the environment to grow, reproduce, and maintain organization.
Enduring understanding 2.B: Growth, reproduction, and dynamic homeostasis require that cells create and maintain internal environments that are different from their external environments.	**Essential knowledge 2.B.1:** Cell membranes are selectively permeable due to their structure.
	Essential knowledge 2.B.2: Growth and dynamic homeostasis are maintained by the constant movement of molecules across membranes.
	Essential knowledge 2.B.3: Eukaryotic cells maintain internal membranes that partition the cell into specialized regions.
Enduring understanding 2.C: Organisms use feedback mechanisms to regulate growth and reproduction, and to maintain dynamic homeostasis.	**Essential knowledge 2.C.1:** Organisms use feedback mechanisms to maintain their internal environments and respond to external environmental changes.
	Essential knowledge 2.C.2: Organisms respond to changes in their external environments.
Enduring understanding 2.D: Growth and dynamic homeostasis of a biological system are influenced by changes in the system's environment.	**Essential knowledge 2.D.1:** All biological systems from cells and organisms to populations, communities, and ecosystems are affected by complex biotic and abiotic interactions involving exchange of matter and free energy.
	Essential knowledge 2.D.2: Homeostatic mechanisms reflect both common ancestry and divergence due to adaptation in different environments.
	Essential knowledge 2.D.3: Biological systems are affected by disruptions to their dynamic homeostasis.
	Essential knowledge 2.D.4: Plants and animals have a variety of chemical defenses against infections that affect dynamic homeostasis.

Enduring understanding 2.E: Many biological processes involved in growth, reproduction, and dynamic homeostasis include temporal regulation and coordination.	**Essential knowledge 2.E.1:** Timing and coordination of specific events are necessary for the normal development of an organism, and these events are regulated by a variety of mechanisms.
	Essential knowledge 2.E.2: Timing and coordination of physiological events are regulated by multiple mechanisms.
	Essential knowledge 2.E.3: Timing and coordination of behavior are regulated by various mechanisms and are important in natural selection.
BIG IDEA 3: Living systems store, retrieve, transmit, and respond to information essential to life processes.	
Enduring understanding 3.A: Heritable information provides for continuity of life.	**Essential knowledge 3.A.1:** DNA, and in some cases RNA, is the primary source of heritable information.
	Essential knowledge 3.A.2: In eukaryotes, heritable information is passed to the next generation via processes that include the cell cycle and mitosis or meiosis plus fertilization.
	Essential knowledge 3.A.3: The chromosomal basis of inheritance provides an understanding of the pattern of passage (transmission) of genes from parent to offspring.
	Essential knowledge 3.A.4: The inheritance pattern of many traits cannot be explained by simple Mendelian genetics.
Enduring understanding 3.B: Expression of genetic information involves cellular and molecular mechanisms.	**Essential knowledge 3.B.1:** Gene regulation results in differential gene expression, leading to cell specialization.
	Essential knowledge 3.B.2: A variety of intercellular and intracellular signal transmissions mediate gene expression.
Enduring understanding 3.C: The processing of genetic information is imperfect and is a source of genetic variation.	**Essential knowledge 3.C.1:** Changes in genotype can result in changes in phenotype.
	Essential knowledge 3.C.2: Biological systems have multiple processes that increase genetic variation.
	Essential knowledge 3.C.3: Viral replication results in genetic variation, and viral infection can introduce genetic variation into the hosts.

Enduring understanding 3.D: Cells communicate by generating, transmitting, and receiving chemical signals.	**Essential knowledge 3.D.1:** Cell communication processes share common features that reflect a shared evolutionary history.
	Essential knowledge 3.D.2: Cells communicate with each other through direct contact with other cells or from a distance via chemical signaling.
	Essential knowledge 3.D.3: Signal transduction pathways link signal reception with cellular response.
	Essential knowledge 3.D.4: Changes in signal transduction pathways can alter cellular response.
Enduring understanding 3.E: Transmission of information results in changes within and between biological systems.	**Essential knowledge 3.E.1:** Individuals can act on information and communicate it to others.
	Essential knowledge 3.E.2: Animals have nervous systems that detect external and internal signals, transmit and integrate information, and produce responses.
BIG IDEA 4: Biological systems interact, and these systems and their interactions possess complex properties.	
Enduring understanding 4.A: Interactions within biological systems lead to complex properties.	**Essential knowledge 4.A.1:** The subcomponents of biological molecules and their sequence determine the properties of that molecule.
	Essential knowledge 4.A.2: The structure and function of subcellular components, and their interactions, provide essential cellular processes.
	Essential knowledge 4.A.3: Interactions between external stimuli and regulated gene expression result in specialization of cells, tissues, and organs.
	Essential knowledge 4.A.4: Organisms exhibit complex properties due to interactions between their constituent parts.
	Essential knowledge 4.A.5: Communities are composed of populations of organisms that interact in complex ways.
	Essential knowledge 4.A.6: Interactions among living systems and with their environment result in the movement of matter and energy.

Enduring understanding 4.B: Competition and cooperation are important aspects of biological systems.	**Essential knowledge 4.B.1:** Interactions between molecules affect their structure and function.
	Essential knowledge 4.B.2: Cooperative interactions within organisms promote efficiency in the use of energy and matter.
	Essential knowledge 4.B.3: Interactions between and within populations influence patterns of species distribution and abundance.
	Essential knowledge 4.B.4: Distribution of local and global ecosystems changes over time.
Enduring understanding 4.C: Naturally occurring diversity among and between components within biological systems affects interactions with the environment.	**Essential knowledge 4.C.1:** Variation in molecular units provides cells with a wider range of functions.
	Essential knowledge 4.C.2: Environmental factors influence the expression of the genotype in an organism.
	Essential knowledge 4.C.3: The level of variation in a population affects population dynamics.
	Essential knowledge 4.C.4: The diversity of species within an ecosystem may influence the stability of the ecosystem.

SCIENCE PRACTICES FOR AP BIOLOGY
SCIENCE PRACTICE 1: The student can use representations and models to communicate scientific phenomena and solve scientific problems.
1.1 The student can *create representations and models* of natural or man-made phenomena and systems in the domain.
1.2 The student can *describe representations and models* of natural or man-made phenomena and systems in the domain.
1.3 The student can *refine representations and models* of natural or man-made phenomena and systems in the domain.
1.4 The student can *use representations and models* to analyze situations or solve problems qualitatively and quantitatively.
1.5 The student can *reexpress key elements* of natural phenomena across multiple representations in the domain.
SCIENCE PRACTICE 2: The student can use mathematics appropriately.
2.1 The student can *justify the selection of a mathematical routine* to solve problems.
2.2 The student can *apply mathematical routines* to quantities that describe natural phenomena.

2.3 The student can *estimate numerically* quantities that describe natural phenomena.
SCIENCE PRACTICE 3: The student can engage in scientific questioning to extend thinking or to guide investigations within the context of the AP course.
3.1 The student can *pose scientific questions.*
3.2 The student can *refine scientific questions.*
3.3 The student can *evaluate scientific questions.*
SCIENCE PRACTICE 4: The student can plan and implement data collection strategies appropriate to a particular scientific question.
4.1 The student can *justify the selection of the kind of data* needed to answer a particular scientific question.
4.2 The student can *design a plan* for collecting data to answer a particular scientific question.
4.3 The student can *collect data* to answer a particular scientific question.
4.4 The student can *evaluate sources of data* to answer a particular scientific question.
SCIENCE PRACTICE 5: The student can perform data analysis and evaluation of evidence.
5.1 The student can *analyze data* to identify patterns or relationships.
5.2 The student can *refine observations and measurements* based on data analysis.
5.3 The student can *evaluate the evidence provided by data sets* in relation to a particular scientific question.
SCIENCE PRACTICE 6: The student can work with scientific explanations and theories.
6.1 The student can *justify claims with evidence.*
6.2 The student can *construct explanations of phenomena based on evidence* produced through scientific practices.
6.3 The student can *articulate the reasons that scientific explanations and theories are refined or replaced.*
6.4 The student can *make claims and predictions about natural phenomena* based on scientific theories and models.
6.5 The student can *evaluate alternative scientific explanations.*
SCIENCE PRACTICE 7: The student is able to connect and relate knowledge across various scales, concepts, and representations in and across domains.
7.1 The student can *connect phenomena and models* across spatial and temporal scales.
7.2 The student can *connect concepts* in and across domain(s) to generalize or extrapolate in and/or across enduring understandings and/or big ideas.

Part II

A Review of Topics with Sample Questions

Part II is keyed to *Campbell Biology*, Tenth Edition, by Reece et al. as well as *Biology in Focus* by Urry et al. The content is divided into 10 broad topics, and within each topic there are bold headings. Each of these is aligned with material in the textbooks by reference to the concepts where it is found. You will find an overview of important information and sample multiple-choice and free-response questions. The content necessary to do well on the AP Biology Exam is included in the bulleted information, but you will need to rely on sample questions and your teacher's instruction to reinforce the science practices. Some of the included material may be outside the scope of the AP exam, but we have included it because your teacher may use it to help you better understand the richness of biology. Answers and explanations to the questions can be found in Part V. Be sure to review the answers thoroughly to prepare yourself for the range of questions you will encounter on the AP Biology Examination.

TOPIC 1

The Chemistry of Life

The Chemical Context of Life

(Biology, *10e: Chapters 2, 3*; Focus, *1e: Chapter 2*)

WHAT'S IMPORTANT TO KNOW?
Some of this material is considered prior knowledge for the AP Biology Examination. However, you will need to know this information to proceed with the required topics, so we include what is most important in this area.

YOU MUST KNOW

- The three subatomic particles and their significance.
- The types of chemical bonds and how they form
- The importance of hydrogen bonding to the properties of water.
- Four unique properties of water, and how each contributes to life on Earth.
- How to interpret the pH scale.
- How changes in pH can alter biological systems.
- The importance of buffers in biological systems.

Matter consists of chemical elements in pure form and in combinations called compounds (Biology, *10e: 2.1*, Focus, *1e: 2.1*)

- **Matter** is anything that takes up space and has mass.
- An **element** is a substance that cannot be broken down to other substances by chemical reactions. *Examples*: gold, copper, carbon, and oxygen.
- A **compound** is a substance consisting of two or more elements combined in a fixed ratio. *Examples*: water (H_2O) and table salt (NaCl).
- **C, H, O, N** make up 96% of living matter. About 25 of the 92 natural elements are known to be essential to life.
- **Trace elements** are those required by an organism in only minute quantities. *Examples*: iron and iodine.

An element's properties depend on the structure of its atoms (Biology, *10e: 2.2*, Focus, *1e: 2.2*)

- **Atoms** are the smallest unit of an element that still retains the property of the element. Atoms are made up of neutrons, protons, and electrons.

- **Protons** are positively charged particles found in the nucleus of the atom.
- **Electrons** are negatively charged particles that are found in electron shells around the nucleus. They determine the chemical properties and reactivity of the element.
- **Neutrons** are particles with no charge. They are found in the nucleus. Their number can vary in the same element, resulting in isotopes.
- **Isotopes** are forms of an element with differing numbers of neutrons. *Example*: ^{12}C and ^{14}C are isotopes of carbon. Both have 6 protons, but ^{12}C has 6 neutrons, whereas ^{14}C has 8 neutrons.
- The **atomic number** is the number of protons an element possesses. This number is unique to every element. (See Figure 1.1.)
- The **mass number** of an element is the sum of its protons and neutrons.

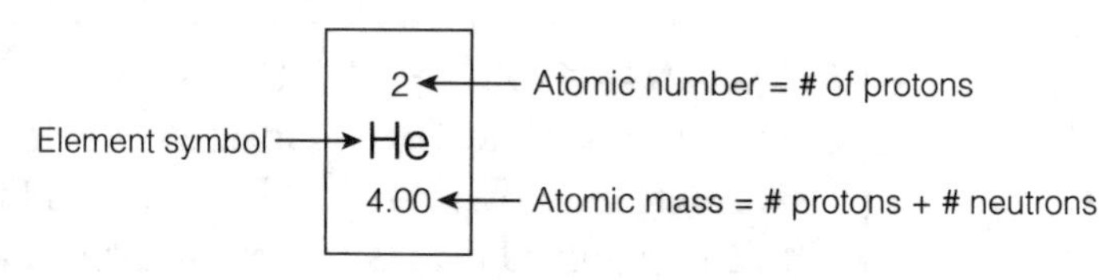

Figure 1.1 An element of the periodic table

The formation and function of molecules depend on chemical bonding between atoms (Biology, *10e: 2.3,* Focus, *1e: 2.3*)

- **Chemical bonds** are defined as interactions between the valence electrons of different atoms. Atoms are held together by chemical bonds to form molecules.
- A **covalent bond** occurs when valence electrons are shared by two atoms.
    - **Nonpolar covalent bonds** occur when the electrons being shared are shared equally between the two atoms. *Examples*: O=O, H–H.
    - Atoms vary in their *electronegativity*, a tendency to attract electrons of a covalent bond. Oxygen and nitrogen are strongly electronegative.
    - In **polar covalent bonds**, one atom has greater electronegativity than the other, resulting in an unequal sharing of the electrons. This results in the region of the oxygen atom being slightly negative, whereas the regions about the hydrogen atoms are slightly positive. *Example*: Refer to Figure 1.2 and note that within each molecule of H_2O the electrons are shared unequally.
- **Ionic bonds** are ones in which two atoms attract valence electrons so unequally that the more electronegative atom steals the electron away from the less electronegative atom.
    - An **ion** is the resulting charged atom or molecule.
    - **Ionic bonds** occur because these ions will be either positively or negatively charged and will be attracted to each other by these opposite charges.
- **Hydrogen bonds** are relatively weak bonds that form between the partial positively charged hydrogen atom of one molecule and the strongly electronegative

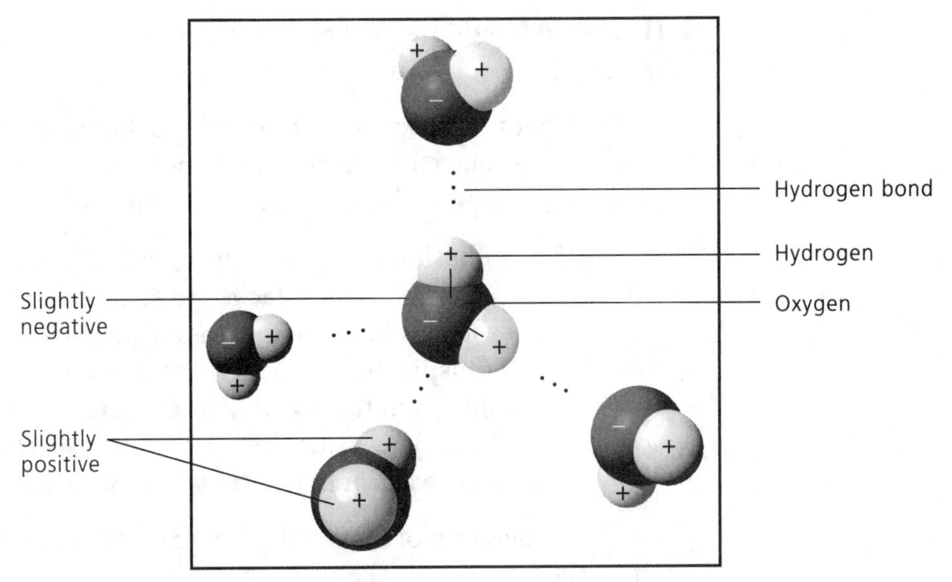

Figure 1.2 Hydrogen bonds between water molecules

oxygen or nitrogen of *another* molecule. Hydrogen bonds play a major role in the three-dimensional shape of proteins and nucleic acids.

- **Van der Waals interactions** are very weak, transient connections that are the result of asymmetrical distribution of electrons within a molecule. These weak interactions contribute to the three-dimensional shape of molecules.

Chemical reactions make and break chemical bonds (Biology, *10e: 2.4,* Focus, *1e: 2.4*)

- A **chemical reaction** shows the **reactants**, which are the starting materials, an arrow to indicate their conversion into the **products**, the ending materials. Example: $6\ CO_2 + 6\ H_2O \longrightarrow C_6H_{12}O_6 + 6\ O_2$.
- The chemical reaction above also shows the number of molecules involved. This is the coefficient in front of each molecule. Note that the number of atoms of each element is the same on each side of the reaction.
- Some chemical reactions are reversible, which is indicated with a double-headed arrow: $3\ H_2 + N_2 \rightleftharpoons 2\ NH_3$.

Water and its properties due to hydrogen bonding helps make life on Earth possible (Biology, *10e: 3.1–3.3,* Focus, *1e: 2.5*)

- The **structure of water** is the key to its special properties. Water is made up of one atom of oxygen and two atoms of hydrogen, bonded to form a molecule.
- Water molecules are **polar**. The oxygen region of the molecule has a partial negative charge, and each hydrogen has a partial positive charge.
- **Hydrogen bonds** form between water molecules. The slightly negative oxygen atom from one water molecule is attracted to the slightly positive hydrogen end of *another* water molecule.
- Each water molecule can form a maximum of four hydrogen bonds at a time. (Refer to Figure 1.2.)

- **Hydrogen bonds** are the key to each of the following properties of water and what makes water so unique.

  1. **Cohesion.** Cohesion is the linking of like molecules. Think "water molecule joined to water molecule" and visualize a water strider walking on top of a pond due to the *surface tension* that is the result of this property.
     - **Adhesion** is the clinging of one substance to another. Think "water molecule attached to some other molecule" such as water droplets adhering to a glass windshield.
     - **Transpiration** is the movement of water molecules up the very thin xylem tubes and their evaporation from the stomata in plants. The water molecules cling to each other by *cohesion* and to the walls of the xylem tubes by *adhesion*.
  2. Moderation of temperature is possible because of water's high specific heat.
     - **Specific heat** is the amount of heat required to raise or lower the temperature of a substance by 1°C. Relative to most other materials, the temperature of water changes less when a given amount of heat is lost or absorbed. This high specific heat makes the temperature of Earth's oceans relatively stable and able to support vast quantities of both plant and animal life.
  3. Insulation of bodies of water by floating ice.
     - Water is less dense as a solid than in its liquid state, whereas the opposite is true of most other substances. Because ice is less dense than liquid water, ice floats. This keeps larger bodies of water from freezing solid, allowing life to exist in ponds, lakes, and even oceans.
  4 Water is an important *solvent*. (The substance that something is dissolved in is called the *solvent*, whereas the substance being dissolved is the *solute*. Together they are called the *solution*.)
     - **Hydrophilic** substances are water-soluble. These include ionic compounds, polar molecules (for example, sugars), and some proteins.
     - **Hydrophobic** substances such as oils are nonpolar and do not dissolve in water.

- Acidic and basic conditions affect living organisms
  - The **pH** scale runs between 0 and 14 and measures the relative acidity and alkalinity of aqueous solutions. (See Figure 1.3.)
  - **Acids** have an excess of H^+ ions and a pH below 7.0. $[H^+] > [OH^-]$

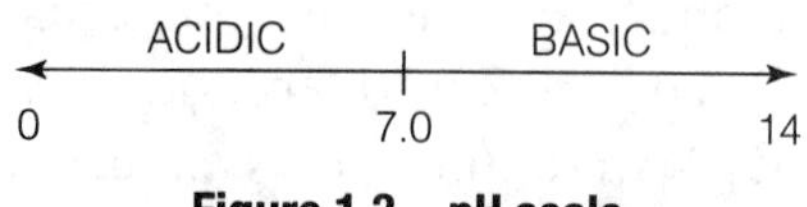

Figure 1.3 pH scale

- **Bases** have an excess of OH^- ions, and pH above 7.0. $[H^+] < [OH^-]$
- Pure water is neutral, which means it has a pH of 7. $[H^+] = [OH^-]$
- **Buffers** are substances that minimize changes in pH. They accept H^+ from solution when they are in excess and donate H^+ when they are depleted. Buffering compounds are essential in living tissues to minimize pH changes.
- **Carbonic acid (H_2CO_3)** is an important buffer in living systems. It moderates pH changes in blood plasma and the ocean.

Carbon and the Molecular Diversity of Life

(Biology, *10e: Chapters 4, 5*, Focus, *1e: Chapter 3*)

YOU MUST KNOW

- The properties of carbon that make it so important.
- The role of **dehydration reactions** in the formation of organic compounds and **hydrolysis** in the digestion of organic compounds.
- How the sequence and subcomponents of the four groups of organic compounds determine their properties.
- The cellular functions of carbohydrates, lipids, proteins, and nucleic acids.
- How changes in these organic molecules would affect their function.
- The four structural levels of proteins and how changes at any level can affect the activity of the protein.
- How proteins reach their final shape **(conformation)**, the **denaturing** impact that heat and pH can have on protein structure, and how these changes may affect the organism.
- Directionality influences structure and function of polymers, such as nucleic acids (5' and 3' ends) and proteins (amino and carboxyl ends).

Carbon and molecular diversity (Biology, *10e: 4.1–4.3*, Focus, *1e: 3.1*)

- The major elements of life are C, H, O, N, S, and P, sometimes recalled with the acronym for a person's name: ***P.S. COHN***.
- All ***organic compounds*** contain carbon, and most also contain hydrogen.
- Carbon is unparalleled in its ability to form molecules that are large, complex, and diverse. Why?
    - It has 4 valence electrons.
    - It can form up to 4 covalent bonds.
    - These can be single, double, or triple covalent bonds.
    - It can form large molecules.
    - These molecules can be chains, ring-shaped, or branched.
- **Isomers** are molecules that have the same molecular formula but differ in their arrangement of these atoms. These differences can result in molecules

that are very different in their biological activities. *Examples*: glucose and fructose (both have the molecular formula of $C_6H_{12}O_6$).

- **Functional groups** attached to the carbon skeleton have diverse properties. The behavior of organic molecules is dependent on the identity of their functional groups.
- Some common functional groups are listed next:

Functional Group Name/Structure	Organic Molecules with the Functional Group and Items of Note about Functional Group
Hydroxyl, —OH	Alcohols such as ethanol, methanol; helps dissolve molecules such as sugars
Carboxyl, —COOH	Carboxylic acids such as fatty acids and sugars; acidic properties because it tends to ionize; source of H^+ ions
Carbonyl, <CO	Ketones and aldehydes such as sugars
Amino, —NH_2	Amines such as amino acids
Phosphate, PO_3	Organic phosphates, including ATP, DNA, and phospholipids
Sulfhydryl, —SH	This group is found in some amino acids; forms disulfide bridges in proteins
Methyl, —CH_3	Addition of a methyl group affects expression of genes

Macromolecules are polymers, built from monomers (Biology, *10e: 5.1,* Focus, *1e: 3.2*)

- **Polymers** are long chain molecules made of repeating subunits called **monomers.** *Examples*: Starch is a polymer composed of glucose monomers. Proteins are polymers composed of amino acid monomers. (See Figure 1.4.)
- **Dehydration reactions** create polymers from monomers. Two monomers are joined by removing one molecule of water. *Example*: $C_6H_{12}O_6 + C_6H_{12}O_6 \rightarrow C_{12}H_{22}O_{11} + H_2O$.
- **Hydrolysis** occurs when water is added to split large molecules. This occurs in the reverse of the above reaction.

Figure 1.4 Synthesis and breakdown of polymers

Carbohydrates serve as fuel and building material (Biology, *10e: 5.2,* Focus, *1e: 3.3*)

- **Carbohydrates** include both simple sugars (glucose, fructose, galactose, etc.) and polymers such as starch made from these and other subunits. All carbohydrates exist in a ratio of 1 carbon:2 hydrogen:1 oxygen or CH_2O.
- **Monosaccharides** are the monomers of carbohydrates. *Examples*: glucose ($C_6H_{12}O_6$) and ribose ($C_5H_{10}O_5$). Notice the 1:2:1 ratio discussed above.
- **Polysaccharides** are polymers of monosaccharides. *Examples*: starch, cellulose, and glycogen.
- Starches and cellulose are both composed of glucose monomers, but the different ring forms (alpha and beta) and linkages between them result in very different functions. For example, you can digest starch, which has 1–4 alpha

linkages, and not cellulose, which has 1–4 beta linkages. Keep this in mind: *Change the structure, change the function*!

- Two functions of polysaccharides are **energy storage** and **structural support**.

  1. **Energy-storage polysaccharides**
     - **Starch** is a storage polysaccharide found in plants (for example, potatoes).
     - **Glycogen** is a storage polysaccharide found in animals, vertebrate muscle cells, and liver cells.
  2. **Structural support polysaccharides**
     - **Cellulose** is a major component of plant cell walls.
     - **Chitin** is found in the exoskeleton of arthropods, such as lobsters and insects, and the cell walls of fungi. It gives cockroaches their "crunch."

Lipids are a diverse group of hydrophobic molecules (Biology, *10e: 5.4,* Focus, *1e: 3.4*)

- Lipids are all **hydrophobic**. They aren't polymers because they are assembled from a variety of components. *Examples*: **waxes, oils, fats,** and **steroids.**
- **Fats** (also called triglycerides) are made up of a **glycerol** molecule and three **fatty acid** molecules.
- **Fatty acids** include hydrocarbon chains of variable lengths. These chains are nonpolar and therefore hydrophobic.
- **Saturated fatty acids**
  - have no double bonds between carbons
  - tend to pack solidly at room temperature
  - are linked to cardiovascular disease
  - are commonly produced by animals
  - *Examples*: butter and lard
- **Unsaturated fatty acids**
  - have some C=C (carbon double bonds); this results in kinks
  - tend to be liquid at room temperature
  - are commonly produced by plants
  - *Examples*: corn oil and olive oil
- **Functions of lipids**
  - **Energy storage.** Fats store twice as many calories/gram as carbohydrates!
  - *Protection* of vital organs and *insulation*. In humans and other mammals, fat is stored in adipose cells.
- **Phospholipids** make up cell membranes. They
  - have a hydrophilic (polar) head that includes a phosphate group.
  - have two fatty acid tails, which are hydrophobic.

- are arranged in a bilayer in forming the cell membrane, with the hydrophilic heads pointing toward the watery cytosol or extracellular environment, and hydrophobic tails sandwiched in between (Figure 1.5).

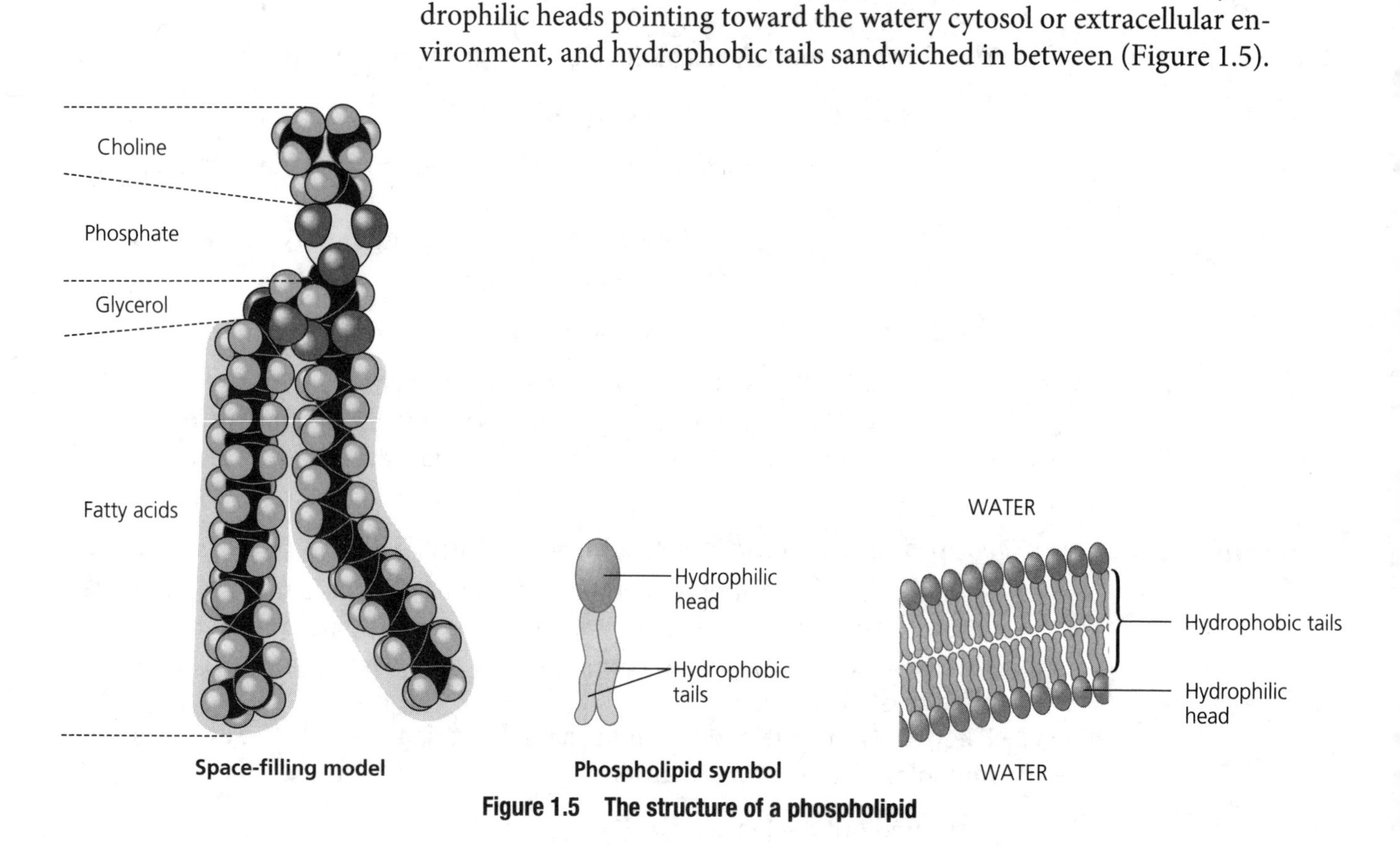

Figure 1.5 The structure of a phospholipid

- **Steroids** are made up of four rings that are fused together.
  - **Cholesterol** is a steroid. It is a common component of cell membranes.
  - **Estrogen** and **testosterone** are steroid hormones.

Proteins include a diversity of structures, resulting in a wide range of functions (Biology, *10e: 5.4,* Focus, *1e: 3.5*)

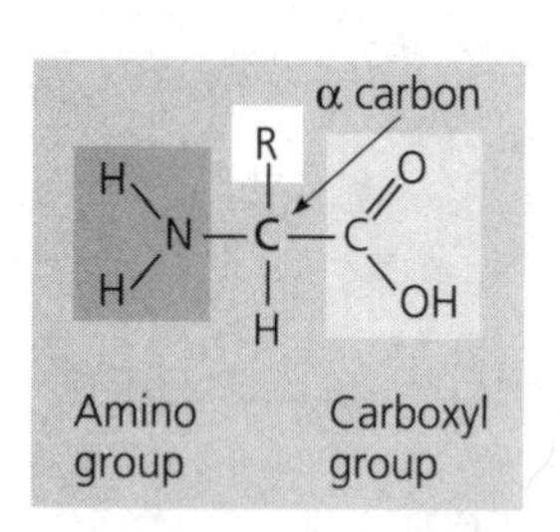

Figure 1.6 The structure of an amino acid

- **Proteins** are polymers made up of amino acid monomers.
- **Amino acids** contain a central carbon bonded to a carboxyl group (COOH) at one end, an amino group (NH_2) at the other end, a hydrogen atom, and an R group (variable group or side chain). (See Figure 1.6.)
- **Peptide bonds** link amino acids. They are formed by dehydration synthesis between the amino and carboxyl groups of adjacent monomers.
- **There are four levels of protein structure** (Figure 1.7):
  - **Primary structure** is the unique sequence in which amino acids are joined.
  - **Secondary structure** refers to one of two three-dimensional shapes that are the result of hydrogen bonding between members of the polypeptide backbone (not the amino acid side chains).
    - **Alpha (α) helix** is a coiled shape, much like a slinky.
    - **Beta (β) pleated sheet** is an accordion shape.

- **Tertiary structure** results in a complex globular shape due to interactions between the side chains (R groups), such as hydrophobic interactions, van der Waals interactions, hydrogen bonds, and disulfide bridges.
  - Globular proteins such as enzymes are held in position by these R group interactions.
- **Quaternary structure** refers to the association of two or more polypeptide chains into one large protein. Hemoglobin is a globular protein with quaternary structure because it is composed of four chains.

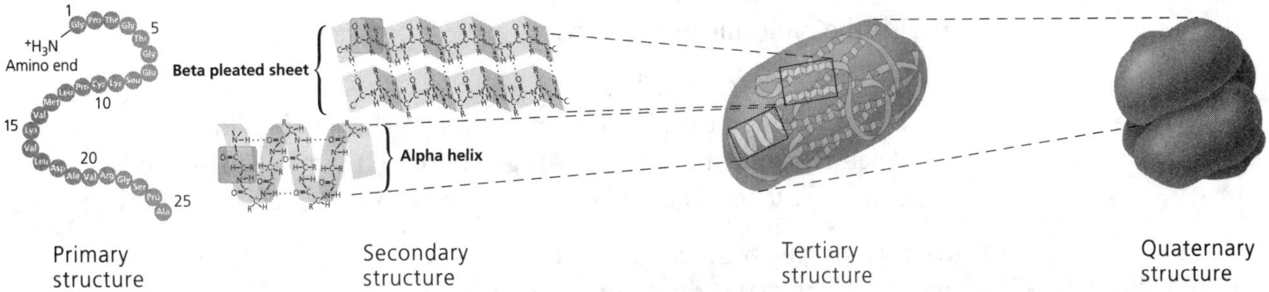

Primary structure

Secondary structure

Tertiary structure

Quaternary structure

Figure 1.7 Levels of protein structure

- **Protein shape is crucial to protein function.** When a protein does not fold properly, its function is changed. This can be the result of a single amino acid substitution, such as that seen in the abnormal hemoglobin typical of sickle-cell disease.

- A protein is **denatured** when it loses its shape and ability to function due to **heat**, a **change in pH**, or some other disturbance. Keep this in mind: *change the structure, change the function*!

> ***CONNECT WITH THE CURRICULUM FRAMEWORK***
>
> Consider each category of organic compound. Describe a specific change that might occur in a molecule in each category, and describe how this might affect the original function of the molecule. Justify your response.

Nucleic acids store, transmit, and help express hereditary information (Biology, *10e: 5.5,* Focus, *1e: 3.6*)

- **DNA** (deoxyribonucleic acid) and **RNA** (ribonucleic acid) are the two nucleic acids. Their monomers are nucleotides.
- **Nucleotides** are made up of three parts (Figure 1.8):
  - **Nitrogenous base** (adenine, thymine, cytosine, and guanine in DNA; adenine, uracil, cytosine, and guanine in RNA)
  - **Pentose** (5-carbon) sugar (deoxyribose in DNA or ribose in RNA)
  - **Phosphate group**

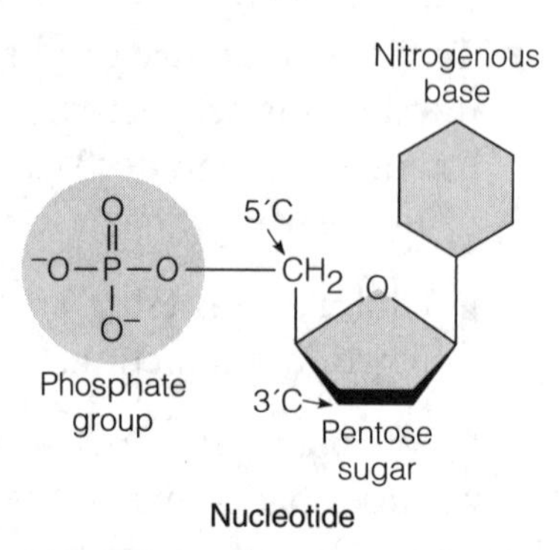

Figure 1.8 The components of nucleic acids

- **DNA** is the molecule of heredity.
  - It is double-stranded helix.
  - Its nucleotides are adenine, thymine, cytosine, and guanine.
  - Adenine nucleotides will hydrogen bond to thymine nucleotides; cytosine nucleotides will hydrogen bond to guanine nucleotides.
- **RNA** is single-stranded. Its nucleotides are adenine, uracil, cytosine, and guanine. Note that it does not have thymine.
- Nucleic acids have directionality, determined by the 3′ and 5′ carbons of the sugar. This will determine the direction of DNA replication and transcription (5′ to 3′).

SUMMARY TABLE

Macromolecules/Polymers	Monomers/Components	Examples	Functions
Carbohydrates	Monosaccharides	Sugars, starch, glycogen, cellulose	Energy, energy storage; structural
Lipids	Fatty acids and glycerol	Fats, oils	Important energy source; insulation; phospholipids of plasma membrane
Proteins	Amino acids	Hemoglobin, pepsin	Enzymes; movement; membrane receptors
Nucleic Acids	Nucleotides (sugar, phosphate group, nitrogenous base)	DNA, RNA	Heredity; code for amino acid sequence

Level 1: Knowledge/Comprehension Questions

The Knowledge/Comprehension questions review essential content knowledge but don't represent the type of question you will see on the AP exam. Level 2, Application/Synthesis, questions are similar to those you will see on the exam.

1. Which list of components characterizes RNA?
   (A) a phosphate group, deoxyribose, and uracil
   (B) a phosphate group, ribose, and uracil
   (C) a phosphate group, ribose, and thymine
   (D) a phosphate group, deoxyribose, and uracil

2. Which of the following molecules would contain a polar covalent bond?
   (A) Cl_2
   (B) NaCl
   (C) H_2O
   (D) CH_4

3. Which of the following statements regarding carbon is *false*?
   (A) Carbon has the capacity to form polar covalent bonds with hydrogen.
   (B) Carbon has the ability to form covalent bonds with up to four other atoms.
   (C) Carbon has the capacity to form single and double bonds.
   (D) Carbon has the ability to bond together to form extensive branched or unbranched "carbon skeletons."

4. Because of the unique properties of water associated with hydrogen bonding, water evaporates from pores on the leaves of plants and draws up water molecules in a continuous chain from the roots up through the vascular tissues of plants. Which of these groups of terms describes the process and properties of water that explain this?
   (A) adhesion, cohesion, and translocation
   (B) adhesion, cohesion, and transcription
   (C) cohesion, hybridization, and transpiration
   (D) cohesion, adhesion, and transpiration

Directions: Questions 5–8 consist of five lettered choices followed by a list of numbered phrases or sentences. For each numbered phrase or sentence, select the one choice that is most closely related to it. Each choice may be used once, more than once, or not at all.

Questions 5–8
   (A) Lipids
   (B) Peptide bonds
   (C) Alpha helix
   (D) Cellulose

5. The major class of biological molecules that are not polymers

6. Linkages between the monomers of proteins

7. A secondary structure of proteins

8. A structural carbohydrate found in plants

9. Hydrolysis is involved in which of the following?
   (A) formation of starch, a polysaccharide, from monomers of glucose
   (B) hydrogen bond formation between nitrogen bases of nucleic acids
   (C) peptide bond formation between amino acids to form a protein
   (D) the digestion of sucrose, a disaccharide to its monomers of glucose and fructose

10. The process by which protein conformation is lost or broken down is
   (A) dehydration synthesis.
   (B) translation.
   (C) denaturation.
   (D) hydrolysis.

11. An organic compound that is composed of carbon, hydrogen, and oxygen in a 1:2:1 ratio is known as a
   (A) lipid.
   (B) carbohydrate.
   (C) protein.
   (D) nucleic acid.

12. If three molecules of a fatty acid that has the formula $C_{16}H_{22}O_2$ are joined to a molecule of glycerol ($C_3H_8O_3$), then the resulting molecule would have the formula
   (A) $C_{48}H_{96}O_6$.
   (B) $C_{48}H_{98}O_8$.
   (C) $C_{51}H_{68}O_6$.
   (D) $C_{51}H_{106}O_8$.

13. Which of the macromolecules below could be structural parts of the cell, enzymes, or involved in cell movement or communication?
   (A) nucleic acids
   (B) proteins
   (C) lipids
   (D) carbohydrates

14. The plasma membrane is composed of several different macromolecules. Which macromolecule serves as the fluid interface between the intracellular and extracellular environments?
    (A) proteins
    (B) phospholipids
    (C) carbohydrates
    (D) cholesterol

15. The partial negative charge at one end of a water molecule is attracted to a partial positive charge of another water molecule. What is this type of attraction called?
    (A) a polar covalent bond
    (B) an ionic bond
    (C) a hydration shell
    (D) a hydrogen bond

16. Polymers of carbohydrates and proteins are all synthesized from monomers by which of the following processes?
    (A) the joining of monosaccharides
    (B) hydrolysis
    (C) dehydration reactions
    (D) ionic bonding of monomers

17. If the pH of a solution is decreased from 7 to 6, it means that the concentration of
    (A) H^+ has decreased to 1/10 of what it was at pH 7.
    (B) H^+ has increased 10 times what it was at pH 7.
    (C) OH^- has increased 10 times what it was at pH 7.
    (D) OH^- has increased by 1/7 of what it was.

18. Recall the structure of a typical amino acid and how the molecule has two distinct "ends." Which two functional groups are always found in amino acids?
    (A) amine and sulfhydryl
    (B) hydroxyl and carboxyl
    (C) carboxyl and amine
    (D) phosphate group and nitrogenous base

Level 2: Application/Analysis/Synthesis Questions

1. The hydrogen bonds shown in this figure are each between

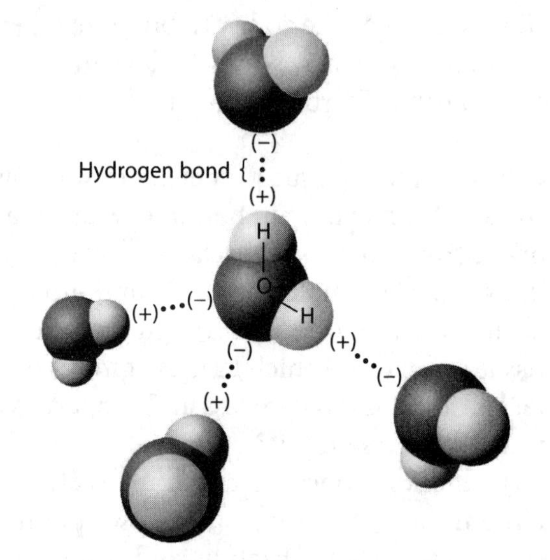

    (A) two hydrogen atoms.
    (B) two oxygen atoms.
    (C) an oxygen and a hydrogen atom of the same water molecule.
    (D) an oxygen and a hydrogen atom of different water molecules.

2. The tremendous variation and unique properties of proteins are most likely a result of
    (A) interactions between R groups of the amino acids.
    (B) hydrogen bonds linking amino acids.
    (C) the sequence of amino acids in the primary structure of the protein.
    (D) peptide bonds linking amino and carboxyl groups.

Questions 3 and 4 refer to the following art.

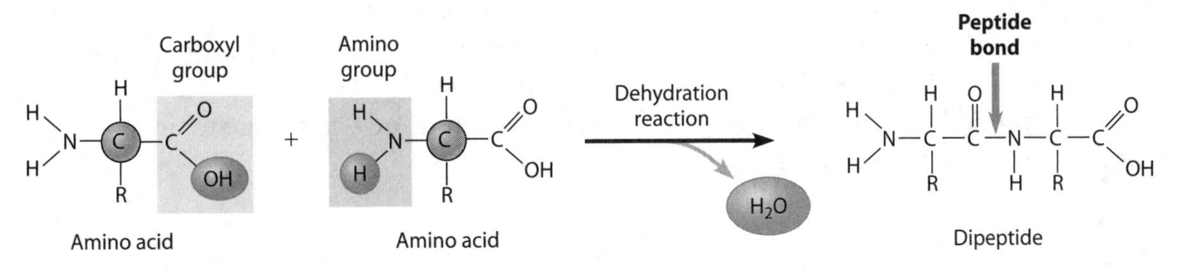

3. How are these two amino acids attached together?
   (A) amino group to amino group
   (B) amino group to carboxylic acid group
   (C) carboxylic acid group to carboxylic acid group
   (D) carbon atom to carbon atom

4. If the dipeptide above were to be digested, how would it be reduced to amino acids?
   (A) by a dehydration reaction which removes water
   (B) by reduction in digestive fluid pH in the stomach
   (C) through the removal of functional groups at either end
   (D) through a hydrolysis reaction in which water is added

After reading the following paragraph, answer questions 5 and 6.

You're the manager of a factory that produces enzyme-washed blue jeans (the enzymes lighten the color of the denim, giving a faded appearance). When the most recent batch of fabric came out of the enzyme wash, however, the color wasn't light enough to meet your standards. Your quality control laboratory wants to do some tests to determine why the wash enzymes didn't perform as expected.

5. Which hypothesis is most likely to be productive for their initial investigation?
   (A) The nucleotide chain of the enzymes may be incorrectly formed.
   (B) The dye in the fabric may have hydrolyzed the fatty acids in the enzymes.
   (C) The polysaccharides in the enzymes may have separated in the wash water.
   (D) The three-dimensional amino acid structure of the enzyme may have been altered.

6. Based on your understanding of enzyme structure, which of the following would you recommend that they also investigate?
   (A) the temperature of the liquid in the washing vat
   (B) washing the blue jeans in smaller vats
   (C) the manufacturer of the fabric
   (D) switching to another product for the enzyme wash

7. The molecular formula for glucose is $C_6H_{12}O_6$. What would be the molecular formula for a polymer made by lining 10 glucose molecules together by dehydration reactions?
   (A) $C_{60}H_{120}O_{60}$
   (B) $C_6H_{12}O_6$
   (C) $C_{60}H_{102}O_{51}$
   (D) $C_{60}H_{100}O_{50}$

8. Which of the following pairs of base sequences could form a short stretch of a normal double helix of DNA?
   (A) 5′-AGCA-3′ with 3′-UCGU-5′
   (B) 5′-AGCT-3′ with 5′-TCGA-3′
   (C) 5′-GCGC-3′ with 5′-TATA-3′
   (D) 5′-ATGC-3′ with 5′-GCAT-3′

9. Water is an excellent solvent. Select the property that justifies this statement.
   (A) As a polar molecule, it can surround and dissolve ionic and polar molecules.
   (B) It forms ionic bonds with ions, hydrogen bonds with polar molecules, and hydrophobic interactions with nonpolar molecules.
   (C) It forms hydrogen bonds with itself so cohesion is possible.
   (D) It is liquid and will adhere to many substances.

Grid-In Questions

These call for a numerical response.

1. If nine molecules of a monosaccharide with the formula $C_6H_{12}O_6$ are assembled to produce a complex carbohydrate, how many atoms of hydrogen will be in the final polymer?

2. Compare the number of H^+ ions in a solution with a pH of 2 to a solution with a pH of 6. If appropriate, include a negative sign in your answer.

Free-Response Question

1. *The selectively permeable plasma membrane is composed of phospholipids and protein, which allow for its unique functions.*

   (a) Describe the structure and properties of phospholipids and *explain* the important roles of phospholipids in the plasma membrane.
   (b) Predict how the normal function of the plasma membrane would be altered if all phospholipids were saturated, resulting in fatty acid tails without kinks or bends. **Explain** the effect this would have on plants located in very cold regions.
   (c) Proteins are an important component of the cell membrane. **Describe** two specific functions of proteins in the membrane.
   (d) Explain the role of each type of protein you select for part (c) based on the structure and properties of a protein.

TOPIC 2

The Cell

A Tour of the Cell

(Biology, *10e: Chapter 6*, Focus, *1e: Chapter 4*)

YOU MUST KNOW

- Three differences between prokaryotic and eukaryotic cells.
- The structure and function of organelles common to plant and animal cells.
- The structure and function of organelles found only in plant cells or only in animal cells.
- How different cell types show differences in subcellular components.
- How internal membranes and organelles contribute to cell functions. (LO 2.13)
- How cell size and shape affect the overall rate of nutrient intake and waste elimination. (LO 6.2)

Eukaryotic cells have internal membranes that compartmentalize their functions (Biology, *10e: 6.2*, Focus, *1e: 4.2*)

- The following table organizes the major characteristics of prokaryotic and eukaryotic cells.

Characteristics	Prokaryotic Cells	Eukaryotic Cells
Plasma membrane	Yes	Yes
Ribosomes	Yes	Yes
Membrane-bound organelles in cytosol	No	Yes
Nucleus	No	Yes
Size	1 μm–10 μm	10 μm–100 μm

- Prokaryotic cells are found in the domains Bacteria and Archaea. Eukaryotic cells belong to the domain Eukarya and include animals, fungi, plants, and protists.
- Three key details to remember about prokaryotes include:
  - The single circular chromosome is found in a region called the nucleoid, but there is no nuclear membrane and therefore no true nucleus.
  - No membrane-bounded organelles are found in the cytosol. (Ribosomes are found, but they are not membrane bound.)
  - From the preceding table, notice how much smaller prokaryotes are than eukaryotes.

- Three corresponding details about eukaryotic cells:
  - A membrane-enclosed nucleus contains the cell's linear chromosomes.
  - Many membrane-bounded organelles are found in the cytoplasm.
  - On average, eukaryotes are much larger than prokaryotes.
- Use Figures 2.1 and 2.2 to locate each component of a plant or animal cell as they are reviewed. Notice if the cell structure is found only in animal cells, only in plant cells, or both plant and animal cells (see Figures 6.8 and 6.9, 10e).
- The **plasma membrane** forms the boundary for a cell. It is selectively permeable and permits the passage of materials into and out of the cell.
- The plasma membrane is made up of *phospholipids*, *proteins*, and associated *carbohydrates*. These molecules determine the functions of the membrane.
- The *surface area-to-volume ratio* becomes less favorable as a cell increases in size. The total volume grows proportionately more than the surface area. Because a cell acquires resources through the plasma membrane, cell size is limited.

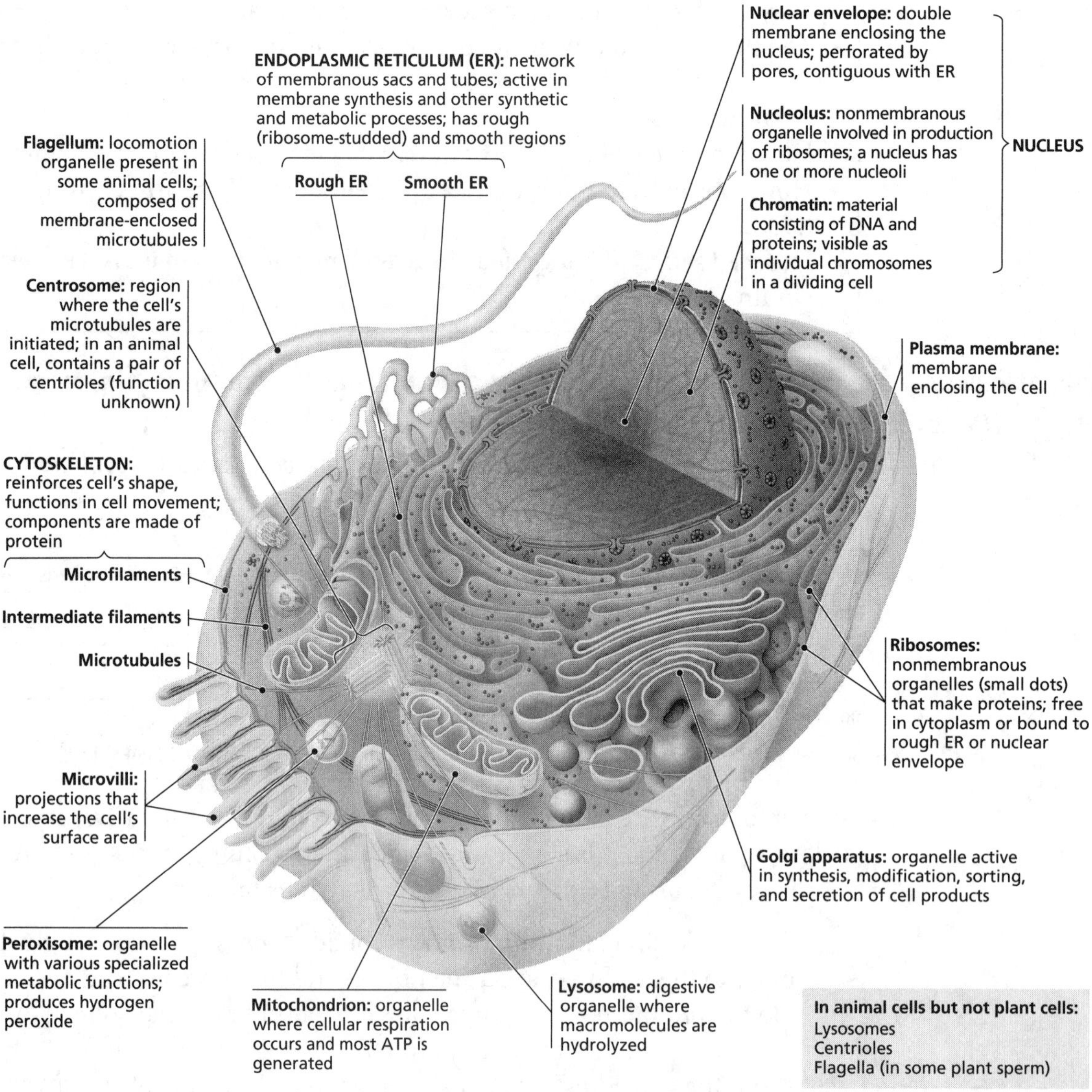

Figure 2.1 Animal Cell Structure

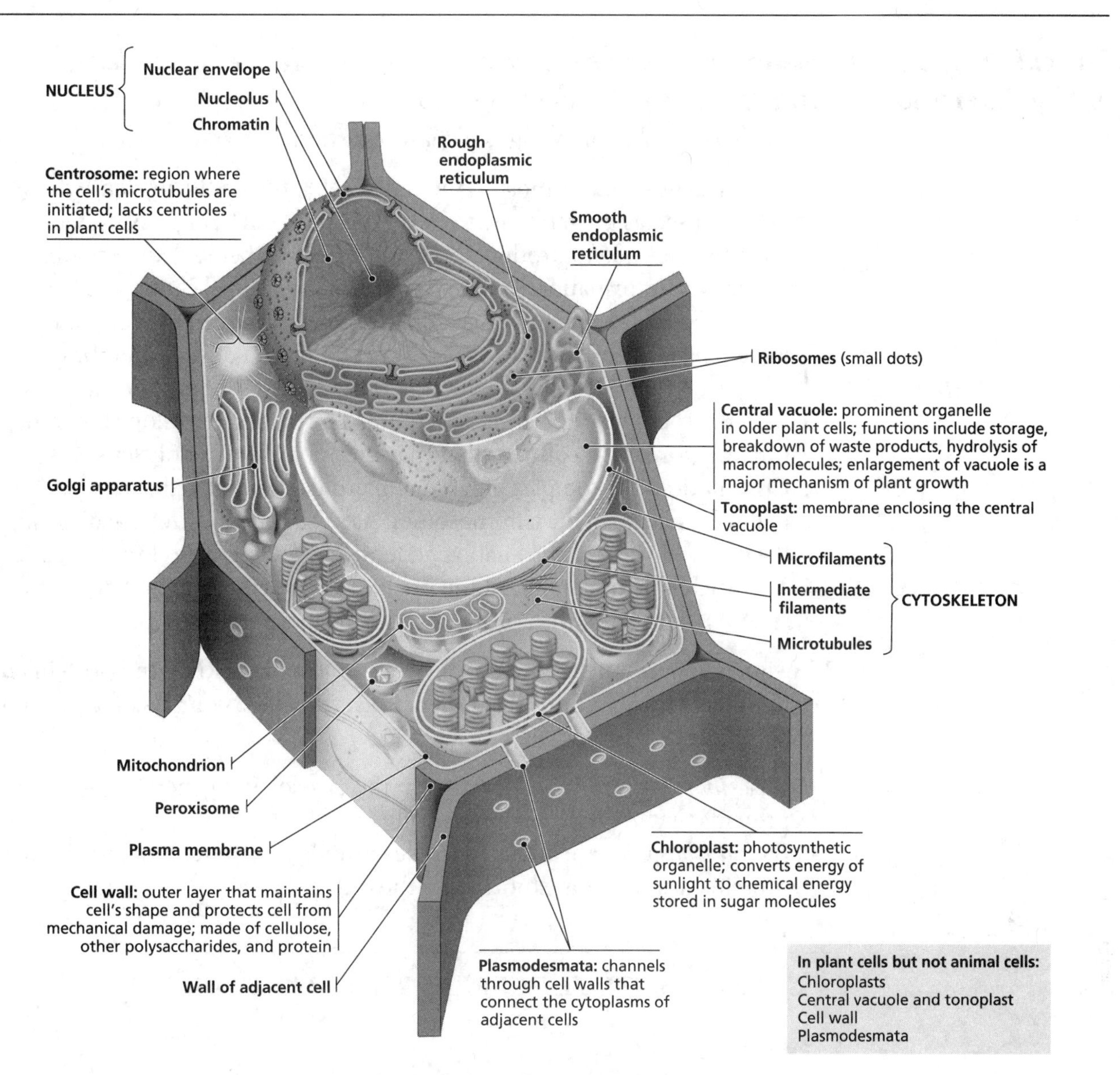

Figure 2.2 Plant Cell Structure

* You could be asked to calculate a surface area-to-volume ratio. Do the grid-in question at the end of this topic for practice. The answers section will have the correct answer and an explanation for how to do the problem.

> ***CONNECT WITH THE CURRICULUM FRAMEWORK***
>
> **Big Idea 2**
>
> Be able to calculate surface area-to-volume ratios for various cell sizes and shapes. Can you use this information to predict relative rates of diffusion into/out of the cell? How do the following structures enhance exchange: root hairs, microvilli, cristae of mitochondria?

The eukaryotic cell's genetic instructions are housed in the nucleus and carried out by the ribosomes (Biology, *10e: 6.3*, Focus, *1e: 4.3*)

- The **nucleus** has the following key characteristics:
  - The nucleus contains most of the cell's DNA. It is in the nucleus where DNA is used as the template to make messenger RNA (mRNA), which contains the code to produce a protein. Because the nucleus contains the genetic information, it is referred to as the control center of the cell.
  - The nucleus is the most noticeable organelle in the cell because of its large relative size. The nucleus is surrounded by a double membrane, the **nuclear envelope**. Note that the nuclear envelope is continuous with the rough endoplasmic reticulum. The nuclear envelope contains **nuclear pores** that control what may enter or leave the nucleus.
  - **Chromatin** is the complex of DNA and protein housed in the nucleus that is formed from the chromosomes. As a cell gets ready for cell division, the diffuse threads of chromatin condense back into visible chromosomes.
  - The **nucleolus** is a region of the nucleus where ribosomal RNA (rRNA) complexes with proteins to form ribosomal subunits.
- **Ribosomes** are protein factories. They are composed of rRNA and protein and are sites of protein synthesis in the cell. Each ribosome consists of a large and a small subunit.
  - *Free ribosomes* are found floating in the cytosol and generally produce proteins that are used within the cell.
  - *Bound ribosomes* are attached to the endoplasmic reticulum and make proteins destined for export from the cell.

> *CONNECT WITH THE CURRICULUM FRAMEWORK*
>
> **Big Idea 2**
>
> Consider what cell features might be present in abundance or absent in certain cells based on their functions. We will test your understanding of this in questions at the end.

The endomembrane system regulates protein traffic and performs metabolic functions in the cell (Biology, *10e: 6.4*, Focus, *1e: 4.4*)

- **Endoplasmic reticulum (ER)** makes up more than half the total membrane structure in many cells. The ER is a network of membranes and sacs whose internal area is called the *cisternal space*. There are two types of ER:
  - **Smooth ER** has three primary functions: synthesis of lipids, metabolism of carbohydrates, and detoxification of drugs and poisons.
  - **Rough ER** is so called because its associated ribosomes make the structure appear rough under the microscope. Ribosomes associated with ER synthesize proteins that are generally secreted by the cell. As

the proteins are produced by the ER-bound ribosomes, the polypeptide chains travel across the ER membrane and into the cisternal space. Within the cisternal space the proteins are packaged into *transport vesicles,* which bud off the ER and move toward the Golgi apparatus.

- The **Golgi apparatus** operates something like the postal system—proteins from the transport vesicles are modified, stored, and shipped. As Figures 2.1 and 2.2 show, the Golgi apparatus consists of flattened sacs of membranes called cisternae, arranged in stacks. Golgi stacks have polarity—the *cis* face receives vesicles, whereas the *trans* face ships vesicles. Products of the ER are modified here, and the Golgi apparatus is extensive in cells specialized for secretion.
- **Lysosomes** are membrane-bound sacs of hydrolytic enzymes that can digest large molecules, including proteins, polysaccharides, fats, and nucleic acids. They have digestive enzymes that break down macromolecules to organic monomers that are released into the cytosol and thus recycled by the cell. The digestive or hydrolytic enzymes work best in the acidic environment found in lysosomes. If a lysosome breaks open or leaks, the enzymes are not very active in the neutral pH of the cell. This is a good example of the importance of cell compartmentalization.
- **Vacuoles** are membrane-bound vesicles. *Food vacuoles* such as those formed by phagocytosis of protists are one example, as are the *contractile vacuoles* that maintain water balance in *Paramecia* and other protists.
- **Central vacuoles** in plant cells may concentrate and contain compounds not found in the cytosol. A large central vacuole is one of the striking differences between plant and animal cells. In plants, a vacuole can make up as much as 80% of the cell.

Mitochondria and chloroplasts change energy from one form to another (Biology, *10e: 6.5,* Focus, *1e: 4.5*)

- **Mitochondria** are the sites of cellular respiration, the metabolic process that uses oxygen to generate ATP by extracting energy from sugars, fats, and other fuels. Mitochondria are found in both plant and animal cells. Study Figure 2.3 to learn the structure.

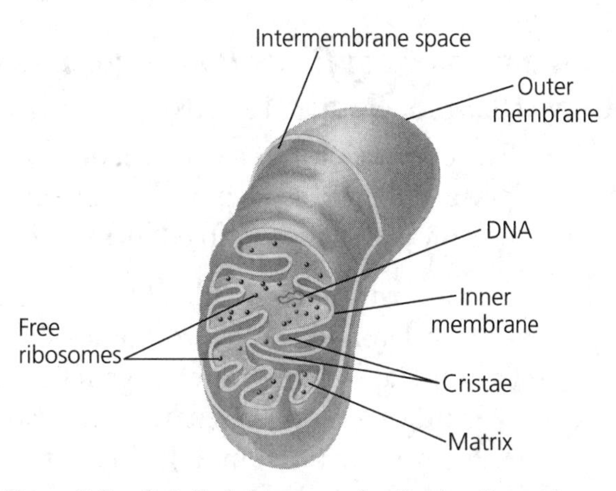

Figure 2.3 Detailed structure of animal cell membrane

- Mitochondria consist of an *outer* and *inner membrane.* The inner membrane is highly folded. These *cristae* (folds) increase the surface area, enhancing the productivity of cellular respiration.
- The inner compartment, the *mitochondrial matrix,* is fluid-filled and many of the reactions of cellular respiration occur here.
- The mitochondrial matrix contains mitochondrial DNA separate from nuclear DNA as well as ribosomes.

- **Chloroplasts,** found in plants and algae, are the sites of photosynthesis.
- The *endosymbiont theory* proposes that both mitochondria and chloroplasts share a similar origin. This theory states that these organelles descended from prokaryotic cells once engulfed by ancestors of eukaryotic cells. There are several lines of evidence for this:

  - Both organelles have a double-membrane structure.
  - Both organelles have their own ribosomes and circular DNA molecules.
  - Both reproduce independently within the cell.

- **Peroxisomes** are single-membrane-bound compartments in the cell responsible for various metabolic functions that involve the transfer of hydrogen from compounds to oxygen, producing hydrogen peroxide (H_2O_2). Peroxisomes break down fatty acids to be sent to the mitochondria for fuel and detoxify alcohol by transferring hydrogen from the poison to oxygen.
- This is an excellent example of how the cell's compartmental structure is crucial to its functions: The enzymes that produce hydrogen peroxide and those that dispose of this toxic compound are separate from other cellular components that could be damaged.

WHAT'S IMPORTANT TO KNOW?
Parts of 10e Concepts 6.6 and 6.7/Focus Concept 4.6 and part of Focus Concept 4.7 on the cytoskeleton and cell junctions are not part of the Essential Knowledges for the AP Biology Exam, although your teacher may select illustrative examples from this material. Knowledge of this information will also help you understand vital processes such as cell motility and division.

The cytoskeleton is a network of fibers that organizes structures and activities in the cell (Biology, *10e: 6.6*, Focus, *1e: 4.6*)

- The **cytoskeleton** is a network of protein fibers that runs throughout the cytoplasm, where it is responsible for support, motility, and regulating some biochemical activities. Three types of fibers make up the cytoskeleton:

  - **Microtubules,** made of the protein tubulin, are the largest of the cytoskeleton fibers. Microtubules shape and support the cell and also serve as tracks along which organelles equipped with *motor molecules* can move. They also separate chromosomes during mitosis and meiosis (forming the spindle) and are the structural components of cilia and flagella (found primarily in animal cells).

- **Microfilaments** are composed of the protein actin. Much smaller than microtubules, microfilaments function in smaller-scale support. When coupled with the motor molecule *myosin*, microfilaments can be involved with movement. *Examples*: amoeboid movement, cytoplasmic streaming, and contraction of muscle cells.
- **Intermediate filaments** are slightly larger than microfilaments and smaller than microtubules. Intermediate fibers are more permanent fixtures in the cell, where they are important in maintaining the shape of the cell and fixing the position of certain organelles.

- **Centrosomes** are a region located near the nucleus from which microtubules grow (the area is also called the microtubule-organizing center). Centrosomes contain centrioles in animal cells.
- **Centrioles** are located within the centrosomes of animal cells, where they replicate before cell division.
- A specialized arrangement of microtubules is responsible for the beating of flagella and cilia.

  - **Flagella** are usually long and few in number. Many unicellular eukaryotic organisms are propelled through the water by flagella, as are the sperm of animals, algae, and some plants.
  - **Cilia** are usually much shorter and more numerous than flagella. Cilia can also be used in locomotion or, when held in place as part of a tissue layer, they can move fluid over the surface of the tissue. For example, the lining of the trachea moves mucus-trapped debris out of the lungs in this manner.

- Although different in length, number per cell, and beating pattern, cilia and flagella share a common ultrastructure. Nearly all eukaryotic cilia and flagella have nine pairs of microtubules surrounding a central core of two microtubules. This arrangement is referred to as the "9 + 2" pattern.

Extracellular components and connections between cells help coordinate cellular activities (Biology, *10e: 6.7*, Focus, *1e: 4.7*)

- The **cell wall** of a plant protects the plant and helps maintain its shape. It is outside the plasma membrane. The primary component of cell walls is the carbohydrate *cellulose*.
- Prokaryotes and fungi also have cell walls, although not formed of cellulose.
- **Plasmodesmata** are channels that perforate adjacent plant cell walls and allow the passage of some molecules from cell to cell.
- **Extracellular matrix (ECM)** of animal cells is situated just external to the plasma membrane; it is composed of glycoproteins secreted by the cell (most prominent of which is collagen). The ECM greatly strengthens tissues and serves as a conduit for transmitting external stimuli into the cell, which can turn genes on and modify biochemical activity.
- Animal cells have three types of intercellular junctions:

  - **Tight junctions** are sections of animal cell membrane where two neighboring cells are fused, making the membranes watertight.

- **Desmosomes** fasten adjacent animal cells together, functioning like rivets to fasten cells into strong sheets.
- **Gap junctions** provide channels between adjacent animal cells through which ions, sugars, communication molecules, and other small molecules can pass.

> ***STUDY TIP*** Know the structure and function of each organelle and whether it is found in a plant cell or an animal cell, or both. Be able to predict and justify how a change in a cellular organelle would affect the function of the entire cell or organism.

Membrane Structure and Function

(Biology, *10e: Chapter 7*; Focus, *1e: Chapter 5*)

YOU MUST KNOW

- Why membranes are selectively permeable.
- The role of phospholipids, proteins, and carbohydrates in membranes.
- How water will move if a cell is placed in an isotonic, hypertonic, or hypotonic solution and be able to predict the effect of different environments on the organism.
- How electrochemical gradients and proton gradients are formed and function in cells.

Cellular membranes are fluid mosaics of lipids and proteins (Biology, *10e: 7.1*, Focus, *1e: 5.1*)

- The cell or **plasma membrane** is **selectively permeable**; that is, it allows some substances to cross it more easily than others.
- Membranes are predominantly made of phospholipids and proteins held together by weak interactions that cause the membrane to be fluid. The *fluid mosaic model* of the cell membrane describes the membrane as fluid, with proteins embedded in or associated with the phospholipid bilayer. Figure 2.4 shows the current model of an animal cell's plasma membrane. Find each part of the membrane as the three primary organic molecules of the membrane are described:
    - The hydrophilic phosphate portions of the phospholipids are oriented toward the aqueous inside and outside environments of cells, whereas the hydrophobic fatty acids face each other in a double layer (the *bilayer)* in the interior.
    - The **phospholipids** in the membrane provide a hydrophobic barrier that separates the cell from its liquid environment. Hydrophilic molecules cannot easily enter the cell, but hydrophobic molecules can enter much more easily; hence, the selectively permeable nature of the membrane.

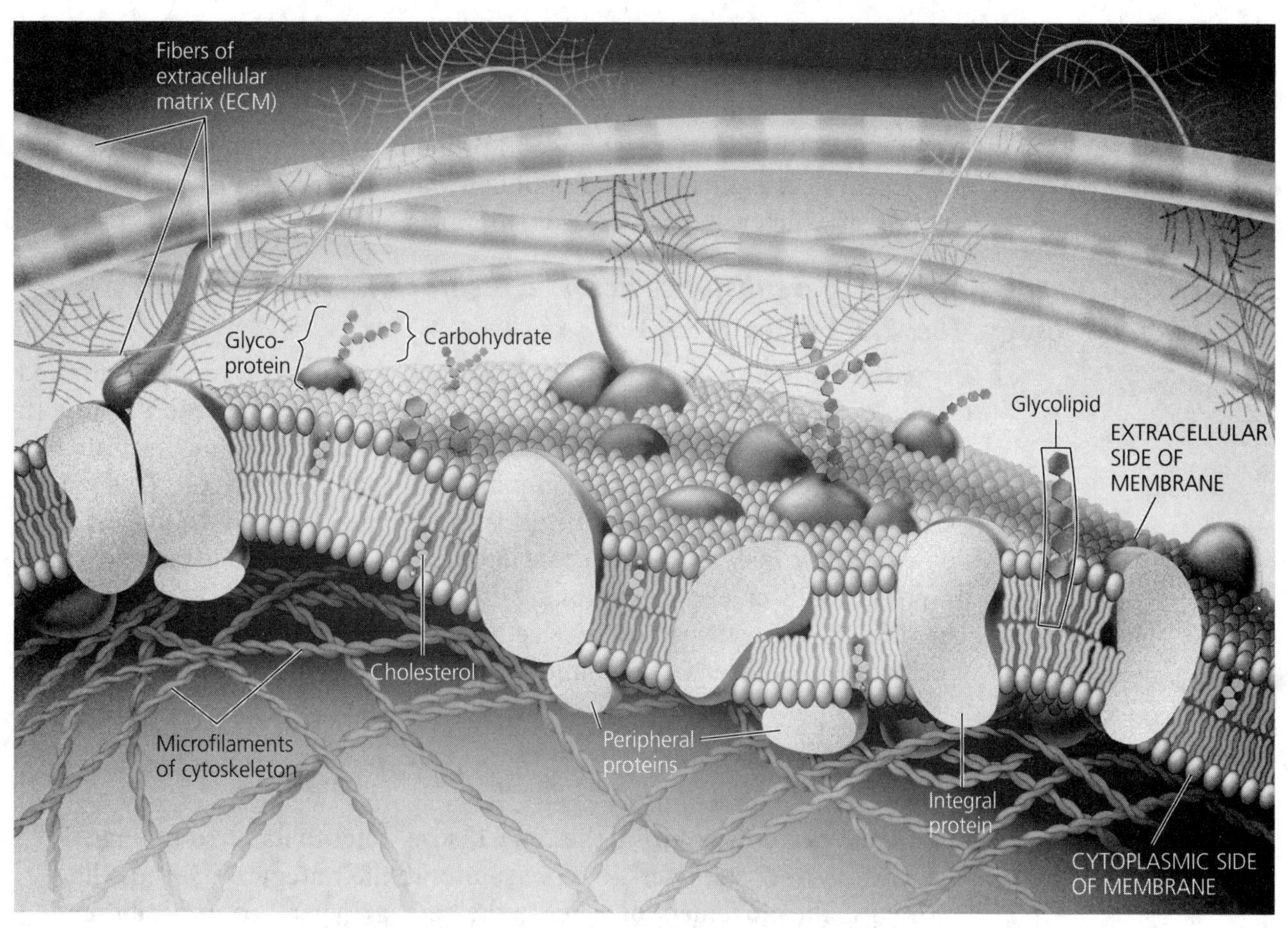

Figure 2.4 Structure of an animal cell's plasma membrane

- There are **proteins** completely embedded in the membrane, including some that span the membrane completely. These may serve as transport channels to move materials across the hydrophobic interior of the phospholipid bilayer, or as molecular receptors to bind to signaling molecules (ligands). Peripheral proteins are loosely bound to the membrane's surface. An example would be a G protein, which can move along the membrane when a ligand binds a G protein-coupled receptor. (See Figure 2.9 on page 60.)
- **Carbohydrates** on the membrane are crucial in cell-cell recognition (which is necessary for proper immune function) and in developing organisms (for tissue differentiation). Cell-surface carbohydrates vary from tissue to tissue and are the reason that blood transfusions must be type-specific.

Membrane structure results in selective permeability (Biology, *10e: 7.2*, Focus, *1e: 5.2*)

- Nonpolar molecules—such as hydrocarbons, carbon dioxide, and oxygen—are hydrophobic and can dissolve in the hydrophobic interior of the phospholipid bilayer and cross the membrane easily.
- The hydrophobic core of the membrane impedes the passage of ions and polar molecules, which are hydrophilic. However, hydrophilic substances can avoid

the lipid bilayer by passing through **transport proteins** that span the membrane (see Figure 2.4).

- Perhaps the most important molecule to move across the membrane is water. Water moves through special transport proteins termed **aquaporins**. Aquaporins greatly accelerate the speed (3 billion water molecules per aquaporin per second!) at which water can cross membranes.

Passive transport is diffusion of a substance across a membrane with no energy investment (Biology, *10e: 7.3*, Focus, *1e: 5.3*)

- In **passive diffusion,** a substance travels from where it is more concentrated to where it is less concentrated, diffusing down its **concentration gradient**. Hydrocarbons, carbon dioxide, and oxygen (O_2) are hydrophobic substances that can pass easily across the cell membrane by passive diffusion. This type of diffusion requires that no work be done, and it relies only on the thermal motion energy intrinsic to the molecule in question. It is called "passive" because the cell expends no energy in moving the substances.
- The diffusion of water across a selectively permeable membrane is **osmosis.** A cell has one of three water relationships with the environment around it.
    - In an **isotonic solution** there will be no net movement of water across the plasma membrane. Water crosses the membrane but at the same rate in both directions.
    - In a **hypertonic solution** the cell will lose water to its surroundings. The *hyper-* prefix refers to more solutes in the water around the cell; hence, the movement of water to the higher (hyper-) concentration of solutes. In this case the cell loses water to the environment, will shrivel, and may die.
    - In a **hypotonic solution** water will enter the cell faster than it leaves. The *hypo-* prefix refers to fewer solutes in the water around the cell; hence, the movement of water into the cell where solutes are more heavily concentrated. In this case the cell will swell and may burst.
- Be sure to watch the wording: Is the cell hypertonic to the solution it is placed in, or is the surrounding solution hypertonic to the cell? In the first example water moves into the cell, whereas in the second water moves out of the cell. Remember this: Water moves *from* a hypotonic solution *to* a hypertonic solution. **Hypo→Hyper**

> ***STUDY TIP*** It is likely that you will do an investigation or lab that focuses on osmosis and diffusion. Work with these ideas until you can predict the direction of water movement based on the concentration of solutes inside and outside the cell.

- **Ions** and **polar molecules** cannot pass easily across the membrane. The process by which ions and hydrophilic substances diffuse across the cell membrane with the help of transport proteins is called **facilitated diffusion**. Transport proteins are specific (like enzymes) for the substances they transport. They work in one of two ways:

- They provide a hydrophilic channel through which the molecules in question can pass.
- They bind loosely to the molecules in question and carry them through the membrane.

Active transport uses energy to move solutes against their gradients (Biology, *10e: 7.4*, Focus, *1e: 5.4*)

- In **active transport**, substances are moved against their concentration gradient—that is, from the region where they are *less* concentrated to the region where they are *more* concentrated. This type of transport requires energy, usually in the form of ATP.
- A common example of active transport is the **sodium-potassium pump**. This transmembrane protein pumps sodium out of the cell and potassium into the cell. The sodium-potassium pump is necessary for proper nerve transmission and is a major energy consumer in your body as you read this.
- The inside of the cell is negatively charged compared with outside the cell. The difference in electric charge across a membrane is expressed in voltage and termed the **membrane potential**. Because the inside of the cell is negatively charged, a positively charged ion on the outside, like sodium, is attracted to the negative charges inside the cell. Thus, two forces drive the diffusion of ions across a membrane:
  - *A chemical force*, which is the ion's concentration gradient.
  - *A voltage gradient* across the membrane, which attracts positively charged ions and repels negatively charged ions.

  This combination of forces acting on an ion forms an **electrochemical gradient**.

> ***STUDY TIP*** Both photosynthesis and cellular respiration, the topics of two upcoming chapters, utilize electrochemical (proton) gradients as potential energy sources to generate ATP. By carefully studying electrochemical gradients now, you will be in a good position to understand more complex processes later.

To review transport, try your hand at the following questions using Figure 2.5.

1. Which figure represents *active transport*? There are two ways you should be able to tell.
2. Which section shows *simple diffusion*?
3. Which section shows *facilitated diffusion* with a *carrier protein*?
4. Which section shows *facilitated diffusion* with a *channel protein*?

(Answers are at the end of this chapter.)

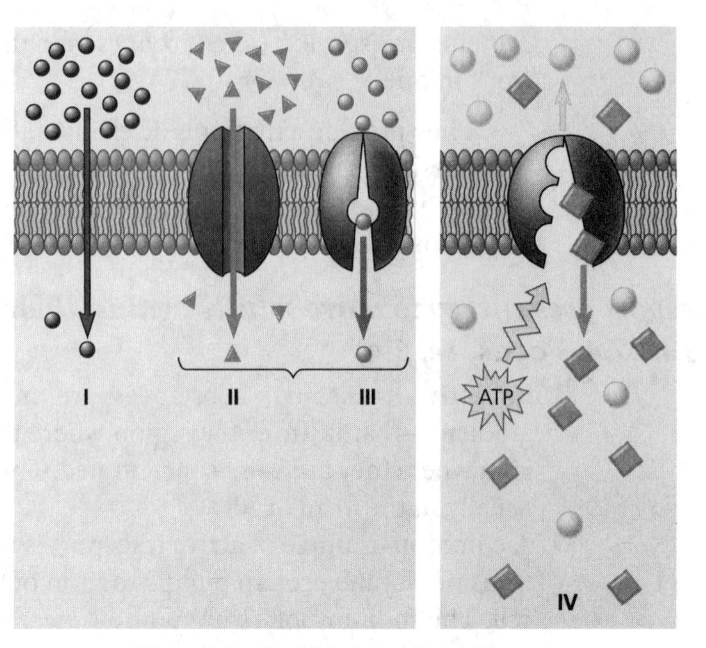

Figure 2.5 Passive and active transport

Bulk transport across the plasma membrane occurs by exocytosis and endocytosis (Biology, *10e: 7.5*, Focus, *1e: 5.5*)

- Large molecules are moved across the cell membrane through exocytosis and endocytosis.
  - In **exocytosis**, vesicles from the cell's interior fuse with the cell membrane, expelling their contents.
  - In **endocytosis**, the cell forms new vesicles from the plasma membrane; this is basically the reverse of exocytosis, and this process allows the cell to take in macromolecules. Examples of endocytosis include the engulfing of foreign particles by white blood cells or amoebas.

Answers to questions for Figure 2.5:

1. IV (It shows the use of ATP and molecules being moved against a concentration gradient.)
2. I
3. III
4. II

Cell Communication

(Biology, *10e: Ch. 11*, Focus, *1e: 5.6*)

YOU MUST KNOW

- The three stages of cell communication: reception, transduction, and response.
- How a receptor protein recognizes signal molecules and starts transduction.
- How a cell signal is amplified by a phosphorylation cascade.
- An example of a second messenger and its role in a signal transduction pathway.
- How a cell response in the nucleus turns on genes, whereas in the cytoplasm it activates enzymes.
- What apoptosis means and why it is important to normal functioning of multicellular organisms.

External signals are converted to responses within the cell (Biology, *10e: 11.1*, Focus, *1e: 5.6*)

- In signaling, animal cells communicate by direct contact or by secreting local regulators, such as growth factors or neurotransmitters. There are three stages of cell signaling.
  - **Reception**—The target cell's detection of a signal molecule coming from outside the cell.
  - **Transduction**—The conversion of the signal to a form that can bring about a specific cellular response.
  - **Response**—The specific cellular response to the signal molecule.

Reception: A signaling molecule binds to a receptor protein, causing it to change shape (Biology,*10e: 11.2*, Focus, *1e: 5.6*)

- The binding between a signal molecule **(ligand)** and a **receptor** is highly specific. A conformational change in a receptor is often the initial transduction of the signal. Receptors are found in two places:
  - **Intracellular receptors** are found inside the plasma membrane in the cytoplasm or nucleus. The signal molecule must cross the plasma membrane and therefore must be hydrophobic, like the steroid hormone testosterone.
  - **Plasma membrane receptors** bind to water-soluble ligands.

> **CONNECT WITH THE CURRICULUM FRAMEWORK**
>
> **Big Idea 3**
>
> You are expected to be able to describe a model that expresses the key elements of a signal transduction pathway leading to a cellular response. Examples you may use include G protein-coupled receptors, ligand-gated ion channels, and receptor tyrosine kinases. Select the one you wish to perfect, and learn it well!

- A **G protein-coupled receptor** is a membrane receptor that works with the help of a **G protein**. Follow Figure 2.6 to review how these receptors work.
  - Step 1: The ligand or signaling molecule has bound to the G protein-coupled receptor. This causes a conformational change in the receptor

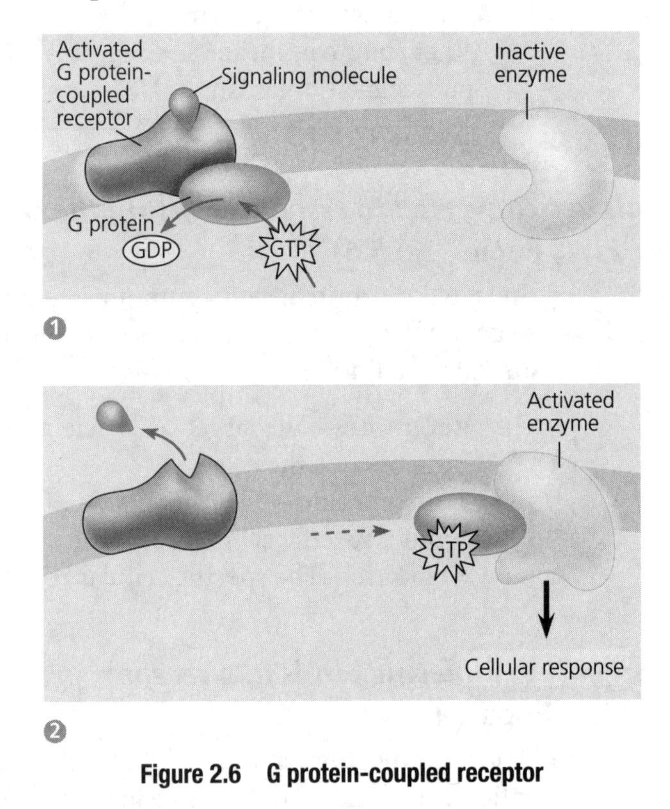

Figure 2.6 G protein-coupled receptor

so that it may now bind to an inactive G protein, causing a GTP to displace the GDP. This activates the G protein.
  - Step 2: The G protein binds to a specific enzyme and activates it. When the enzyme is activated, it can trigger the next step in a pathway leading to a cellular response. All the molecular shape changes are temporary. To continue the cellular response, new signal molecules are required.

- The **receptor tyrosine kinases** are a second type of membrane protein. Follow Figure 2.7 to review how they function *(not discussed in* Focus *but explained next).*

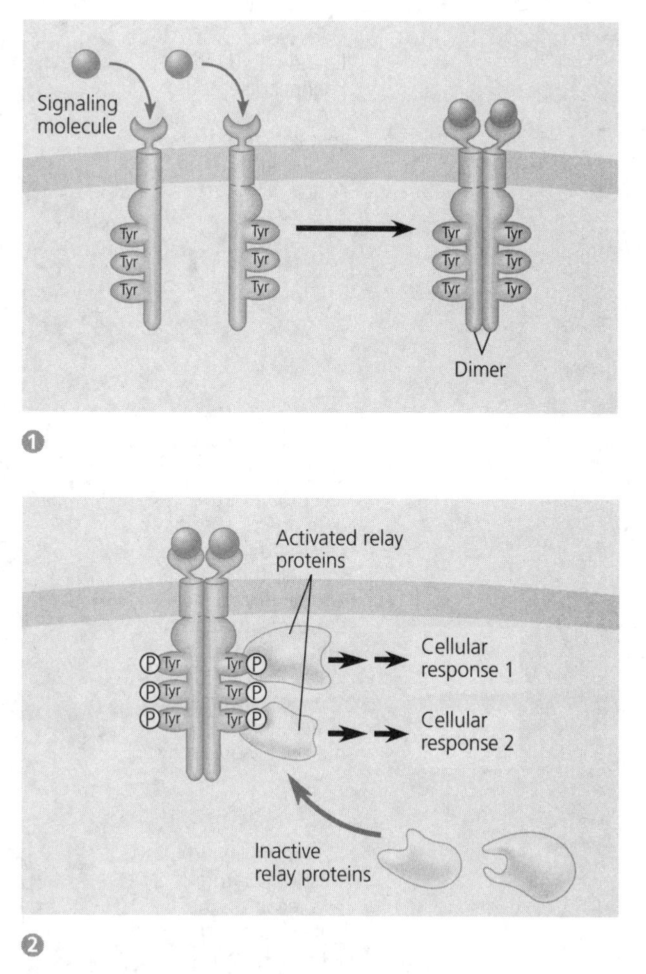

Figure 2.7 Receptor tyrosine kinase

- Step 1 shows the binding of signal molecules to the receptors and the subsequent formation of a dimer. In the dimer configuration each tyrosine kinase adds a phosphate from an ATP molecule.
- Step 2 shows the fully activated receptor protein as it initiates a unique cellular response for each phosphorylated tyrosine.
- The ability of a single ligand to activate multiple cellular responses is a key difference between G protein-coupled receptors and receptor tyrosine kinases.

- Specific signal molecules cause **ligand-gated ion channels** in a membrane to open or close, regulating the flow of specific ions. When the ligand opens the gate and ions flow into the cell, the change in ion concentration can result in a change in cell activity. See Figure 2.8.

Transduction: Cascades of molecular interactions relay signals from receptors to target molecules in the cell (Biology, *10e: 11.3*, Focus, *1e: 5.6*)

- Signal transduction pathways often involve a **phosphorylation cascade**. Because the pathway is usually a multistep one, the possibility of greatly amplifying the signal exists. At each step enzymes called **protein kinases** phosphorylate

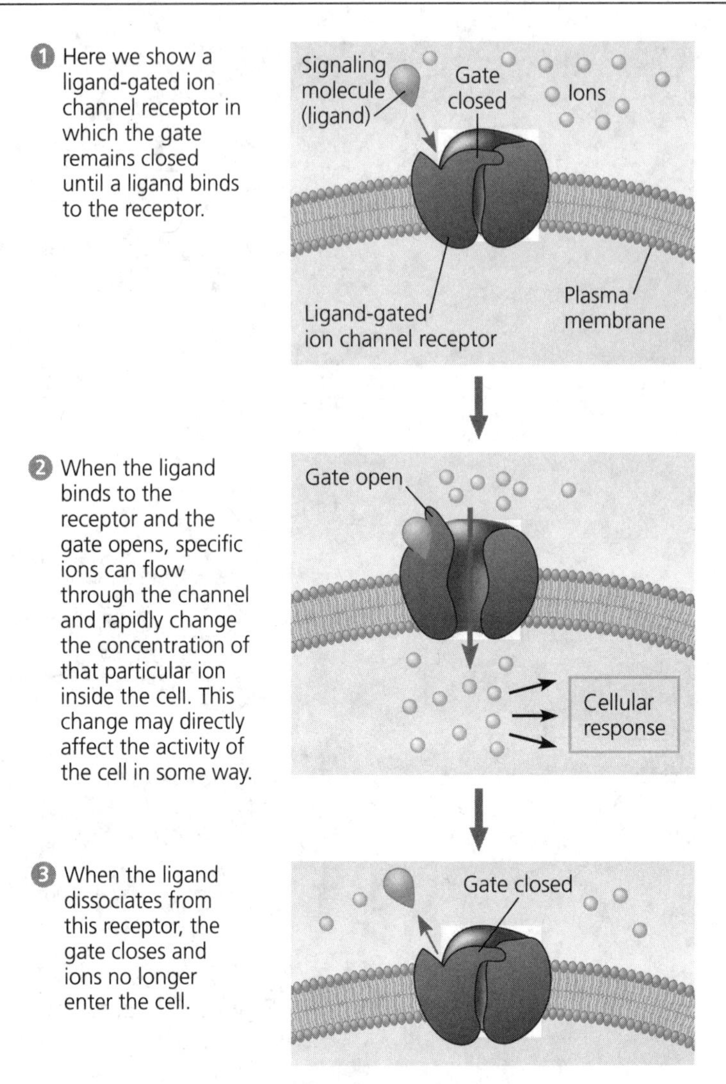

Figure 2.8 Ion channel receptors

and thereby activate many proteins at the next level. This cascade of phosphorylation greatly enhances the signal, allowing for a large cellular response.

- **Protein phosphatases** are enzymes that remove phosphate groups and inactive protein kinases. Thus, the signal can be turned *on* by kinases and *off* by phosphatases.
- Not all components of signal transduction pathways are proteins. Many signaling pathways involve small, nonprotein water-soluble molecules or ions called **second messengers**. *Calcium ions* and *cyclic AMP* are two common second messengers. The second messengers, once activated, can initiate a phosphorylation cascade resulting in a cellular response.

Response: Cell signaling leads to regulation of transcription or cytoplasmic activities (Biology, *10e: 11.4*, Focus, *1e: 5.6*)

- Many signaling pathways ultimately regulate protein synthesis, usually by turning specific genes on or off in the nucleus. Often, the final activated molecule in a signaling pathway functions as a transcription factor.

- In the cytoplasm, signaling pathways often regulate the activity of proteins rather than their synthesis. For example, the final step in the signaling pathway may affect the activity of enzymes or cause cytoskeleton rearrangement.

Apoptosis integrates multiple cell-signaling pathways (Biology, *10e: 11.5*, Focus, *1e: 16.1*)

- An elaborate example of cell signaling is a program of controlled cell suicide called **apoptosis**. During apoptosis the cell is systematically dismantled and digested. This protects neighboring cells from damage that would occur if a dying cell merely leaked out its digestive and other enzymes.
  - Apoptosis is triggered by signals that activate a cascade of "suicide" proteins in the cells.
  - In vertebrates apoptosis is a normal part of development and is essential for a normal nervous system, for the operation of the immune system, and for normal morphogenesis of hands and feet in humans.

STUDY TIP Cell signaling has emerged as an important topic that explains cell interactions. Devote study time to understanding how reception, transduction, and response collectively result in cell communication.

The Cell Cycle

(Biology, *10e: Chapter 12*; Focus, *1e: Chapter 9*)

YOU MUST KNOW

- The structure of the duplicated chromosome.
- The events that occur in the cell cycle (G_1, S, and G_2).
- The role of cyclins and cyclin-dependent kinases in the regulation of the cell cycle.
- Ways in which the normal cell cycle is disrupted to cause cancer, or halted in certain specialized cells.
- The features of mitosis that result in the production of genetically identical daughter cells including replication, alignment of chromosomes (metaphase), and separation of chromosomes (anaphase).

Most cell division results in genetically identical daughter cells (Biology, *10e: 12.1*, Focus, *1e: 9.1*)

- The **cell cycle** is the life of a cell from the time it is first formed from a dividing parent cell until its own division into two cells.

- A cell's endowment of DNA, its genetic information, is called its **genome**. Before the cell can divide, the cell's genome must be copied.
  - All eukaryotic organisms have a characteristic number of chromosomes in their cell nuclei. As an example, human **somatic cells** (all body cells except gametes) have 46 chromosomes, which is the diploid chromosome number. *Mitosis* is the process by which somatic cells divide, forming daughter cells that contain the same chromosome number as the parent cell.
  - Human **gametes**—sperm and egg cells—are haploid and have half the number of chromosomes as a diploid cell. Human gametes have 23 chromosomes. A special type of cell division called meiosis (the topic of Chapter 13) results in gametes.
- When the chromosomes are replicated, each duplicated chromosome consists of two **sister chromatids** attached by a **centromere**. Figure 2.9 will help you visualize this arrangement.
  - The two sister chromatids have identical DNA sequences.
  - Later, in the process of cell division, the two sister chromatids will separate and move into two new cells. Once the sister chromatids separate, they are considered individual chromosomes.

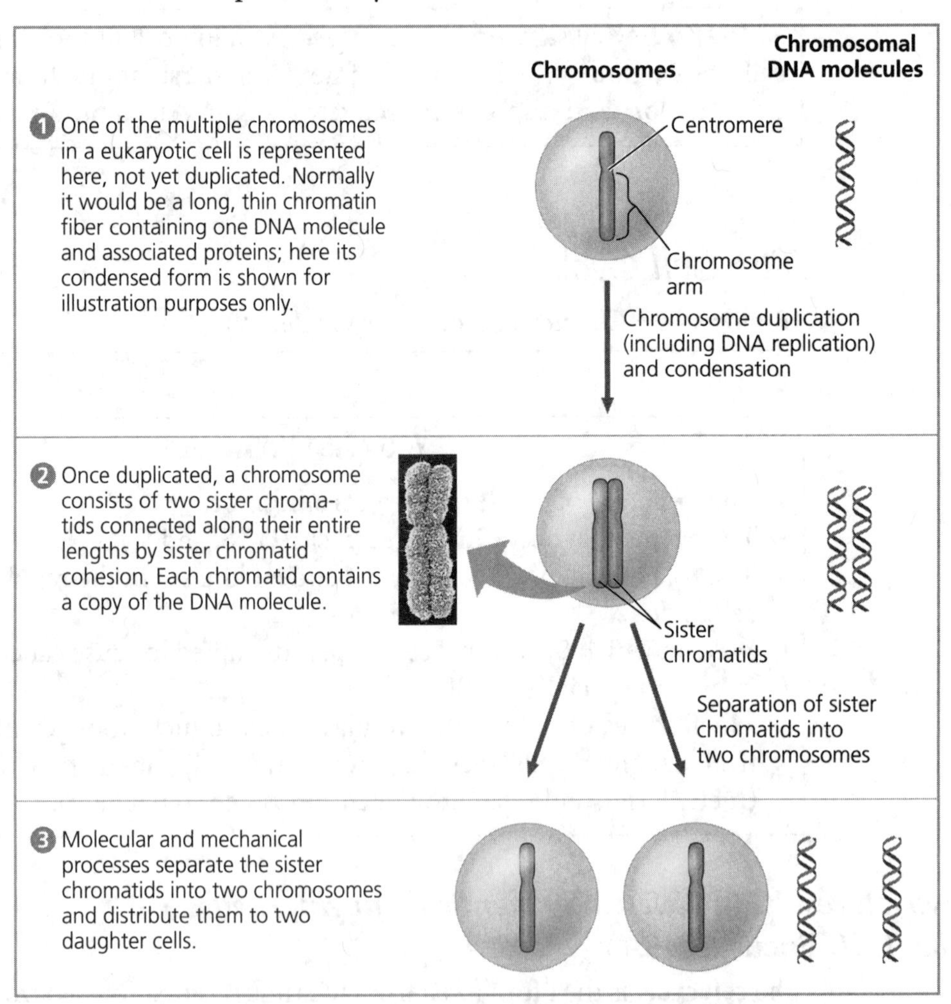

Figure 2.9 Chromosome duplication and distribution during cell division

- **Mitosis** is the division of the cell's nucleus. It may be followed by **cytokinesis**, which is the division of the cell's cytoplasm. Where there was one cell, there are now two, each the genetic equivalent of the parent cell.

The mitotic phase alternates with interphase in the cell cycle (Biology, *10e: 12.2*, Focus, *1e: 9.2*)

- The primary events of **interphase**, which is 90% of the cell cycle, follow:
  - In the **G_1 phase** the cell grows while carrying out cell functions unique to its cell type.
  - In the **S phase** the cell continues to carry out its unique functions but does one other important process—it duplicates its chromosomes. This means it faithfully makes a copy of the DNA that makes up the cell's chromosomes.
  - The **G_2 phase** is the period after the chromosomes have been duplicated and just before mitosis. The cell continues to grow and carry out its functions during this time.

WHAT'S IMPORTANT TO KNOW?
The Curriculum Framework says that you do not have to know the specific phases of mitosis. However, many students find it useful to know the phases in order to organize their understanding of the process.

- Mitosis can be broken down into five phases, not including cytokinesis. At each stage, find the specific references in Figure 2.10. To simplify your studying, key features of each phase are given. Focus on how each step contributes to the distribution of identical genetic information to two daughter cells.
  - **Prophase:**

    1. The chromatin becomes more tightly coiled into discrete chromosomes.
    2. The nucleoli disappear.
    3. The mitotic spindle (consisting of microtubules extending from the two centrosomes) begins to form in the cytoplasm.

  - **Prometaphase:**

    1. The nuclear envelope begins to fragment, allowing the microtubules to attach to the chromosomes.
    2. The two chromatids of each chromosome are held together by the centromere. The centromere contains protein kinetochores on each chromatid, which is where the microtubules will attach.

  - **Metaphase:**

    1. The microtubules move the chromosomes to the metaphase plate at the equator of the cell. The microtubule complex is referred to as the *spindle*.
    2. The centrioles have migrated to opposite poles in the cell, riding along on the developing spindle.

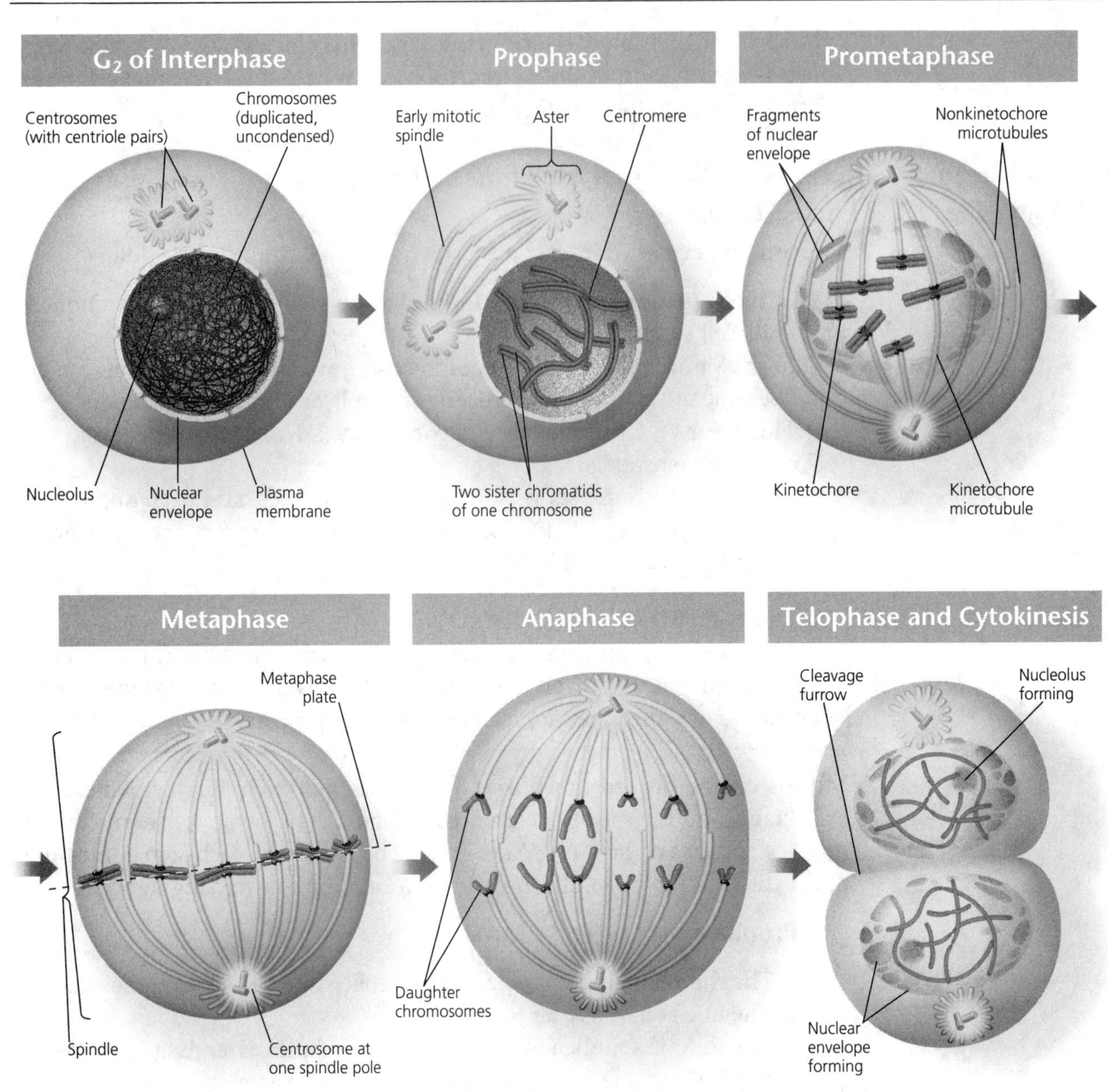

Figure 2.10 The mitotic division of an animal cell

- **Anaphase:**

  1. Sister chromatids begin to separate, pulled apart by motor molecules interacting with kinetochore microtubules.
  2. The cell elongates, as the nonkinetochore microtubules ratchet apart, again with the help of motor molecules.
  3. By the end of anaphase, the opposite ends of the cell both contain complete and equal sets of chromosomes.

- **Telophase:**

  1. The nuclear envelopes re-form around the sets of chromosomes located at opposite ends of the cell.

2. The chromatin fiber of the chromosomes becomes less condensed.
3. **Cytokinesis** begins, during which the cytoplasm of the cell is divided. In animal cells, a **cleavage furrow** forms that eventually divides the cytoplasm; in plant cells, a **cell plate** forms that divides the cytoplasm.
4. Prokaryotes replicate their genome by **binary fission** rather than mitosis.

The eukaryotic cell cycle is regulated by a molecular control system (Biology, *10e: 12.3*, Focus, *1e: 9.3*)

- The steps of the cycle are controlled by a **cell cycle control system**. This control system moves the cell through its stages by a series of **checkpoints**, during which molecular signals tell the cell either to continue dividing or to stop.
- The major cell cycle checkpoints include the **G_1 phase checkpoint**, **G_2 phase checkpoint**, and **M phase checkpoint**.
- The **G_1 phase checkpoint** seems to be most important. If the cell gets the go-ahead signal at this checkpoint, it usually completes the whole cell cycle and divides. If it does not receive the go-ahead signal, it enters a nondividing phase called the *G_0 phase*.
- Most mature human cells remain in G_0 and never receive the molecular signal to divide (pass through the G_1 checkpoint). Muscle and nerve cells never divide, but liver cells can respond to signals, moving from G_0 back to the cell cycle at G_1.
- **Kinases** are the protein enzymes that control the cell cycle. They exist in the cells at all times but are active only when they are connected to **cyclin** proteins. Thus, they are called **cyclin-dependent kinases (Cdks)**. Specific kinases give the go-ahead signals at the G_1 and G_2 checkpoints.
- As a specific example, cyclin molecules combine with Cdk molecules, producing enough molecules of **MPF** to pass the G_2 checkpoint and initiate the events of mitosis. (MPF promotes mitosis—think of it as **M**itosis **P**romoting **F**actor.)
- How does the cell stop cell division? During anaphase, MPF switches itself off by starting a process that leads to the destruction of cyclin molecules. Without cyclin molecules Cdk molecules become inactive, bringing mitosis to a close.
- Normal cell division has two key characteristics:
  - **Density-dependent inhibition**—The phenomenon in which crowded cells stop dividing.
  - **Anchorage dependency**—Normal cells must be attached to a substratum, like the extracellular matrix of a tissue, to divide.
- Knowing the features of normal cell division is important because cancer cells exhibit neither density-dependent inhibition nor anchorage dependency. Cancer is covered in more depth in the section on molecular genetics, but here are several important points:
  - **Transformation** is the process that converts a normal cell to a cancer cell.

- A **tumor** is a mass of abnormal cells within otherwise normal tissue. If the abnormal cells remain at the original site, the lump is called a **benign tumor**. A **malignant tumor** becomes invasive enough to impair the functions of one or more organs. An individual with a malignant tumor is said to have cancer.
- **Metastasis** occurs when cells separate from a malignant tumor and enter blood or lymph vessels and travel to other parts of the body.

For Additional Review

Compare the process of meiosis with the process of mitosis. In your comparison, include a study of the change in chromosomal number through the cell, the purposes of each process within an organism, and the starting material and product for each. Note: The details of meiosis are covered in the first section of Topic 4.

Level 1: Knowledge/Comprehension Questions

The Knowledge/Comprehension questions review essential content knowledge but don't represent the type of question you will see on the AP exam. Level 2, Application/Synthesis, questions are similar to those you will see on the exam.

1. Salivary glands produce a large quantity of enzymes, which are transported out of the secretory cells. Which of the following organelles would be in abundance in these cells?
   (A) nuclei
   (B) ribosomes
   (C) smooth ER
   (D) lysosomes

2. Prokaryotic and eukaryotic cells have all of the following structures in common EXCEPT
   (A) a plasma membrane.
   (B) cytoplasm.
   (C) linear chromosomes.
   (D) ribosomes.

Directions: Questions 3–6 consist of four lettered choices followed by a list of numbered phrases or sentences. For each numbered phrase or sentence, select the one choice that is most closely related to it. Each choice may be used once, more than once, or not at all in each group.

Questions 3–6

(A) Mitochondria
(B) Golgi apparatus
(C) Lysosomes
(D) Smooth endoplasmic reticulum

3. An organelle that is characterized by extensive, folded membranes and is abundant in cells that detoxify poisons, such as liver cells

4. An organelle with a *cis* and *trans* face, which acts as the packaging and secreting center of the cell

5. These organelles are not found in red blood cells but are present in large numbers in muscle cells

6. Large membrane-bound structures that contain hydrolytic enzymes and that are found predominantly in animal cells

7. Insulin is a protein synthesized in the cytoplasm of pancreas cells and then transported to the plasma membrane where it enters the bloodstream. Which of the following summarizes the pathway for insulin through a pancreatic cell?
   (A) ribosome → Golgi body → vesicle → plasma membrane
   (B) smooth ER → nucleus → vesicle → plasma membrane
   (C) rough ER → Golgi body → vesicle → plasma membrane
   (D) nucleus → vesicle → rough ER → plasma membrane

8. Which of the following component molecules of the plasma membrane is most important in the reception phase of cell signaling?
   (A) protein
   (B) phospholipids
   (C) cholesterol
   (D) carbohydrates

9. Which of the following uses passive transport, without protein channels, to move materials across the cell membrane?
   (A) the depolarization of a muscle cell
   (B) the uptake of glucose by the microvilli of cells lining the stomach
   (C) the movement of insulin across the cell membrane
   (D) the movement of carbon dioxide across the cell membrane

10. Large molecules are moved out of the cell by which of the following processes?
   (A) endocytosis
   (B) exocytosis
   (C) translocation
   (D) passive diffusion

11. In cell signaling, how is the flow of specific ions regulated?
   (A) opening and closing of ligand-gated ion channels
   (B) transduction
   (C) cytoskeleton rearrangement
   (D) endocytosis

12. Which of the following processes could result in the net movement of a substance into a cell if the substance is more concentrated in the cell than in the surroundings?
   (A) active transport
   (B) facilitated diffusion
   (C) diffusion
   (D) osmosis

13. What is a G protein?
   (A) a specific type of membrane receptor protein
   (B) a protein on the cytoplasmic side of a membrane that becomes activated by a receptor protein
   (C) a membrane-bound enzyme that converts ATP to cAMP
   (D) a guanine nucleotide that converts between GDP and GTP to activate and inactivate relay proteins

14. Which of the following can activate a protein by transferring a phosphate group to it?
   (A) cAMP
   (B) G protein
   (C) protein phosphatase
   (D) protein kinase

15. Many signal transduction pathways use second messengers to
   (A) transport a signal through the lipid bilayer portion of the plasma membrane.
   (B) relay a signal from the outside to the inside of the cell.
   (C) relay the message from the inside of the membrane throughout the cytoplasm.
   (D) diffuse directly into the nucleus, turning on genes.

16. Which of the following signal molecules pass through the plasma membrane and bind to intracellular receptors that move into the nucleus and function as transcription factors to regulate gene expression?
   (A) epinephrine
   (B) neurotransmitter released into synapse between nerve cells
   (C) yeast mating factors alpha and **a**
   (D) testosterone, a steroid hormone

Directions: Questions 17–22 consist of four lettered choices followed by a list of numbered phrases or sentences. For each numbered phrase or sentence, select the one choice that is most closely related to it. Each choice may be used once, more than once, or not at all in each group.

Questions 17–22

(A) Synthesis (S-phase)
(B) Gap 1
(C) Cytokinesis
(D) Anaphase

17. If this did not occur, daughter cells would have two complete sets of chromosomes.

18. Division of the cytoplasm of the cell.

19. The genetic material of the cell replicates to prepare for cell division.

20. The cell is carrying out its typical functions, not engaged in producing molecules necessary for division.

21. Failure of this stage to occur prior to mitosis would result in daughter cells without sufficient DNA.

22. Sister chromatids begin to separate.

23. The fluid mosaic model describes the plasma membrane as consisting of
    (A) a phospholipid bilayer with embedded carbohydrates.
    (B) two layers of phospholipids with cholesterol sandwiched between them.
    (C) carbohydrates and phospholipids that can drift in the membrane.
    (D) diverse proteins embedded in a phospholipid bilayer.

24. Small, nonpolar, hydrophobic molecules such as fatty acids
    (A) easily pass through a membrane's lipid bilayer.
    (B) very slowly diffuse through a membrane's lipid bilayer.
    (C) require transport proteins to pass through a membrane's lipid bilayer.
    (D) are actively transported across cell membranes.

25. Which of the following statements regarding the cell-cycle control system is *false*?
    (A) The cell-cycle control system receives messages from outside the cell that influence cell division.
    (B) The cell cycle control system triggers and controls major events in the cell cycle.
    (C) The cell cycle control system includes fluctuating levels of cyclins and cyclin-dependent kinases.
    (D) The cell cycle control system operates independently of the growth factors.

Level 2: Application/Analysis/Synthesis Questions

1. Cells of the pancreas will incorporate radioactively labeled amino acids into proteins. This "tagging" of newly synthesized proteins enables a researcher to track their location. In this case, we are tracking an enzyme secreted by pancreatic cells. What is its most likely pathway?
    (A) ER → lysosomes → vesicles that fuse with plasma membrane
    (B) Golgi apparatus→ ER → lysosome
    (C) nucleus → ER → Golgi apparatus
    (D) ER → Golgi apparatus → vesicles that fuse with plasma membrane

2. If the S phase were eliminated from the cell cycle, the daughter cells would
   (A) have half the genetic material found in the parent cell.
   (B) be generally identical to each other.
   (C) be genetically identical to the parent.
   (D) synthesize the missing genetic material on their own.

3. Which figure depicts an animal cell placed in a solution hypotonic to the cell?

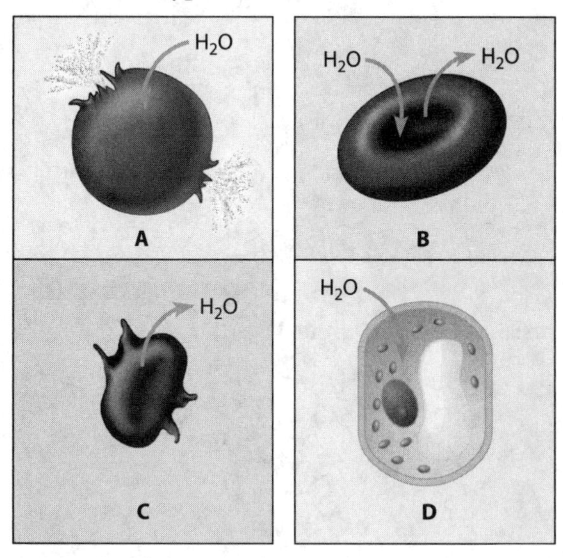

   (A) cell A
   (B) cell B
   (C) cell C
   (D) cell D

4. Which two figures show a cell that is hypertonic to its environment?
   (A) cells A and B
   (B) cells A and C
   (C) cells A and D
   (D) cells C and D

After reading the paragraphs, answer the question(s) that follow.

Americans spend up to $100 billion annually for bottled water (41 billion gallons). The only beverages with higher sales are carbonated soft drinks. Recent news stories have highlighted the fact that most bottled water comes from municipal water supplies (the same source as your tap water), although it may undergo an extra purification step called reverse osmosis.

Imagine two tanks that are separated by a membrane that is permeable to water but not to the dissolved minerals present in the water. Tank A contains tap water and Tank B contains the purified water. Under normal conditions, the purified water would cross the membrane to dilute the more concentrated tap water solution. In the reverse osmosis process, pressure is applied to the tap water tank to force the water molecules across the membrane into the pure water tank.

5. After the reverse osmosis system has been operating for 30 minutes, the solution in Tank A would
   (A) be hypotonic to Tank B.
   (B) be isotonic to Tank B.
   (C) be hypertonic to Tank B.
   (D) move by passive transport to Tank B.

6. If you shut the system off and pressure was no longer applied to Tank A, you would expect
   (A) the water to flow from Tank A to Tank B.
   (B) the water to reverse flow from Tank B to Tank A.
   (C) the water to flow in equal amounts in both directions.
   (D) the water to flow against the concentration gradient.

7. Earl Sutherland received the Nobel Prize for his discovery of cAMP as a second messenger. Which observation suggested to Sutherland the involvement of a second messenger in epinephrine's effect on liver cells?
   (A) Enzymatic activity was proportional to the amount of calcium added to a cell-free extract.
   (B) Receptor studies indicated that epinephrine was a ligand.
   (C) Glycogen breakdown was observed only when epinephrine was administered to intact cells.
   (D) Glycogen breakdown was observed when epinephrine and glycogen phosphorylase were combined.

8. Protein phosphorylation is commonly involved with all of the following EXCEPT
   (A) regulation of transcription by extracellular signaling molecules.
   (B) enzyme activation.
   (C) activation of G protein-coupled receptors.
   (D) activation of receptor tyrosine kinases.

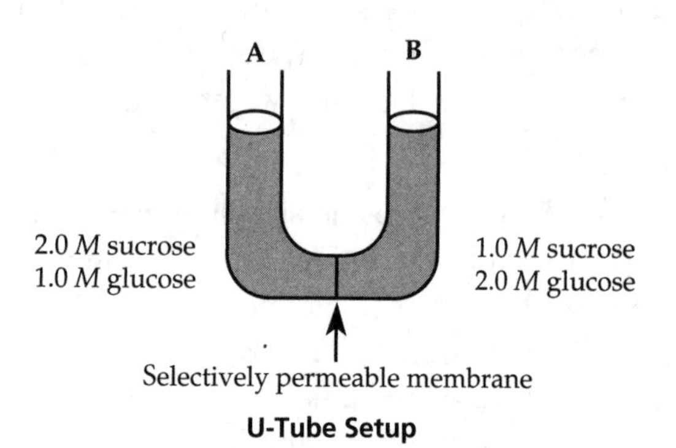

U-Tube Setup

9. The drawing above shows two solutions of glucose and sucrose in a U-tube containing a semipermeable membrane (which allows the passage of sugars). Which of the following accurately describes what will take place next?
   (A) Glucose will diffuse from side A to side B.
   (B) Sucrose will diffuse from side B to side A.
   (C) No net movement of molecules will occur.
   (D) Glucose will diffuse from side B to side A.

10. Look at the same figure, but let's assume the membrane is not freely permeable to both glucose ($C_6H_{12}O_6$) and sucrose ($C_{12}H_{22}O_{11}$) but only to the smaller of the two sugars. Now what results would you predict?
    (A) Glucose will diffuse from side A to side B.
    (B) Glucose will diffuse from side B to side A.
    (C) Sucrose will diffuse from side A to side B.
    (D) Sucrose will diffuse from side B to side A.

11. Using the same figure and assumption as for question 10, what effect will the movement of the sugar have on the water level in the tubes?
    (A) The water level will go up in side A.
    (B) The water level will go up in side B.
    (C) No net movement of water molecules will occur.
    (D) The membrane is not permeable to water, only to the sugar molecules.

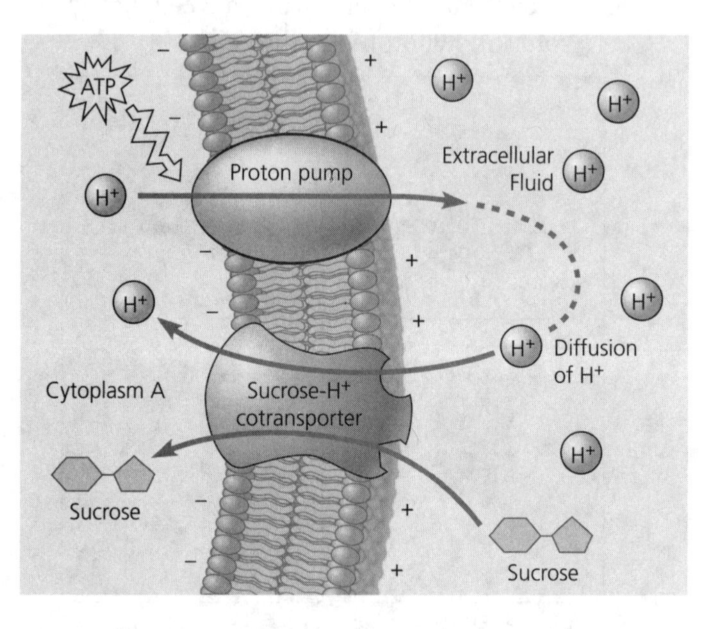

12. An ATP-powered pump that transports a specific solute can indirectly drive the active transport of another solute in a mechanism called cotransport. This is shown in the figure above. The cotransporter protein is able to use the diffusion of H+ ions down their electrochemical gradient into the cell to drive the uptake of sucrose into the cell. By what process are H+ ions being moved in this figure?
    (A) active transport
    (B) simple diffusion
    (C) facilitated diffusion
    (D) exocytosis

13. Based on the figure above, how might the rate of sucrose into the cell be increased?
    (A) Decrease the extracellular sucrose concentration.
    (B) Decrease the extracellular pH.
    (C) Increase the cytoplasmic pH.
    (D) Make the membrane more permeable to hydrogen ions.

14. White blood cells (WBCs) are more resistant to lysis than red blood cells (RBCs). When looking at a sample of blood for WBCs, what could you do to reduce interference from RBCs?
    (A) Mix the blood in a salty solution to cause the RBCs to lyse.
    (B) Mix the blood in an isotonic solution and allow the WBCs to float to the top.
    (C) Mix the blood in a hypotonic solution, which will cause the RBCs to lyse.
    (D) Mix the blood in a hypertonic solution, which will cause the RBCs to lyse.

After reading the paragraph, answer the question(s) that follow.

The skin is the body's largest organ. It's made up of many different types of cells. Oils, produced by the sebaceous glands, prevent the skin from drying and splitting. The protein melanin, produced by melanocytes in the epidermis, protects the skin from the harmful effects of ultraviolet radiation. Sweat, released through ducts to the skin surface, helps to cool the body. The types of cells that produce these compounds have different numbers of specific organelles, depending on their function.

15. Based on their function, you would expect melanocytes in the skin to have a higher than usual number of
    (A) lysosomes.
    (B) chloroplasts.
    (C) Golgi bodies.
    (D) microtubules.

16. If the melanin production were not sufficient to prevent UV damage to the DNA, which of the following would you expect?
    (A) decreased rate of melanin production
    (B) apoptosis and lysis of damaged cells
    (C) decreased cell divisions
    (D) peeling of the skin due to sebaceous glands shutting down

17. The endosymbiosis hypothesis is supported by all of the following pieces of evidence, *except* the fact that
    (A) mitochondria have circular DNA like prokaryotes.
    (B) mitochondria use ATP like prokaryotes.
    (C) chloroplasts have ribosomes like prokaryotes.
    (D) chloroplasts reproduce through a splitting process like certain prokaryotes.

Grid-In Questions

These call for a numerical response.

1. Simple cuboidal epithelial cells line the ducts of certain human exocrine glands. Various materials are transported into or out of the cell by diffusion. (The formula for the surface area of a cube is $6 \times S^2$, and the formula for the volume of a cube is S^3, where S = the length of a side of the cube.) What would be the volume of the larger cell in μm^3?

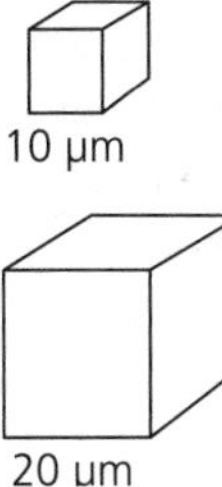

2. How many of the small cell could fit inside the larger cell?

3. Potato cores were placed in solutions of varying concentrations and were found to neither gain nor lose mass in a sucrose solution of 0.32M. Use this information to calculate the solute potential of the potato cores. The temperature of the solution is 22° C. (Refer to Appendix A, AP Biology Equations and Formulas.)

Free-Response Question

1. *Prokaryotic and eukaryotic cells are physiologically different in many ways, but both represent functional, evolutionarily successful cells.*

   (a) It has been theorized that the organelles of eukaryotic cells evolved from prokaryotes living symbiotically within a larger cell. **Explain three** lines of evidence that support this theory.

   (b) **Create** a model to show the path of a protein in a eukaryotic cell from its formation to its secretion from the cell.

> ***CONNECT WITH THE CURRICULUM FRAMEWORK***
>
> A scientific explanation has three distinct features:
>
> 1. A claim
> 2. Evidence for the claim
> 3. Justification for the claim

TOPIC 3

The Energy of Life

An Introduction to Metabolism

(Biology, *10e: Chapter 8*, Focus, *1e: Chapter 6*)

YOU MUST KNOW

- Examples of endergonic and exergonic reactions.
- The key role of ATP in energy coupling.
- That enzymes work by lowering the energy of activation.
- The catalytic cycle of an enzyme that results in the production of a final product.
- Factors that change enzyme shape and how they influence enzyme activity.
- How the shape of enzymes, their active sites, and interaction with specific molecules affect their function.
- How feedback inhibition is used to maintain appropriate levels of enzymes in a pathway.

An organism's metabolism transforms matter and energy, subject to the laws of thermodynamics (Biology, *10e: 8.1*, Focus, *1e: 6.1*)

- **Metabolism** is the totality of an organism's chemical reactions. Metabolism as a whole manages the material and energy resources of the cell.
- A **catabolic pathway** leads to the release of energy by the breakdown of complex molecules to simpler compounds. *Example*: Catabolic pathways occur when your digestive enzymes break down food and release energy.
- **Anabolic pathways** consume energy to build complicated molecules from simpler ones. *Example*: Anabolic pathways occur when your body links together amino acids to form muscle protein in response to physical exercise.
- **Energy** is defined as the capacity to do work. Anything that is moving is said to possess **kinetic energy**. An object at rest can possess **potential energy** if it has stored energy as a result of its position or structure. **Chemical energy**, a form of potential energy, is stored in molecules, and the amount of chemical energy a molecule possesses depends on its chemical bonds.
- **Thermodynamics** is the study of energy transformations that occur in matter.
  - The **first law of thermodynamics** states that the energy of the universe is constant and that energy *can* be transferred and transformed, but it *cannot* be created or destroyed.

- The **second law of thermodynamics** states that every energy transfer or transformation increases the **entropy**, or the amount of disorder or randomness, in the universe.

The free-energy change of a reaction tells us whether or not the reaction occurs spontaneously (Biology, *10e: 8.2*, Focus, *1e: 6.2*)

- **Free energy** is defined as the part of a system's energy that is able to perform work when the temperature of a system is uniform.
- ΔG is a symbol for a change in free energy.
  - An **exergonic reaction** is one in which energy is released. Exergonic reactions occur spontaneously (that does not necessarily mean quickly) and release free energy to the system. $\Delta G < 0$.
  - An **endergonic reaction** is one that requires energy in order to proceed. Endergonic reactions absorb free energy; that is, they require free energy from the system. $\Delta G > 0$.
  - Is the breakdown of glucose in cellular respiration exergonic or endergonic? (ΔG is –686 kcal/mol.)

ATP powers cellular work by coupling exergonic reactions to endergonic reactions (Biology, *10e: 8.3*, Focus, *1e: 6.3*)

- A key feature in the way cells manage their energy resources to do cell work is **energy coupling**, the use of an exergonic process to drive an endergonic one.
- The primary source of energy for cells in energy coupling is **ATP (adenosine triphosphate)**. Study Figure 3.1 and note that ATP is made up of the nitrogenous base adenine, bonded to ribose and a chain of three phosphate groups. When a phosphate group is hydrolyzed, energy is released in an exergonic reaction.

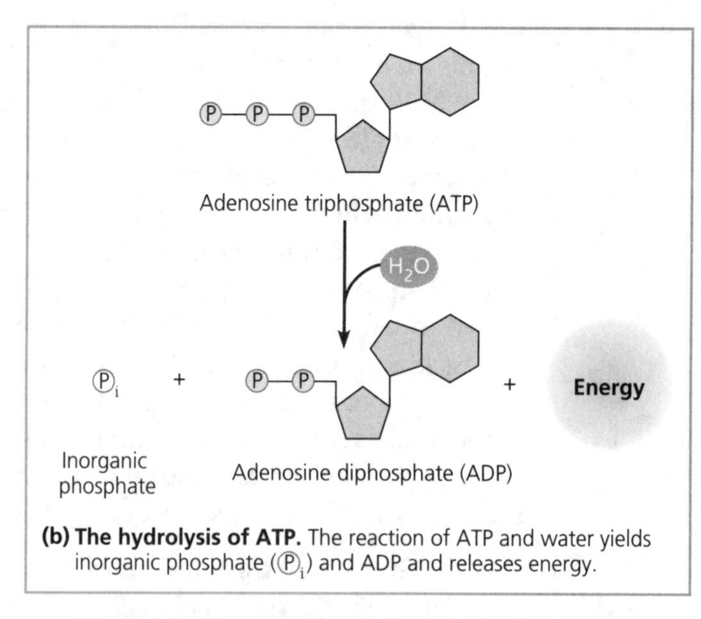

(b) The hydrolysis of ATP. The reaction of ATP and water yields inorganic phosphate (P_i) and ADP and releases energy.

Figure 3.1 The structure and hydrolysis of adenosine triphosphate (ATP)

- Work in the cell is done by the release of a phosphate group from ATP. The exergonic release of the phosphate group is used to do the endergonic work of the cell. When ATP transfers one phosphate group through hydrolysis, it becomes **ADP (adenosine diphosphate)**.

> ***CONNECT TO THE CURRICULUM FRAMEWORK***
>
> Essential Knowledge 2.A.1 is about the constant input of free energy required by all living systems. The preceding section about entropy, ATP/ADP, and endergonic/exergonic reactions is an important part of your study. Take your time to understand this material well!

Enzymes speed up metabolic reactions by lowering energy barriers (Biology, *10e: 8.4*, Focus, *1e: 6.4*)

- **Catalysts** are substances that can change the rate of a reaction without being altered in the process.
- **Enzymes** are macromolecules that are biological catalysts.
- The **activation energy** of a reaction is the amount of energy it takes to start a reaction—the amount of energy it takes to break the bonds of the reactant molecules. Enzymes speed up reactions *by lowering the activation energy* of the reaction—but without changing the free-energy change of the reaction. The reactant that the enzyme acts on is called a **substrate**. Figure 3.2 graphically depicts how enzymes function.

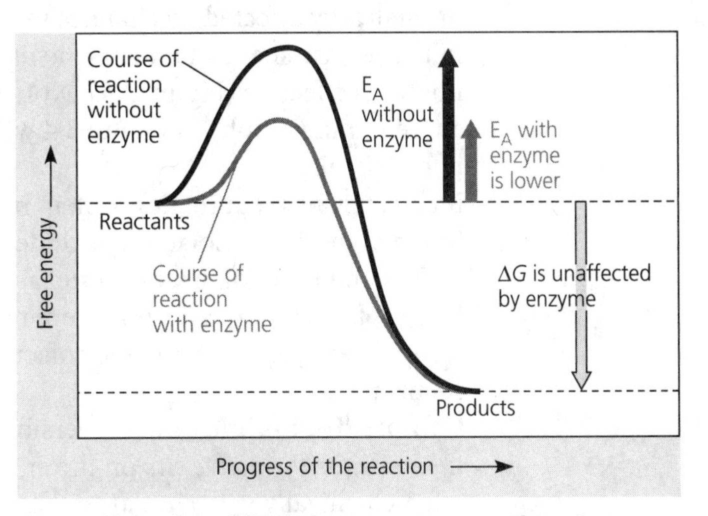

Figure 3.2 Effect of an enzyme on reaction rate

- The **active site** is the part of the enzyme that binds to the substrate. The enzyme and substrate form a complex called an **enzyme-substrate complex** that is generally held together by weak interactions. The substrate is then converted into **products**, and the products are released from the enzyme. Use Figure 3.3 to locate each step in the catalytic cycle of an enzyme.

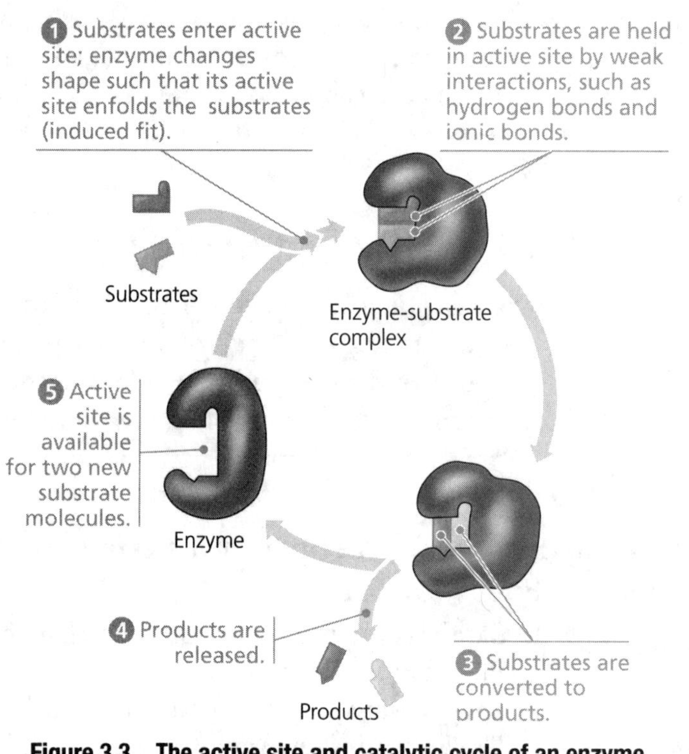

Figure 3.3 The active site and catalytic cycle of an enzyme

- The activity of an enzyme can be affected by several factors:
  - Protein enzymes have complicated three-dimensional shapes that are dramatically affected by changes in pH and temperature. Changes in the precise shape of an enzyme usually mean the enzyme will not be as effective (Figure 3.4). Note how the rate of the reaction is altered in the graphs in Figure 3.4 when temperature and pH are not optimal.
  - Many enzymes require nonprotein helpers, termed **cofactors**, to function properly. Cofactors include metal ions like zinc, iron, and copper and function in some crucial way to allow catalysis to occur. If the cofactor is organic, it is more properly referred to as a coenzyme. **Coenzymes** are organic cofactors; vitamins are examples of coenzymes.
  - **Competitive inhibitors** are reversible inhibitors that *compete with the substrate for the active site* on the enzyme. Competitive inhibitors are often chemically very similar to the normal substrate molecule and reduce the efficiency of the enzyme as it competes for the active site.
  - **Noncompetitive inhibitors** do not directly compete with the substrate molecule; instead, they impede enzyme activity by binding to another part of the enzyme. This causes the enzyme to change its shape, rendering the active site nonfunctional.

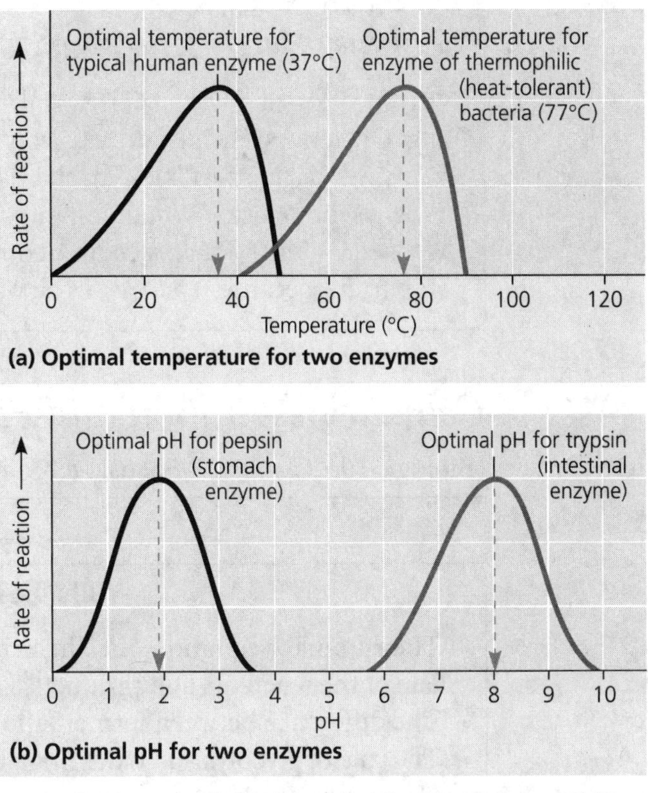

Figure 3.4 Environmental factors affecting enzyme activity

> ***STUDY TIP*** Enzymes are a key topic in biology. It is likely that you will do an investigation with enzymes. Know why certain factors such as pH and temperature affect enzyme action. Be able to pose a question about enzyme function, design an experiment to test this question, predict results, and analyze data from an enzyme experiment.

Regulation of enzyme activity helps control metabolism (Biology, *10e: 8.5*, Focus, *1e: 6.5*)

- Many enzyme regulators bind to an **allosteric** site on the enzyme, which is a specific binding site, but not the active site. Once bound, the shape of the enzyme is changed, and this can either stimulate or inhibit enzyme activity.
- The end product on an enzymatic pathway can switch off its pathway by binding to the allosteric site of an enzyme in the pathway. This type of allosteric inhibition is termed **feedback inhibition**. Feedback inhibition increases the efficiency of the pathway by turning it off when the end product accumulates in the cell.

CONNECT WITH THE CURRICULUM FRAMEWORK

Big Idea 2 deals with energy and its transformations. Several Learning Objectives (LOs) address how changes in molecular concentrations will trigger feedback mechanisms. What is the effect of a change? How is homeostasis maintained? What predictions can you make? (LOs 2.15-2.18) The preceding concept on enzyme regulation is very important because it addresses all of these LOs.

Cellular Respiration and Fermentation

(Biology, *10e: Chapter 9*, Focus, *1e: Chapter 7*)

YOU MUST KNOW

- The summary equation of cellular respiration including the source and fate of the reactants and products.
- The difference between fermentation and cellular respiration.
- The role of glycolysis in oxidizing glucose to two molecules of pyruvate.
- How pyruvate is moved from the cytosol into the mitochondria and introduced into the citric acid cycle.
- How electrons from NADH and $FADH_2$ are passed to a series of electron acceptors to produce ATP by chemiosmosis.
- The roles of the mitochondrial membrane, proton (H^+) gradient, and ATP synthase in generating ATP.

TIP FROM THE READERS

Oxidation-reduction reactions, fermentation, cellular respiration, and photosynthesis are among the most technically challenging sections of the course. Here, we focus on the major steps of each of the processes as well as the results. Questions on the AP Biology Exam are likely to focus on the energy transfers of photosynthesis and respiration—not on the exact reactions that create the products, nor are you expected to know the names of the enzymes involved in the process. As you work through these chapters, compare and contrast the two fundamental cell processes.

Catabolic pathways yield energy by oxidizing organic fuels (Biology, *10e: 9.1*, Focus, *1e: 7.1*)

- **Catabolic pathways** occur when molecules are broken down and their energy is released. You should know these two catabolic pathways:
  - **Fermentation** is the partial degradation of sugars that occurs without the use of oxygen.

- **Aerobic respiration** is the most prevalent and efficient catabolic pathway in which oxygen is consumed as a reactant along with the organic fuel.

- Carbohydrates, fats, and proteins can all be broken down to release energy in cellular respiration. However, glucose is the primary molecule that is used in cellular respiration. The standard way of representing the process of cellular respiration shows glucose being broken down in the following reaction:

$$C_6H_{12}O_6 + 6\ O_2 \rightarrow 6\ CO_2 + 6\ H_2O + \text{Energy (686 kcal/mol of glucose)}$$

- The exergonic release of energy from glucose is used to phosphorylate ADP to ATP. Life processes constantly consume ATP; cellular respiration burns fuels and uses the energy to regenerate ATP.
- The reactions of cellular respiration are of a type termed **oxidation-reduction (redox)** reactions. In redox reactions electrons are transferred from one reactant to another.
  - The loss of one or more electrons from a reactant is called **oxidation**. When a reactant is *oxidized*, it loses electrons and, consequently, energy.
  - The gain of one or more electrons is **reduction**. When a reactant is *reduced*, it gains electrons and, therefore, energy.
- At key steps in cellular respiration, electrons are stripped from glucose. Each electron travels with a proton, thereby forming a hydrogen atom. The hydrogen atoms are not transferred directly to oxygen, as the formula might suggest, but instead are usually passed to an electron carrier, the coenzyme **NAD^+** (a derivative of the B vitamin niacin). Within the cell **NAD^+** accepts two electrons, plus the stabilizing hydrogen ion, to form **NADH**. Note that NADH has been reduced and therefore has gained energy.
- Figure 3.5 shows the three stages of cellular respiration. Each stage is separately featured in the next three concepts. Use this figure to begin to develop an overall concept of the process of cellular respiration.

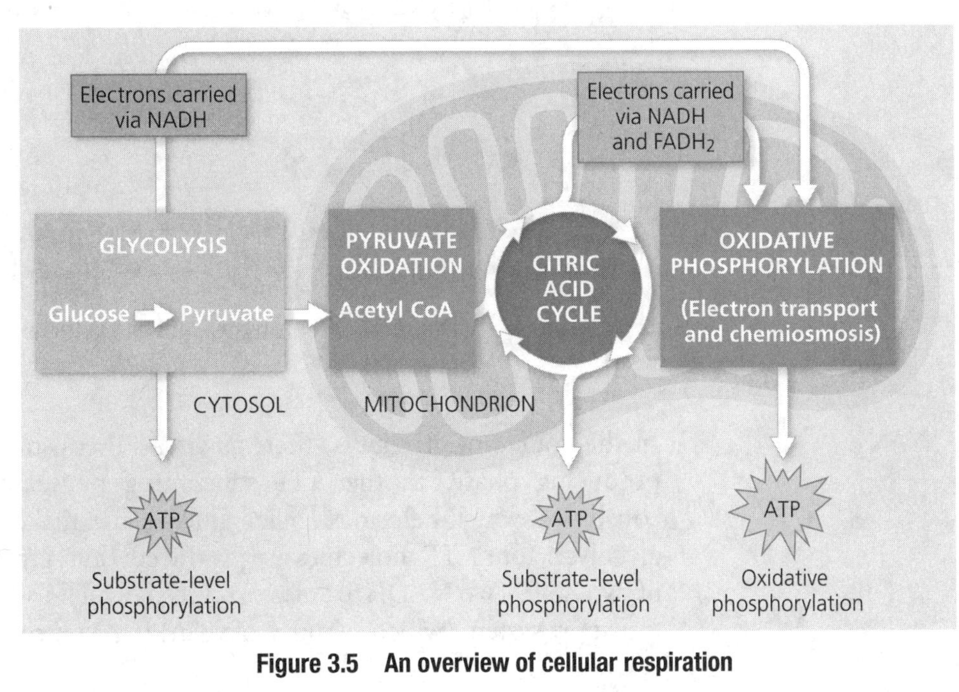

Figure 3.5 An overview of cellular respiration

Glycolysis harvests chemical energy by oxidizing glucose to pyruvate (Biology, *10e: 9.2*, Focus, *1e: 7.2*)

- In **glycolysis** (which occurs in the cytosol), the degradation of glucose begins as it is broken down into two pyruvate molecules. The six-carbon glucose molecule is split into two three-carbon sugars through a long series of steps.

> ***STUDY TIP*** Use Figure 3.6 as a guide to the important features of glycolysis. It is not necessary to know the enzymes and reactions for each step in glycolysis.

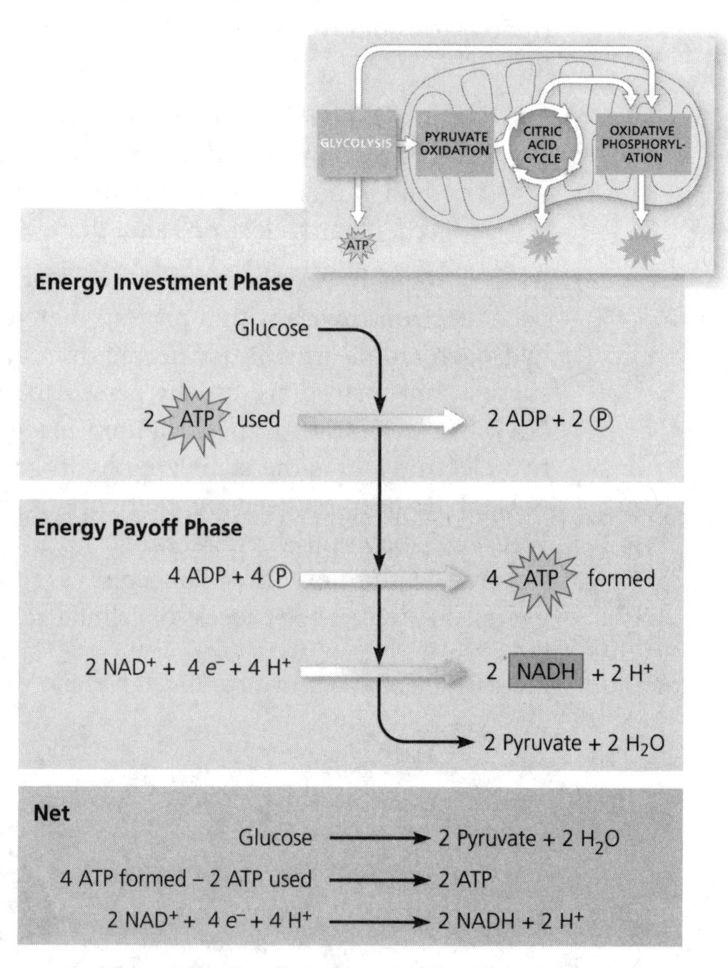

Figure 3.6 The energy input and output of glycolysis

- In the course of glycolysis, there is an ATP-consuming phase and an ATP-producing phase. In the ATP-consuming phase, two ATP molecules are consumed, which helps destabilize glucose and make it more reactive. Later in glycolysis, four ATP molecules are produced; thus, glycolysis results in a net gain of two ATP. Two NADH are also produced, which will be utilized in the electron transport system (which requires oxygen) to produce ATP (see Figure 3.5).

- Notice the *net* energy gain in glycolysis as indicated in Figure 3.6—two ATP molecules and two NADH molecules. Most of the potential energy of the glucose molecule still resides in the two remaining pyruvates, which will now feed into the citric acid cycle, as discussed in the next concept.

After pyruvate is oxidized, the citric acid cycle completes the energy-yielding oxidation of organic molecules (Biology, *10e: 9.3*, Focus, *1e: 7.3*)

- After glycolysis, **pyruvate is oxidized to acetyl CoA**. This junction between glycolysis and the citric acid cycle is shown in Figure 3.7. Note the following steps in this figure:
  - A transport protein moves pyruvate from the cytosol into the matrix of the mitochondria.
  - In the matrix an enzyme complex catalyzes three reactions: a CO_2 **is removed**, electrons are stripped from pyruvate to convert NAD^+ to NADH, and coenzyme A joins with the remaining two-carbon fragment to form acetyl CoA.

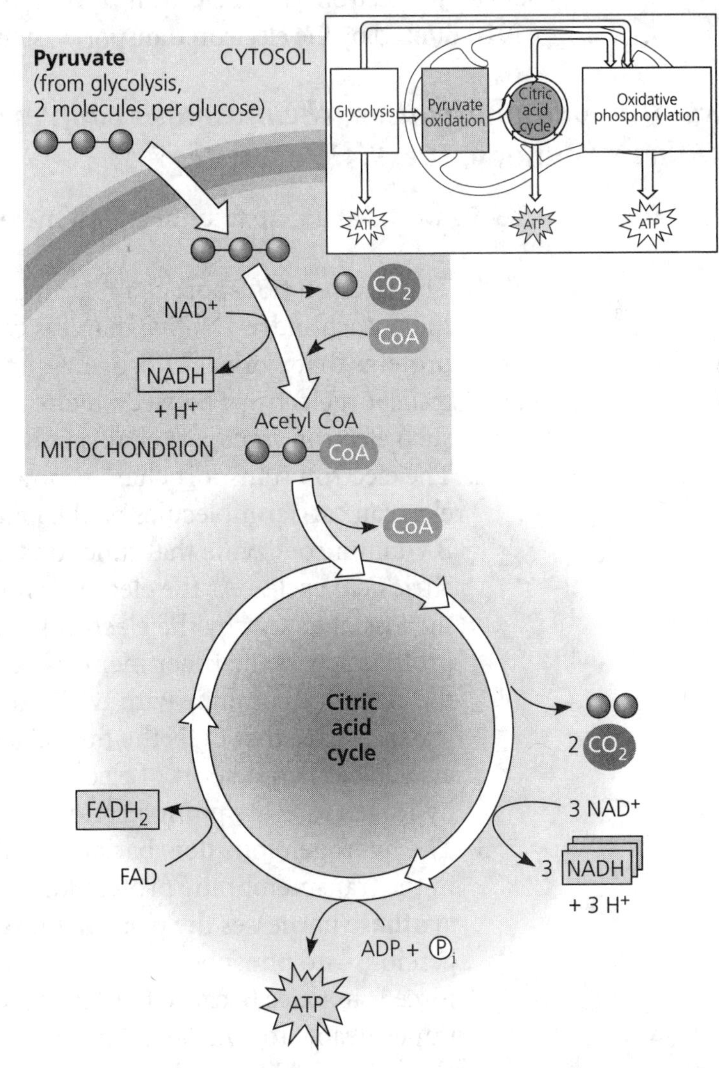

Figure 3.7 An overview of pyruvate oxidation and the citric acid cycle

- Two acetyl CoA molecules are produced per glucose. Acetyl CoA now enters the enzymatic pathway termed the citric acid cycle.

- In the **citric acid cycle** (which occurs in the mitochondrial matrix), the job of breaking down glucose is completed with CO_2 released as a waste product. Each turn of the citric acid cycle requires the input of one acetyl CoA. The citric acid cycle must make two turns before the glucose is completely oxidized.
- Study Figure 3.7 and note that the citric acid cycle results in the following:
  - Each turn of the citric acid cycle produces **2 CO_2, 3 NADH, 1 $FADH_2$,** and **1 ATP**.
  - Because each glucose yields *two* pyruvates, the *total* products of the citric acid cycle are usually listed as the result of two cycles:

4 CO_2, 6 NADH, 2 $FADH_2$, and **2 ATP**

- Note that at the end of the citric acid cycle the six original carbons in glucose have been released as CO_2. (You are exhaling this gas as you study.) Only two ATP molecules, however, have been produced. Where is all the energy? The energy is held in the electrons in the electron carriers, NADH and $FADH_2$. These electrons will be utilized by the electron transport system, explained in the next concept.

During oxidative phosphorylation, chemiosmosis couples electron transport to ATP synthesis (Biology, *10e: 9.4*, Focus, *1e: 7.4*)

- Use Figure 3.8 as a map to understand the process of electron transport.

1. The electron transport chain is embedded in the inner membrane of the mitochondria. Notice that it is composed of three transmembrane proteins that work as hydrogen pumps and two carrier molecules that transport electrons between hydrogen pumps. There are thousands of such electron transport chains in the inner mitochondrial membrane.
2. The electron transport chain is powered by electrons from the electron carrier molecules NADH and $FADH_2$ ($FADH_2$ is also a B vitamin coenzyme that functions as an electron acceptor in the citric acid cycle). As the electrons flow through the electron chain, the loss of energy by the electrons is used to power the pumping of protons across the inner membrane. At the end of the electron chain, the electrons combine with two hydrogen ions and oxygen to form water. Notice that O_2 is the final electron acceptor, and when it is not available, the transport of electrons comes to a screeching halt! No hydrogen ions are pumped and no ATP is produced.
3. The hydrogen ions flow back down their gradient through a channel in the transmembrane protein known as ATP synthase. **ATP synthase** harnesses the **proton-motive force**—the gradient of hydrogen ions—to phosphorylate ADP, forming ATP. The proton-motive force is in place because the inner membrane of the mitochondria is impermeable to hydrogen ions. Like water behind a dam with its only exit being a spillway, electrons are held behind the inner membrane with their only exit, ATP synthase.

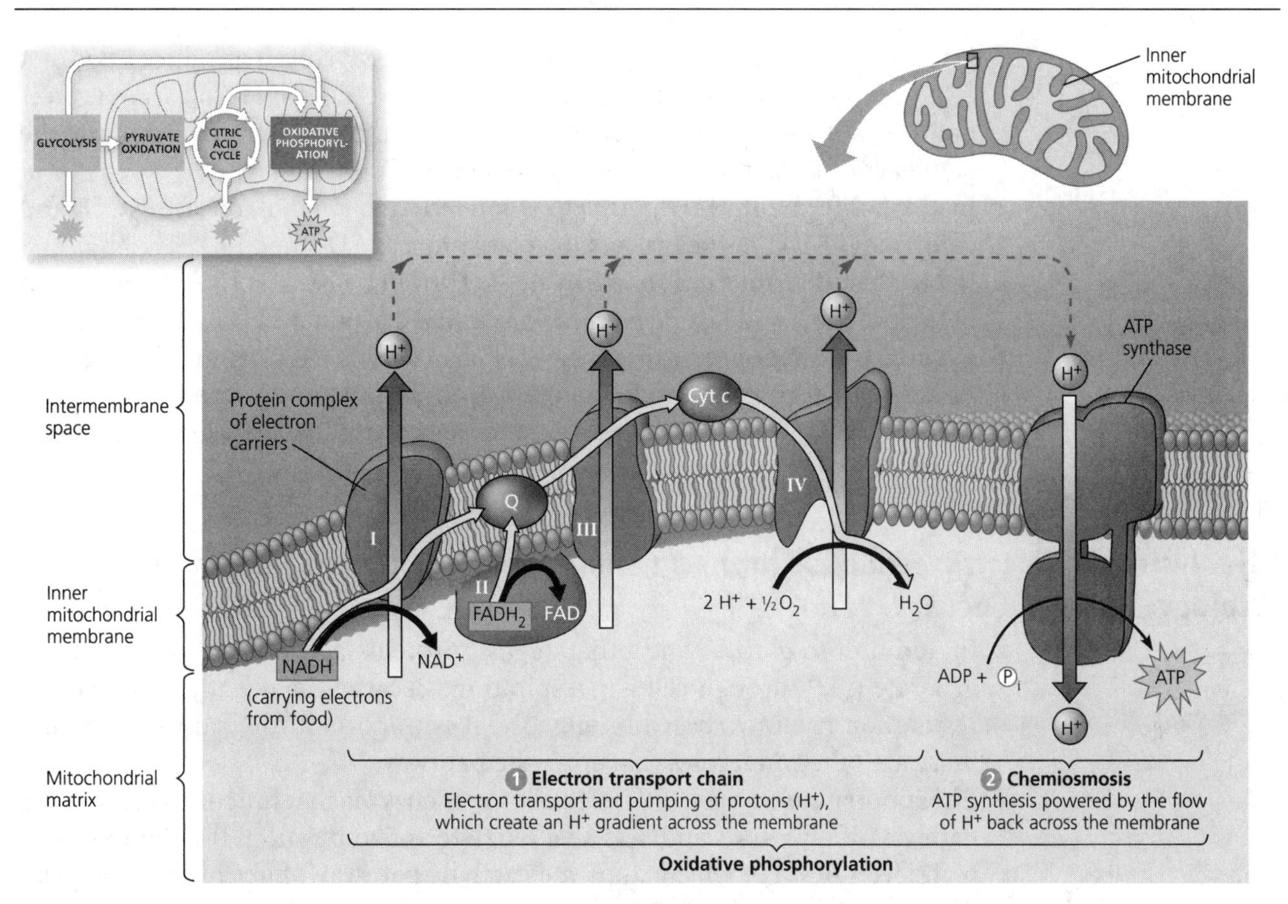

Figure 3.8 ATP production by chemiosmosis

4. This process is referred to as chemiosmosis. **Chemiosmosis** is an energy-coupling mechanism that uses energy stored in the form of an H^+ gradient across a membrane to drive cellular work (ATP synthesis in our example). The electron transport chain and chemiosmosis compose **oxidative phosphorylation**. This specific term is used because ADP is phosphorylated and oxygen is necessary to keep the electrons flowing.
5. The ATP yield per molecule of glucose is 30 to 32 ATP. Oxidative phosphorylation produces 26 to 28 of the total. (You may notice that this number is less than you might have learned in an earlier text, but it more accurately reflects the current biochemical analysis.)

> ***STUDY TIP*** Sketch this process and explain it verbally. This is a fundamental biological process that you should understand.

Fermentation and anaerobic respiration enable cells to produce ATP without the use of oxygen (Biology, *10e: 9.5*, Focus, *1e: 7.5*)

- **Anaerobic respiration** by certain prokaryotes generates ATP without oxygen using an electron transport chain.
- **Fermentation** is an expansion of glycolysis in which ATP is generated by substrate-level phosphorylation.

- Fermentation consists of glycolysis (recall that glycolysis produces two net ATP molecules) and reactions that regenerate NAD^+. In glycolysis oxygen is not needed to accept electrons; NAD^+ is the electron acceptor. Therefore, the pathways of fermentation must regenerate NAD^+.
- In **alcohol fermentation**, pyruvate is converted to ethanol, releasing CO_2 and oxidizing NADH in the process to create more NAD^+.
- In **lactic acid fermentation**, pyruvate is reduced by NADH (NAD^+ is formed in the process), and lactate is formed as a waste product.
- **Facultative anaerobes** can make ATP by aerobic respiration if oxygen is present but can switch to fermentation under anaerobic conditions. On a cellular level, your muscle cells operate as facultative anaerobes. **Obligate anaerobes** cannot survive in the presence of oxygen.

Glycolysis and the citric acid cycle connect to many other metabolic pathways (Biology, *10e: 9.6*, Focus, *1e: 7.6*)

- In addition to glucose and other sugars, proteins and fats are often used to generate ATP through cellular respiration. Because the equation for aerobic respiration is always based on glucose, it is important for you to know that there are other molecules that enter the pathway.
- **Phosphofructokinase** (PFK) is an allosteric enzyme that functions early in the pathway of glycolysis and acts as a regulator of respiration. It is inhibited by high levels of ATP, which stops the catalytic pathway of glycolysis. Adequate ATP? Breakdown of glucose → pyruvate is not required.
- By controlling the rate of the entire process of cellular respiration, phosphofructokinase is considered the pacemaker of respiration. Consider this as an excellent example of regulation of a process by negative feedback.

Photosynthesis

(Biology, *10e: Chapter 10*, Focus, *1e: Chapter 8*)

YOU MUST KNOW

- The summary equation of photosynthesis including the source and fate of the reactants and products.
- How leaf and chloroplast anatomy relate to photosynthesis.
- How photosystems convert solar energy to chemical energy.
- How linear electron flow in the light reactions results in the formation of ATP, NADPH, and O_2.
- How the formation of a proton gradient in the light reactions is used to form ATP from ADP + inorganic phosphate by ATP synthase.
- How the Calvin cycle uses the energy molecules of the light reactions (ATP and NADPH) to produce carbohydrates (G3P) from CO_2.

Photosynthesis converts light energy to the chemical energy of food (Biology, *10e: 10.1*, Focus, *1e: 8.1*)

- Before you look at the molecular details of photosynthesis, it is important to think of photosynthesis in an ecological context.
  - Life on Earth is solar powered by autotrophs. **Autotrophs** are "self-feeders"; they sustain themselves without eating anything derived from other organisms. Autotrophs are the ultimate source of organic compounds and are therefore known as *producers.*
  - **Heterotrophs** live on compounds produced by other organisms and are thus known as *consumers.* Animals immediately come to mind as heterotrophs, but also remember that decomposers like fungi and many prokaryotes are heterotrophs. Heterotrophs are dependent on the process of photosynthesis for both food and oxygen.
- **Chloroplasts** are the specific sites of photosynthesis in plant cells.
  - Use Figure 3.9 to become familiar with the structure of chloroplasts. An envelope of two membranes encloses the **stroma**, which is a dense fluid-filled area. Within the stroma is a vast network of interconnected membranous sacs called **thylakoids**. The thylakoids segregate the stroma from another compartment, the **thylakoid space**.
  - Note that the thylakoids set up compartments separate from the stroma. This will allow a proton gradient to be established.
  - **Chlorophyll** is located in the thylakoid membranes and is the light-absorbing pigment that drives photosynthesis and gives plants their green color.
- The exterior of the lower epidermis of a leaf contains many tiny pores called **stomata**, through which carbon dioxide enters and oxygen and water vapor exit the leaf. The loss of water through open stomata is *transpiration.*

> ***STUDY TIP*** Practice drawing a chloroplast in the margin of this page. Label its parts and know what major events occur in each region.

- The overall reaction of photosynthesis looks like this:

$$6\ CO_2 + 6\ H_2O + \text{Light energy} \rightarrow C_6H_{12}O_6 + 6\ O_2$$

  - Notice that the overall chemical change during photosynthesis is the reverse of the one that occurs during cellular respiration.
  - All the oxygen you breathe was formed in the process of photosynthesis when a water molecule was split! Water is split for its electrons, which are transferred along with hydrogen ions from water to carbon dioxide, reducing it to sugar. This process requires energy (an endergonic process), which is provided by the sun.

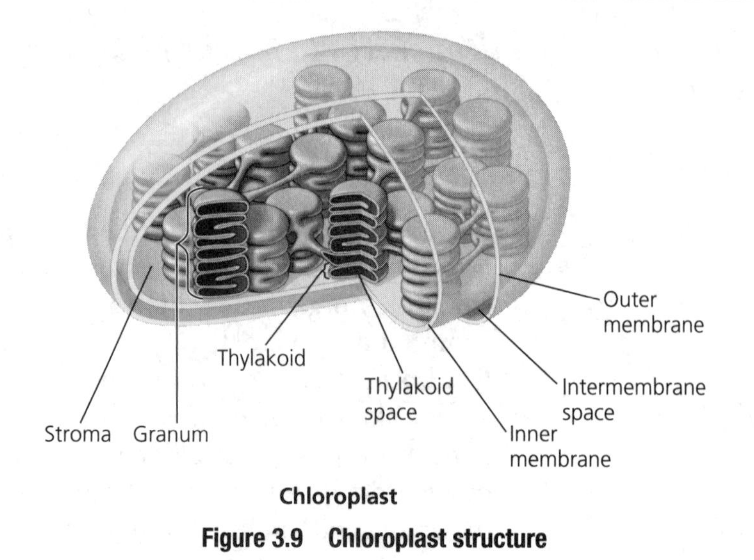

Figure 3.9 Chloroplast structure

- Photosynthesis is a chemical process that requires two stages to complete.
  - The **light reactions** occur in the thylakoid membranes where solar energy is converted to chemical energy. The net products of the light reactions are **NADPH** (which stores electrons), **ATP**, and **oxygen**. Here are the primary events.

    1. Light energy is absorbed by chlorophyll, which drives the transfer of electrons from water to $NADP^+$, forming **NADPH**.
    2. Water is split during these reactions, and $\mathbf{O_2}$ is released.
    3. **ATP** is generated, using chemiosmosis to power the addition of a phosphate group to ADP, a process called **photophosphorylation**.

  - The **Calvin cycle** occurs in the stroma, where CO_2 from the air is incorporated into organic molecules in **carbon fixation**. The Calvin cycle uses the fixed carbon plus NADPH and ATP from the light reactions in the formation of new sugars.

> ***STUDY TIP*** Use Figure 3.10 to help in understanding where reactions occur and the overall purpose of the two stages of photosynthesis. If you understand the big picture, the details will be easier to comprehend.

The light reactions convert solar energy to the chemical energy of ATP and NADPH (Biology, *10e: 10.2,* Focus, *1e: 8.2*)

- Not surprisingly, light is an important concept in photosynthesis. It is the primary energy source for life on Earth.
  - Light is electromagnetic energy, and it behaves as though it is made up of discrete particles, called **photons**—each of which has a fixed quantity of energy.
  - Substances that absorb light are called **pigments**, and different pigments absorb light of different wavelengths. Chlorophyll is a

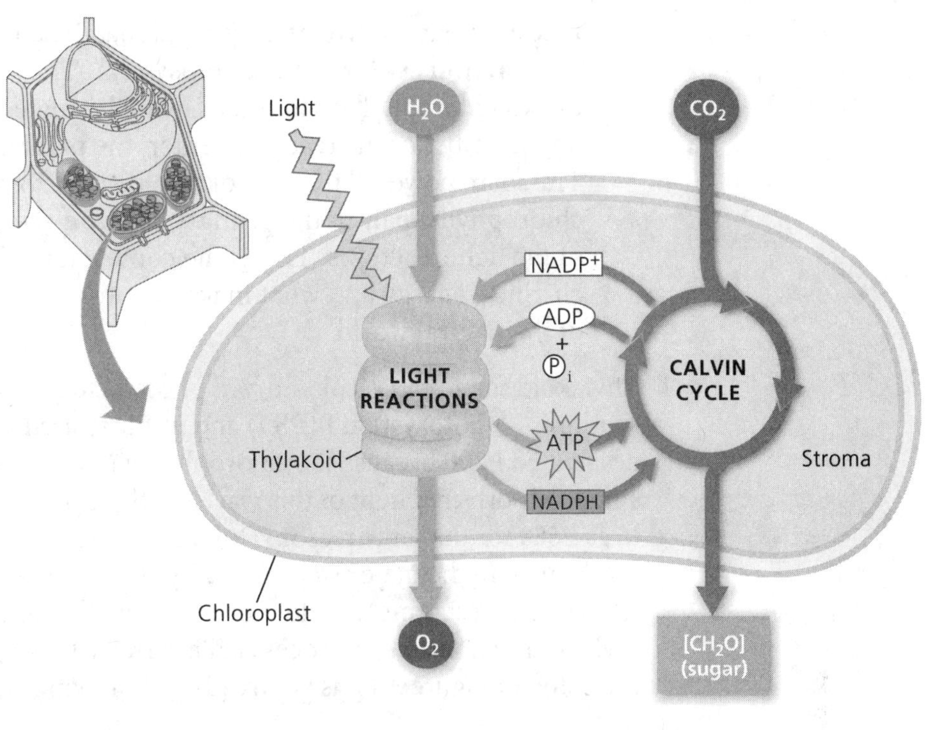

Figure 3.10 Overview of photosynthesis

pigment that absorbs violet-blue and red light while transmitting and reflecting green light. This is why we see summer leaves as green.

- A graph plotting a pigment's light absorption versus wavelength is called an **absorption spectrum**. The absorption spectrum of chlorophyll provides clues to the effectiveness of different wavelengths for driving photosynthesis. This is confirmed by an action spectrum.
- An **action spectrum** for photosynthesis graphs the effectiveness of different wavelengths of light in driving the process of photosynthesis. Note examples of both of these graphs in your text, Figure 10.10 in Campbell *Biology, 10e,* and Figure 8.9 in Campbell *Biology in Focus.*
- The action spectrum confirms that plants use energy from red and blue light (which is absorbed) and very little energy from green light (which is reflected).

- Photons of light are absorbed by certain groups of pigment molecules in the thylakoid membrane of chloroplasts. These groups are called **photosystems** and consist of two parts: a light-harvesting complex and a reaction center.
  - The **light-harvesting complex** is made up of many chlorophyll and carotenoid molecules (accessory pigments in the thylakoid membrane); this allows the complex to gather light effectively. When chlorophyll absorbs light energy in the form of photons, one of the molecule's electrons is raised to an orbital of higher potential energy. The chlorophyll is then said to be in an "excited" state.

- Like a human "wave" at a sports arena, the energy is transferred to the **reaction center** of the photosystem. The reaction center consists of two chlorophyll *a* molecules, which donate the electrons to the second member of the reaction center, the **primary electron acceptor**. The solar-powered transfer of an electron from the reaction-center chlorophyll *a* pair to the primary electron acceptor is the first step of the light reactions. This is the conversion of light energy to chemical energy and what makes photoautotrophs the producers of the natural world.

- Thylakoid membranes contain two photosystems that are important to photosynthesis—**photosystem I (PS I)** and **photosystem II (PS II)**. PS I is sometimes designated P700 because the chlorophyll *a* in the reaction center of this photosystem absorbs red light of this wavelength the best; PS II is sometimes referred to as P680 for the same reason. Don't let switches in designation be confusing.
- Following are the major steps of the light reactions of photosynthesis. The key to the light reactions is a flow of electrons through the photosystems in the thylakoid membrane, a process called **linear (noncyclic) electron flow**. Find each step in Figure 3.11 as you read the following summary:

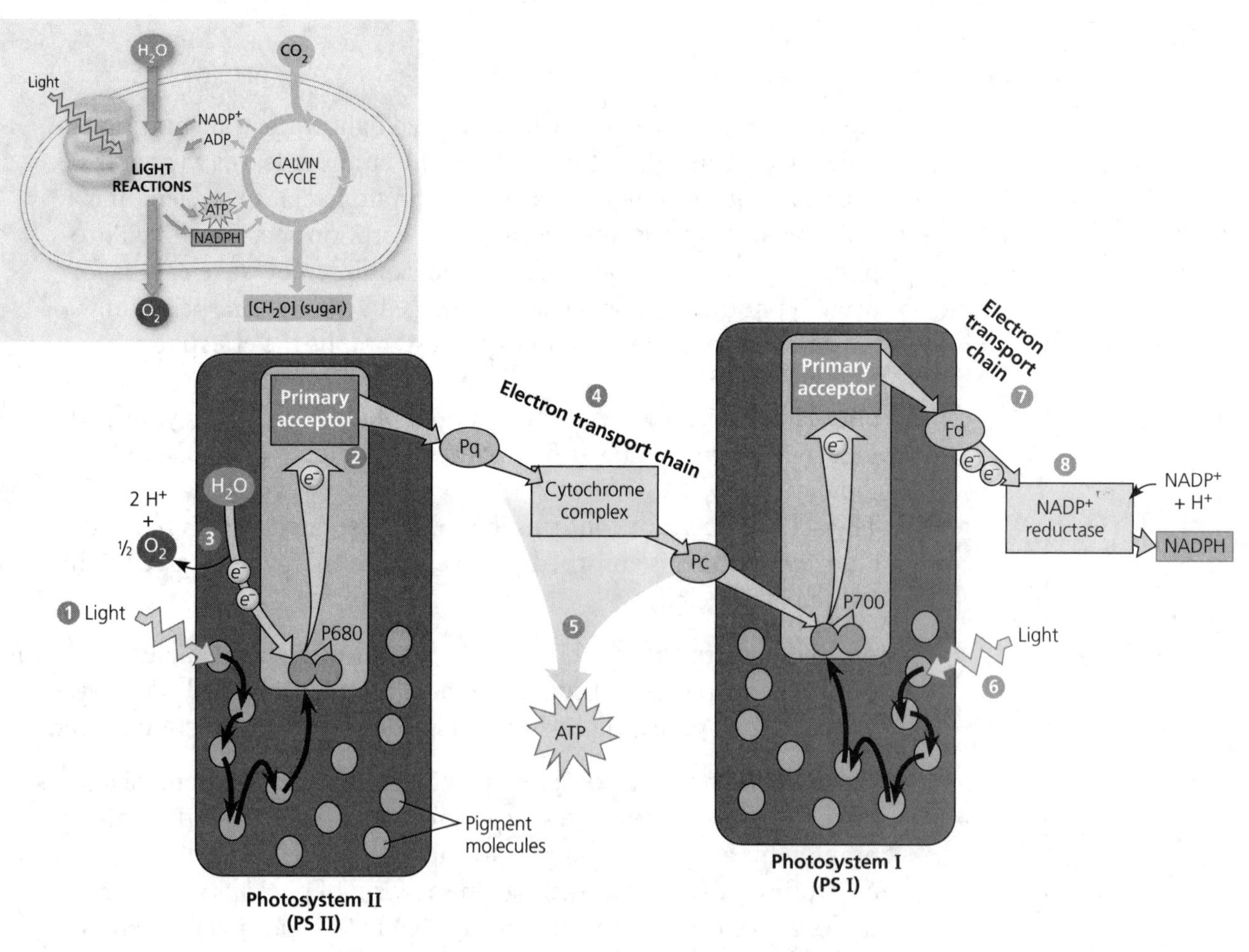

Figure 3.11 Noncyclic electron flow

1. Photosystem II absorbs light energy, allowing the P680 reaction center of two chlorophyll *a* molecules to donate an electron to the primary electron acceptor. The reaction-center chlorophyll is oxidized and now requires an electron.
2. An enzyme splits a water molecule into two hydrogen ions, two electrons, and an oxygen atom. The electrons are supplied to the P680 chlorophyll *a* molecules. The oxygen combines with another oxygen molecule, forming the O_2 that will be released into the atmosphere.
3. The original excited electron passes from the primary electron acceptor of photosystem II to photosystem I through an electron transport chain (similar to the electron chain in cellular respiration).
4. The energy from the transfer of electrons down the electron transport chain is used to pump protons, creating a gradient that is used in chemiosmosis to phosphorylate ADP to ATP. ATP will be used as energy in the formation of carbohydrates in the Calvin cycle.
5. Meanwhile, light energy has also activated PS I, resulting in the donation of an electron to its primary electron acceptor. The electrons just donated by PS I are replaced by the electrons from PS II. (Keep in mind that the ultimate source of electrons is water.)
6. The primary electron acceptor of photosystem I passes the excited electrons along to another electron transport chain, which transmits them to $NADP^+$, which is reduced to NADPH—the second of the two important light-reaction products. The high-energy electrons of NADPH are now available for use in the Calvin cycle.

- Chloroplasts and mitochondria generate ATP by the same basic mechanism: chemiosmosis. Examining Figure 3.12 will quickly demonstrate the same basic chemiosmotic plan as cellular respiration. Use Figure 3.12 to illustrate the following:
    - An electron transport chain uses the flow of electrons to pump protons across the thylakoid membrane from the stroma into the thylakoid space.
    - A proton-motive force is created within the thylakoid space that can be utilized by ATP synthase to phosphorylate ADP to ATP. This occurs when protons (H^+) flow out of the thylakoid space, down their electrochemical gradient, through ATP synthase and into the stroma. Notice that the proton-motive force is generated in three places: (1) hydrogen ions from water; (2) hydrogen ions pumped across the membrane by the cytochrome complex; (3) the removal of a hydrogen ion from the stroma when $NADP^+$ is reduced to NADPH.
    - Although similar, chemiosmosis in cellular respiration and photosynthesis are not identical. In addition to some spatial differences, the key conceptual difference is that mitochondria use chemiosmosis to transfer chemical energy from food molecules to ATP, whereas chloroplasts transform light energy into chemical energy in ATP. This is the essence of the difference between a consumer and a producer.

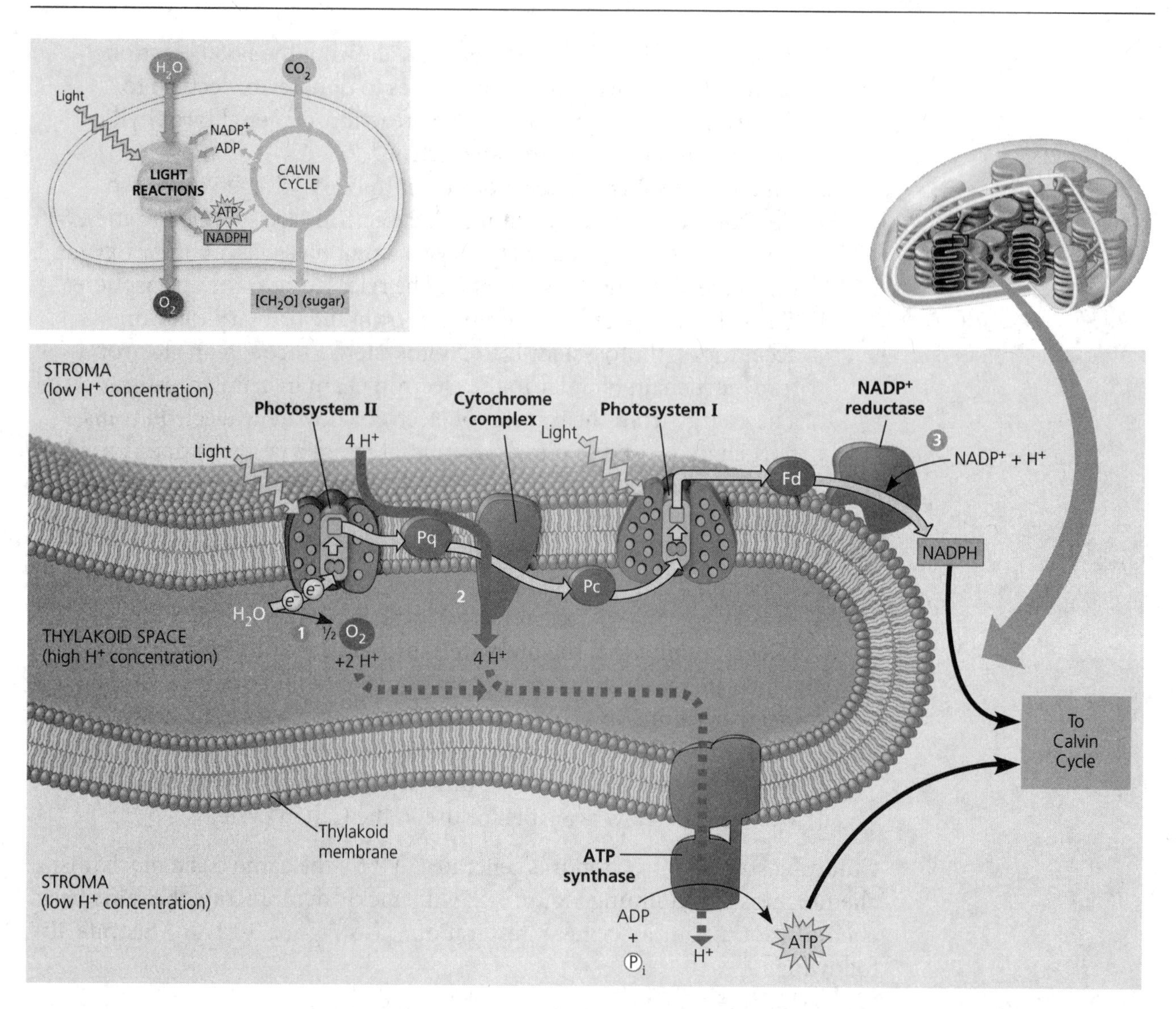

Figure 3.12 Light reactions and chemiosmosis

> ***CONNECT WITH THE CURRICULUM***
>
> Create a model that shows electron transport establishing an H^+ gradient across either the inner membrane of the mitochondria or a thylakoid membrane and use it to explain the production of ATP. Be able to predict how changes in the model would affect the output of ATP.

The Calvin cycle uses the chemical energy of ATP and NADPH to convert CO_2 to sugar (Biology, *10e: 10.3*, Focus, *1e: 8.3*)

- In the course of the **Calvin cycle**, carbon enters in the form of CO_2 and leaves in the form of a sugar. The cycle spends ATP as an energy source and consumes NADPH as reducing power for adding high-energy electrons to make the sugar. Use Figure 3.13 to chart each step summarized in the outline that follows. You must note that in order to net one molecule of G3P, the cycle must go through three rotations and fix three molecules of CO_2.

> ***STUDY TIP*** Try sketching the Calvin cycle as you read these steps. When you are done, you should have a figure much like Figure 3.13. You do not have to create the figure from memory, but you should be able to explain the process!

- These are the major steps of the Calvin cycle. It is not important that you memorize the intermediate organic molecules, but it is important to understand the conceptual scheme of reducing CO_2 to a sugar.

    1. Three CO_2 molecules are attached to three molecules of the five-carbon sugar **ribulose bisphosphate (RuBP)**. These reactions are catalyzed by the enzyme **rubisco** (probably the most common protein in the biosphere) and produce an unstable product that immediately splits into two 3-carbon compounds called 3-phosphoglycerate.

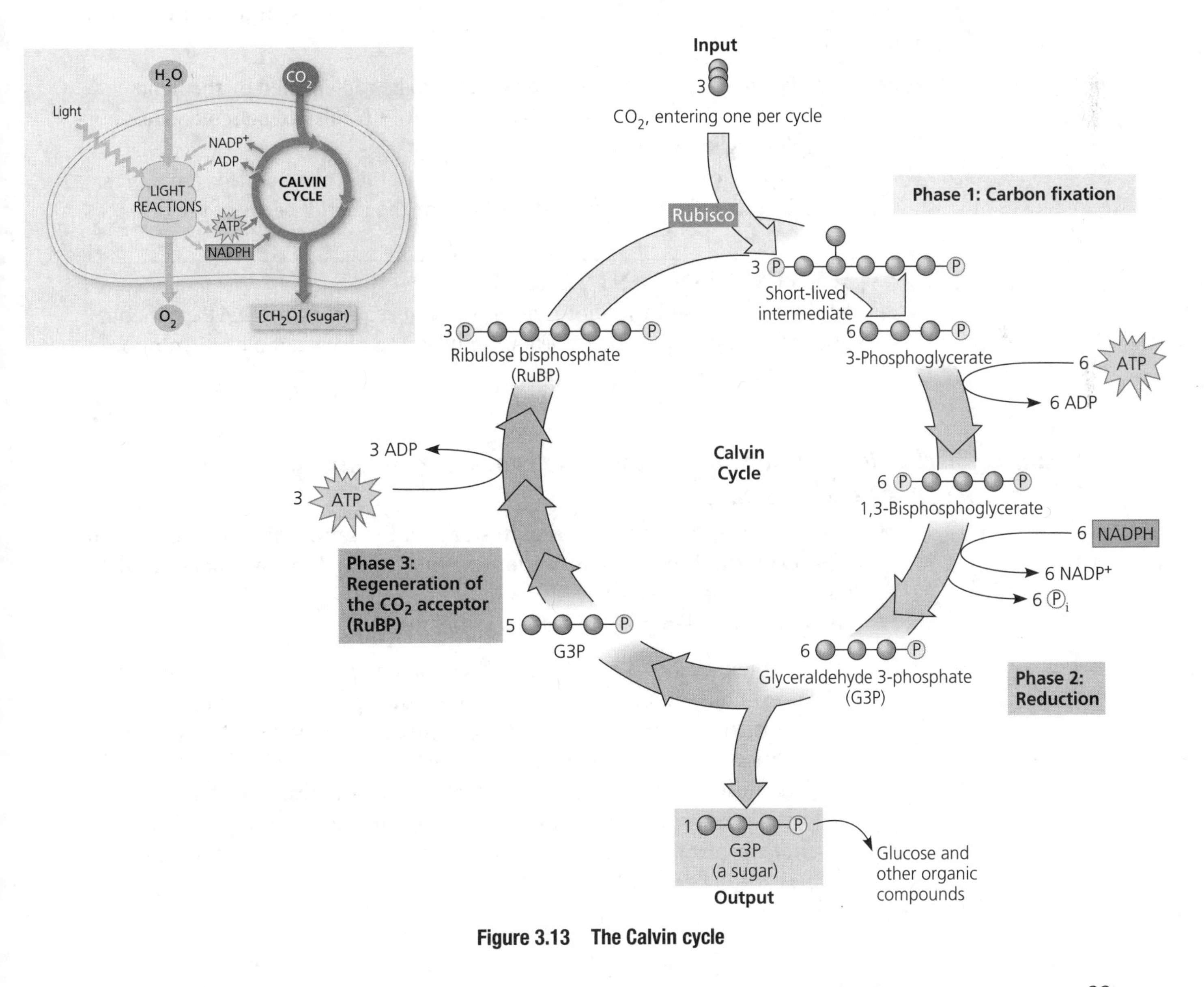

Figure 3.13 The Calvin cycle

At this point carbon has been fixed—the incorporation of CO_2 into an organic compound.

2. The 3-phosphoglycerate molecules are phosphorylated (using ATP from the light reactions) to become 1,3-bisphosphoglycerate.
3. Next, six NADPH (from the light reactions) reduce the six 1,3-bisphosphoglycerate molecules to six **glyceraldehyde 3-phosphate (G3P)**.
4. *One* G3P leaves the cycle to be used by the plant cell. (Two G3P molecules can combine to form glucose, which is generally listed as the final product of photosynthesis.)
5. Finally, the three beginning RuBPs are generated as the *five* G3Ps remaining are reworked into three of the starting molecules, with the expenditure of three ATP molecules. Notice that the 5G3Ps molecules have 15 carbons, which are rearranged to 3 RuBPs with 15 carbons. With the generation of new RuBPs the Calvin cycle begins again.

- In the Calvin cycle, the formation of one net G3P requires the following energy molecules:
  - *Nine molecules of ATP* are consumed (to be replenished by the light reactions) along with *six molecules of NADPH* (also to be replenished by the light reactions).
  - One of the six G3P molecules produced in the Calvin cycle is a net gain and will be used for biosynthesis or the energy needs of the cell.

WHAT'S IMPORTANT TO KNOW?
Knowledge of C_4/CAM photosynthesis is not required for the AP exam, but your teacher may use the information here as a nice example of evolutionary adaptations to arid conditions.

Alternative mechanisms of carbon fixation have evolved in hot, arid climates (Biology, *10e: 10.4*, Focus, *1e: 8.3*)

- Because CO_2 enters the leaf through stomata, the same pores through which water exits the leaf in transpiration, having open stomata quickly lead to dehydration in arid habitats.
- The specific problem for C_3 plants is as follows:
  - On hot, dry days C_3 plants must keep their stomata closed to conserve water—thus, no CO_2 uptake. Therefore, they produce less sugar because the declining levels of CO_2 in the leaf starve the Calvin cycle.
  - Additionally, the enzyme rubisco can bind O_2 in place of CO_2. This causes the oxidation or breakdown of RuBP, resulting in a loss of energy and carbon for the plant—a metabolic process called **photorespiration**. Photorespiration can drain away as much as 50% of the carbon fixed by the Calvin cycle!

- How can hot, arid regions have any plants? They have metabolic adaptations (as well as structural ones covered in Topic 8) that reduce photorespiration. The two most important of these adaptations are C_4 and CAM plants.
- C_4 plants have two strategies for reducing photorespiration.

    1. **Structural strategy.** In C_3 photosynthesis the light reactions and the Calvin cycle occur in the same cell. This puts the production of O_2 in close proximity to rubisco, leading to photorespiration. In C_4 plants the two stages of photosynthesis are separated spatially into different cells. One cell specializes in the light reactions, while the other type of cell specializes in the Calvin cycle. This spatially separates the two phases of photosynthesis, thus reducing photorespiration.
    2. **Biochemical strategy.** C_4 plants have an extra enzyme, PEP carboxylase, to fix carbon. PEP carboxylase does not combine with O_2, so it helps to further reduce photorespiration. PEP carboxylase acts as a CO_2 pump, helping to keep the concentration of CO_2 higher in the cells specializing in the Calvin cycle, again reducing photorespiration. Notice that C_4 plants still do the Calvin cycle and still use rubisco but have structural and biochemical adaptations that allow photosynthesis to occur even on hot, dry days.

- **CAM photosynthesis** is another adaptation to hot, dry climates.

    - CAM stands for *crassulacean acid metabolism,* referring to the family Crassulaceae in which the process was first discovered. (This is a group with succulent plants like the jade plant.)
    - CAM plants keep their stomata closed during the day to prevent excessive water loss. Of course, this also prevents gas exchange. At night, the stomata open and CO_2 is fixed in organic acids and stored in vacuoles. In the morning when the stomata close, the plant cells release the stored CO_2 from the acids and proceed with photosynthesis.
    - In CAM plants the two stages of photosynthesis are separated temporally.

- In both C_4 and CAM photosynthesis, CO_2 is first transformed into an organic intermediate before it enters the Calvin cycle. All of the processes—C_3, C_4, and CAM photosynthesis—use the Calvin cycle; they just have different methods for getting there.

> ***TIP FROM THE READERS***
> Be able to *compare* the process of chemiosmosis in the mitochondrion and the chloroplast. *Explain* how the H^+ gradient is established and *describe* the orientation of the ATP synthase molecules.

Level 1: Knowledge/Comprehension Questions

The Knowledge/Comprehension questions review essential content knowledge but don't represent the type of question you will see on the AP exam. Level 2, Application/Synthesis, questions are similar to those you will see on the exam.

1. The graph below most accurately depicts the energy changes that take place in which of the following types of reactions?

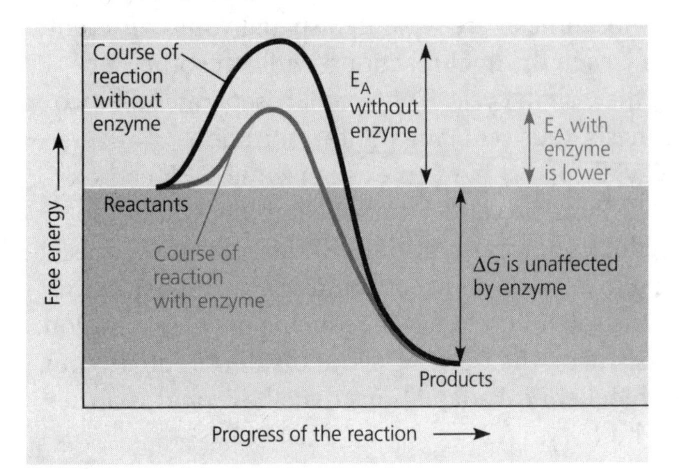

   (A) hypothermic
   (B) hyperthermic
   (C) endergonic
   (D) exergonic

2. Enzymes are organic catalysts. How do they increase the rate of chemical reactions?
   (A) by decreasing the free-energy change of the reaction
   (B) by increasing the free-energy change of the reaction
   (C) by raising the activation energy of the reaction
   (D) by lowering the activation energy of the reaction

Directions: Questions 3–6 consist of four lettered choices followed by a list of numbered phrases or sentences. For each numbered phrase or sentence, select the one choice that is most closely related to it. Each choice may be used once, more than once, or not at all.

Questions 3–6

(A) Allosteric interactions
(B) Feedback inhibition
(C) Competitive inhibitor
(D) Noncompetitive inhibitor

3. Describes a process by which an enzyme's function at the active site may be either activated or inhibited by the binding of a regulatory molecule at a separate site

4. A molecule that is so similar to the normal substrate that it competes for the active site of the enzyme

5. The process by which a metabolic pathway is shut off by the product it produces

6. A molecule that binds to the enzyme at a site other than the active site, causing the enzyme to change shape and be unable to bind substrate

7. The purpose of cellular respiration in a eukaryotic cell is to
   (A) synthesize carbohydrates from CO_2.
   (B) synthesize fats and proteins from CO_2.
   (C) break down carbohydrates to provide energy for the cell in the form of ATP.
   (D) provide oxygen to the cell.

8. One glucose molecule provides enough carbons for two trips through the citric acid cycle. How many molecules of ATP are directly produced in two trips through the citric acid cycle?
   (A) one
   (B) two
   (C) three
   (D) four

Directions: Questions 9–12 consist of four lettered choices followed by a list of numbered phrases or sentences. For each numbered phrase or sentence, select the one choice that is most closely related to it. Each choice may be used once, more than once, or not at all in each group.

Questions 9–12

(A) Chemiosmosis
(B) Electron transport chain
(C) Fermentation
(D) Glycolysis

9. The process by which glucose is split into pyruvate producing two ATP and two NADH

10. The process by which a proton (hydrogen ion) gradient is used to produce ATP

11. A process by which glucose is split to produce two ATP and lactic acid

12. A series of membrane-embedded electron carriers that ultimately create the hydrogen ion gradient to drive the synthesis of ATP

13. Groups of photosynthetic pigment molecules organized in the thylakoid membrane are called
(A) photosystems.
(B) carotenoids.
(C) chlorophyll.
(D) grana.

14. The main products of the light reactions of photosynthesis are
(A) NADPH and $FADH_2$.
(B) NADPH and ATP.
(C) ATP and $FADH_2$.
(D) ATP and CO_2.

15. The major product of the Calvin cycle is
(A) rubisco.
(B) ribulose bisphosphate.
(C) pyruvate.
(D) glyceraldehyde 3-phosphate.

16. Which of the following statements correctly links a process to its result?
(A) The light reactions convert solar energy to chemical energy in the form of ATP and glucose.
(B) The Calvin cycle uses ATP and NADPH to convert CO_2 to sugar.
(C) The Calvin cycle occurs in the thylakoid membrane and results in the capture of light energy.
(D) In chemiosmosis, electron transport chains pump protons (H^+) across a membrane from a region of high H^+ concentration to a region of low H^+ concentration.

17. Which of the following is mismatched with its location?
(A) photosystems—thylakoid membrane
(B) electron transport chain—thylakoid membrane
(C) Calvin cycle—stroma
(D) ATP synthesis—double membrane surrounding chloroplast

18. When light energy boosts electrons from the chlorophyll known as P680 to the primary electron acceptor, P680 has its electron "holes" filled by electrons from
(A) photosystem I.
(B) photosystem II.
(C) water.
(D) CO_2.

19. Photorespiration occurs when the levels of carbon dioxide are relatively low, and the enzyme rubisco catalyzes its release from the plant, rather than fixing it into an organic acid. CAM plants avoid photorespiration by
(A) fixing CO_2 into organic acids during the night; these acids then release CO_2 during the day.
(B) performing the Calvin cycle at night.
(C) fixing CO_2 into four-carbon compounds in the mesophyll, which immediately release CO_2 in the bundle-sheath cells.
(D) keeping their stomata open during the day.

20. How many "turns" of the Calvin cycle are required to produce one molecule of glucose?
    (A) 1
    (B) 2
    (C) 3
    (D) 6

21. What are the final electron acceptors for the electron transport chains in the light reactions of photosynthesis and in cellular respiration?
    (A) O_2 in both
    (B) CO_2 in both
    (C) $NADP^+$ in the light reactions and O_2 in respiration
    (D) H_2O in the light reactions and O_2 in respiration

22. In this drawing of a chloroplast, which structure represents the location of the Calvin cycle?

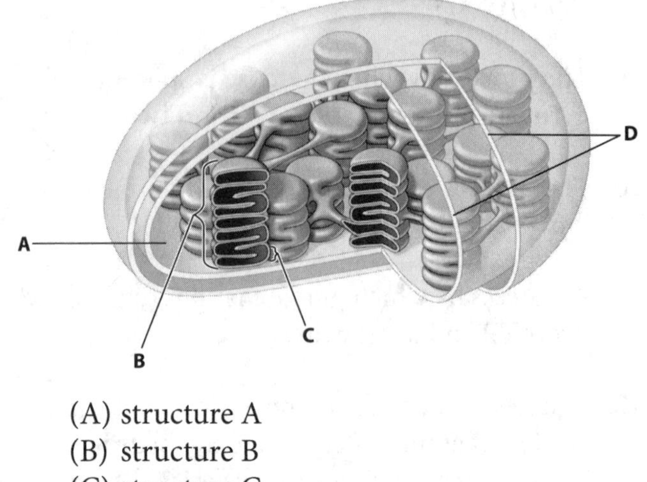

    (A) structure A
    (B) structure B
    (C) structure C
    (D) structure D

Level 2: Application/Analysis/Synthesis Questions

Use the following information to answer questions 1 and 2.

You're conducting an experiment to determine the effect of different wavelengths of light on the absorption of carbon dioxide as an indicator of the rate of photosynthesis in aquatic ecosystems. If the rate of photosynthesis increases, the amount of carbon dioxide in the environment will decrease and vice versa. You've added an indicator to each solution. When the carbon dioxide concentration decreases, the color of the indicator solution also changes.

Small aquatic plants are placed into three containers of water mixed with carbon dioxide and indicator solution. Container A is placed under normal sunlight, B under green light, and C under red light. The containers are observed for a 24-hour period.

1. Based on your knowledge of the process of photosynthesis, the plant in the container placed under red light would probably
    (A) absorb no CO_2.
    (B) absorb the same amount of CO_2 as the plants under both the green light and normal sunlight.
    (C) absorb less CO_2 than the plants under the green light.
    (D) absorb more CO_2 than the plants under the green light.

2. Carbon dioxide absorption is an appropriate indicator of photosynthesis because
    (A) CO_2 is needed to produce sugars in the Calvin cycle.
    (B) CO_2 is needed to complete the light reactions.
    (C) plants produce oxygen gas by splitting CO_2.
    (D) the energy in CO_2 is used to produce ATP and NADPH.

3. In the figure below, which step of the citric acid cycle requires both NAD^+ and ADP as reactants?
   (A) step 1
   (B) step 2
   (C) step 3
   (D) step 4

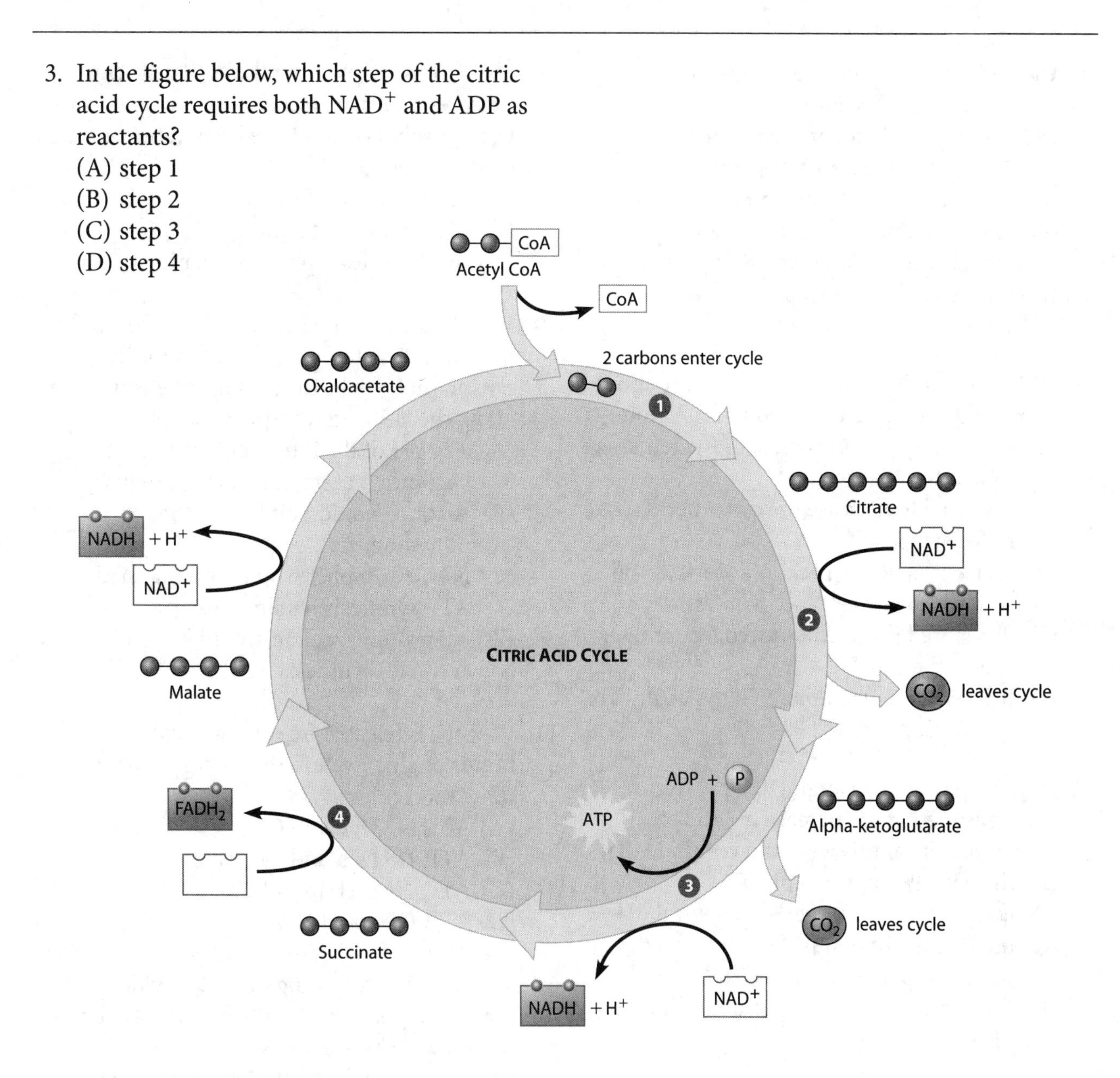

After reading the paragraph, answer question 4.

As a scientist employed by the FDA, you've been asked to sit on a panel to evaluate a pharmaceutical company's application for approval of a new weight loss drug called Fat Away. The company has submitted a report summarizing the results of their animal and human testing. In the report, it was noted that Fat Away works by affecting the electron transport chain. It decreases the synthesis of ATP by making the mitochondrial membrane permeable to H^+, which allows H^+ to leak from the intermembrane space to the mitochondrial matrix. This effect leads to weight loss.

4. Fat Away prevents ATP from being made by
   (A) destroying the H^+ gradient that allows ATP synthase to work.
   (B) preventing glycolysis from occurring.
   (C) preventing the conversion of NADH to NAD^+.
   (D) slowing down the citric acid cycle.

5. Catabolic and anabolic pathways are often coupled in a cell because
   (A) the intermediates of a catabolic pathway are used in the anabolic pathway.
   (B) both pathways use the same enzymes.
   (C) the free energy released from one pathway is used to drive the other pathway.
   (D) their enzymes are controlled by the same activators and inhibitors.

6. When a cell breaks down glucose, only about 34% of the energy is captured in ATP molecules. What happens to most of the remaining 66% of the energy?
   (A) It is used to increase the order necessary for life to exist.
   (B) It is lost as heat, in accordance with the second law of thermodynamics.
   (C) It is stored in starch or glycogen for later use by the cell.
   (D) It is released when the ATP molecules are hydrolyzed.

7. In an experiment, changing the pH from 7 to 6 resulted in an increase in product formation. From this we could conclude that
   (A) the enzyme's optimal pH is 6.
   (B) this enzyme works best at a neutral pH.
   (C) the temperature must have increased when the pH was changed to 6.
   (D) the enzyme was in a more active shape at pH 6.

8. When substance A was added to an enzyme reaction, product formation decreased. The addition of more substrate did not increase product formation. From this we conclude that substance A could be
   (A) product molecules.
   (B) an allosteric enzyme.
   (C) a competitive inhibitor.
   (D) a noncompetitive inhibitor.

9. The process in photosynthesis that bears the most resemblance to chemiosmosis and oxidative phosphorylation in cell respiration is called
   (A) glycolysis.
   (B) linear electron flow.
   (C) ATP synthase coupling.
   (D) substrate-level phosphorylation.

10. Which of the following statements correctly describes a metabolic effect of cyanide, a poison that blocks the passage of electrons along the electron transport chain?
    (A) The pH of the intermembrane space becomes much lower than normal.
    (B) Alcohol would build up in the mitochondria.
    (C) NADH supplies would be exhausted, and ATP synthesis would cease.
    (D) No proton gradient would be produced, and ATP synthesis would cease.

11. Glycolysis releases free energy held in the bonds of glucose, and this energy is held in these molecules:
    (A) ATP, NADH, and CO_2
    (B) ATP, NADH, and pyruvate
    (C) ATP, NADPH, and RUBP
    (D) ATP, CO_2, and H_2O

12. Which of the following provides evidence that glycolysis is one of the first metabolic pathways to have evolved?
    (A) It relies on fermentation, which is characteristic of archaea and bacteria.
    (B) It is found only in prokaryotes, whereas eukaryotes use mitochondria to produce ATP.
    (C) It produces ATP only by oxidative phosphorylation and does not involve redox reactions.
    (D) It is nearly universal, occurs in the cytosol, and does not involve O_2.

13. Bacterial production of the enzymes needed for the synthesis of the amino acid tryptophan declines with increasing levels of tryptophan and increases as tryptophan levels decline. This is an example of
   (A) competitive inhibition.
   (B) noncompetitive inhibition.
   (C) feedback inhibition.
   (D) irreversible inhibition.

14. Brown fat, which is found in newborn infants and hibernating mammals, has uncoupler proteins that, when activated, make the inner mitochondrial membrane leaky to H^+. What is the function of brown fat?
   (A) It produces more ATP than does regular fat and is also found in the flight muscles of ducks and geese, providing more energy for long-distance migrations.
   (B) It lowers the pH of the intermembrane space, which results in the production of more ATP per gram than is produced by the oxidation of glucose or regular fat tissue.
   (C) Because it dissipates the proton gradient, it generates heat through cellular respiration without producing ATP, thereby raising the body temperature of hibernating mammals or newborn infants.
   (D) Its main function is insulation in endothermic animals where brown fat is common.

15. The release of the energy stored in glucose involves a series of reactions and processes. Which of the following correctly states the location and function of the process named?
   (A) Citric acid cycle releases carbon dioxide from organic intermediates and synthesizes ATP from ADP via substrate level phosphorylation.
   (B) Glycolysis releases free energy from glucose to produce ATP by oxidative phosphorylation.
   (C) Electron transport occurs in the cytoplasm and harvests the electrons from NADH and $FADH_2$ to establish the proton gradient.
   (D) Chemiosmosis occurs when an enzyme in the inner mitochondrial membrane produces carbon dioxide and water from NADH and ATP.

Grid-In Questions

These call for a numerical response.

1. The following chart shows the energy products produced in various stages of the breakdown of glucose.

Process	ATP Produced	$NADH/FADH_2$ Produced
Glycolysis	2	2 NADH
Pyruvate Oxidation *(per molecule of pyruvate)*		1 NADH
Citric Acid (Krebs) Cycle *per molecule of pyruvate*	1	3 NADH 1 $FADH_2$

Each molecule of NADH results in approximately 2.5 molecules of ATP, whereas each molecule of $FADH_2$ results in approximately 1.5 molecules of ATP when these molecules are fed into the electron transport chain.

1. What is the difference in the total number of ATP molecules produced between three molecules of glucose that undergo fermentation compared to three molecules of glucose that undergo aerobic respiration?
2. Glutamine is formed from glutamic acid by adding an ammonium molecule to glutamine. The overall reaction is endergonic, requiring 3.4 kcal/mole. The energy for the reaction comes from the exergonic splitting of a phosphate from ATP to form ADP, which releases 7.3 kcal/mole. What is the free energy change for this coupled reaction?

An enzyme in the liver removes a phosphate group from glucose so the glucose molecule can enter the bloodstream, providing energy for cellular respiration to the cells of the body. The rate of enzyme activity can be monitored by measuring the phosphate concentration over time. In this experiment, liver cells were placed in a phosphate solution, and every 5 minutes cells were removed and the intracellular concentration of phosphate was measured. What is the rate of phosphate formation per minute from 15 to 20 minutes?

3.

Time (min)	Phosphate Concentration in Liver Cells (μmol/mL)
0	0
5	10
10	90
15	180
20	270

Free-Response Question

1. *Organisms capture and store free energy for use in biological processes. Each of the following parts plays a role in the transfer of energy.*

   (a) **Describe** how a photosystem converts light energy to chemical energy.
   (b) **Explain** how glycolysis releases free energy from glucose.
   (c) **Describe** the role of water in both cellular respiration and photosynthesis.

TOPIC 4

Mendelian Genetics

Meiosis and Sexual Life Cycles

(Biology, *10e: Chapter 13*, Focus, *1e: Chapter 10*)

YOU MUST KNOW

- The differences between asexual and sexual reproduction.
- The role of meiosis and fertilization in sexually reproducing organisms.
- The importance of homologous chromosomes to meiosis.
- How the chromosome number is reduced from diploid to haploid in meiosis.
- Three events that occur in meiosis but not mitosis.
- The importance of crossing over, independent assortment, and random fertilization to increasing genetic variability.

Offspring acquire genes from parents by inheriting chromosomes (Biology, *10e, 13.1,* Focus, *1e: 10.1*)

- **Genes** are segments of DNA that code for the basic units of heredity and are transmitted from one generation to the next. In animals and plants, reproductive cells that transmit genes from one generation to the next are called **gametes.**
- A **locus** (plural, *loci*) is the location of a gene on a chromosome. See Figure 4.1a.
    - In **asexual reproduction** a single parent is the sole parent and passes copies of all its genes to its offspring. In asexual reproduction the new offspring arise by mitosis and have virtually exact copies of the parent's genome. An individual that reproduces asexually gives rise to a **clone**, a group of genetically identical individuals.
    - In **sexual reproduction**, two individuals (parents) contribute genes to the offspring. This form of reproduction results in greater genetic variation in the offspring than asexual reproduction.

Fertilization and meiosis alternate in sexual life cycles (Biology, *10e: 13.2,* Focus, *1e: 10.2*)

- A **life cycle** is the generation-to-generation sequence of stages in the reproductive history of an organism, from conception to production of its own offspring.
- **Somatic cells** are any cells in the body that are not gametes. Each somatic cell in humans has 46 chromosomes. Liver cells and neurons are somatic cells.

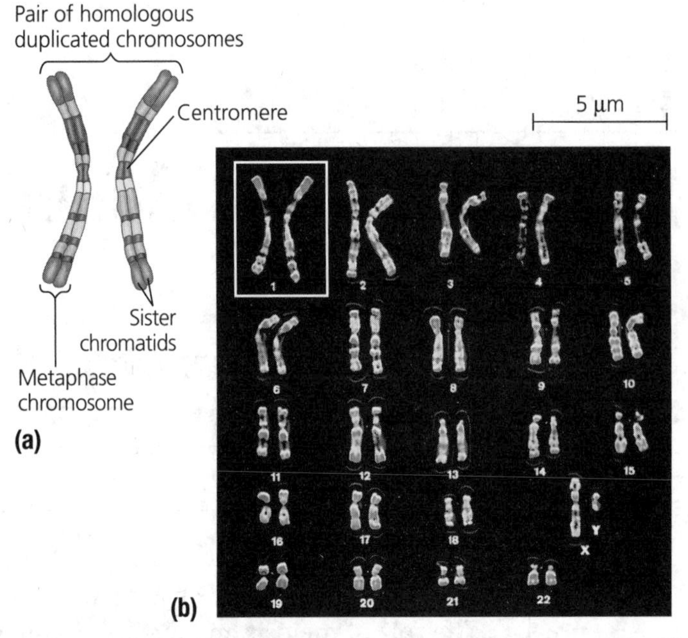

Figure 4.1 Metaphase chromosome (a) and human karyotype (b)

- The **karyotype** of an organism refers to a picture of its complete set of chromosomes, arranged in pairs of homologous chromosomes from the largest pair to the smallest pair. Figure 4.1b is a karyotype made from a human somatic cell. Notice that the 46 chromosomes are paired into 23 homologous chromosomes.
- In **homologous chromosomes** both chromosomes of each pair carry genes that control the same inherited characteristics. If a gene for eye color is found at a specific locus on one chromosome, its homologs will have the same gene at the same locus.
  - Homologous chromosomes are similar in length and centromere position, and they have the same staining pattern.
  - One homologous chromosome from each pair is inherited from each parent; in other words, half of the set of 46 chromosomes in your somatic cells was inherited from your mother, and the other half was inherited from your father.
- Exceptions to the rule that all chromosomes are part of a homologous pair may be found with the **sex chromosomes**—in humans, it is the **X** and **Y**. Human females have a homologous pair of chromosomes, XX, but males have one X chromosome and one Y chromosome. Nonsex chromosomes; that is, all the chromosomes except the X and Y are called **autosomes**.
- *What sex did the somatic cell come from that was used to make the karyotype in Figure 4.1?*
- **Gametes**—meaning sperm and ova (eggs)—are haploid cells. Haploid cells contain half the number of chromosomes of somatic cells. In humans, gametes contain 22 autosomes plus a single sex chromosome (X in female, either X or Y in male), giving them a haploid number of 23. The haploid number of chromosomes is symbolized by *n*.

- **Meiosis** and **fertilization** are the key events in sexually reproducing life cycles. The human life cycle in Figure 4.2 is typical of a sexually reproducing animal. Note the key points in the figure as you read about the life cycle.

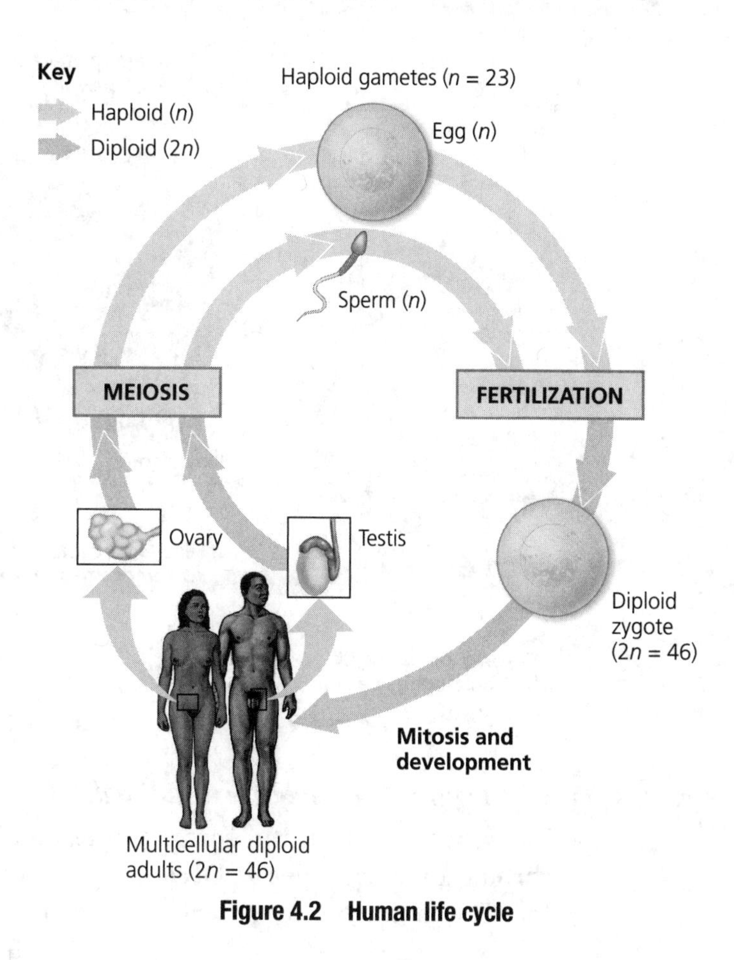

Figure 4.2 Human life cycle

- During **fertilization** (the combination of a sperm cell and an egg cell), one haploid gamete from the father fuses with one haploid gamete from the mother. The result is a fertilized egg called a **zygote**. It is **diploid** (has two sets of chromosomes) and may be symbolized by $2n$.
- **Meiosis** is the type of cell division that reduces the numbers of sets of chromosomes from two to one. Fertilization restores the diploid number as the gametes are combined. *Fertilization and meiosis alternate in the life cycles of sexually reproducing organisms.*

Meiosis reduces the number of chromosome sets from diploid to haploid (Biology, *10e: 13.3,* Focus, *1e: 10.3*)

- Meiosis and mitosis look similar—both are preceded by the replication of the cell's DNA, for instance, but in meiosis this replication is followed by *two* stages of cell division, meiosis I and meiosis II.
- The final result of meiosis is *four* **daughter cells**, each of which has *half* as many chromosomes as the parent cell.

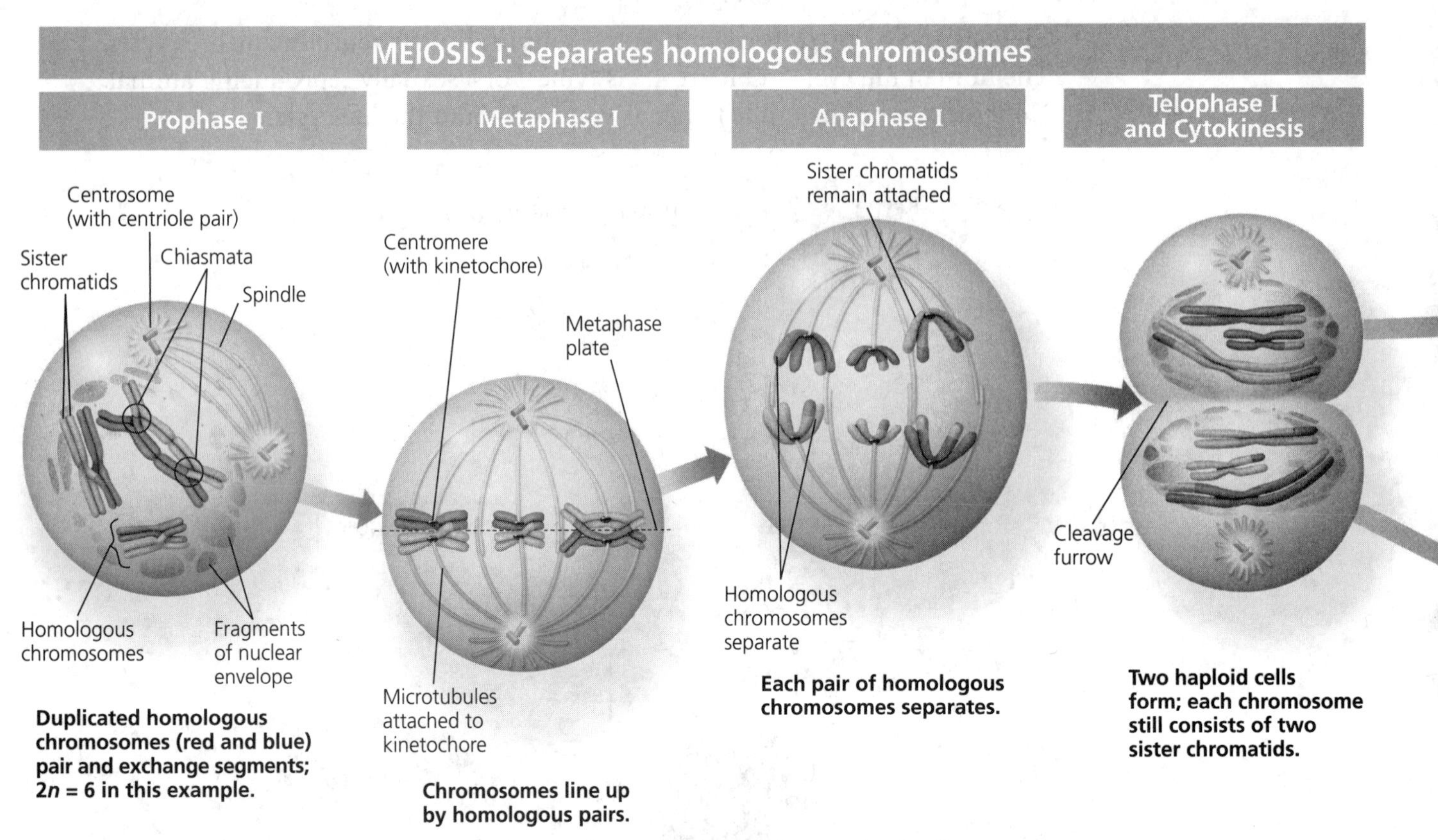

Figure 4.3 Meiosis in an animal cell

Carefully follow the stages in Figure 4.3 as they are explained:

- **Interphase:** Each of the chromosomes makes a copy of itself; that is, each chromosome replicates its DNA, roughly doubling the amount of DNA in the cell. The centrosome also divides during this phase.

> *STUDY TIP* Understanding prophase I is critical to understanding meiosis. Study the unique events of prophase I carefully!

- **Meiosis I:** The first cellular division in meiosis is referred to as meiosis I. Meiosis I begins with a diploid cell.
  - **Prophase I:** The chromosomes condense, resulting in two sister chromatids attached at their centromeres.
  - **Synapsis** occurs—that is, the joining of homologous chromosomes along their length. This newly formed structure is called a *tetrad* and precisely aligns the homologous chromosomes gene by gene. This perfect alignment is necessary for the next step—crossing over.
  - In **crossing over** the DNA from one homolog is cut and exchanged with an exact portion of DNA from the other homolog. Essentially, a small part of the DNA from one parent is exchanged with the DNA from the other parent. *The result of crossing over is to increase genetic variation.* Where crossing over has occurred (two to three times per homologous pair), crisscrossed regions termed **chiasmata** form, which hold the homologs together until anaphase I.

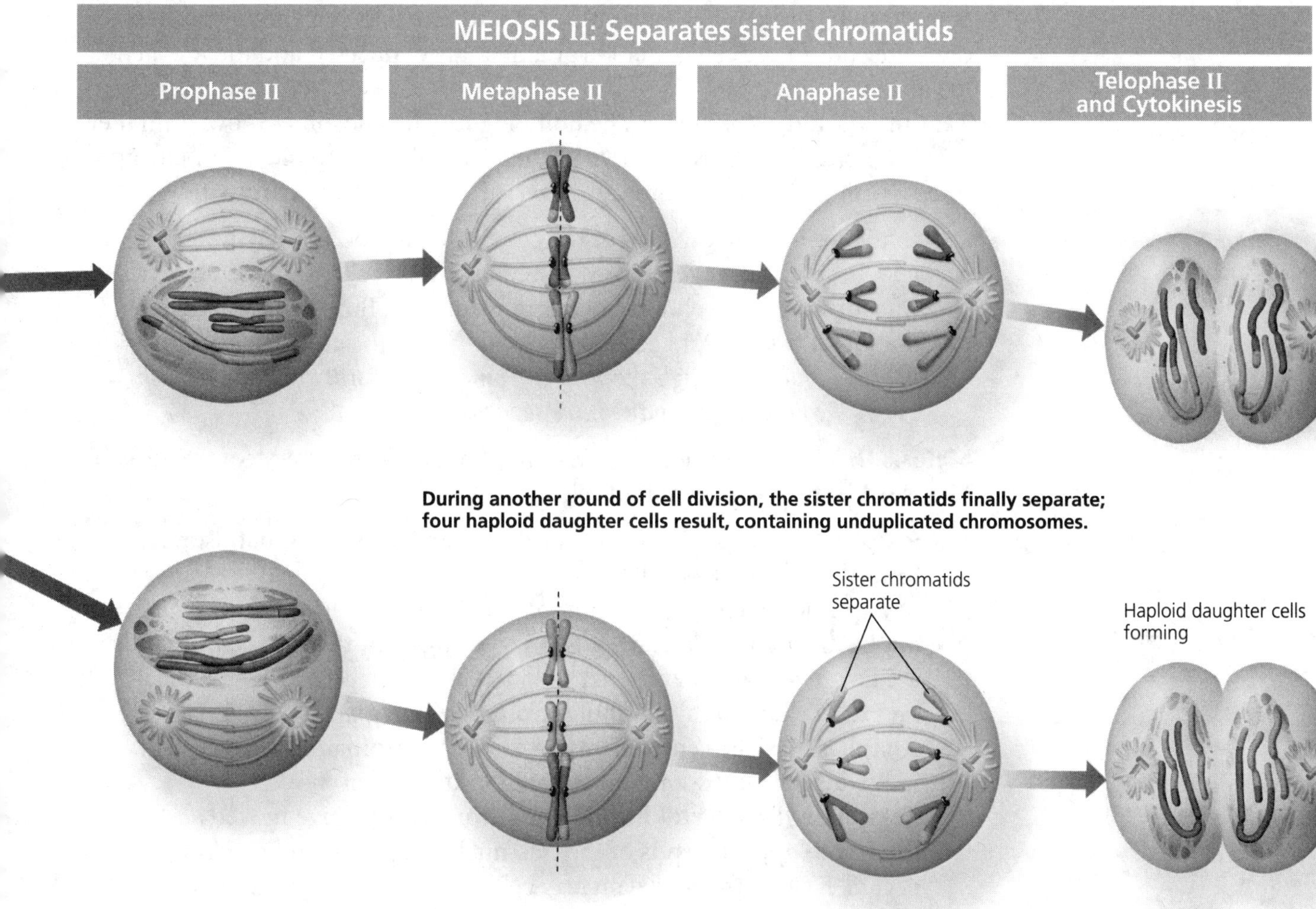

MEIOSIS II: Separates sister chromatids

Prophase II | Metaphase II | Anaphase II | Telophase II and Cytokinesis

During another round of cell division, the sister chromatids finally separate; four haploid daughter cells result, containing unduplicated chromosomes.

Sister chromatids separate

Haploid daughter cells forming

- After crossing over, the spindle poles move away from each other, the nuclear envelope disintegrates, and the spindle microtubules attach to the kinetochores forming on the chromosomes. The microtubules then begin to move the chromosomes to the metaphase plate of the cell.

> ***ORGANIZE YOUR THOUGHTS***
>
> In Prophase I:
>
> 1. Synapsis occurs, forming tetrads.
> 2. Crossing over occurs between *homologous chromosomes* in the tetrads.
> 3. Crossing over increases genetic variation.
> 4. Areas of crossing over form chiasmata.
> 5. The nuclear envelope disintegrates, allowing the spindle to attach to the homologs.

- **Metaphase I:** At this point in meiosis the homologous pairs of chromosomes are lined up at the metaphase plate. How many chromosomes and how many homologous pairs are found in the cell in Figure 4.3? (There are four chromosomes and two homologous pairs.)

- **Anaphase I:** The spindle apparatus helps to move the chromosomes toward opposite ends of the cell; sister chromatids stay connected and move together toward the poles.
- **Telophase I** and **cytokinesis:** The homologous chromosomes move until they reach the opposite poles. Each pole, then, contains a haploid set of chromosomes, with each chromosome still consisting of two sister chromatids.
  - Cytokinesis is the division of the cytoplasm and occurs during telophase. A **cleavage furrow** occurs in animal cells, and **cell plates** (the forming new cell wall) occur in plant cells. Both result in the formation of two haploid cells.*Note carefully that the daughter cells are now haploid—although the sister chromatids are still attached to each other, the homologous pairs have separated.*
- **Meiosis II:** The second cellular division in meiosis is referred to as meiosis II. Meiosis II begins with a haploid cell.
  - **Prophase II:** A spindle apparatus forms, and sister chromatids move toward the metaphase plate.
  - **Metaphase II:** The haploid number of chromosomes is now arrayed on the metaphase plate. Because of crossing over, the sister chromatids are not genetically identical. The kinetochores of each sister chromatid are attached to microtubules from opposite poles.
  - **Anaphase II:** The centromeres of the sister chromatids separate and individual chromosomes move to opposite ends of the cell.
  - **Telophase II** and **cytokinesis:** The chromosomes have moved all the way to opposite ends of the cell, nuclei reappear, and cytokinesis occurs. Each of the four daughter cells has the haploid number of chromosomes and is genetically different from the other daughter cells and from the parent cell.

STUDY TIP Be prepared to cite the following three examples when asked to explain differences between mitosis and meiosis.

- Three events occur during meiosis I that do not occur during mitosis.
  1. Synapsis and crossing over do not occur during mitosis.
  2. At metaphase I, paired homologous chromosomes (tetrads) are positioned on the metaphase plate, rather than individual replicated chromosomes, as in mitosis.
  3. At anaphase I, duplicated chromosomes of each homologous pair separate, but the sister chromatids of each duplicated chromosome stay attached. In mitosis, the chromatids separate.

TIP FROM THE READERS

When does the cell go from diploid to haploid? Be sure you know and understand this! Refer again to Figure 4.3 and you should see that chromosome number is reduced during **meiosis I**. Each chromosome consists of two sister chromatids but the homologous pairs have separated.

Genetic variation produced in sexual life cycles contributes to evolution (Biology, *10e: 13.4,* Focus, *1e: 10.4*)

> ***STUDY TIP*** There are three important processes that contribute to variation. They are given next. Be able to list and explain them.

- **Crossing over:** During prophase I the exchange of genetic material on homologous chromosomes between nonsister chromatids occurs. Use Figure 4.3 to help make this unique feature of meiosis clear. Notice that all four chromatids that make up the tetrad are different due to crossing over. In metaphase II when sister chromatids separate, each chromatid is unique, thus increasing variation.
- **Independent assortment of chromosomes:** In metaphase I, when the homologous chromosomes are lined up on the metaphase plate, they can pair up in any combination, with any of the homologous pairs facing either pole. This means that there is a 50% chance that a particular daughter cell will get a maternal chromosome or a paternal chromosome from each of the homologous pairs.
- **Random fertilization:** Because each egg and sperm is different, as a result of independent assortment and crossing over, each combination of egg and sperm is unique.

Mendel and the Gene Idea

(Biology, *10e: Chapter 14*, Focus, *1e: Chapter 11*)

> **YOU MUST KNOW**
>
> - Terms associated with genetics problems: P, F_1, F_2, dominant, recessive, homozygous, heterozygous, phenotype, and genotype.
> - How to derive the proper gametes when working a genetics problem.
> - The difference between an allele and a gene.
> - How to read a pedigree.
> - How to use data sets to determine Mendelian patterns of inheritance.

Mendel used the scientific approach to identify two laws of inheritance (Biology, *10e: 14.1,* Focus, *1e: 11.1*)

- True-breeding parents in a genetic cross are called the **P (parental) generation**; their offspring are called the **F_1 (first filial) generation**. If the F_1 population is crossed, their offspring are called the **F_2 (second filial) generation.**
- The following are four related concepts that make up Mendel's model explaining the 3:1 inheritance pattern that he observed among F_2 offspring.

  1. **Alternative versions of genes account for variations in inherited characteristics among offspring.** For example, consider flower color in peas. The gene for flower color in pea plants comes in two versions: white and purple. These alternative versions of the gene, called **alleles**, are the result of slightly different DNA sequences.

2. **For each character, every organism inherits one allele from each parent.**
3. **If the two alleles are different, then the *dominant allele* will be fully expressed in the offspring, whereas the *recessive allele* will have no noticeable effect on the offspring.**
4. **The two alleles for each character separate during gamete production.** If the parent has two of the same alleles, then the offspring will all get that version of the gene, but if the parent has two different alleles for a gene, each offspring has a 50% chance of getting one of the two alleles. This is Mendel's **law of segregation**.

STUDY TIP Use Figure 4.4 to find each of the four basic concepts of Mendel's model.

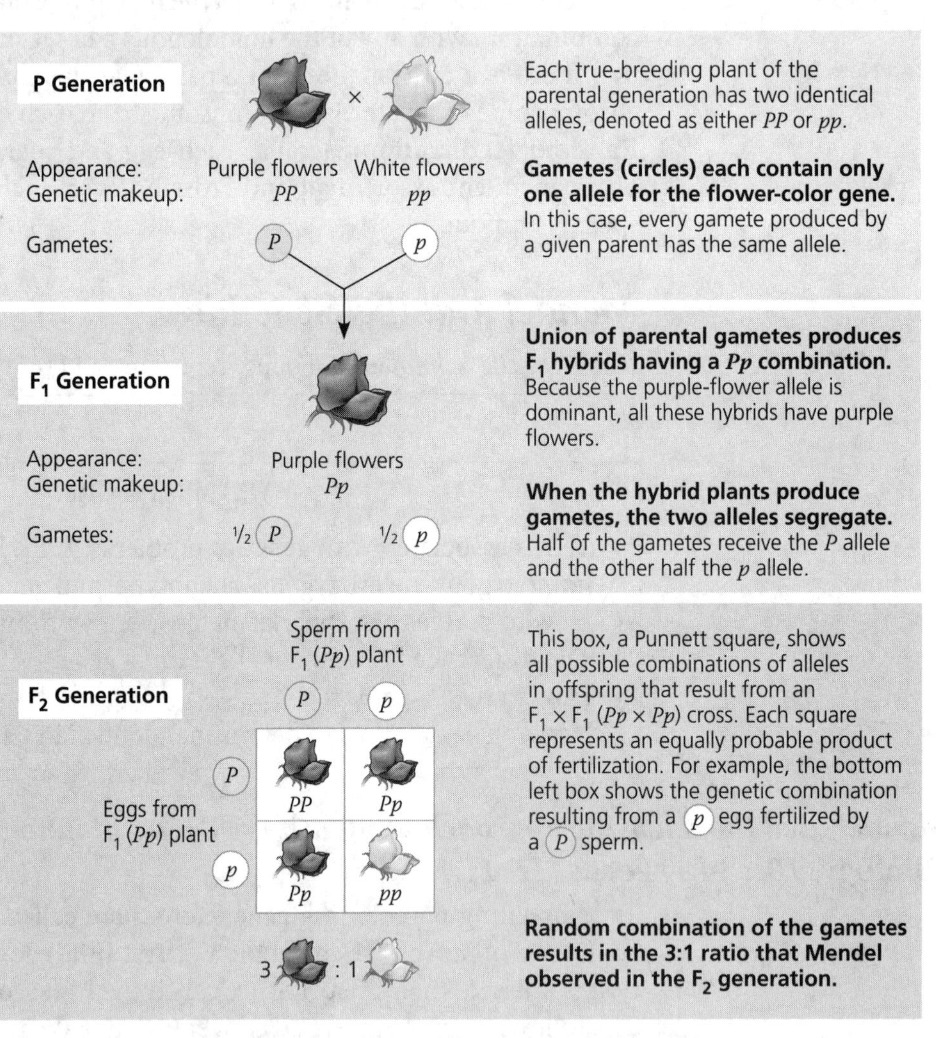

Figure 4.4 Mendel's law of segregation

- **Law of independent assortment** was Mendel's second law. It states that each pair of alleles will segregate (separate) independently during gamete formation. This occurs during anaphase I of meiosis.
- **Homozygous** organisms have two of the same alleles for a particular trait. If the dominant allele for a trait is designated as *R* (dominant traits are generally

capitalized), and the recessive allele is designated *r* (recessive traits are generally not capitalized), then an individual could be homozygous for the dominant trait (*RR)* or homozygous for the recessive trait (*rr*).

- A **heterozygous** organism has two different alleles for a trait (*Rr*).
- **Phenotype** refers to an organism's expressed physical traits, and **genotype** refers to an organism's genetic makeup. For example, the phenotype of a seed might be round, and its genotype could be *RR* or *Rr*.
- A **testcross** is done to determine if an individual showing a dominant trait is homozygous or heterozygous. A test cross is always done between the unknown genotype and a *homozygous recessive* individual. If the unknown parent is homozygous dominant (*RR* × *rr*) all the offspring will show the dominant trait, but if the unknown parent is heterozygous (*Rr* × *rr*) some of the offspring will show the recessive trait.
- A **monohybrid cross** is a cross involving the study of only one character (for example, flower color), whereas a **dihybrid cross** is a cross intended to study two characters (for example, flower color and seed shape).
- The diagram that follows shows the results of a dihybrid cross. In this case, in the parental generation two homozygous plants are crossed: one homozygous dominant for yellow and round seeds (*YYRR*) and one homozygous recessive for green and wrinkled seeds (*ppyy*). The only gamete type the first parent can produce is *YR*, and the only gamete the second parent can produce is *yr*. The F_1 generation, therefore, is composed of individuals with genotype *YyRr*. Crossing *YyRr* with a second *YyRr* gives an F_2 generation that completes the cross and looks like Figure 4.5.

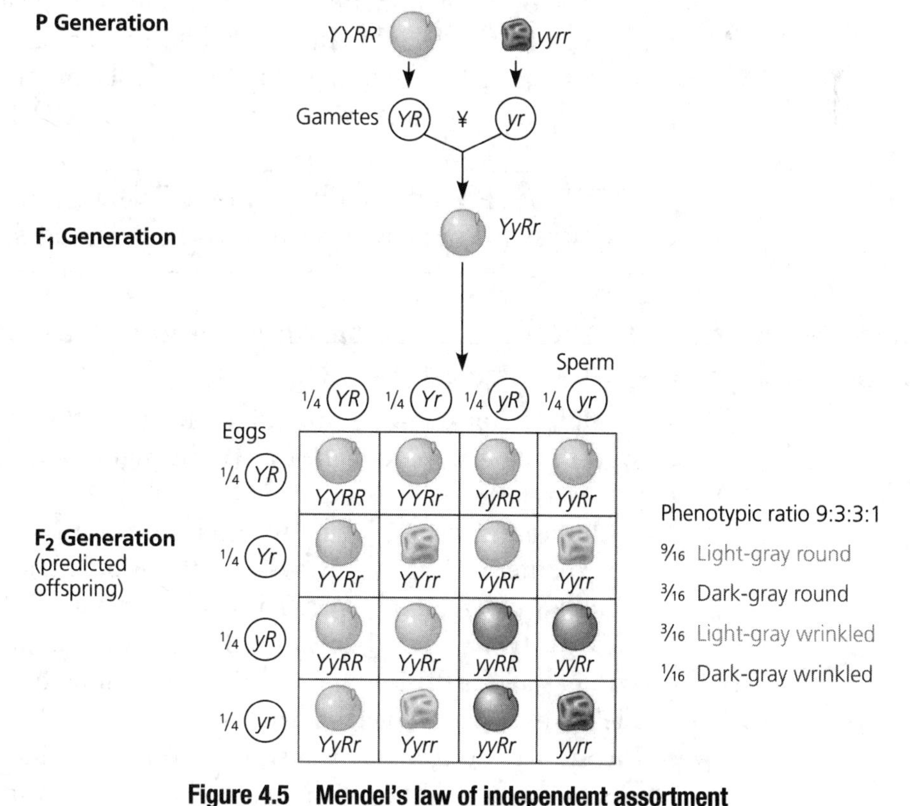

Figure 4.5 Mendel's law of independent assortment

> ***STUDY TIP*** When working genetics problems, be sure you have written the genotypes of the parents correctly and derived the gametes correctly. Study Figure 4.5, paying special attention to how the gametes were obtained for the F_2 generation. If you are having trouble doing genetics problems work through **TIPS FOR GENETICS PROBLEMS** and do the problems in **TEST YOUR UNDERSTANDING** at the end of the *Mendel and the Gene Idea* (Biology, *10e: Chapter 14*, Focus, *1e: Chapter 11*). All end-of-chapter problems are worked in Appendix A of your textbook (page 290 in Biology, *10e*, and page 225 in Focus, *1e*).

The laws of probability govern Mendelian inheritance (Biology, *10e: 14.2*, Focus, *1e: 11.2*)

- Understanding how to predict offspring of genetic crosses involves familiarity with the basic laws of probability. There are two laws that you will use directly in solving genetics problems:
  - **The rule of multiplication:** When calculating the probability that two or more independent events will occur together in a specific combination, multiply the probabilities of each of the two events. Thus, the probability of a coin landing face up two times in two flips is $\frac{1}{2} \times \frac{1}{2} = \frac{1}{4}$. If you cross two organisms with the genotypes *AABbCc* and *AaBbCc*, the probability of an offspring having the genotype *AaBbcc* is $\frac{1}{2} \times \frac{1}{2} \times \frac{1}{4} = \frac{1}{16}$.
  - **The rule of addition:** When calculating the probability that any of two or more mutually exclusive events will occur, you need to add together their individual probabilities. For example, if you are tossing a die, what is the probability that it will land on either the side with 4 spots or the side with 5 spots? ($\frac{1}{6} + \frac{1}{6} = \frac{1}{3}$)

> ***STUDY TIP*** What are the chances of event 1 ***and*** event 2? Multiply them. What are the chances of event 1 ***or*** event 2? Add them.

Inheritance patterns are often more complex than predicted by simple Mendelian genetics (Biology, *10e: 14.3*, Focus, *1e: 11.3*)

- **Complete dominance** is dominance in which the heterozygote and the homozygote for the dominant allele are indistinguishable. A *Yy* yellow seed is just as yellow as a *YY* yellow seed.
- **Codominance** occurs when two alleles are dominant and affect the phenotype in two different but equal ways. The traditional example for this type of dominance is human blood types. The alleles for A and B blood are dominant to the allele for type O blood, but A and B are codominant to each other. A person who has alleles for both A and B blood will be blood type AB because these alleles are each completely expressed.
- **Incomplete dominance** is a type of dominance in which the F_1 hybrids have an appearance that is in between that of the two parents. For example, if

two plants, one with white flowers and one with red flowers, were crossed and all of the offspring had pink flowers, you could conclude that the trait for flower color exhibits incomplete dominance. Breeding two of the hybrids with incomplete dominance gives a flower ratio of 1 red:2 pink:1 white.

- **Multiple alleles** occur when a gene has more than two alleles. Again, a good example of this is seen in human blood types. There are three alleles for human blood types: I^A, I^B, and *I*, but one person only receives any combination of two alleles.
- **Pleiotropy** is the property of a gene that causes it to have multiple phenotypic effects. For example, sickle-cell disease has multiple symptoms all caused by a single defective gene.
- In **epistasis**, a gene at one locus alters the effects of a gene at another locus. For example, an individual may have genes for heavy skin pigmentation, but if a separate gene that produces the pigment is defective, the genes for pigment deposition will not be expressed. This would lead to a condition known as albinism. *(For those who read the Curriculum Framework closely, you will note epistasis and pleiotropy are not required information—but we think it helps tell the complete story.)*
- In **polygenic inheritance**, two or more genes have an additive effect on a single character in the phenotype (such as height or skin color in humans). When several genes are involved, the phenotype usually is described by a bell-shaped curve, with fewer individuals at each extreme and most individuals clustered in the middle.

Many human traits follow Mendelian patterns of inheritance (Biology, *10e: 14.4,* Focus, *1e: 11.4*)

- A **pedigree** is a diagram that shows the relationship between parents and offspring across two or more generations. (See Figure 4.6.) In a typical pedigree, circles represent females, and squares represent males. White open circles or squares indicate that the individual did not or does not express a particular trait, whereas the shaded ones indicate that the individual expresses or expressed that trait. Through the patterns they reveal, pedigrees can help determine the genome of individuals that comprise them; pedigrees can also help predict the genome of future offspring.
- **Recessively inherited disorders** require two copies of the defective gene for the disorder to be expressed. Examples include the following:
  - **Cystic fibrosis** is caused by a mutation in an allele that codes for a cell membrane protein that functions in the transport of chloride ions into and out of cells. The resulting high extracellular levels of chloride cause mucus to be thicker and stickier, leading to organ malfunction and recurrent bacterial infections.
  - **Tay-Sachs** disease is caused by an allele that codes for a dysfunctional enzyme, which is unable to break down certain lipids in the brain. As these lipids accumulate in the brain cells, the child suffers from blindness, seizures, and degeneration of brain function, leading to death.

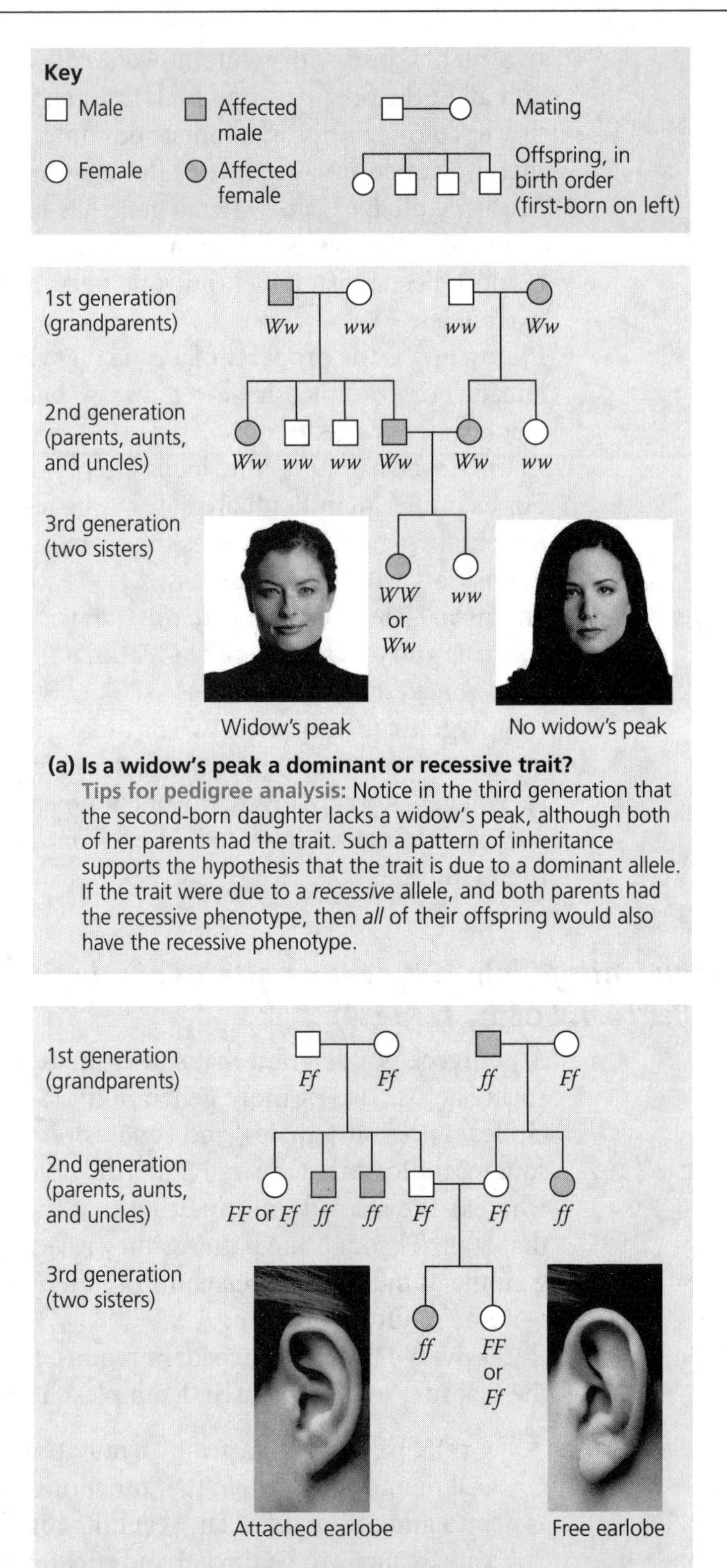

(a) Is a widow's peak a dominant or recessive trait?

Tips for pedigree analysis: Notice in the third generation that the second-born daughter lacks a widow's peak, although both of her parents had the trait. Such a pattern of inheritance supports the hypothesis that the trait is due to a dominant allele. If the trait were due to a *recessive* allele, and both parents had the recessive phenotype, then *all* of their offspring would also have the recessive phenotype.

(b) Is an attached earlobe a dominant or recessive trait?

Tips for pedigree analysis: Notice that the first-born daughter in the third generation has attached earlobes, although both of her parents lack that trait (they have free earlobes). Such a pattern is easily explained if the attached-lobe phenotype is due to a recessive allele. If it were due to a *dominant* allele, then at least one parent would also have had the trait.

Figure 4.6 Human pedigree analysis

- **Sickle-cell disease** is caused by an allele that codes for a mutant hemoglobin molecule that forms long rods when the oxygen levels in the blood are low. These long rods cause the red blood cell to sickle, clogging small blood vessels and leading to pain, organ damage, and even paralysis.

- **Lethal dominant alleles** require only one copy of the allele in order for the disorder to be expressed. Usually, only lethal alleles that act late in life are passed on.
- **Huntington's disease** is caused by a lethal dominant allele. It is a degenerative disease of the nervous system, which usually doesn't affect the individual until he or she is over 40 years old.
- Genetic testing may be used on a fetus to detect certain genetic disorders. Two common tests are amniocentesis and chorionic villus sampling (CVS).
  - **Amniocentesis** occurs when the physician removes amniotic fluid from around the fetus. The amniotic fluid can be utilized to detect some genetic disorders, and the cells in the fluid can be cultured for a karyotype.
  - **Chorionic villus sampling** involves using a narrow tube inserted through the cervix to suction out a tiny sample of the placenta that contains only fetal cells. A karyotype can immediately be developed from these cells.

The Chromosomal Basis of Inheritance

(Biology, *10e: Chapter 15*, Focus, *1e: Chapter 12*)

YOU MUST KNOW

- How the chromosome theory of inheritance connects the physical movement of chromosomes in meiosis to Mendel's laws of inheritance.
- The unique pattern of inheritance in sex-linked genes.
- How alteration of chromosome number or structurally altered chromosomes (deletions, duplications, etc.) can cause genetic disorders.
- How genomic imprinting and inheritance of mitochondrial DNA are exceptions to standard Mendelian inheritance.

Mendelian inheritance has its physical basis in the behavior of chromosomes (Biology, *10e: 15.1*, Focus, *1e: 12.1*)

- The **chromosome theory of inheritance** states that genes have specific locations (called *loci*) on chromosomes and that it is chromosomes that segregate and assort independently. It is important to connect this physical movement of chromosomes in meiosis to Mendel's laws of inheritance. If you are having trouble visualizing this, carefully work through Figure 15.2 in *Biology*, 10e, or Figure 12.2 in *Focus*, 1e.

- A **sex-linked gene** is one located on a sex chromosome (X or Y in humans). After the chromosome theory of inheritance was formed, *Thomas Hunt Morgan* discovered the existence of sex-linked genes.

Sex-linked genes exhibit unique patterns of inheritance (Biology, *10e: 15.2,* Focus, *1e: 12.2*)

- In humans, there are two types of sex chromosomes, X and Y. A person who inherits two X chromosomes usually develops as a female, whereas a person who inherits one X and one Y normally develops as a male.
- Sex-linked genes may be either X-linked or Y-linked. Figure 4.7 shows the unique pattern of inheritance in X-linked genes. In addition to tracking the gene from one generation to the next, it is also necessary to track the sex of the offspring. Figure 4.7 is based on work with *Drosophila* (fruit flies) performed by Thomas Hunt Morgan, but the pattern is the same in humans.

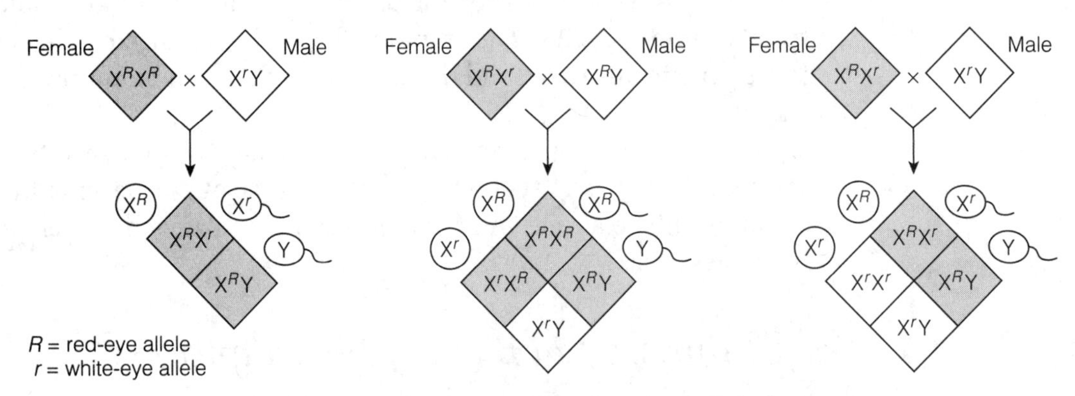

Figure 4.7 Patterns of inheritance with sex-linked traits

- Each egg or ovum contains an X chromosome; there are two types of sperm: those with an X chromosome and those with a Y chromosome. In fertilization, there is a 50% chance that a sperm carrying an X or Y will reach and penetrate the egg first. Thus, gender is determined by chance and by the male sperm cell in humans.
- Fathers pass X-linked genes on to their daughters but not to their sons; fathers pass the Y chromosome to all their sons.
- Females will express an X-linked trait exactly like any other trait, but males, with only one X chromosome, will express the allele on the X chromosome they inherited from their mother. The terms *homozygous* and *heterozygous* do not apply to a male pattern of sex-linked genes.
- The vast majority of genes on the X chromosome are not related to sex.
- Several X-linked disorders have medical significance:
    - **Duchenne muscular dystrophy** is an X-linked disorder characterized by a progressive weakening of the muscles and loss of coordination. Affected individuals rarely live past their early 20s.
    - **Hemophilia** is an X-linked disorder characterized by having blood with an inability to clot normally, caused by the absence of proteins required for blood clotting.

- **X-inactivation** regulates gene dosage in females. Although female mammals inherit two X chromosomes, one of the X chromosomes (randomly chosen) in each cell of the body becomes inactivated during embryonic development by **methylation**. As a result, males and females have the same effective dose of genes with loci on the X chromosome.
- The inactive chromosome condenses into a **Barr body**, which lies along the inside of the nuclear envelope. Still, females are not usually affected as heterozygote carriers of problematic alleles, because half of their sex chromosomes are normal and produce the necessary proteins.

Linked genes tend to be inherited together because they are located near each other on the same chromosome (Biology, *10e: 15.3*, Focus, *1e: 12.3*)

- **Linked genes** are located on the same chromosome and therefore tend to be inherited together during cell division.
- **Genetic recombination** is the production of offspring with a new combination of genes inherited from the parents. This is due to crossing over during prophase I of meiosis. Many genetic crosses yield some offspring with the same phenotype as one of the parents (these offspring are referred to as **parental types**) and some offspring with phenotypes different from either parent (these offspring are referred to as **recombinants**).
- **Crossing over** can explain why some linked genes get separated during meiosis. During meiosis, unlinked genes follow independent assortment because they are located on different chromosomes. Linked genes are located on the same chromosome and would not be predicted to follow independent assortment. However, sometimes genetic crosses give results that seem to indicate that some independent assortment has occurred, even when genes are on the same chromosome. These results are not caused by independent assortment but can be explained by crossing over. Research further indicates that the farther apart two genes are on a chromosome, the higher the probability that crossing over will occur between them. The likelihood of crossing over between different genes on the same chromosome is expressed as a percent.
- A **linkage map** is a genetic map that is based on the percentage of crossover events.
- A **map unit** is equal to a 1% recombination frequency. Map units are used to express relative distances along the chromosome. Grid-in question 1 is a linkage map problem. Try this problem, then check in the Answers section to see how you did.

Alterations of chromosome number or structure cause some genetic disorders (Biology, *10e: 15.4*, Focus, *1e: 12.4*)

- **Nondisjunction** occurs when the members of a pair of homologous chromosomes do not separate properly during meiosis I, or sister chromatids don't separate properly during meiosis II.
- As a result of nondisjunction, one gamete receives two copies of the chromosome, whereas the other gamete receives none. If the faulty gametes engage in fertilization, the offspring will have an incorrect chromosome number. This is known as **aneuploidy**.

- Fertilized eggs that have received three copies of the chromosome in question are said to be **trisomic**; those that have received just one copy of a chromosome are said to be **monosomic** for the chromosome.
- **Polyploidy** is the condition of having more than two complete sets of chromosomes, forming a 3*n* or 4*n* individual. Rare in animals, this condition is fairly frequent in plants.
- Errors in meiosis or damaging agents like radiation can cause portions of a chromosome to be lost or rearranged, resulting in the following mutations:
    - A **deletion** occurs when a chromosomal fragment is lost, resulting in a chromosome with missing genes.
    - A **duplication** occurs when a chromosomal segment is repeated.
    - An **inversion** occurs when a chromosomal fragment breaks off and reattaches to its original position but backward, so that the part of the fragment that was originally at the attachment point is now at the end of the chromosome.
    - A **translocation** occurs when the deleted chromosome fragment joins a *nonhomologous* chromosome.
- Human disorders caused by chromosomal alterations include the following:
    - **Down syndrome:** An aneuploid condition that is the result of having an extra chromosome 21 (trisomy 21). Down syndrome includes characteristic facial features, short stature, heart defects, and developmental delays.
    - **Klinefelter syndrome:** An aneuploid condition in which a male possesses the sex chromosomes XXY (an extra X). Klinefelter males have male sex organs but are sterile.
    - **Turner syndrome:** A monosomic condition in which the female has just one sex chromosome, often designated XO. Turner syndrome females are sterile because the reproductive organs do not mature. Turner syndrome is the only known viable monosomy in humans.

Some inheritance patterns are exceptions to standard Mendelian inheritance (Biology, *10e: 15.5*)

- In mammals, geneticists have identified traits that differ, depending on which parent passed along the allele for those traits. This phenomenon is called **genomic imprinting**. The phenotypic effect of a gene may depend on which allele is inherited from which parent.
    - Genomic imprinting occurs during gamete formation and results in the silencing of a particular allele of certain genes. The offspring expresses only one allele of an imprinted gene, hence the exception to Mendelian inheritance. Over 60 imprinted genes have been identified, with hundreds more suspected.
- Genes that are present in mitochondria and plastids are inherited only from the mother because the zygote's cytoplasm comes only from the egg. You inherited your mitochondrial DNA only from your mother; your mother inherited her mitochondrial DNA only from her mother. Your mitochondrial DNA is your maternal grandmother's!

Level 2: Application/Analysis/Synthesis Questions **(All questions for this topic are Level 2.)**

1. A couple has six children, all daughters. If the woman has a seventh child, what is the probability that the seventh child will be a daughter?
   (A) $\frac{6}{7}$
   (B) $\frac{1}{7}$
   (C) $\frac{1}{36}$
   (D) $\frac{1}{2}$

2. If genes *R* and *S* are on two different chromosomes and the probability of allele *R* segregating into a gamete is ¼, while the probability of allele *S* segregating into a gamete is ½, what is the probability that both will segregate into the same gamete?
   (A) $\frac{1}{4} \times \frac{1}{2}$
   (B) $\frac{1}{4} \div \frac{1}{2}$
   (C) $\frac{1}{4} + \frac{1}{2}$
   (D) $\frac{1}{4} + \frac{1}{2}$

3. In llamas, coat color is controlled by a gene that exists in two allelic forms. If a homozygous yellow llama is crossed with a homozygous brown llama, the offspring have gray coats. If two of the gray-coated offspring were crossed, what percentage of their offspring would have brown coats?
   (A) 0%
   (B) 25%
   (C) 50%
   (D) 75%

4. Which of the following descriptions does not apply to meiosis?
   (A) During the first division, the sister chromatids travel to opposite ends of the cell.
   (B) The chromosome number of the daughter cells is half that of the parent cell.
   (C) The homologous chromosomes synapse and form tetrads prior to the first division.
   (D) The cytoplasm of the cell and all its organelles are divided approximately in half.

5. In rabbits, the gene for short hair (*S*) is dominant, and the gene for long hair (*s*) is recessive. The gene for green eyes (*G*) is dominant, and the gene for blue eyes (*g*) is recessive. A cross between two rabbits produces a litter of six short-haired rabbits with green eyes, and two short-haired rabbits with blue eyes. What is the most likely genotype of the parent rabbits in this cross?
   (A) *ssgg* × *ssgg*
   (B) *SSGG* × *SSGG*
   (C) *SsGg* × *SsGg*
   (D) *SsGg* × *SSGg*

6. In humans, hemophilia is an X-linked recessive trait. If a man and a woman have a son who has hemophilia, which of the following is definitely *true*?
   (A) The mother carries an allele for hemophilia.
   (B) The father carries an allele for hemophilia.
   (C) The boy's paternal grandfather has hemophilia.
   (D) Both parents carry an allele for hemophilia.

7. Which of the following explains a significantly low rate of crossing over between two genes?
   (A) The genes are located very close together on the same chromosome.
   (B) They are located on separate but homologous chromosomes.
   (C) The genes code for proteins that have similar functions.
   (D) The genes code for proteins that have very different functions.

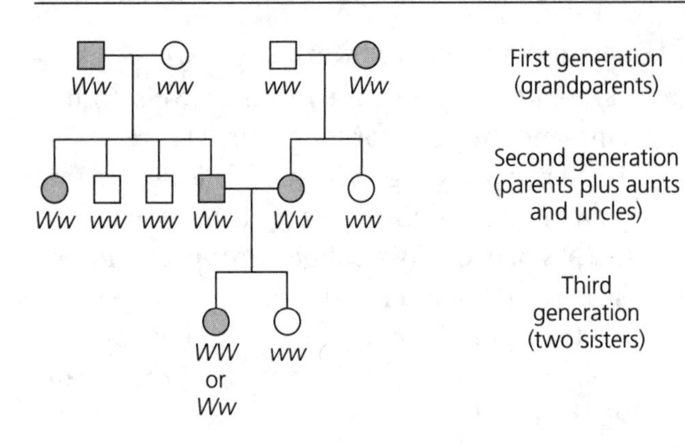

8. In the pedigree above, circles represent females and squares represent males; those who express a particular trait are shaded, whereas those who do not are not shaded. Which pattern of inheritance best explains the pedigree for this trait?
   (A) X-linked recessive
   (B) codominant
   (C) autosomal recessive
   (D) autosomal dominant

9. If a woman has type A blood, and her child has type O blood, the father must have which of the following blood types?
   (A) A, B, or O
   (B) AB or A
   (C) AB or B
   (D) O only

10. If a person with type O blood were to have children with a person with type AB blood, which of the following represents the best prediction of the ratio of possible blood types for their offspring?
   (A) 3 I^Ai:1 I^Bi
   (B) 1 I^Ai:1 I^Bi
   (C) 1 I^Ai:2 I^AI^B:1 I^Bi
   (D) 9 I^AI^B:3 I^AIi:3 I^Bi:1 ii

11. Two yellow mice with the genotype *Yy* are mated. After many offspring, ⅔ are yellow and ⅓ are not yellow (a 2:1 ratio). Mendelian genetics dictates that this cross should produce offspring that were ¼ *YY* (yellow), ½ *Yy* (yellow), and ¼ *yy* (not yellow). What is the most likely conclusion from this experiment?
   (A) The mice did not bear enough offspring for the ratio calculation to be specific.
   (B) *Y* is lethal in the homozygous form and caused death early in development.
   (C) Nondisjunction occurred.
   (D) A mutation masked the effects of the *Y* allele.

12. All of the following contribute to genetic recombination or variation in offspring EXCEPT
   (A) random fertilization.
   (B) independent assortment.
   (C) crossing over.
   (D) gene linkage.

13. In cucumbers, warty (*W*) is dominant over dull (*w*), and green (*G*) is dominant over orange (*g*). A cucumber plant that is homozygous for warty and green is crossed with one that is homozygous for dull and orange. The F_1 generation is then crossed. If a total of 144 offspring is produced in the F_2 generation, which of the following is closest to the number of dull green cucumbers expected?
   (A) 3
   (B) 10
   (C) 28
   (D) 80

14. Chromosome number is reduced to haploid in meiosis. What event must occur to restore the diploid chromosome number?
   (A) synapsis during prophase
   (B) fertilization
   (C) mitosis
   (D) DNA replication

15. Which of the following events is typical of the second meiotic division?
    (A) Paternal and maternal chromosomes assort randomly.
    (B) The chromosome number becomes haploid.
    (C) Crossing over between nonsister chromatids occurs.
    (D) Sister chromatids migrate to separate poles.

16. A cell with a diploid number of 6 could produce gametes with how many different combinations of maternal and paternal chromosomes?
    (A) 6
    (B) 8
    (C) 12
    (D) 64

17. The DNA content of a diploid cell is measured in the G_1 phase. *After* meiosis I, the DNA content of one of the two cells produced would be
    (A) equal to that of the G_1 cell.
    (B) twice that of the G_1 cell.
    (C) one-half that of the G_1 cell.
    (D) one-fourth that of the G_1 cell.

18. Which of the following is true for both meiosis II and mitosis?
    (A) Sister chromatids separate.
    (B) Homologous chromosomes separate.
    (C) DNA replication precedes the division.
    (D) Haploid cells are produced.

19. Given the following recombination frequencies, what is the correct order of the genes on the chromosome? A-B, 8 map units; A-C, 28 map units; A-D, 25 map units; B-C, 20 map units; B-D 33 map units
    (A) A-B-C-D
    (B) D-C-A-B
    (C) A-D-C-B
    (D) D-A-B-C

20. X-linked conditions are more common in men than in women because
    (A) the genes associated with the X-linked conditions are linked to the X chromosome, which determines maleness.
    (B) men need to inherit only one copy of the recessive allele for the condition to be fully expressed.
    (C) women simply do not develop the disease regardless of their genetic composition.
    (D) the sex chromosomes are more active in men than in women.

21. If these four cells resulted from cell division of a single cell with diploid chromosome number $2n = 4$, what best describes what just occurred?

Gametes
$n + 1$ $n - 1$ n n
Number of chromosomes

    (A) normal meiosis
    (B) translocation
    (C) inversion
    (D) nondisjunction

After reading the paragraph, answer questions 22 and 23 that follow.

A woman has been trying to conceive for several years unsuccessfully. At a fertility clinic, they discover that she has blocked fallopian tubes. Using modern technologies, some of her eggs are removed, fertilized with her husband's sperm, and implanted into her uterus. The procedure is successful, but the couple discovers that their new son is colorblind and has type O blood. The woman claims that the child can't be theirs because she has type A blood and her husband has type B. Also, neither parent is colorblind, although one grandparent (the woman's father) is also colorblind.

22. As a genetic counselor, you would explain to the parents that
    (A) the eggs must have been accidentally switched because the baby's blood type has to match one of his parents'.
    (B) each parent could have contributed one recessive allele, resulting in type O blood.
    (C) the eggs must have been accidentally switched because a type A parent and a type B parent can have any type children except O.
    (D) it is possible for the baby to have type O blood because type O is inherited through a dominant allele.

23. In regard to the baby's colorblindness, a sex-linked recessive trait, you explain that
    (A) colorblindness often appears randomly, even if neither parent is colorblind.
    (B) the baby's father must have a recessive allele for colorblindness.
    (C) because colorblindness is sex-linked, a son can inherit colorblindness if his mother has the recessive colorblindness allele.
    (D) the eggs must have been accidentally switched because males inherit sex-linked traits only from their fathers.

24. Independent orientation of chromosomes at metaphase I results in an increase in the number of
    (A) gametes.
    (B) homologous chromosomes.
    (C) possible combinations of characteristics.
    (D) sex chromosomes.

Additional Sample Problems

1. A black guinea pig crossed with an albino guinea pig produces 12 black offspring. When the albino is crossed with a second black one, 7 blacks and 5 albinos are obtained. What is the best explanation for this genetic situation? Write genotypes for the parents, gametes, and offspring.

2. In pea plants, pod color may be green (G) or yellow (g), while the pod shape may be inflated (I) or constricted (i). Two pea plants heterozygous for the characters of pod color and pod shape are crossed. Determine the phenotypic ratios of the offspring.

3. In some plants, a true-breeding, red-flowered strain gives all pink flowers when crossed with a white-flowered strain: C^RC^R (red) $\times$ C^WC^W (white) $\rightarrow$ C^RC^W (pink). The placement of the flower can be determined by the dominant allele for an axial flower (A), whereas a terminal flower is determined by the recessive allele (a). What will be the phenotypic ratio of the F_1 generation resulting from the following cross: axial-red (both genes are homozygous) $\times$ terminal white? What will the phenotypic ratio be in the F_2?

Grid-In Questions

These call for a numerical response

1. A yeast cell in the early portion of interphase of meiotic cell division has 24 fg of DNA (fg $= 1 \times 10^{-15}$ grams). If the yeast cell completes meiotic division to form four haploid cells, how many fg of DNA would be expected in each haploid cell?
2. In pea plants $T =$ Tall, $t =$ dwarf, $R =$ Round seeds, and $r =$ wrinkled seeds. If a *TtRr* plant is crossed with a *Ttrr* plant, what fraction of the offspring will be tall and wrinkled?
3. In fruit flies gray body is dominant to black body and normal wings are dominant to vestigial wings. Flies heterozygous for both gray bodies and normal wings were crossed with flies that had black bodies and vestigial wings. The following results were obtained:

Phenotype	Number of flies
Gray body/normal wings	482
Black body/vestigial wings	472
Black body/vestigial wings	103
Black body/ normal wings	92

The results indicate that the genes for wings and body color are on the same chromosome. The recombinant offspring are a result of crossing over. How many map units (expressed as a percent) apart are the two genes? The formula for calculating recombination frequency is:

Recombination Frequency = number of recombinants/total number of offspring × 100

TOPIC 5

Molecular Genetics

The Molecular Basis of Inheritance

(Biology, *10e: Chapter 16*, Focus, *1e: Chapter 13*)

YOU MUST KNOW

- The structure of DNA.
- The knowledge about DNA gained from the work of Griffith; Avery, MacLeod, and McCarty; Hershey and Chase; Wilkins and Franklin; and Watson and Crick. (LO 3.2)
- Replication is semiconservative and occurs 5′ to 3′.
- The roles of DNA polymerase, ligase, helicase, and topoisomerase in replication. (LO 3.3)
- The general differences between bacterial chromosomes and eukaryotic chromosomes.
- How DNA packaging can affect gene expression.

DNA is the genetic material (Biology, *10e: 16.1*, Focus, *1e: 13.1*)

- Once chromosomes were known to carry genes, the next question became which of the two organic compounds that make chromosomes, DNA or protein, was the genetic material?
  - Fredrick Griffith studied two strains of the bacterium *Streptococcus pneumoniae*. In mice, bacteria of the smooth strain caused pneumonia, whereas the rough strain was nonpathogenic. However, when Griffith mixed heat-killed smooth strain bacteria (nonpathogenic) and rough strains (nonpathogenic) and injected mice with the mixture, the mice developed pneumonia and died! Griffith concluded that the living rough strain had been transformed into the pathogenic smooth form by a heritable agent. The question remained: What was the heritable agent, protein or DNA?
  - O. T. Avery, along with his colleagues McCarty and MacLeod, dedicated 14 years to identifying the "transforming agent" from Griffith's work. In 1944 they concluded that DNA was the transforming factor. Their results were greeted with interest but skepticism. Scientists had difficulty believing that a macromolecule as simple as DNA could be the genetic material and not the much more complex protein molecule.

- In 1952 **Alfred Hershey and Martha Chase** provided an unambiguous answer to the DNA or protein question utilizing *bacteriophages*—viruses that infect bacteria. Bacteriophages were excellent organisms for this study, in part because they are made of only two organic compounds, DNA and protein. Hershey and Chase used a radioactive isotope of phosphorus to tag the DNA in one culture of bacteriophages and radioactive sulfur to tag the protein in a second culture. Their results clearly showed that only the DNA entered bacteria infected by the virus; the radioactive protein never entered the cell. This research convinced scientists that DNA must be the genetic material.
- The next big question centered on the structure of DNA. Would the structure of DNA give any clues as to how it functioned as the genetic material?
    - **James Watson and Francis Crick** were the first to solve the puzzle of the structure of DNA. Critical to their success was the work of Rosalind Franklin and Maurice Wilkins, both working in the field of X-ray crystallography.
    - **X-ray crystallography** is a process used to visualize molecules three-dimensionally. X-rays are diffracted as they pass through the molecule, and they bounce back to produce patterns that can be interpreted through mathematical equations. Through this technique, a rough blueprint of the molecule was formed and the helical structure was deduced.
- Watson and Crick's model determined four major features of DNA. Find each major point by following Figure 5.1 as the model is explained.

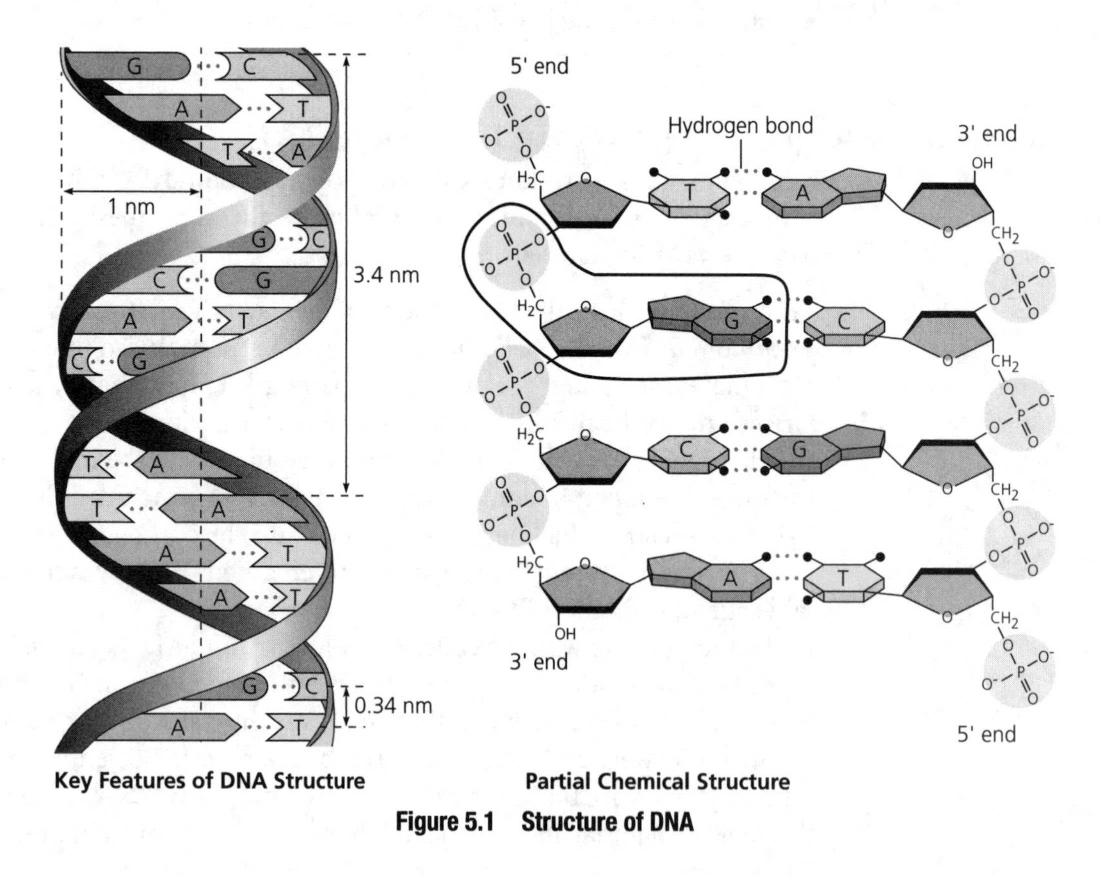

Figure 5.1 Structure of DNA

- DNA is a **double helix**, which can be described as a twisted ladder with rigid rungs. The side, or backbone, is made up of sugar-phosphate components, whereas the rungs are made up of pairs of nitrogenous bases.
- Notice that a single nucleotide is circled in Figure 5.1. It is composed of a sugar (deoxyribose) attached to a phosphate and a nitrogen base.
- The nitrogenous bases of DNA are adenine (A), thymine (T), guanine (G), and cytosine (C). In DNA, adenine pairs only with thymine, and guanine pairs only with cytosine.
- Notice that the chain on the right side of the model runs in one direction, while the left side of the chain runs in the opposite, upside-down direction. The strands are termed **antiparallel**. The left side runs 5′ to 3′ and the opposite strand runs 3′ to 5′. (Recall that the carbons are numbered, and you will see that the number 5 carbon and number 3 carbon and the resultant nucleotides are flipped relative to each other.) Nucleic acid strands are always antiparallel, whether they are DNA/DNA or DNA/RNA or RNA/RNA interactions.

Many proteins work together in DNA replication and repair (Biology, *10e: 16.2,* Focus, *1e: 13.2*)

> ***CONNECT WITH THE CURRICULUM FRAMEWORK***
>
> **Big Idea 3**
>
> Be able to use a model to illustrate how genetic information is copied for transmission between generations. Know the roles of the enzymes involved in DNA replication.

- **Replication** is the making of DNA from an existing DNA strand. DNA replication is *semiconservative*. This means that at the end of replication, each of the daughter molecules has one old strand, derived from the parent strand of DNA, and one strand that is newly synthesized. Study Figure 5.2 to see the pattern of semiconservative replication.

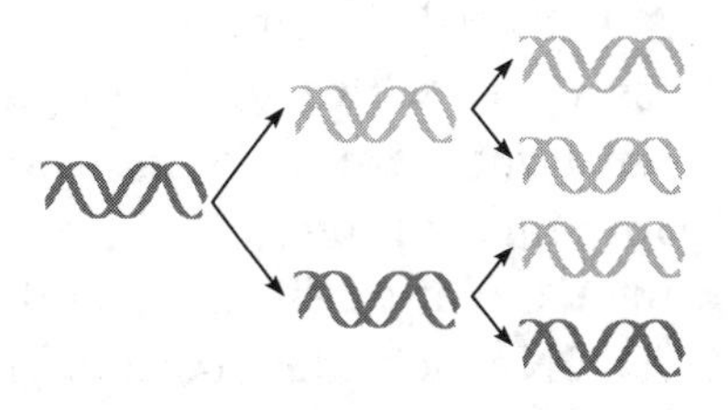

Figure 5.2 Semiconservative replication

- The replication of DNA includes six major points:
  1. The replication of DNA begins at sites called the *origins of replication*.
  2. Initiation proteins bind to the origin of replication and separate the two strands, forming a *replication bubble*. DNA replication then proceeds in both directions along the DNA strand until the molecule is copied.
  3. A group of enzymes called **DNA polymerases** catalyzes the elongation of new DNA at the replication fork.
  4. DNA polymerase adds nucleotides to the growing chain one by one, working in a 5′ to 3′ direction, matching adenine with thymine and guanine with cytosine.
  5. Recall that the strands of DNA are antiparallel. This means that DNA replication occurs continuously along the 5′ to 3′ strand, which is called the **leading strand**. The strand that runs 3′ to 5′ is copied in series of segments and termed the **lagging strand**. Read steps 1–3 in Figure 5.3 to visualize this process.
  6. The lagging strand is synthesized in separate pieces called **Okazaki fragments**, which are then sealed together by **DNA ligase** (steps 4–7, Figure 5.3), forming a continuous DNA strand.
- There are several factors contributing to the accuracy of DNA replication:
  - The specificity of base pairing (A = T, G = C)
  - **Mismatch repair**, in which special repair enzymes fix incorrectly paired nucleotides
  - **Nucleotide excision repair**, in which incorrectly placed nucleotides are excised or removed by enzymes termed **nucleases**, and the gap left over is filled in with the correct nucleotides
- The fact that DNA polymerase can add nucleotides only to the 3′ end of a molecule means that it would have no way to complete the 5′ end of the DNA molecule at the end of the chromosome. Every time the chromosome is replicated for mitosis, a small portion of the tip of the chromosome is removed. To avoid losing the terminal genes, the linear ends of eukaryotic chromosomes are "capped" with **telomeres**, short, repetitive nucleotide sequences that do not contain genes. This means that any given cell has a finite number of times it can divide before essential information is lost. In tumor cells, such as those in HeLa cells, a mutation activates an enzyme called *telomerase,* which prevents this degradation and renders the cells "immortal."

A chromosome consists of a DNA molecule packed together with proteins (Biology, *10e: 16.3,* Focus, *1e: 13.3*)

- A bacterial chromosome is one double-stranded, circular DNA molecule associated with a small amount of protein.
- Eukaryotic chromosomes are linear DNA molecules associated with large amounts of protein.
- In eukaryotic cells, DNA and proteins are packed together as **chromatin**.
- As DNA becomes more highly packaged, it becomes less accessible to transcription enzymes, which reduces gene expression. In interphase cells, most

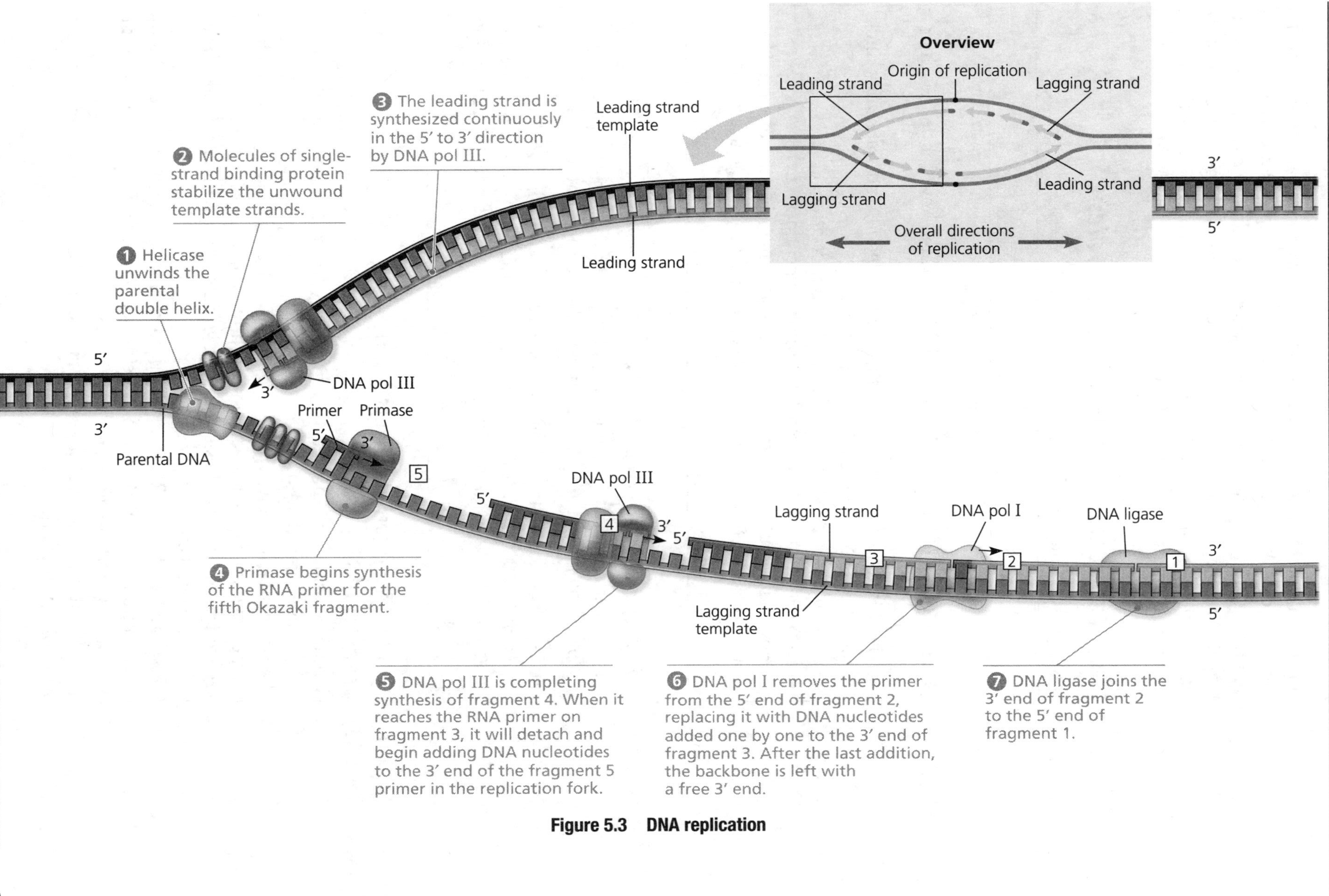

Figure 5.3 DNA replication

chromatin is in the highly extended form **(euchromatin)** and is available for transcription. When the euchromatin condenses to chromosomes during mitotic division, the more condensed chromatin **(heterochromatin)** is no longer available for transcription. Heterochromatin is largely inaccessible to transcription enzymes and, thus, generally is not transcribed. Barr bodies are another example of heterochromatin.

Gene Expression: From Gene to Protein

(Biology, *10e: Chapter 17*, Focus, *1e: Chapter 14*)

YOU MUST KNOW

- How RNA and DNA are similar and different, and how this defines their roles.
- The differences between *replication, transcription*, and *translation* and the role of DNA and RNA in each process.
- How eukaryotic cells modify RNA after transcription.
- How genetic material is translated into polypeptides. (LO 3.4)
- How mutations can change the amino acid sequence of a protein and be able to predict how a mutation can result in changes in gene expression. (LO 3.6)

TIP FROM THE READERS

This is the central chapter for molecular genetics. It is one of the top five chapters you must know to perform well on the AP exam!

Be sure you know the difference between *replication* (DNA to DNA), *transcription* (DNA to RNA), and *translation* (RNA to protein). In essay questions that use these terms, often 20% of the students confuse the processes!

***Genes specify proteins via transcription and translation* (Biology, *10e: 17.1*, Focus, *1e: 14.1*)**

- **Gene expression** is the process by which DNA directs the synthesis of proteins (or, in some cases, RNAs).
- The **one gene–one polypeptide hypothesis** states that each gene codes for a polypeptide, which can be—or can constitute a part of—a protein.
- **Transcription** is the synthesis of RNA using DNA as a template. It takes place in the nucleus of eukaryotic cells.
- **Messenger RNA**, or **mRNA**, is produced during transcription. It carries the genetic message of DNA to the protein-making machinery of the cell in the cytoplasm, the *ribosome.*
- In eukaryotes, transcription results in pre-mRNA, which undergoes **RNA processing** to yield the final mRNA.

- In prokaryotes, transcription results directly in mRNA, which is not processed. Transcription and translation can occur simultaneously.
- **Translation** is the production of a polypeptide chain using the mRNA transcript and occurs at the ribosomes.
- The instructions for building a polypeptide chain are written as a series of three-nucleotide groups; this is called a *triplet code.*
- During transcription, only one strand of the DNA is transcribed, and it is called the **template strand**. The mRNA that is produced is said to be *complementary* to the original DNA strand. The mRNA base triplets are called **codons.** *They are written in the 5′ to 3′ direction.*
- The genetic code is *redundant*, meaning that more than one codon codes for each of the 20 amino acids. The codons are read based on a consistent reading frame—the groups of three must be read in the correct groupings in order for translation to be successful.

Transcription is the DNA-directed synthesis of RNA: a closer look (Biology, *10e: 17.2,* Focus, *1e: 14.2*)

- **RNA polymerase** is an enzyme that separates the two DNA strands and connects the RNA nucleotides as they base-pair along the DNA template strand.
- The RNA polymerases can add RNA nucleotides only to the 3′ end of the strand, so RNA elongates in the 5′ to 3′ direction. As RNA nucleotides are added, remember that uracil replaces thymine when base pairing to adenine.
- The DNA sequence at which RNA polymerase attaches is called the **promoter,** whereas the DNA sequence that signals the end of transcription is called the **terminator**.
- A **transcription unit** is the entire stretch of DNA that is transcribed into an RNA molecule. A transcription unit may code for a polypeptide or an RNA, like transfer RNA or ribosomal RNA.
- There are three main stages of transcription:

1. **Initiation:** In bacteria, RNA polymerase recognizes and binds to the promoter. In eukaryotes, RNA polymerase II, the specific RNA polymerase that transcribes mRNA, cannot bind to the promoter without supporting help from proteins known as transcription factors. **Transcription factors** assist the binding of RNA polymerase to the promoter and, thus, the initiation of transcription. The whole complex of RNA polymerase II and transcription factors is called a **transcription initiation complex** (Figure 5.4).

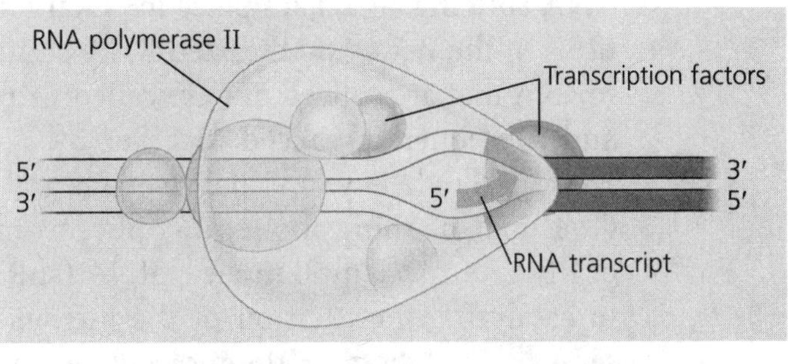

Figure 5.4 Transcription initiation complex

2. **Elongation:** RNA polymerase moves along the DNA, continuing to untwist the double helix. RNA nucleotides are continually added to the 3′ end of the growing chain. As the complex moves down the DNA strand, the double helix re-forms, with the new RNA molecule straggling away from the DNA template. Find these key steps in Figure 5.5.

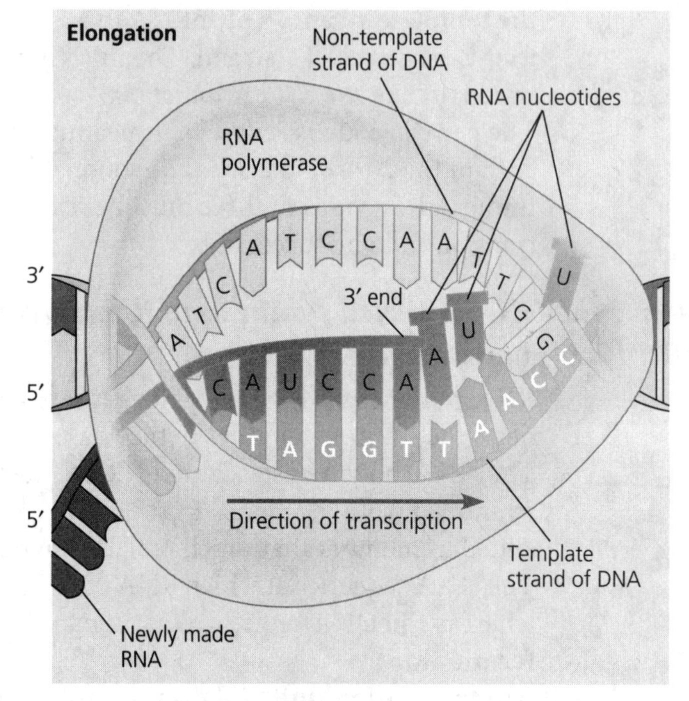

Figure 5.5 Elongation stage of transcription

3. **Termination:** After RNA polymerase transcribes a terminator sequence in the DNA, the RNA transcript is released, and the polymerase detaches.

Eukaryotic cells modify RNA after transcription (Biology, *10e: 17.3,* Focus, *1e: 14.3*)

- In eukaryotes, there are a couple of key post-transcriptional modifications to RNA—the addition of a **5′ cap** and the addition of a **poly-A tail**.
- The 5′ cap and the poly-A tail facilitate the export of mRNA from the nucleus, help protect the mRNA from degradation by enzymes, and facilitate the attachment of the mRNA to the ribosome.
- **RNA splicing** also takes place in eukaryotic cells. In RNA splicing, large portions of the newly synthesized RNA strand are removed. The sections of the mRNA that are spliced out are called **introns**, and the sections that remain—and subsequently spliced together by a *spliceosome*—are called **exons**. Use Figure 5.6 to help you visualize exons and introns.
- One amazing thing about how spliceosomes work is the role of a special kind of RNA, termed **small nuclear RNA** (snRNA). The snRNA plays a major role in catalyzing the excision of the introns and joining of exons. When RNA serves a catalytic role, the molecule is termed a **ribozyme**. For many years it

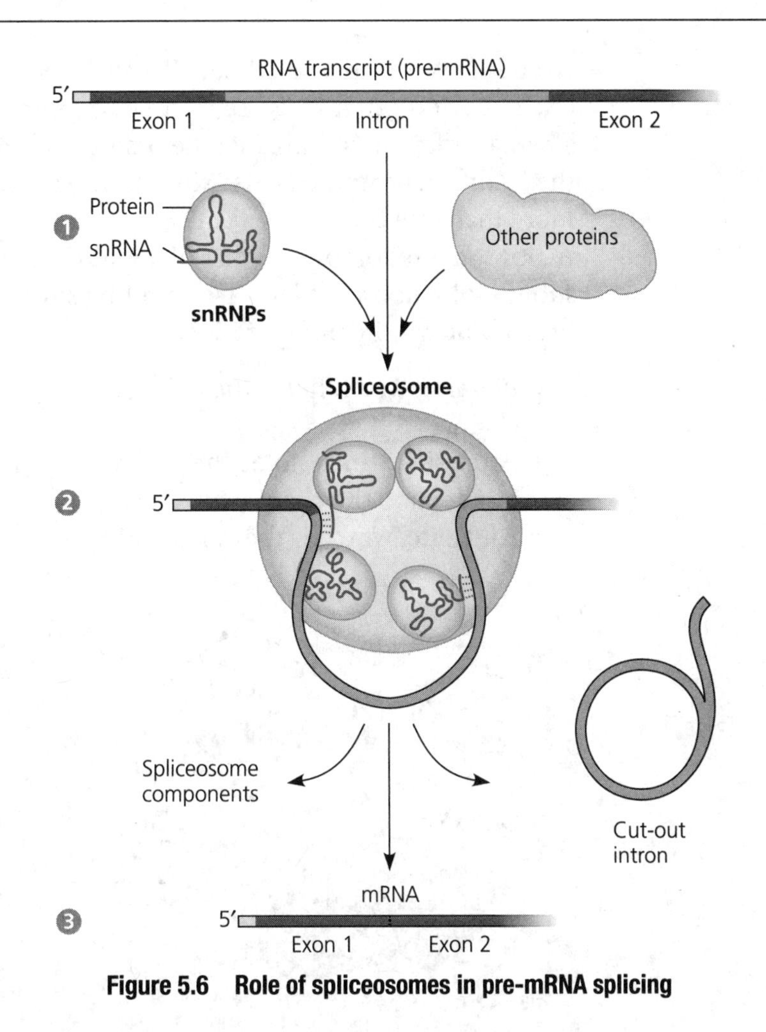

Figure 5.6 Role of spliceosomes in pre-mRNA splicing

was thought that only proteins could be catalytic, but the discovery of ribozymes totally changed that idea!

- Another rethinking that has taken place came with the realization that we have only about 20,000 genes to make approximately 100,000 polypeptides. One gene can often make more than one polypeptide. An intron removed in the production of one polypeptide can be an exon in a second polypeptide made from the same gene! Alternative RNA splicing allows for different combinations of exons, resulting in more than one polypeptide per gene.

Translation is the RNA-directed synthesis of a polypeptide: a closer look (Biology, *10e: 17.4,* Focus, *1e: 14.4*)

- In addition to mRNA, two additional types of RNA play important roles in translation: transfer RNA (tRNA) and ribosomal RNA (rRNA).
- The **tRNA** functions in transferring amino acids from a pool of amino acids in the cell's cytoplasm to a ribosome. The ribosome accepts the amino acid from the tRNA and incorporates the amino acid into a growing polypeptide chain.
- Each type of tRNA is specific for a particular amino acid; at one end it loosely binds the amino acid, and at the other end it has a nucleotide triplet called an **anticodon**, which allows it to pair specifically with a complementary codon on the mRNA.

- A **codon** is an mRNA triplet. Because there are four different nucleotides (A, T, C, and G), taking them three at a time results in 64 different codons. The 64 different codons include 3 stop codons and 61 codons that code for the 20 different amino acids. Most amino acids can therefore be designated by more than one codon.
- A ribosome is composed of **rRNA** and protein and has 2 subunits. The large subunit of a ribosome has three binding sites for tRNA molecules (locate each tRNA binding site in Figure 5.7).
  - A **P site**, which holds the tRNA that carries the growing polypeptide chain.
  - An **A site**, which holds the tRNA that carries the amino acid that will be added to the chain next.
  - An **E site**, which is the exit site for each tRNA.

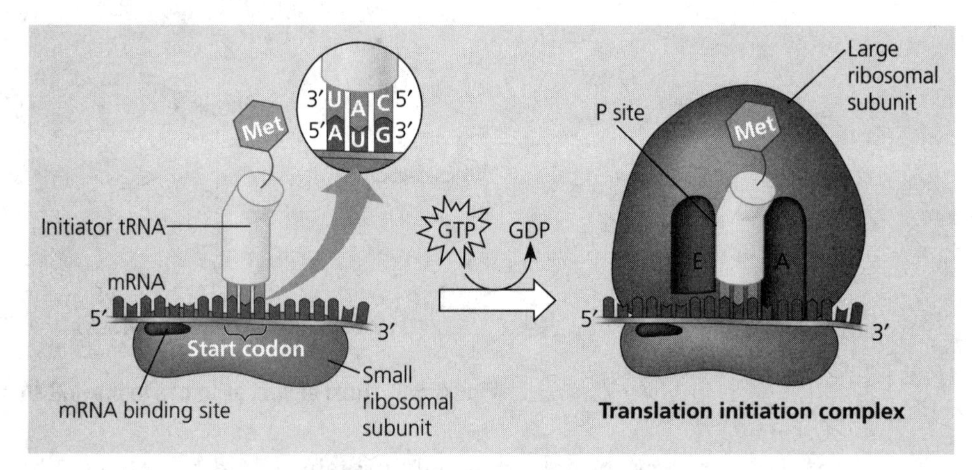

Figure 5.7 Initiation stage of translation

- Translation, like transcription, can be divided into three stages:
  1. **Initiation:** Organize initiation into these three steps. Use Figure 5.7 to find each step.
     A. A small ribosomal subunit binds to mRNA in such a way that the first codon of the mRNA strand, which is always AUG, is placed in the proper position.
     B. The tRNA with anticodon UAC, which carries the amino acid methionine, hydrogen bonds to the first codon (initiation factors are proteins that assist in holding all this together).
     C. Large subunit of ribosome attaches, allowing the tRNA with methionine to attach to the P site. Notice that the A site is now available to the tRNA that will bring the second amino acid.
  2. **Elongation:** Read about the three steps of elongation—codon recognition, peptide bond formation, and translocation—using Figure 5.8. This is an important idea; follow the explanations and the diagrams carefully.

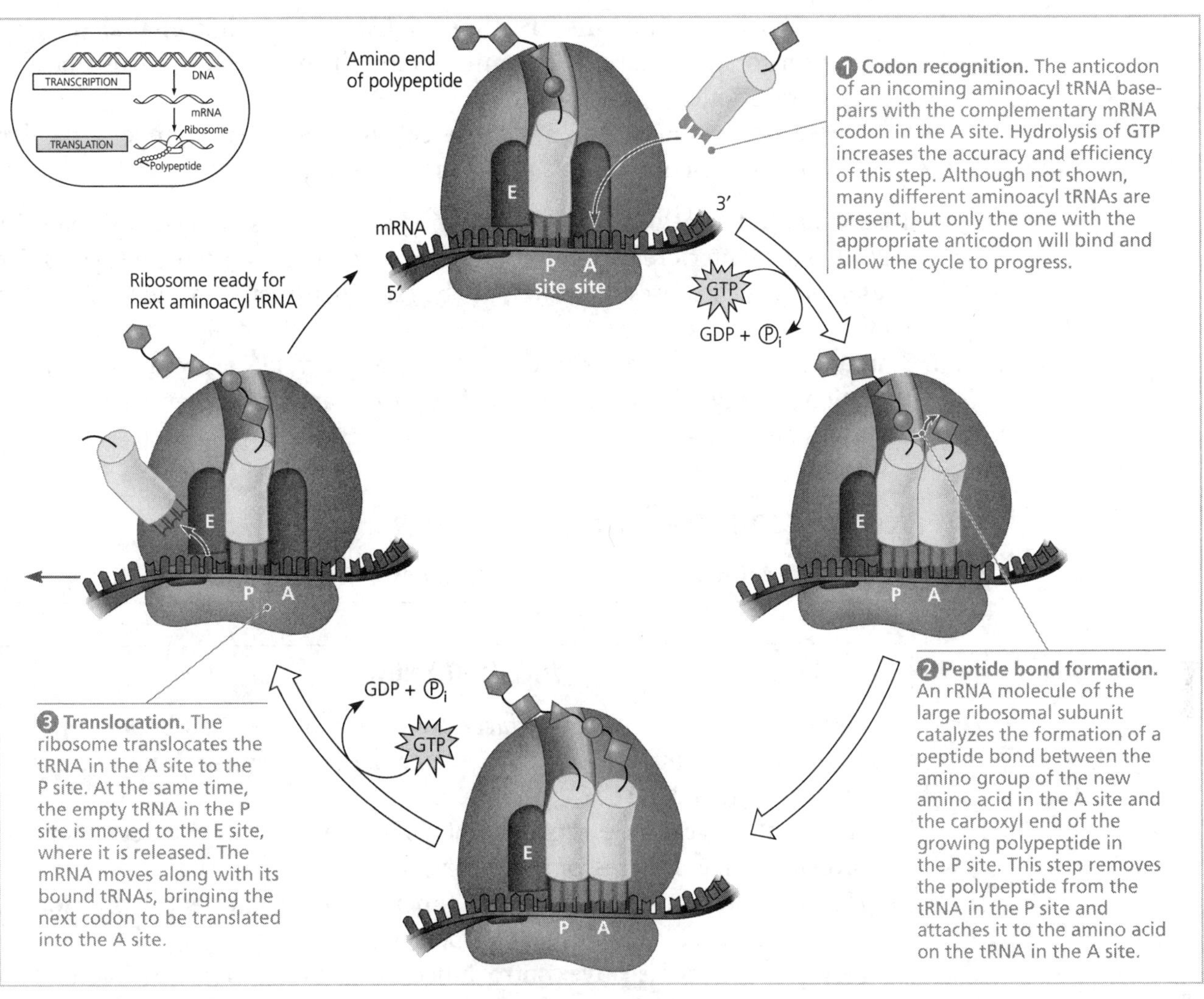

Figure 5.8 Protein synthesis

3. **Termination:** A stop codon in the mRNA is reached and translation stops. A protein called release factor binds to the stop codon, and the polypeptide is freed from the ribosome.

- Polypeptides then fold to assume their specific conformation, and they are almost always modified further to render them functional. The destination of a protein is often determined by the sequence of about 20 amino acids at the leading end of the polypeptide chain. The **signal peptide**, the sequence of the leading 20 or so amino acids, serves as a sort of cellular zip code, directing proteins to their final destination.

Mutations of one or a few nucleotides can affect protein structure and function (Biology, *10e: 17.5*, Focus, *1e: 14.5*)

- Mutations are alterations in the genetic material of the cell; **point mutations** are alterations of just one nucleotide base pair of a gene. They come in two basic types:
  - A **nucleotide-pair substitution** is the replacement of one nucleotide and its partner with another pair of nucleotides.

- **Missense mutations** are those substitutions that enable the codon to still code for an amino acid, although it might not be the correct one.
- **Nonsense mutations** are those substitutions that change a regular amino acid codon into a stop codon, ceasing translation.

- **Insertions** and **deletions** refer to the additions and losses of nucleotide pairs in a gene. If they interfere with the codon groupings, they can cause a **frameshift mutation**, which causes the mRNA to be read incorrectly on each remaining codon.
- **Mutagens** are substances or forces that interact with DNA in ways that cause mutations. X-rays and other forms of radiation are known mutagens, as are certain chemicals.

Regulation of Gene Expression

(Biology, *10e: Chapter 18*, Focus, *1e: Chapter 15*)

YOU MUST KNOW

- Genes can be activated by *inducer* molecules, or they can be inhibited by the presence of a *repressor* as they interact with regulatory proteins or sequences. (EK 3.B.1)
- A *regulatory gene* is a sequence of DNA that codes for a regulatory protein such as a repressor protein.
- How the components of an operon function to regulate gene expression in both repressible and inducible operons.
- How positive and negative control function in gene expression. (EK 3.B.1)
- The impact of DNA methylation and histone acetylation on gene expression.
- How timing and coordination of specific events are regulated in normal development, including pattern formation and induction. (EK 2.E.1)
- The role of microRNAs in control of cellular functions.
- The role of gene regulation in embryonic development and cancer.

***Bacteria often respond to environmental change by regulating transcription* (Biology, *10e: 18.1*, Focus, *1e: 15.1*)**

- In bacteria, genes are often clustered into units called *operons,* which allow the expression of several related genes to be controlled as a unit. Figure 5.9 shows a repressible operon with an inactive repressor. Locate each part of the operon and the regulatory gene as you read the accompanying text.
- An **operon** consists of three parts:
  - An **operator** that controls the access of RNA polymerase to the genes. The operator is found within the promoter site or between the promoter and the protein coding genes of the operon.

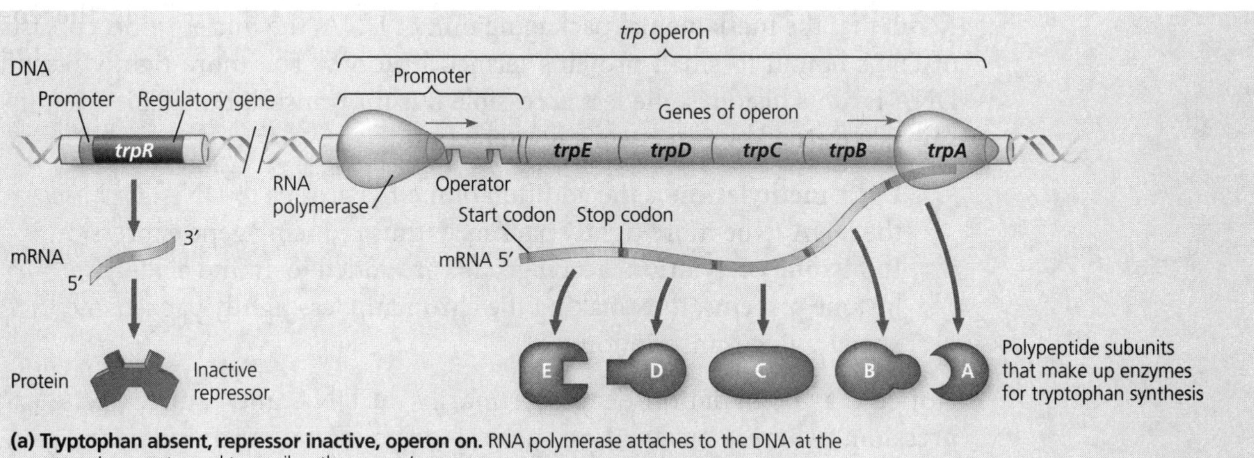

Figure 5.9 A repressible operon

- The **promoter**, which is where RNA polymerase attaches.
- The **genes of the operon**. This is the entire stretch of DNA required for all the enzymes produced by the operon.

- Located some distance from the operon is a regulatory gene. **Regulatory genes** produce repressor proteins that may bind to the operator site. When a regulatory protein occupies the operator site, RNA polymerase is blocked from the genes of the operon. In this situation the operon is off.
- A **repressible operon** is normally *on* but can be inhibited. This type of operon is normally anabolic, building an essential organic molecule. The repressor protein produced by the regulatory gene is inactive. If the organic molecule being produced by the operon is provided to the cell, the molecule can act as a **corepressor** and bind to the repressor protein, activating it. The activated repressor protein binds to the operator site, shutting down the operon. This type of repressible operon is shown in Figure 5.9. Figure 18.3b in *Campbell Biology,* 10e, or Figure 15.3b in *Focus,* 1e, show the corepressor turning off the operon.
- An **inducible operon** is normally *off* but can be activated. This type of operon is normally catabolic, breaking down food molecules for energy. The repressor protein produced by the regulatory gene is active. To turn an inducible operon on, a specific small molecule, called an **inducer**, binds to and inactivates the repressor protein. With the repressor out of the operator site, RNA polymerase can access the genes of the operon. The *lac* operon is an example of an inducible operon and can be seen in Figure 18.4 in *Campbell Biology,* 10e, or Figure 15.4 in *Focus,* 1e.

Eukaryotic gene expression is regulated at many stages (Biology, *10e: 18.2,* Focus, *1e: 15.2*)

- The expression of eukaryotic genes can be turned off and on at any point along the pathway from gene to functional protein. Further, the differences between cell types are not caused by different genes being present but by **differential gene expression**, the expression of different genes by cells with the same genome.

- Recall that the fundamental packaging unit of DNA, the nucleosome, consists of DNA bound to small proteins termed histones. The more tightly bound DNA is to its histones, the less accessible it is for transcription. This relationship is governed by two chemical interactions:
  - **DNA methylation** is the addition of methyl groups to DNA. It causes the DNA to be more tightly packaged, thus reducing gene expression.
  - In **histone acetylation**, acetyl groups are added to amino acids of histone proteins, thus making the chromatin less tightly packed and encouraging transcription.
- Notice that methylation occurs primarily on DNA and reduces gene expression, whereas acetylation occurs on histones and increases gene expression.
- **Epigenetic inheritance** is the inheritance of traits transmitted by mechanisms not directly involving the nucleotide sequence. The DNA sequence is not changed, just its expression.
- *Transcription initiation* is another important control point in gene expression. At this stage, DNA control elements that bind transcription factors are involved in regulation.
- The **transcription initiation complex** greatly enhances gene expression. Study Figure 5.10 (see next page) and note its essential elements. DNA sequences far from the gene, termed **enhancer regions**, are bound to the promoter region by proteins termed **activators**.
- The control of gene expression may also occur after transcription and just after translation, when proteins are processed.
- Coordinately controlled genes, such as the genes coding for the enzymes of a metabolic pathway, are expressed together. This is possible even though the genes in a given pathway may be scattered on different chromosomes. All of the genes that code for the enzymes of the pathway share the same control elements. In general, eukaryotes do not have operons.

> ***CONNECT WITH THE CURRICULUM FRAMEWORK***
>
> **Big Idea 3**
>
> The control of gene expression is an extremely important topic, so spend enough time to understand different ways genes are controlled in both prokaryotes (with operons) and eukaryotes. Any factor that changes gene expression will affect cells, the organism, and populations. (LOs 3.18–3.23)

Noncoding RNAs play multiple roles in controlling gene expression (Biology, *10e: 18.3*, Focus, *1e: 15.3*)

- Recent research has indicated that large, diverse populations of RNA molecules in the cell play crucial roles in regulating gene expression. Small molecules of single-stranded RNA can complex with proteins and influence gene expression.

- Two types of RNA, *micro RNAs* (*miRNA*) and *small interfering RNAs* (*siRNAs*), can bind to mRNA and degrade the mRNA or bind to mRNA and block its translation. The blocking of gene expression in this manner is called RNA interference (RNAi).

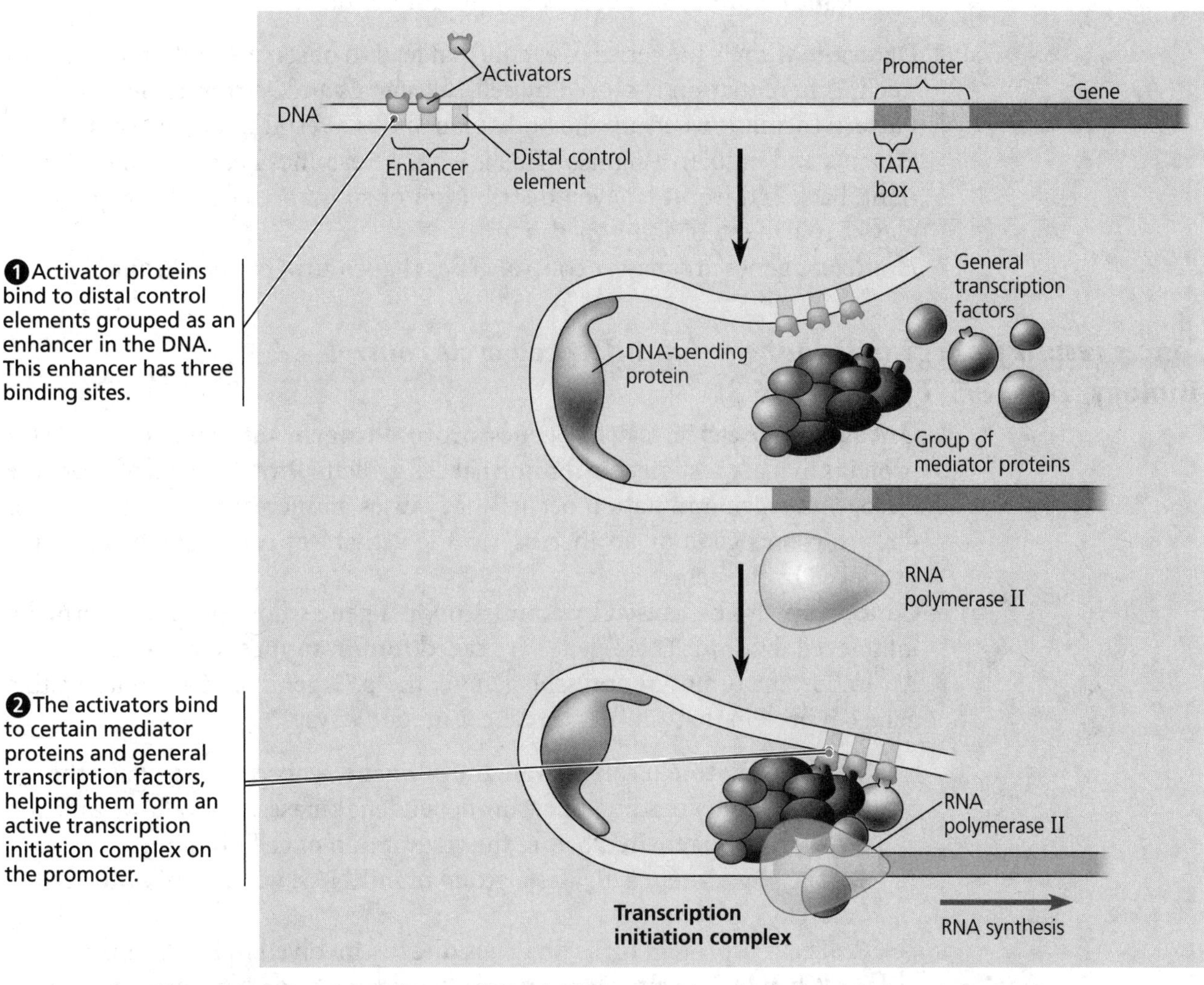

Figure 5.10 Formation of transcription initiation complex plays a key role in gene expression

A program of differential gene expression leads to the different cell types in a multicellular organism (Biology, *10e: 18.2,* Focus, *1e:16.1*)

- The zygote undergoes transformation through three interrelated processes:

  1. **Cell division** is the series of mitotic divisions that increases the number of cells.
  2. **Cell differentiation** is the process by which cells become specialized in structure and function.
  3. **Morphogenesis** gives an organism its shape.

- What controls differentiation and morphogenesis?

    1. **Cytoplasmic determinants** are maternal substances in the egg that influence the course of early development. These are distributed unevenly in the early cells of the embryo and result in different effects.
    2. **Cell-cell signals** result from molecules, such as growth factors, produced by one cell influencing neighboring cells, a process called **induction**, which causes cells to differentiate.

- **Determination** is the series of events that lead to observable differentiation of a cell. Differentiation is caused by cell-cell signals and is irreversible.
- **Pattern formation** sets up the body plan and is a result of cytoplasmic determinants and inductive signals. This is what determines head and tail, left and right, back and front. Uneven distribution of substances called **morphogens** plays a role in establishing these axes.
- **Homeotic genes** are master control genes that control pattern formation.

Cancer results from genetic changes that affect cell cycle control **(Biology, *10e: 18.5*, Focus, *1e: 16.3*)**

- **Oncogenes** are cancer-causing genes; **proto-oncogenes** are genes that code for proteins that are responsible for normal cell growth. Proto-oncogenes become oncogenes when a mutation occurs that causes an increase in the product of the proto-oncogene or an increase in the activity of each protein molecule produced by the gene.
- **Cancer** can also be caused by a mutation in a gene whose products normally inhibit cell division. These genes are called **tumor-suppressor genes.**
- An important tumor-suppressor gene is the ***p53* gene**. The product of this gene is a protein that suppresses cancer in four ways:

    1. The *p53* protein can activate the *p21* gene, whose product halts the cell cycle by binding to cyclin-dependent kinases. This allows time for DNA to be repaired before the resumption of cell division.
    2. The *p53* protein activates a group of miRNAs, which inhibit the cell cycle.
    3. The *p53* protein turns on genes directly involved in DNA repair.
    4. When DNA damage is too great to repair, the *p53* protein activates "suicide" genes whose products cause cell death, a process termed **apoptosis**.

- The multistep model of cancer development is based on the idea that cancer results from the accumulation of mutations that occur throughout life. The longer we live, the more mutations that are accumulated and the more likely that cancer might develop.
- *Embryonic development* represents what happens when gene regulation proceeds correctly and *cancer* shows what can happen when gene regulation goes awry.

CONNECT WITH THE CURRICULUM FRAMEWORK

Big Idea 3

Intercellular and intracellular signals mediate gene expression, and this underlies many diverse processes including cancer, development in the embryo, seed germination and fruit ripening, and morphological changes such as that seen in yeast cells when mating pheromones are present. (EK 3.B. 2) The importance of this topic to your understanding of biology cannot be overstated.

Viruses

(Biology, *10e: Chapter 19*, Focus, *1e: Chapter 17*)

YOU MUST KNOW

- The components of a virus.
- The differences between lytic and lysogenic cycles.
- How viruses can introduce genetic variation into host organisms. (LO 3.29)
- Mechanisms that introduce genetic variation into viral populations. (LO 3.30)

A virus consists of a nucleic acid surrounded by a protein coat (Biology, *10e: 19.1*, Focus, *1e: 17.1*)

- Smaller than ribosomes, the tiniest viruses are about 20 nm across.
- The two essential components of a virus are a protein shell or **capsid** that surrounds **the genetic material** (either double- or single-stranded DNA *or* double- or single-stranded RNA).
- Many viruses found in animals have membranous **viral envelopes** that surround the capsid and aid the viruses in infecting their hosts. (See Figure 5.11.)
- **Bacteriophages**, or **phages**, are viruses that infect bacterial cells.

Viruses replicate only in host cells (Biology, *10e: 19.2*, Focus, *1e: 17.2*)

- Viruses have a limited **host range**. This means they can infect only a very limited variety of hosts. *Example*: Human cold virus infects only cells of the upper respiratory tract.
- Viral reproduction occurs only in host cells. Two variations have been studied in bacteriophages. Read the following, then study Figure 5.12.

    1. The **lytic cycle** ends in the death of the host cell by rupturing it (lysis). In this cycle, a bacteriophage injects its DNA into a host cell and takes over the host cell's machinery to synthesize new copies of the viral DNA as well as protein coats. These self-assemble, and the bacterial cell is lysed, releasing multiple copies of the virus.

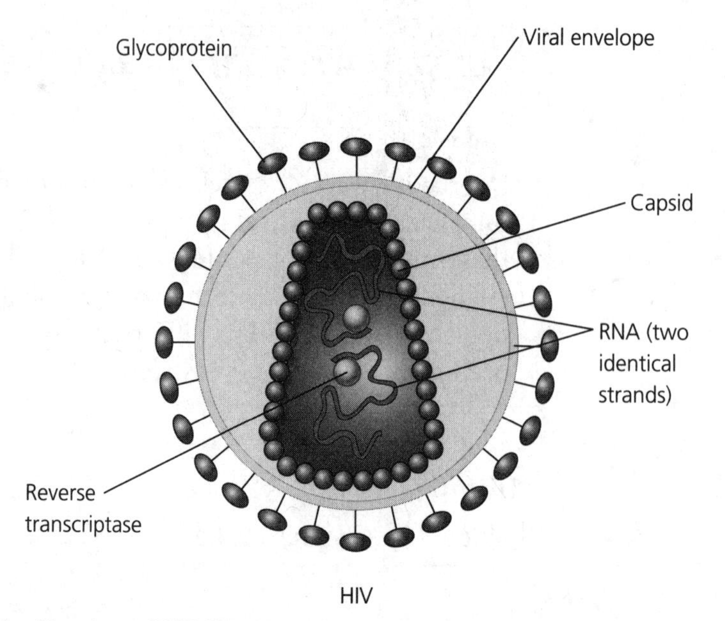

Figure 5.11 Structure of HIV. Note the components of a typical virus: nucleic acid, capsid. This virus also has an envelope, and reverse transcriptase

2. In the **lysogenic cycle** the bacteriophage's DNA becomes incorporated into the host cell's DNA and is replicated along with the host cell's genome. The viral DNA is known as a **prophage**. Under certain conditions, the prophage will enter the lytic cycle, described on the previous page.

- **Retroviruses** are RNA viruses that use the enzyme **reverse transcriptase** to transcribe DNA from an RNA template. The new DNA then permanently integrates into a chromosome in the nucleus of an animal cell. The host transcribes the viral DNA into RNA that may be used to synthesize viral proteins or may be released from the host cell to infect more cells. *Example*: HIV is a retrovirus.
- Viruses have the ability to introduce genetic change into organisms as well as to undergo rapid genetic change themselves. Moving from one host to another, viruses may pick up pieces of the first host's DNA and carry it to the next cell to be infected. This is very common in bacteria infected by viruses, where the process is called **transduction**.
- RNA viruses lack replication error-checking mechanisms and thus have higher rates of mutation. Mutations may accumulate rapidly and give rise to diverse clones of the virus within one organism, as occurs in humans with AIDS, or result in new genetic strains that may cause disease. This rapid mutation of viruses explains why there is no vaccine against the common cold.

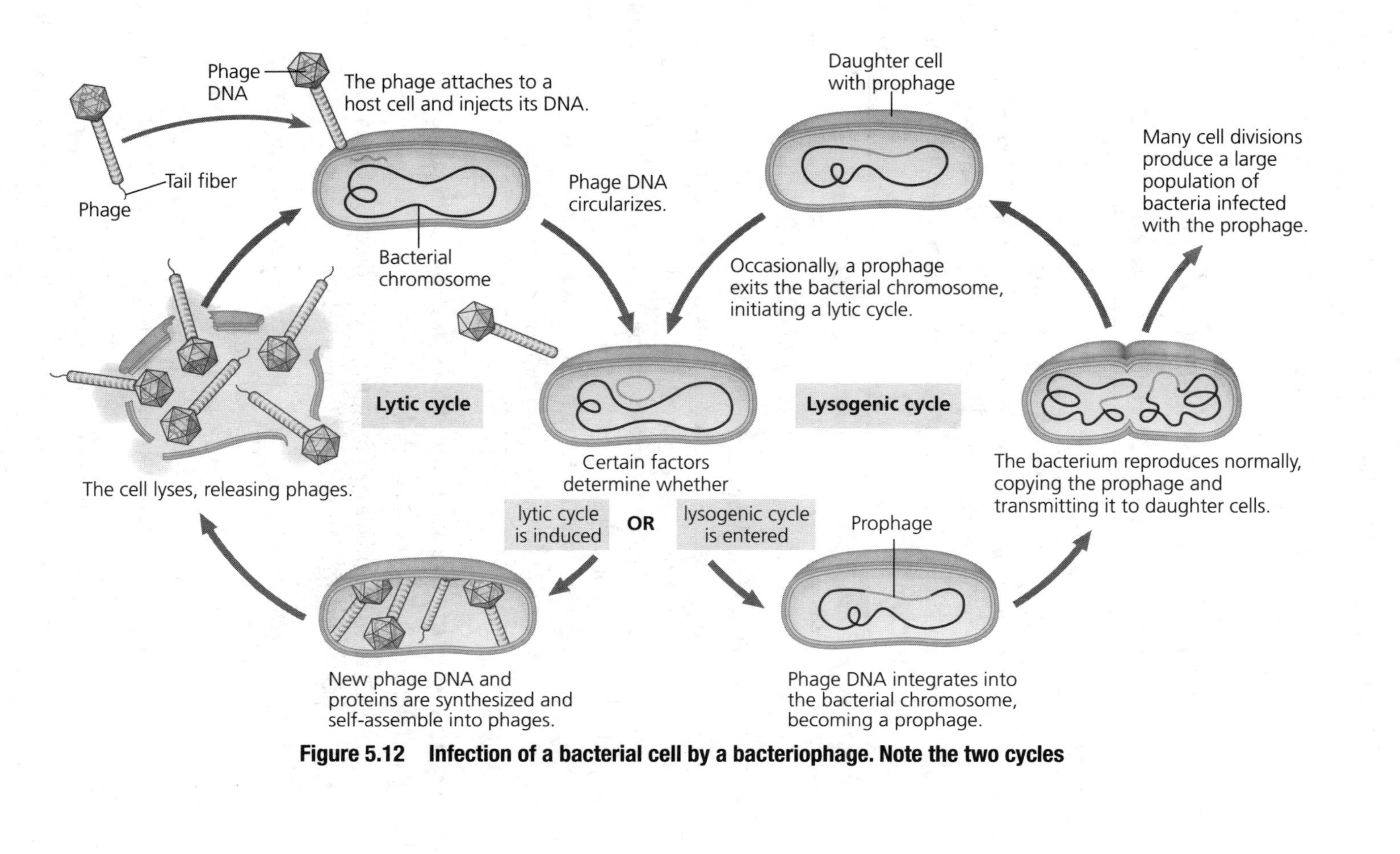

Figure 5.12 Infection of a bacterial cell by a bacteriophage. Note the two cycles

Prions are formidable pathogens (Biology, *10e: 19.3*)

- **Prions** are misfolded, infectious proteins that cause the misfolding of normal proteins in the brains of various animal species. Their damage to the brain accumulates over time and eventually leads to death. *Examples* of diseases caused by prions include mad cow disease and, in humans, Creutzfeldt-Jakob disease.

DNA Tools and Biotechnology

(Biology, *10e: Chapter 20*, Focus, *1e: Concepts 13.4, 15.4, 16.2*)

WHAT'S IMPORTANT TO KNOW?
The Curriculum Framework expects that you be familiar with techniques of modern biotechnology as well as know an example of a product of genetic engineering. Your teacher may not cover all of the possibilities described in this chapter.

YOU MUST KNOW

- The terminology of biotechnology.
- How plasmids are used in bacterial transformation to clone genes.
- The key ideas that make PCR possible and applications of this technology.
- How gel electrophoresis can be used to separate DNA fragments or protein molecules.
- Information that can be determined from DNA gel results, such as fragment sizes and RFLP analysis.

DNA sequencing and cloning are valuable tools in genetic engineering and biological inquiry (Biology, *10e: 20.1*, Focus, *1e: 13.4*)

- The key to unlocking the concepts of biotechnology is to understand the terms. Know the following commonly used terms:
  - **Genetic engineering** is the process of manipulating genes and genomes.
  - **Biotechnology** is the process of manipulating organisms or their components for the purpose of making useful products.
  - **Recombinant DNA** is DNA that has been artificially made, using DNA from different sources—and often different species. An example is the introduction of a human gene into an *E. coli* bacterium.
  - **Gene cloning** is the process by which scientists can produce multiple copies of specific segments of DNA that they can then work with in the lab. Many bacteria have DNA outside the main circular chromosome in plasmids. A **plasmid** is a small, circular extra-chromosomal loop of DNA. Plasmids are often used in biotechnology.

- **Restriction enzymes** are used to cut strands of DNA at specific locations (called **restriction sites**). They are mostly derived from bacteria where they serve the important function of protection against invading viruses.
- When a DNA molecule is cut by restriction enzymes, the result will always be a set of **restriction fragments**, which may have at least one single-stranded end, called a **sticky end**. Sticky ends can form hydrogen bonds with complementary single-stranded pieces of DNA. These unions can be sealed with the enzyme **DNA ligase**. Follow this important step in *Biology,* 10e, Figure 20.6 or *Focus,* 1e, Figure 13.23.

- Follow the steps that may occur to clone a gene in Figure 5.13.

1. *Identify and isolate the gene of interest and a* ***cloning vector***. The vector will carry the DNA sequence to be cloned and is often a bacterial plasmid.

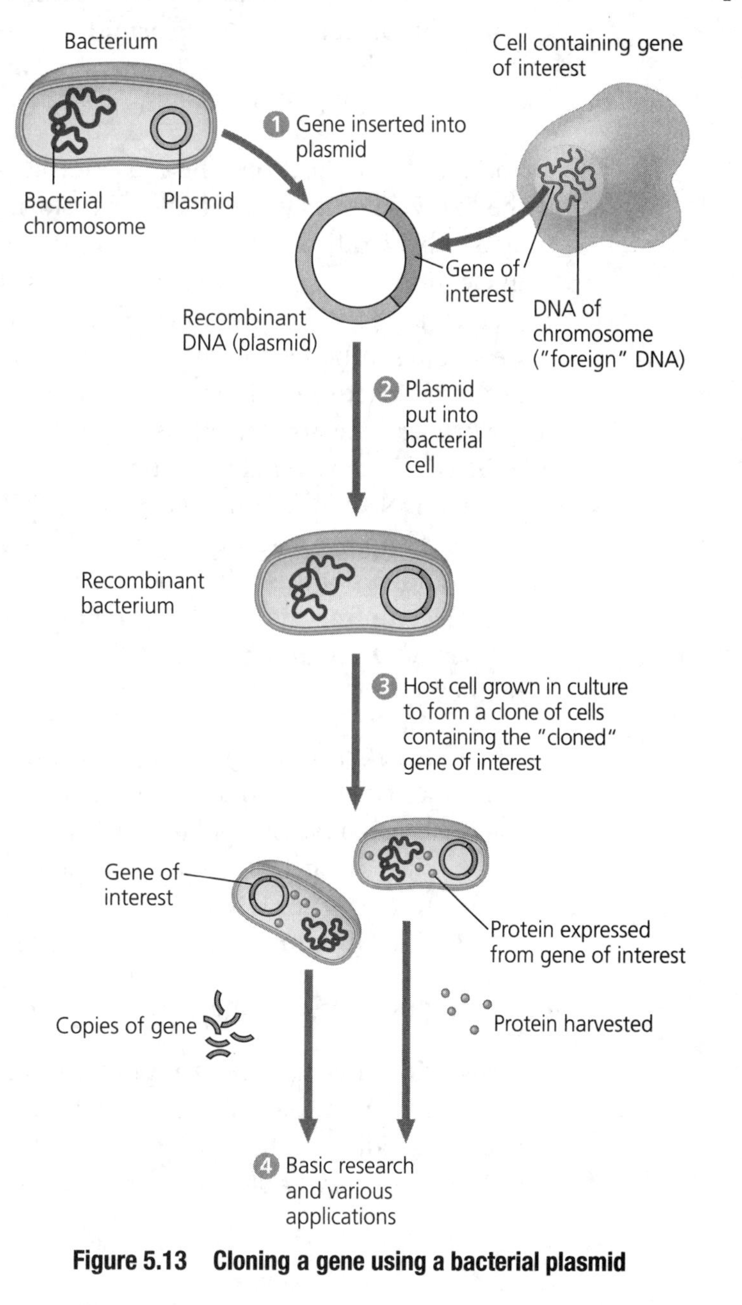

Figure 5.13 Cloning a gene using a bacterial plasmid

2. *Cut both the gene of interest and the vector with the same restriction enzyme.* This gives the plasmid and the human gene matching sticky ends.
3. *Join the two pieces of DNA.* Form recombinant plasmids by mixing the plasmids with the DNA fragments. The human DNA fragments can be sealed into the plasmid using DNA ligase.
4. *Get the vector carrying the gene of interest into a host cell.* The plasmids are taken up by the bacterium by *transformation*. The process of transformation is a key part of Investigation 8.
5. *Select for cells that have been transformed.* The bacterial cells carrying the clones must be identified or selected. This can be done by linking the gene of interest to an antibiotic resistance gene or a *reporter gene* such as green fluorescent protein. In AP Investigation 8, we use an ampicillin-resistant plasmid. Any bacterial cells that do not pick up the plasmid by transformation will be killed when grown on agar with the antibiotic ampicillin.

- For certain applications the next step is finding the gene of interest among the many colonies present after transformation. A process known as **nucleic acid hybridization** can be used to find the gene. If we know at least part of the nucleotide sequence of the gene of interest, we can synthesize a **probe** complementary to it. For example, if the known sequence is G-G-C-T-A-A, then we would synthesize the complementary probe C-C-G-A-T-T. If we make the probe radioactive or fluorescent, the probe will be easy to track, taking us to the proper gene of interest.
- **PCR** (polymerase chain reaction) is a method used to amplify a particular piece of DNA without the use of cells. PCR is used to amplify DNA when the source is impure or scanty (as it would be at a crime scene). Figure 5.14 shows the basic steps of the PCR procedure.
- **Gel electrophoresis** is a lab technique used to separate macromolecules, primarily DNA and proteins. The principles of this separation of DNA include:

1. An *electric current* is applied to the field. DNA is negatively charged and migrates to the positive electrode.
2. A gel made of a polymer is used as a matrix to separate molecules by size. The gel allows smaller molecules to move more easily than larger fragments of DNA.
3. The DNA must be stained or tagged for visualization.

Follow Figure 5.15 (on page 144) as the specific steps are shown.

- **Restriction fragment length polymorphisms (RFLPs)** result from small differences in DNA sequences and can be detected by electrophoresis. The difference in banding patterns after electrophoresis allows for diagnosis of disease or is used to answer paternity and identity questions.

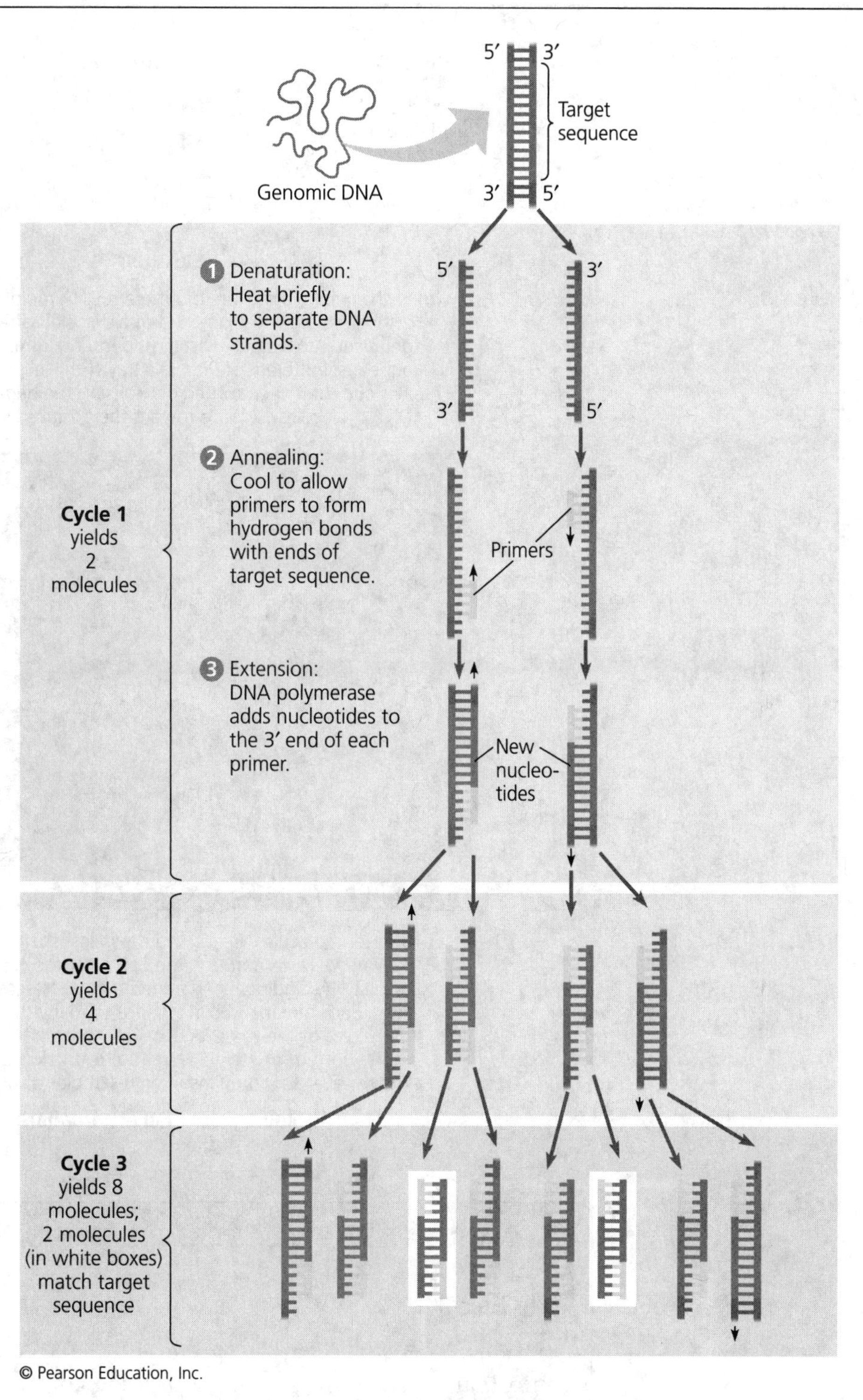

Figure 5.14 Polymerase chain reaction (PCR)

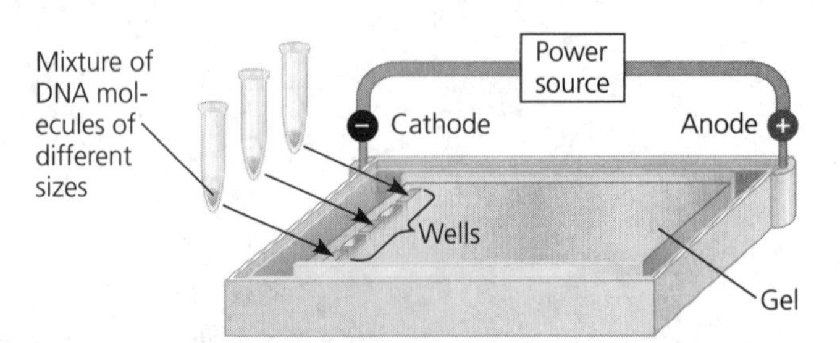

(a) Each sample, a mixture of different DNA molecules, is placed in a separate well near one end of a thin slab of agarose gel. The gel is set into a small plastic support and immersed in an aqueous, buffered solution in a tray with electrodes at each end. The current is then turned on, causing the negatively charged DNA molecules to move toward the positive electrode.

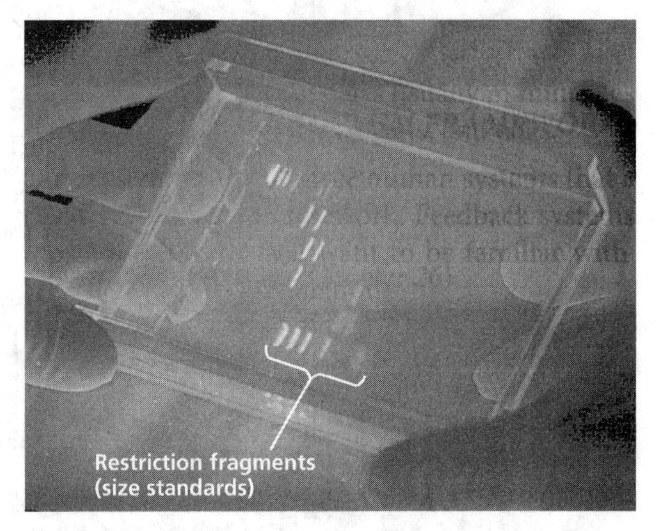

(b) Shorter molecules are slowed down less than longer ones, so they move faster through the gel. After the current is turned off, a DNA-binding dye is added that fluoresces pink in UV light. Each pink band corresponds to many thousands of DNA molecules of the same length. The horizontal ladder of bands at the bottom of the gel is a set of restriction fragments of known sizes for comparison with samples of unknown length.

Figure 5.15 Gel electrophoresis

> ### *CONNECT WITH THE CURRICULUM FRAMEWORK*
>
> **Big Idea 3**
>
> Be able to justify the claim that humans can manipulate heritable information by identifying *at least two* commonly used technologies. (LO 3.6) Investigation 8 is Bacterial Transformation and Investigation 9 is Restriction Enzyme Analysis of DNA. Be sure that you thoroughly understand both of these processes, and then we suggest that you be able to describe PCR. The focus here is how the technology can be used in genetic manipulation, so carefully consider applications of technology. Also, be able to discuss ethical implications of specific applications of biotechnology.

- The process just described leads to a genomic library. A **genomic library** is a set of thousands of recombinant plasmid clones, each of which has a piece of the original genome being studied. A **cDNA library** is made up of complementary DNA made from mRNA transcribed by reverse transcriptase. This technique rids the gene of introns but may not contain every gene in the organism.

Biologists use DNA technology to study gene expression and function (Biology, *10e: 20.2*, Focus, *1e: 15.4*)

- Genome-wide studies of gene expression are made possible by the use of **DNA microarray assays**. DNA microarray chips work as follows:

  1. Small amounts of single-stranded DNA (ssDNA) fragments representing different genes are fixed to a glass slide in a tight grid, termed a *DNA chip*.
  2. The mRNA molecules from the cells being tested are isolated and used to make cDNA using reverse transcriptase, then tagged with a fluorescent dye.
  3. The cDNA bonds to the ssDNA on the chip, indicating which genes are "on" in the cell (actively producing mRNA). This enables researchers, for example, to see differences in gene expression between breast cancer tumors and noncancerous breast tissue.

Cloned organisms and stem cells are useful for basic research and other applications (Biology, *10e: 20.3*, Focus, *1e: 16.2*)

- In animal cloning the nucleus of an egg is removed and replaced with the diploid nucleus of a body cell, a process termed *nuclear transplantation*. The ability of a body cell to successfully form a clone decreases with embryonic development and cell differentiation.
- The major goal of most animal cloning is reproduction, but not for humans. In humans, the major goal is the production of **stem cells**. A stem cell can both reproduce itself indefinitely and, under the proper conditions, produce other specialized cells. Stem cells have enormous potential for medical applications.
- **Embryonic stem cells** are *pluripotent*, which means "capable of differentiating into many different cell types." The ultimate aim is to use them for the repair of damaged or diseased organs, such as insulin-producing pancreatic cells for people with diabetes or certain kinds of brain cells for people with Parkinson's disease.

The practical applications of DNA technology affect our lives in many ways (Biology, *10e: 20.4*)

There are many different uses for DNA technology, some of which are as follows:

- **Diagnosis of disease:** A number of diseases can be detected by RFLP analysis (for example, cystic fibrosis, sickle-cell disease) or through amplification of blood samples to test for viruses (for example, HIV).
- **Gene therapy:** This is alteration of an afflicted individual's genes. Gene therapy holds great potential for treating disorders traceable to a single defective gene, such as cystic fibrosis.

- **The production of pharmaceuticals:** Gene splicing and cloning can be used to produce large amounts of particular proteins in the lab (for example, human insulin and growth hormone).
- **Transgenic** animals are created when eggs are fertilized *in vitro* and then a desired gene is cloned and inserted into the nucleus of the embryo. If successful, the transgenic animal will express the "foreign" gene, which might be for a human protein that can be produced in large quantities. For example, goats are used to express human antithrombin in milk.
- **Forensic applications:** DNA samples taken from the blood, skin cells, or hair of alleged criminal suspects can be compared to DNA collected from the crime scene. *Genetic profiles* can be compared and used to identify persons at that crime scene.
- **Environmental cleanup:** Scientists engineer metabolic capabilities into microorganisms, which are then used to treat environmental problems, such as removing heavy metals from toxic mining sites.
- **Agricultural applications:** Certain genes that produce desirable traits have been inserted into crop plants to increase their productivity or efficiency. An organism that has acquired by artificial means one or more genes from another species or variety is termed a **genetically modified (GM) organism**. Currently, a debate is in progress over the safety of GM organisms.

Genomes and Their Evolution

(Biology, *10e: Chapter 21*; Focus *1e, Chapter 18*)

YOU MUST KNOW

- How prokaryotic genomes compare to eukaryotic genomes.
- Applications of bioinformatics to medicine, evolution, and health.
- The activity and role of transposable elements and retrotransposons in generating genetic diversity
- How evo-devo relates to our understanding of the evolution of genomes.
- The role of homeotic genes and homeoboxes in developmental patterns and sequences.

Scientists use bioinformatics to analyze genomes and their functions (Biology, *10e: 21.2,* Focus, *1e: 18.2*)

- **Bioinformatics** is the use of computers, software, and mathematical models to process and integrate the incredible volume of data from sequencing projects such as the Human Genome Project. In addition to DNA sequences, protein interactions are analyzed in an approach called *proteomics.*
- *Systems biology* aims to model the behavior of entire biological systems and is enhanced by bioinformatics. This has many applications, including medical ones—for example, in the understanding and treatment of cancers.

> ***CONNECT WITH THE CURRICULUM FRAMEWORK***
>
> Investigation 3 uses BLAST, a primary tool of bioinformatics. If you did not do this investigation review the information about it on p. 302. Be sure to understand this powerful tool of modern science!

Genomes vary in size, number of genes, and gene density (Biology, *10e: 21.3*, Focus, *1e: 18.3*)

- More than 3,700 genomes (about 3,500 of them prokaryotes) have now been sequenced with thousands more in progress. In general, bacteria and archaea have fewer genes than eukaryotes, and the number of genes in eukaryotic genomes is less than was expected.
- There does not seem to be any correlation between the complexity of an organism and its number of genes. The 1 mm nematode *C. elegans* and humans both have between 20,000 and 21,000 genes! The human genome is able to function with relatively few genes by utilizing alternative splicing of RNA transcripts. Recall that this process results in more than one functional protein from a single gene.

Multicellular eukaryotes have much noncoding DNA and many multigene families (Biology, *10e: 21.4*, Focus, *1e: 18.4*)

- Only a tiny part of the human genome—1.5%—codes for proteins or is transcribed into rRNAs or tRNAs. Much of the rest is **repetitive DNA**, sequences that are present in multiple copies in the genome.
- **Transposable elements** make up much of the repetitive DNA. These are stretches of DNA that can move from one location to another in the genome with the aid of an enzyme, *transposase.*
- Transposons can interrupt normal gene function if inserted in the middle of a functional gene, or alter gene expression if inserted into a regulatory element. Although these effects may be harmful or lethal, over many generations some may have small beneficial effects.
- Transposons can account for multiple copies of genes and the resulting genetic diversity provides raw material for natural selection.
- **Multigene families** are collections of two or more identical or very similar genes. A classic example is the *human alpha-globin and beta-globin* gene families. Here, the genes for different human globins are on different chromosomes. This allows for different forms of the beta-globin gene to function at different times in the human life cycle. For example, the embryonic and fetal forms of hemoglobin have a higher affinity for oxygen that the adult forms, ensuring the efficient transfer of oxygen from mother to fetus.

Duplication, rearrangement, and mutation of DNA contribute to genome evolution (Biology, *10e: 21.5*, Focus, *1e: 18.5*)

- How might genes with novel functions evolve? Duplication events can lead to the evolution of genes with related functions, such as those of the alpha-globin and beta-globin gene families. Mutations and transpositions can occur, and nonfunctional *pseudogenes* may be found in the clusters. Ultimately, new genes with new functions may occur.

Comparing genome sequences provides clues to evolution and development (Biology, *10e: 21.6*, Focus, *1e: 18.6*)

- Determining which genes have remained similar, that is, are *highly conserved*, in distantly related species can help clarify evolutionary relationships among species that diverged from each other long ago.
- **Evo-devo** is a field of biology that compares developmental processes to understand how they may have evolved and how changes can modify existing organismal features or lead to new ones.
- **Homeotic** genes are master regulatory genes that control placement and spatial organization of body parts by controlling the developmental fate of groups of cells.
- A **homeobox** is a widely conserved 180-nucleotide sequence found within homeotic genes. When we say that a sequence is *widely conserved*, this means that it is found in many groups (for example, fungi, animals, and plants) with very few differences. This hints at the relatedness and common evolution of all life-forms.

Level 1: Knowledge/Comprehension Questions

The Knowledge/Comprehension questions review essential content knowledge but don't represent the type of question you will see on the AP exam. Level 2, Synthesis/Application, questions are similar to those you will see on the exam.

1. Which of the following has a DNA genome surrounded by a protein coat?
   (A) a retrovirus
   (B) a virus
   (C) a eukaryotic cell
   (D) a prokaryotic cell

2. Which of the following is a laboratory application of restriction enzymes?
   (A) to block the replication of DNA
   (B) to artificially force the transcription of DNA
   (C) to prevent the translation of mRNA
   (D) to cut DNA molecules at specific locations

Directions: Questions 3–6 consist of four lettered choices followed by a list of numbered phrases or sentences. For each numbered phrase or sentence, select the one choice that is most closely related to it. Each choice may be used once, more than once, or not at all.

Questions 3–6

(A) Transcription
(B) Translation
(C) Transposon
(D) DNA methylation

3. A mobile segment of DNA that travels from one location on a chromosome to another, making it an important element of genetic change

4. The addition of chemical groups (CH_4, for example) to certain bases of DNA after DNA synthesis; this is thought to be an important control mechanism for gene expression

5. The synthesis of polypeptides from the genetic information coded in mRNA

6. The synthesis of RNA from a DNA template

7. Which of these processes can cause an increase in the rate of transcription?
   (A) methylation of DNA, particularly the cytosine bases
   (B) acetylation of histone proteins
   (C) binding of the operator in an operon
   (D) negative feedback from accumulating product

8. When a virus infects a bacterium by injecting its DNA and the genome of the virus is incorporated into the host cell, this is
   (A) the lytic cycle of a phage.
   (B) the lysogenic cycle of a phage.
   (C) transposition by a virus.
   (D) retroviral infection.

9. In genetic engineering, DNA ligase is used for which of the following purposes?
   (A) to act as a probe for locating cloned genes
   (B) to create breaks in DNA in order to allow foreign DNA fragments to be inserted
   (C) to seal up nicks created in newly created recombinant DNA
   (D) to ensure that "sticky ends" of like DNA fragments do not re-anneal

Directions: Questions 10–13 consist of four lettered choices followed by a list of numbered phrases or sentences. For each numbered phrase or sentence, select the one choice that is most closely related to it. Each choice may be used once, more than once, or not at all.

Questions 10–13
   (A) tRNA
   (B) mRNA
   (C) Poly-A tail
   (D) RNA polymerase

10. An example of a post-transcriptional modification

11. Binds to the promoter on DNA to initiate transcription

12. Binds to free amino acids in the cytoplasm

13. Travels out of the nucleus and into the cytoplasm where it serves as a template in translation

14. After eukaryotic transcription takes place, mRNA undergoes several modifications before leaving the nucleus to take part in translation. One of these is the cutting out of nonessential sections of mRNA and the subsequent splicing together of stretches of mRNA necessary for the final functional molecule. Which of the following mRNA sections are spliced together into the finished mRNA molecule?
   (A) introns
   (B) exons
   (C) genes
   (D) ribozymes

Level 2: Application/Analysis/Synthesis Questions

1. A goat can produce milk containing the same polymers present in the silk produced by spiders when particular genes from a spider are inserted into the goat's genome. Which of the following reasons describes why this is possible?
   (A) Goats and spiders share a common ancestor and, thus, produce similar protein excretions.
   (B) Goats retain all the genes that were found in ancestral species and will express these proteins when activated by a gene insertion.
   (C) The proteins in goats' milk and spiders' silk have the same amino acid sequence.
   (D) The universal nature of the genetic code allows for the production of identical proteins from different organisms as long as the DNA sequence is identical.

2. Enzymes and energy are required for the synthesis of DNA and RNA. Which enzyme is correctly linked with its function?
   (A) Telomerase ensures that the transcription machinery starts with the codon AUG.
   (B) DNA ligase joins together Okazaki fragments from the lagging strand.
   (C) DNA polymerase oversees the binding of mRNA to the ribosome.
   (D) Helicase alters the new DNA strands so it will coil into a double helix.

3. PCR (polymerase chain reaction) allows target segments of DNA to be produced quickly by using heat to denature and separate strands of DNA, then cooling to allow primers to attach to the single strands and DNA polymerase to synthesize new strands of the target DNA by adding nucleotides to the 3′ end of each primer on the two strands. Why was this process considered so important that it earned Kary Mullis a Nobel Prize?
   (A) It is useful for isolating the source of DNA.
   (B) It facilitates the introduction of plasmid DNA into a host cell.
   (C) It enables selection of transformed cells that have taken up foreign DNA.
   (D) It can produce multiple copies of even minute samples of DNA.

Questions 4 and 5 refer to an experiment that was performed to separate linear DNA fragments from three samples radioactively labeled with ^{32}P. *The fragments were then separated using gel electrophoresis. The visualized bands are depicted below:*

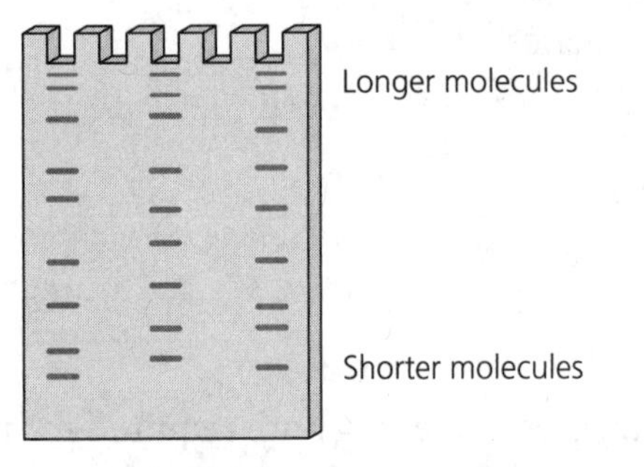

Completed gel

4. When the electric field was applied, the fragments of DNA in each of the three samples migrated to different locations along the gel. Which of the following explains the pattern shown here?
   (A) The fragments differed in their levels of radioactivity.
   (B) The fragments differed in their charges—some were positively charged, whereas others were negatively charged.
   (C) The fragments differed in size.
   (D) The fragments differed in solubility.

5. How many sites on DNA were cut by the particular restriction enzyme used in Sample 1 (the lane on the far left)?
   (A) 6
   (B) 7
   (C) 8
   (D) 9

6. A bacterium is infected with an experimentally constructed bacteriophage composed of the T2 phage protein coat and T4 phage DNA. The new phages produced by lysis of infected cells would have
   (A) T2 protein and T4 DNA.
   (B) T2 protein and T2 DNA.
   (C) T4 protein and T4 DNA.
   (D) T4 protein and T2 DNA.

7. RNA retroviruses such as HIV discharge both strands of RNA and molecules of reverse transcriptase into cells they infect. What is the purpose of the reverse transcriptase?
   (A) The host cells rapidly destroy the viral RNA so more must constantly be made.
   (B) The host cells lack enzymes that can replicate the viral genome.
   (C) The reverse transcriptase is required to translate viral mRNA into proteins.
   (D) The reverse transcriptase binds to active sites of the lytic enzymes of the host cells.

8. A mutation is any change in a DNA sequences. Of the following mutation events, which one would have the most significant effect on the resulting polypeptide?
   (A) A nucleotide and its complementary base are replaced, resulting in a new amino acid in translation.
   (B) A nucleotide pair is added into a gene near its terminal end, altering the reading frame of the genetic message.
   (C) A nucleotide pair substitution occurs, which results in the introduction of a "stop" codon.
   (D) A nucleotide is altered that results in no change in the amino acid sequence due to redundancy.

9. In analyzing the number of different bases in a DNA sample, which result would be consistent with the base-pairing rules?
   (A) A = G
   (B) A + G = C + T
   (C) A + T = G + T
   (D) A = C

10. The segment of DNA shown below has restriction sites I and II, which create restriction fragments a, b, and c. Which of the following gels produced by electrophoresis would represent the separation and identity of these fragments?

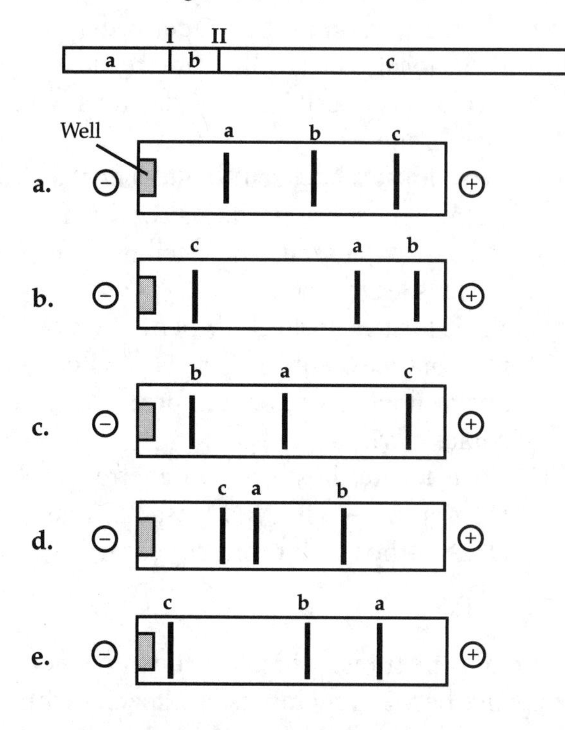

11. Which of the following is a difficulty in getting prokaryotic cells to express eukaryotic genes?
    (A) The signals that control gene expression are different for prokaryotes, thus requiring prokaryotic promoter regions to be added to the vector.
    (B) The genetic code differs because prokaryotes substitute the base uracil for thymine.
    (C) Prokaryotic cells do not recognize eukaryotic stop and start codons.
    (D) The ribosomes of prokaryotes are not large enough to handle long eukaryotic genes.

12. Microarrays use single-stranded pieces of DNA of known genes, which are placed on a slide to determine whether these genes are being expressed in a particular tissue. For example, if the cells come from seedlings that are germinated and kept in the dark, the chlorophyll genes would not be active. Which of the following would you wash across the slide to determine that the chlorophyll genes are active in a plant grown in light?
    (A) single-stranded cDNA made from the mRNA transcribed from the chlorophyll genes
    (B) a portion of the amino acid sequences of the chlorophyll molecules
    (C) mRNA transcribed from the plant grown in the dark
    (D) cell fragments that include ribosomes and both tRNA and rRNA

13. The human genome appears to have only about as many genes as the simple nematode worm, *C. elegans*. Which of the following best explains how the more complex humans can have relatively few genes?
    (A) Human genes have unusually long introns involved in the regulation of gene expression.
    (B) More than one polypeptide can be produced from a gene by alternative splicing.
    (C) The human genome has a high proportion of noncoding DNA.
    (D) The human genome has a large number of SNPs (single nucleotide polymorphisms), which increases genetic variability.

14. The human globin genes are located on two different chromosomes and include variants that are expressed at different times in the life cycle as well as a number of nonfunctional pseudogenes. Which of the following would best describe this gene family?
   (A) All the functional genes in the family have the same transcription control elements.
   (B) This gene family would be equivalent to the operons of prokaryotes.
   (C) Mutations in the hemoglobin gene led to its translocation to two different chromosomes.
   (D) Identical or similar genes have evolved by gene duplication.

After reading the paragraphs, answer the question(s) that follow.

Exposure to the HIV virus doesn't necessarily mean that a person will develop AIDS. Some people have genetic resistance to infection by HIV. Dr. Stephen O'Brien from the U.S. National Cancer Institute has recently identified a mutant form of a gene, called *CCR5*, which can protect against HIV infection. The mutation probably originated in Europe among survivors of the bubonic plague. The mutated gene prevents the plague bacteria from attaching to cell membranes and, therefore, from entering and infecting body cells.

Although the HIV virus is very different from the bacterium that causes the plague, both diseases affect the exact same cells and use the same method of infection. The presence of the mutated gene in descendants of plague survivors helps prevent them from contracting AIDS. Pharmaceutical companies are using this information as the basis for a new approach to AIDS prevention. This would be very important in areas of the world where the mutation is scarce or absent, such as Africa.

15. The most likely method by which the mutated *CCR5* gene prevents AIDS is by
   (A) blocking the protein channels that allow the virus to enter the cell.
   (B) directing the formation of lytic enzymes packaged in the lysosomes.
   (C) being presented on immune system cells to target the virus for attack and destruction.
   (D) coding for a mutant receptor protein on the cell membrane, thus not allowing the entry of the virus.

16. Which of the following shows the steps of a retroviral infection in the proper order?
   (A) Virus locates host cell → enters nucleus → alters host cell DNA → destroys cell membrane.
   (B) Virus locates host cell → alters host cell DNA → host cell produces copies of virus → copies enter host cell nucleus → nucleus leaves cell.
   (C) Virus locates host cell → penetrates cell membrane → enters nucleus → alters host cell DNA → host cell produces copies of virus.
   (D) Virus locates host cell → penetrates cell membrane → viral RNA used to make DNA → host cell produces copies of virus.

17. Lateral gene transfer involves the movement of genes between members of a species without sexual reproduction. In bacteria, there are three mechanisms for this: conjugation, transformation, and transduction. Predict how these processes are most likely to affect bacteria.
   (A) Their DNA content will be reduced.
   (B) The amount of RNA in the cytoplasm will be increased.
   (C) Their genetic diversity will be increased.
   (D) Their metabolic needs will be modified to give a selective advantage.

After reading the paragraph, answer questions 18 and 19.

Four decades after the end of the Vietnam War, the remains of an Air Force pilot were discovered and returned to the United States. A search of Air Force records identified three families to which the remains might possibly belong. Each family had a surviving twin of a missing service member. A DNA comparison was done, and the following STR profiles were obtained from the remains of the pilot and the surviving twins from the three families.

	Air Force Pilot	Family 1	Family 2	Family 3
1		—		
2	—	—		—
3	—			—
4		—		
5	—	—	—	—
6			—	
7	—	—	—	—
8			—	
9		—		
10	—			—
11			—	
12		—		
13	—		—	—

18. In order to match the pilot's remains to the correct family using DNA profiling,
    (A) the majority of the STR bands must match.
    (B) each of the 13 STR bands must match.
    (C) the bands for site 13 must match.
    (D) bands 5 and 7 must match.

19. Based on analysis of the STR sites shown, does the missing pilot belong to any of these three families?
    (A) No, none of the families match.
    (B) Yes, family 1 matches.
    (C) Yes, family 2 matches.
    (D) Yes, family 3 matches.

20. RNA processing involves a number of important steps before the final transcript is ready to leave the nucleus. Of the following, which does not represent an event in this process?
    (A) Exons are cut out before mRNA leaves the nucleus.
    (B) Nucleotides may be added at both ends of the RNA.
    (C) Ribozymes may function in RNA splicing.
    (D) RNA splicing can be catalyzed by spliceosomes.

21. What is the basis for the difference in how the leading and lagging strands of DNA molecules are synthesized?
    (A) The origins of replication occur only at the $5'$ end.
    (B) Helicases and DNA polymerase work at the $5'$ end.
    (C) DNA polymerase can join new nucleotides only to the $3'$ end of a growing strand.
    (D) DNA ligase works only in the $3' \rightarrow 5'$ direction.

22. Below are amino acid sequences for six species for critical parts of the *FOXP2* gene. Each letter represents a single amino acid. The data can be used to determine the relatedness of the six species. Assuming that mutation rates are equal for all species over time and based only on these data, which number represents the species *least* related to the other species?

    1. ATETI . . . PKSSD . . . TSSTT . . . NARRD
    2. ATETI . . . PKSSE . . . TSSTT . . . NARRD
    3. ATETI . . . PKSSD . . . TSSTT . . . NARRD
    4. ATETI . . . PKSSD . . . TSSNT . . . SARRD
    5. ATETI . . . PKSSD . . . TSSTT . . . NARRD
    6. VTETI . . . PKSSD . . . TSSTT . . . NARRD

    (A) Species 2
    (B) Species 3
    (C) Species 4
    (D) Species 6

Grid-In Questions

These call for a numerical response.

1. A linear strip of DNA has two restriction sites for the restriction enzyme *EcoR1* and three restriction sites for the restriction enzyme *Hind III.* If the strip of DNA was incubated with the restriction enzymes then the cut DNA was collected and run on an electrophoretic gel, how many bands would be expected on the gel?
2. In *E. coli* the DNA is 24% adenine. Based on this, what percentage of this DNA is guanine?

Free-Response Question

1. *Genes are located on chromosomes and are the basic unit of heredity that is passed on from parent to child through generations.*

(a) For each of the following mutations, describe how the genetic information is altered and predict one specific effect that could occur with the error.

- nondisjunction
- base-pair deletion resulting in a frameshift within an intron
- base-pair substitution resulting in a different amino acid

(b) **Explain** why it is more common for human males to be colorblind than females.

(c) Proper gene dosages are critical to normal development and function. Discuss the mechanism that maintains proper gene dosages related to the X chromosome in females.

TOPIC 6

Mechanisms of Evolution

Descent with Modification

(Biology, *10e: Chapter 22,* Focus, *1e: Chapter 19*)

YOU MUST KNOW

- How Lamarck's view of the mechanism of evolution differed from Darwin's.
- Several examples of evidence for evolution and how they each support how organisms have changed over time.
- The difference between structures that are homologous and those that are analogous, and how this relates to evolution.
- The role of adaptations, variation, time, reproductive success, and heritability in evolution.

The Darwinian revolution challenged traditional views of a young Earth inhabited by unchanging species (Biology, *10e: 22.1;* Focus, *1e: 19.1*)

- Historical Setting
  - **Carolus Linnaeus** (1707–1778): Grouped similar species into increasingly general categories, reflecting what he considered the pattern of their creation.
    - Developed **taxonomy**, the branch of biology dedicated to the naming and classification of all forms of life.
    - Developed **binomial nomenclature**, a two-part naming system that includes the organism's genus and species.
  - **Georges Cuvier** (1769–1832): French geologist opposed to the idea of evolution.
    - Advocated the principle that events in the past occurred suddenly, as with catastrophes, and by different mechanisms than those occurring today. This explained boundaries between strata and location of different species.
- **Charles Lyell** (1797–1875): English geologist and friend of Charles Darwin.
  - Developed the idea that the geologic processes that have shaped the planet have been uniform over a long period of time and not by a series of catastrophes occurring over a short period of time.

- *Importance*: ***The Earth must be very old.*** An old Earth has time for evolution; a very young Earth does not. Lyell gave Darwin the gift of time.
- Lyell's *Principles of Geology* was studied by Darwin during his journeys.

- **Jean-Baptiste de Lamarck** (1744–1829): Developed an early theory of evolution based on two principles:
  - **Use and disuse** is the idea that parts of the body that are used extensively become larger and stronger, while those that are not used deteriorate.
  - **Inheritance of acquired characteristics** assumes that characteristics acquired during an organism's lifetime could be passed on to the next generation. *Example*: A weightlifter's child could be born with a more muscular anatomy.
  - *Importance:* Lamarck recognized that species evolve and the match of organisms to their environment occurs through gradual evolutionary change. His explanatory mechanism, however, was flawed.

Descent with modification by natural selection explains the adaptations of organisms and the unity and diversity of life (Biology, *10e: 22.2,* Focus, *1e: 19.2*)

- **Charles Darwin's** voyage on the HMS *Beagle* from 1831 to 1836 was the impetus for the development of his theory of evolution by natural selection.
- Darwin's mechanism for evolution was natural selection. Recall that Lamarck's mechanism was the inheritance of acquired characteristics.
- **Natural selection** explains how adaptations arise.
  - **Adaptations** are heritable characteristics that enhance organisms' ability to survive and reproduce in specific environments. *Example*: Desert foxes have large ears, which radiate heat. Arctic foxes have small ears, which conserve body heat.
- Darwin's theory of natural selection involves these important points:
  1. Individuals in a population vary in their traits, many of which are heritable.
  2. A population can produce far more offspring than can survive. With more individuals than the environment can support, competition is inevitable.
  3. Individuals with inherited traits that are better suited to the local environment are more likely to survive and reproduce than individuals less well-suited. This is sometimes phrased as "differential reproductive success."
  4. Evolution occurs as the unequal reproductive success of individuals ultimately leading to adaptations to their environment. Over time, natural selection can increase the match between organisms and their environment.

If an environment changes, or if individuals move to a new environment, natural selection may result in adaptation to these new conditions, sometimes giving rise to new species in the process.

> ***TIP FROM THE READERS***
> When explaining evolution, avoid any language that makes it sound like a species "needs" a feature and so it evolves. Don't write "survival of the fittest" but instead describe how selection favors a feature that results in leaving more offspring. Evolution is not goal oriented.

- **Artificial selection** is the process by which species are modified by humans. *Example*: Selective breeding for milk or meat production; development of dog breeds.
- *Individuals do not evolve.* ***Populations*** evolve.

Evolution is supported by an overwhelming amount of scientific evidence (Biology, *10e: 22.3*, Focus, *1e: 19.3*)

> ***ORGANIZE YOUR THOUGHTS***
>
> EVIDENCE FOR EVOLUTION is seen in the following ways:
>
> **1.** Direct observations
> **2.** Homology
> **3.** The fossil record
> **4.** Biogeography

- Evidence for Evolution
    1. **Direct Observations of Evolutionary Change**
        - Insect populations can rapidly become resistant to pesticides such as DDT.
        - Evolution of drug-resistant viruses and antibiotic-resistant bacteria.
    2. **Homology and Convergent Evolution**
        - **Homology**: Characteristics in related species can have an underlying similarity even though they have very different functions. *Similarity resulting from common ancestry is known as homology.*
        - **Homologous structures** are anatomical signs of evolution. *Examples*: Forelimbs of mammals that are now used for a variety of purposes, such as flying in bats or swimming in whales, but were present and used in a common ancestor for walking. (See *Biology,* 10e, Figure 22.15, page 473, or *Focus,* 1e, Figure 19.17, page 375.)
        - **Embryonic homologies**: Comparison of early stages of animal development reveals many anatomical homologies in embryos that are not visible in adult organisms. *Examples*: All vertebrate embryos have a post-anal tail and pharyngeal pouches.

- **Vestigial organs** are structures of marginal, if any, importance to the organism. They are remnants of structures that served important functions in the organisms' ancestors. *Example*: Remnants of the pelvis and leg bones are found in some snakes.
- **Molecular homologies** are shared characteristics on the molecular level. *Examples*: All life-forms use the same genetic language of DNA and RNA. Amino acid sequences coding for hemoglobin in primate species shows great similarity, thus indicating a common ancestor.
- **Convergent evolution** explains why distantly related species can resemble one another. Convergent evolution has taken place when two organisms developed similarities as they adapted to similar environmental challenges—not because they evolved from a common ancestor. The likenesses that result from convergent evolution are considered **analogous** rather than homologous. Think of it like this: Similar problems have similar solutions. Here are some examples:

    1. The torpedo shapes of a penguin, dolphin, and shark are the solution to movement through an aqueous environment.
    2. Sugar gliders (marsupial mammals) and flying squirrels (eutherian mammals) occupy similar niches in their respective habitats.

STUDY TIP Homologous structures show evidence of common ancestry (whale fin/bat wing).

Analogous structures are similar solutions to similar problems but do *not* indicate close relatedness (bird wing/butterfly wing).

3. **The Fossil Record:** *Fossils provide evidence for the theory of evolution.*

    - Fossils are remains or traces of organisms from the past. They are found in sedimentary rock. **Paleontology** is the study of fossils.
    - Fossils show that evolutionary changes have occurred over time and the origin of major new groups of organisms.
    - Darwin's theory of evolution through natural selection explains the succession of forms in the fossil record. Transitional fossils have been found that link ancient organisms to modern species, just as Darwin's theory predicts.

4. **Biogeography:** The geographic distribution of species.

    - Species in a discrete geographic area tend to be more closely related to each other than to species in distant geographic areas. *Example*: In South America, desert animals are more closely related to local animals in other habitats than they are to the desert animals of Asia. This reflects evolution, not creation.
    - **Continental drift** and the break-up of *Pangaea* can explain the similarity of species on continents that are distant today.
    - **Endemic species** are found at a certain geographic location and nowhere else. *Example*: Marine iguanas are endemic to the Galápagos.

ORGANIZE YOUR THOUGHTS

1. Evolution is change in species over time.
2. There is overproduction of offspring, which leads to competition for resources.
3. Heritable variations exist within a population.
4. These variations can result in differential reproductive success.
5. Over generations, this can result in changes in the genetic composition of the population.

And remember . . . Individuals do not evolve! ***Populations*** *evolve.*

The Evolution of Populations

(Biology, *10e: Chapter 23*, Focus, *1e: Chapter 21*)

YOU MUST KNOW

- How mutation and sexual reproduction each produce genetic variation.
- The conditions for Hardy-Weinberg equilibrium.
- How to use the Hardy-Weinberg equation to calculate allele frequencies to test whether a population is evolving. (LO 1.4)
- What effects genetic drift, migration or selection may have on a population, and analyze data to justify your predictions. (LO 1.6–1.8)

Genetic variation makes evolution possible (Biology, *10e: 23.1*, Focus, *1e: 21.1*)

- Phenotypic variation reflects genetic variation. As you look at your classmates, their phenotypic variation may be caused by the either-or differences of a single gene or the range of variation typical of multiple genes. How many students have attached versus free ear lobes (an either-or difference as the results of one gene); what is the range of height in your class (multiple genes yielding variation along a continuum)?
- **Mutations** are the only source of *new* genes and *new* alleles.
  - Only mutations in cell lines that produce gametes can be passed to offspring.
- **Point mutations** are changes in one nucleotide base in a gene. They can have significant impact on phenotype, as in sickle-cell disease.
- **Chromosomal mutations** delete, disrupt, duplicate, or rearrange many loci at once. They are usually harmful but not always. Gene duplications (often from mistakes in crossing over) can result in an expanded genome with new genes that may accumulate mutations over generations and take on new functions.
- However, **most of the genetic variations** within a population are due to the sexual recombination of alleles that already exist in a population.

- Sexual reproduction shuffles existing alleles and deals them at random to produce individual genotypes. Recall that there are *three mechanisms* for this shuffling of alleles:
  - **Crossing over** during prophase I of meiosis.
  - **Independent assortment** of chromosomes during meiosis (2^{23} different combinations possible in the formation of human gametes!)
  - **Fertilization** ($2^{23} \times 2^{23}$ different possible combinations for human sperm and egg)

The Hardy-Weinberg equation can be used to test whether a population is evolving (Biology, *10e: 23.2,* Focus, *1e: 21.2*)

- **Population genetics** is the study of how populations change genetically over time.
- **Population:** A group of individuals of the same species that live in the same area and interbreed, producing fertile offspring.
- **Gene pool:** All of the alleles at all loci in all the members of a population.
- In diploid species, each individual has two alleles for a particular gene, and the individual may be either heterozygous or homozygous.
- If all members of a population are homozygous for the same allele, the allele is said to be **fixed**. Only one allele exists at that particular locus in the population. For example, the fruit fly is heterozygous for 1,920 of its 13,700 genes—the remaining 11,780 are fixed. It follows that the greater the number of fixed alleles, the lower the species' genetic diversity.

> ***ORGANIZE YOUR THOUGHTS***
>
> **Five Conditions for Hardy-Weinberg Equilibrium**
>
> 1. No change in allelic frequency due to mutation.
> 2. Random mating.
> 3. No natural selection.
> 4. The population size must be extremely large. (No genetic drift.)
> 5. No gene flow. (Emigration, immigration, transfer of pollen, etc.)

- The **Hardy-Weinberg principle** is used to describe a population that is *not* evolving. It states that the frequencies of alleles and genes in a population's gene pool will remain constant over the course of generations unless they are acted upon by forces *other* than Mendelian segregation and the recombination of alleles. The population is at **Hardy-Weinberg equilibrium**.
- You will find additional discussion of this very important topic in Investigation 2, beginning on page 297.
- However, it is unlikely that all the conditions for Hardy-Weinberg equilibrium will be met. Allele frequencies change. Populations evolve. This can be tested by applying the **Hardy-Weinberg equation**.
- Note that the dominant phenotype includes both p^2 and $2pq$. For example, if brown eyes are dominant a person with brown eyes could be homozygous (p^2) or heterozygous ($2pq$). Because the dominant phenotype includes two possible genotypes (and two mathematical variables in p and q), we must look to the recessive phenotype to determine q. The recessive phenotype, blue eyes in our example, is q^2 and every individual with blue eyes has the same genotype (bb),

EQUATION FOR HARDY-WEINBERG EQUILIBRIUM

Consider a gene locus that exists in two allelic forms, *A* and *a*, in a population.

Let p = the frequency of A, the dominant allele

and q = the frequency of a, the recessive allele.

All the dominant alleles plus all the recessive alleles will equal 100% of the alleles for this gene, or, expressed mathematically: p + q = 1. If this simple binomial is expanded we get the Hardy-Weinberg equation:

$$p^2 + 2pq + q^2 = 1$$

The three terms of this binomial expansion indicate the frequencies of the three genotypes:

p^2 = frequency of *AA* (homozygous dominant)
$2pq$ = frequency of *Aa* (heterozygous)
q^2 = frequency of *aa* (homozygous recessive)

If we know the frequency of one of the alleles, we can calculate the frequency of the other allele:

$p + q = 1$, so
$p = 1 - q$
$q = 1 - p$

which is not the case with brown eyes. It is therefore possible to take the square root of the frequency of the recessive phenotype in the population to find q. For example, if 36% of the population has blue eyes, or $q^2 = 0.36$, then $q = 0.6$ and $p = 0.4$. *Here is the rule to remember: Always find* q *first!*

TRY THIS FOR PRACTICE!

Suppose in a plant population that red flowers (*R*) are dominant to white flowers (*r*). In a population of 500 individuals, 25% show the recessive phenotype. How many individuals would you expect to be homozygous dominant and heterozygous for this trait? **(The answer is below in bold. Cover this solution and try it first!)**

1. Let p = frequency of the dominant allele (*R*) and q = frequency of the recessive allele (*r*)
2. q^2 = frequency of the homozygous recessive = 25% = 0.25. Since $q^2 = 0.25$, $q = 0.5$
3. Now, $p + q = 1$, so $p = 0.5$
4. Homozygous dominant individuals are *RR* or $p^2 = 0.5^2 = 0.25$ and will represent (0.25)(500) = **125 individuals.**
5. The heterozygous individuals are calculated from $2pq$ = (2) (0.5) (0.5) = 0.5. In a population of 500 individuals, **250 individuals** (0.5 × 500) will be heterozygous dominant.

> ***TIP FROM THE READERS***
> With the emphasis on mathematics you can expect to see Hardy-Weinberg questions on future AP exams. Practice these problems! Investigation 2, Hardy-Weinberg Modeling, requires your deep understanding of how changes in allelic frequency can be measured. Practice problems can be found at the end of this chapter, on page 300 in Part III, the end of your textbook chapter, and the Science Skill activity in your text.

Natural selection, genetic drift, and gene flow can alter allele frequencies in a population (Biology, *10e: 23.3,* Focus, *1e: 21.3*)

- Let's back up before we go on. In nature, is it likely that all the conditions for HW equilibrium will be met? NO! Therefore, populations are evolving, which means their allelic frequencies are changing. This concept will look at specific factors that will alter allelic frequency. Note that each of them represents an exception to one of the five conditions for HW equilibrium listed earlier.
- Mutations can alter gene frequency but are rare.
- **The three major factors** that alter allele frequencies and bring about most evolutionary change are *natural selection, genetic drift,* and *gene flow.*
  - **Natural selection** results in alleles being passed to the next generation in proportions different from their relative frequencies in the present generation. Individuals with variations that are better suited to their environment tend to produce more offspring than those with variations that are less suited.
  - **Genetic drift** is the unpredictable fluctuation in allele frequencies from one generation to the next. The smaller the population, the greater the chance is for genetic drift. This is a *random, nonadaptive* change in allele frequencies. Genetic drift can lead to a loss of genetic diversity, even causing some genes to become fixed in the new population. Two examples follow.
    - **Founder effect:** A few individuals become isolated from a larger population and establish a new population whose gene pool is not reflective of the source population.
    - **Bottleneck effect:** A sudden change in the environment (for example, an earthquake, flood, or fire) drastically reduces the size of a population. The few survivors that pass through the restrictive bottleneck may have a gene pool that no longer reflects the original population's gene pool. *Example*: The population of California condors was reduced to nine individuals. This represents a bottlenecking event.
- **Gene flow** occurs when a population gains or loses alleles by genetic additions or subtractions from the population. This results from the movement of fertile individuals or gametes. *Gene flow tends to reduce the genetic differences between populations*, thus making populations more similar.

Natural selection is the only mechanism that consistently causes adaptive evolution (Biology, *10e: 23.4,* Focus, *1e: 21.4*)

- **Relative fitness** refers to the contribution an organism makes to the gene pool of the next generation relative to the contributions of other members. Fitness does *not* indicate strength or size. It is measured only by reproductive success.
- Natural selection acts more directly on the phenotype and indirectly on the genotype and can alter the frequency distribution of heritable traits in three ways (Figure 6.1).
  - **Directional selection** *Example*: Large black bears survived periods of extreme cold better than smaller ones and so became more common during glacial periods.
  - **Disruptive selection** *Example*: A population has individuals with either large beaks or small beaks but few with the intermediate beak size. Apparently the intermediate beak size is not efficient in cracking either the large or small seeds that are common.

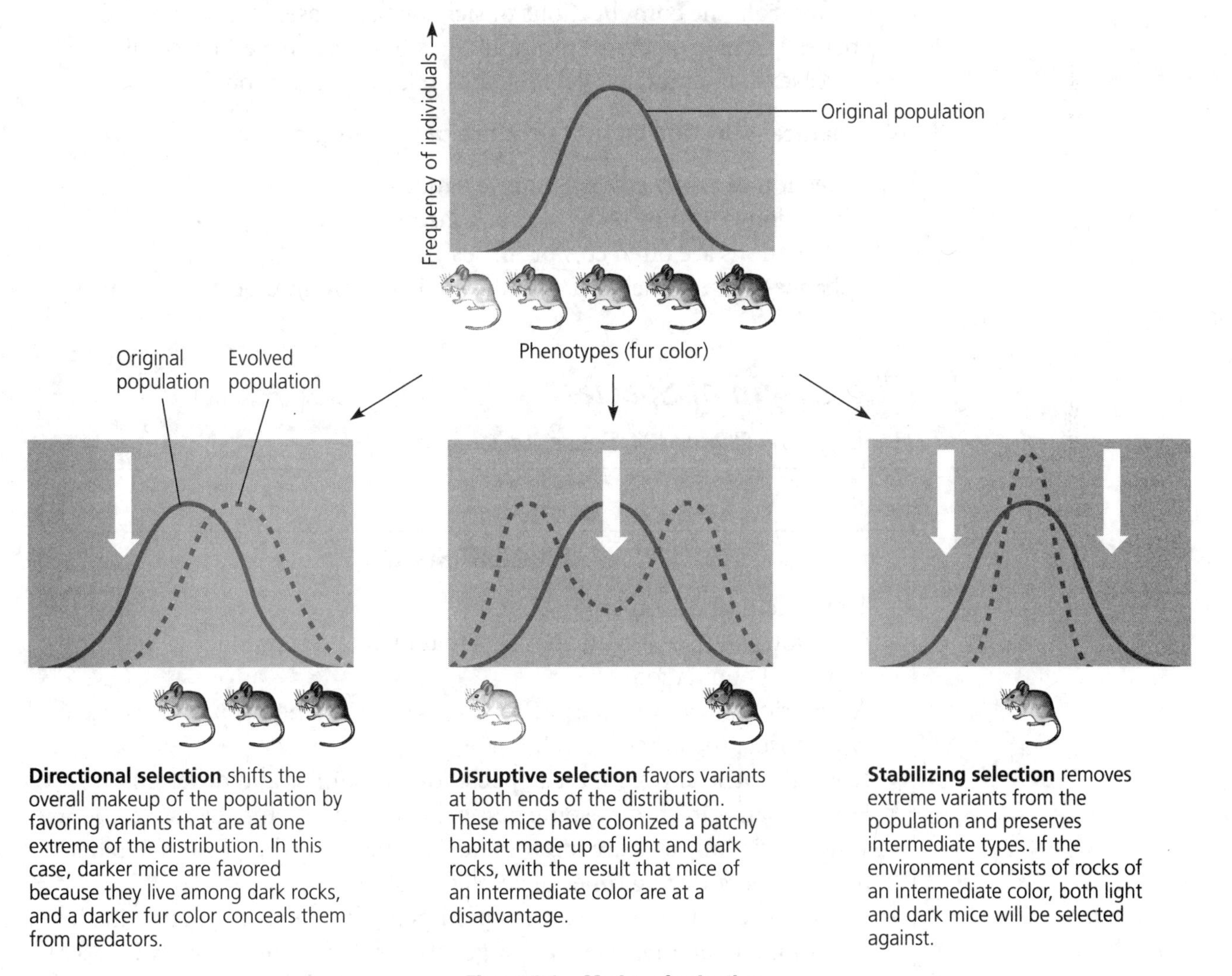

Directional selection shifts the overall makeup of the population by favoring variants that are at one extreme of the distribution. In this case, darker mice are favored because they live among dark rocks, and a darker fur color conceals them from predators.

Disruptive selection favors variants at both ends of the distribution. These mice have colonized a patchy habitat made up of light and dark rocks, with the result that mice of an intermediate color are at a disadvantage.

Stabilizing selection removes extreme variants from the population and preserves intermediate types. If the environment consists of rocks of an intermediate color, both light and dark mice will be selected against.

Figure 6.1 Modes of selection

- **Stabilizing selection** *Example*: Birth weights of most humans lie in a narrow range, as those babies who are very large or very small have higher mortality.

- **Sexual selection** is a form of natural selection in which individuals with certain inherited characteristics are more likely than other individuals to obtain mates. It can result in **sexual dimorphism**, a difference between the two sexes in secondary sexual characteristics such as differences in size, color, ornamentation, and behavior.
- **How is genetic variation preserved in a population?**
  - **Diploidy:** Because most eukaryotes are diploid, they are capable of hiding genetic variation (recessive alleles) from selection.
  - **Heterozygote advantage:** Individuals who are heterozygous at a certain locus have an advantage for survival. *Example*: In sickle-cell disease, individuals homozygous for normal hemoglobin are more susceptible to malaria, whereas homozygous recessive individuals suffer from the complications of sickle-cell disease. Heterozygotes benefit from protection from malaria and do not have sickle-cell disease, so the mutant allele remains relatively common.
- **Why natural selection cannot produce perfect organisms:**
  - Selection can only edit existing variations.
  - Evolution is limited by historical constraints.
  - Adaptations are often compromises.
  - Chance, natural selection, and the environment interact.

The Origin of Species

(Biology, *10e: Chapter 24*, Focus, *1e: Chapter 22*)

YOU MUST KNOW

- The biological concept of species.
- Prezygotic and postzygotic barriers that maintain reproductive isolation in natural populations.
- A description of similar species that are maintained separate by each type of isolating barrier.
- How allopatric and sympatric speciation are similar and different.
- How a change in chromosome number can lead to sympatric speciation.
- Why speciation rates are often rapid in situations when adaptive radiation occurs or during times of ecological stress.
- The connection between a change in gene frequency, a change in the environment, natural selection or genetic drift and speciation. (LO 1.24)
- How punctuated equilibrium and gradualism describe two different tempos of speciation.

The biological species concept emphasizes reproductive isolation (Biology, *10e: 24.1*, Focus, *1e: 22.1*)

- **Speciation** is the process by which new species arise.
- **Microevolution** is change in the genetic makeup of a population from generation to generation. It refers to adaptations that are confined to a single gene pool.
- **Macroevolution** refers to the broad pattern of evolutionary change above the species level, such as the appearance of feathers and other such novelties, used to define higher taxa.
- The **biological species concept** defines a species as a group of populations whose members have the potential to interbreed in nature and produce viable, fertile offspring but are unable to produce viable, fertile offspring with members of other groups.
- **Reproductive isolation** is defined as the existence of biological barriers that impede members of two species from producing viable, fertile hybrids.
- **Prezygotic and postzygotic** are two types of barriers that prevent members of different species from producing offspring that can also successfully reproduce. Examples of prezygotic barriers, those that prevent mating or hinder fertilization if mating has occurred, include the following:

    - **Habitat isolation:** Two species can live in the same geographic area but not in the same habitat; this will prevent them from mating because they will not encounter each other.
    - **Behavioral isolation:** Some species use certain signals or types of behavior to attract mates, and these signals are unique to their species. Members of other species do not respond to the signals; thus, mating does not occur.
    - **Temporal isolation:** Species may breed at different times of day, different seasons, or different years, and this can prevent them from mating.
    - **Mechanical isolation:** Species may be anatomically incompatible.
    - **Gametic isolation:** Even if the gametes of two species do meet, they might be unable to fuse to form a zygote.

  Examples of postzygotic barriers, those that prevent a hybrid zygote from developing into a fertile adult, include the following:

    - **Reduced hybrid viability:** When a zygote *is* formed, genetic incompatibility may cause development to cease.
    - **Reduced hybrid fertility:** Even if the two species produce a viable offspring, reproductive isolation is still occurring if the offspring is sterile and can't reproduce.
    - **Hybrid breakdown:** Sometimes two species mate and produce viable, fertile hybrids; however, when the hybrids mate, their offspring are weak or sterile.

Speciation can take place with or without geographic separation (Biology, *10e: 24.2*, Focus, *1e: 22.2*)

- In **allopatric speciation** a population forms a new species because it is geographically isolated from the parent population. When the population is geographically isolated gene flow is interrupted, resulting in reproductive isolation.

- Some **geologic events or processes** that can fragment a population, resulting in geographic isolation of new populations, include the emergence of a mountain range, the formation of a land bridge, or evaporation in a large lake that produces several small lakes.
- Small, newly isolated populations undergo allopatric speciation more frequently because they are more likely to have their gene pools significantly altered. Speciation is confirmed when individuals from the new population are unable to mate successfully with individuals from the parent population.

- A second type of speciation is **sympatric speciation**, in which a small part of a population forms a new species without being geographically separated from the parent population. It can result from part of the population switching to a new habitat; switching to a different resource such as food; or an accident during cell division, resulting in extra sets of chromosomes (polyploidy).
    - An example of a mechanism that can lead to sympatric speciation in plants is the formation of **polyploid** plants through nondisjunction in meiosis. For example, these plants may have a $4n$ chromosome number instead of the normal $2n$ number. They cannot breed with diploid members and produce fertile offspring. Although the formation of new species due to the formation of polyploids is rare in animals, it is very common in plants (80% of plant species were involved in a polyploid event).
- **Adaptive radiation** occurs when many new species arise from a single common ancestor. In adaptive radiations the new species fill different ecological niches in their communities. Catastrophes such as volcanoes, landslides, or mass extinctions open new niches.

Speciation can occur rapidly or slowly and can result from changes in few or many genes (Biology, *10e: 24.4,* Focus, *1e: 22.4*)

- **Gradualism** proposes that species descended from a common ancestor and gradually diverge more and more in morphology as they acquire unique adaptations.
- **Punctuated equilibrium** is a term used to describe periods of apparent stasis punctuated by sudden change observed in the fossil record.

> ***TIP FROM THE READERS***
> Remember this: Individuals do not evolve! They do not "struggle to survive." They cannot change their genetic makeup in response to a catastrophe. The individual lives or dies. Those that live reproduce and pass on adaptive heritable variations. INDIVIDUALS DO NOT EVOLVE! ONLY POPULATIONS CAN EVOLVE!

The History of Life on Earth

(Biology, *10e: Chapter 25*, Focus, *1e: Chapters 23, 24.1*)

YOU MUST KNOW

- A scientific hypothesis about the origin of life on Earth.
- The age of the Earth and when prokaryotic and eukaryotic life emerged.
- Characteristics of the early planet and its atmosphere.
- How Miller and Urey tested the Oparin-Haldane hypothesis and what they learned.
- Methods used to date fossils and rocks and how fossil evidence contributes to our understanding of changes in life on Earth.
- Evidence for endosymbiosis.
- How continental drift can explain the current distribution of species (biogeography).
- How extinction events open habitats that may result in adaptive radiation.

Conditions on early Earth made the origin of life possible (Biology, *10e: 25.1*, Focus, *1e: 24.1*)

- The current hypothesis about how life arose consists of four main stages:

  1. Abiotic (nonliving) synthesis of small organic molecules, such as amino acids and nitrogenous bases.
  2. The joining of these small molecules into macromolecules, such as proteins and nucleic acids.
  3. The packaging of these molecules into **protocells**, membrane-enclosed droplets, whose internal chemistry differed from that of the external environment.
  4. The origin of self-replicating molecules that made inheritance possible.

- Earth was formed about **4.6 billion years ago**, and life on Earth emerged about **3.8 billion years ago**. For the first three-quarters of Earth's history, all of its living organisms were microscopic and primarily unicellular.
- Hypothetical early conditions of Earth have been simulated in laboratories, and organic molecules have been produced.

  - **Oparin** and **Haldane** hypothesized that the early atmosphere, thick with water vapor, nitrogen, carbon dioxide, methane, ammonia, hydrogen, and hydrogen sulfide, provided with energy from lightning and UV radiation, could have formed organic compounds, a primitive "soup" from which life arose.
  - **Miller** and **Urey** tested this hypothesis and produced a variety of amino acids. Miller-Urey-type experiments show that the abiotic synthesis of organic molecules is possible under various assumptions about the composition of the early atmosphere.

- It is hypothesized that **self-replicating RNA** (not DNA) was the **first genetic material.** RNA, which plays a central role in protein synthesis, can also carry out a number of enzyme-like catalytic functions. These RNA catalysts are called **ribozymes.**

The fossil record documents the history of life (Biology, *10e: 25.2,* Focus, *1e: 23.1*)

- The **fossil record** is the sequence in which fossils appear in the layers of sedimentary rock that constitute Earth's surface. **Paleontologists** study the fossil record. Fossils, which may be remnants of dead organisms or impressions they left behind, are most often found in sedimentary rock formed from layers of minerals settling out of water. The fossil record is incomplete because it favors organisms that existed for a long time, were relatively abundant and widespread, and had shells or hard bony skeletons.
- Rocks and fossils are dated several ways:
  - **Relative dating** uses the order of rock strata to determine the relative age of fossils. The oldest fossils are deposited in the lower strata.
  - **Radiometric dating** uses the decay of radioactive isotopes to determine the age of the rocks or fossils. It is based on the rate of decay, or **half-life** of the isotope. The half-life is the time necessary for 50% of the parent isotope to decay. For example, the half-life of radioactive carbon-14 is 5,730 years (the common isotope carbon-12 is not radioactive). Living organisms accumulate both carbon-12 and carbon-14 in a known ratio. Once the organism dies no more carbon is obtained, but the carbon-14 decays into nitrogen-14. Measuring the carbon-14-to-carbon-12 ratio in a fossil can reveal its age. For example, if the fossil ratio of carbon-14 to carbon-12 is one-eighth the ratio of present-day individuals, the fossil is about 17,190 years old (5780 $\times$ 3 half-lives). See the grid-in questions at the end of this topic for a half-life problem.

Key events in life's history include the origins of single-celled and multicelled organisms and the colonization of land (Biology, *10e: 25.3,* Focus, *1e: 23.1, 25.1*)

- The earliest living organisms were **prokaryotes.**
- About 2.7 billion years ago, **oxygen** began to accumulate in Earth's atmosphere as a result of photosynthesis. The rise of oxygen doomed many prokaryotic groups, but others evolved in the new oxygen-rich environment, including the evolution of groups capable of cellular respiration.
- **Eukaryotes** appeared about 2.1 billion years ago.
  - The **endosymbiotic hypothesis** proposes that mitochondria and plastids (chloroplasts) were formerly small prokaryotes that began living within larger cells.
  - **Evidence** for this hypothesis includes the following:
    - Both organelles have enzymes and transport systems homologous to those found in the plasma membranes of living prokaryotes.
    - Both replicate by a splitting process similar to prokaryotes.
    - Both contain a single, circular DNA molecule, not associated with histone proteins.

- Both have their own ribosomes, which can translate their DNA into proteins.

- **Multicellular eukaryotes** evolved about 1.2 billion years ago.
- The **colonization of land** occurred about 500 million years ago, when **plants, fungi, and animals** began to appear on Earth.

The rise and fall of groups of organisms reflect differences in speciation and extinction rates (Biology, *10e: 25.4,* Focus, *1e: 23.2*)

- **Continental drift** is the movement of Earth's continents on great plates that float on the hot, underlying mantle. The San Andreas Fault marks where two plates are sliding past each other. Where plates have collided, mountains are uplifted.
  - Continental drift can help explain the disjunct geographic distribution of certain species, such as a fossil freshwater reptile found in both Brazil in South America and Ghana in west Africa, today widely separated by ocean.
  - Continental drift can explain why no eutherian (placental) mammals are indigenous to Australia.
- **Mass extinctions**, loss of large numbers of species in a short period, have resulted from global environmental changes that have caused the rate of extinction to increase dramatically.
  - By removing large numbers of species, a mass extinction can drastically alter a complex ecological community. Evolutionary lineages can disappear. *Example*: The dinosaurs were lost in a mass extinction 65 million years ago. Mass extinctions cause many ecological niches to be vacated. After mass extinctions those niches can be filled by the evolution of new species in an adaptive radiation.
  - Mass extinctions open niches that new species may occupy. For examples, the rise of mammals occurred following the loss of dinosaurs. This is an example of adaptive radiation.
- **Adaptive radiations** are periods of evolutionary change in which groups of organisms form many new species whose adaptations allow them to fill different ecological niches. After each of the five major extinctions an adaptive radiation occurred. Adaptive radiations also occur after major evolutionary innovations, such as seeds in plants or feathers in birds, or newly colonized areas where new species face little competition. *Example:* The Galápagos finch species are the result of an adaptive radiation due to the creation of new niches when volcanic action formed new land.

Major changes in body form can result from changes in the sequences and regulation of developmental genes (Biology, *10e: 25.5,* Focus, *1e: 23.3*)

- **Evo-devo** is a field of study in which evolutionary biology and developmental biology converge. (Evo = from evolution, Devo = from development) This field is illuminating how slight genetic divergences can be magnified into major morphological differences between species.

- Evolutionary novelty can arise when structures that originally played one role gradually acquire a different one. Structures that evolve in one context but become co-opted for another function are sometimes called **exaptations**. For example, it is possible that feathers of modern birds were co-opted for flight after functioning in some other capacity, such as thermoregulation.
- **Heterochrony** is an evolutionary change in the rate or timing of developmental events. Changing relative rates of growth even slightly can change the adult form of organisms substantially, thus contributing to the potential for evolutionary change. The increased rate of growth in bat finger bones provide the underlying support for bat wings, whereas the decreased rate of growth in leg and pelvic bones in whales led to the loss of hind limbs.
- **Homeotic genes** are master regulatory genes that determine the location and organization of body parts. Homeotic genes affect where a pair of wings will develop or how a flower's parts are arranged.
- ***Hox*** genes are one class of homeotic genes. Changes in *Hox* genes and in the genes that regulate them can have a profound effect on morphology, thus contributing to the potential for evolutionary change. An example is seen in the variable expression of a *Hox* gene in a snake limb bud and a chicken leg bud, resulting in no legs in the snake and a skeletal extension in the chicken.

Evolution is not goal oriented (**Biology, *10e: 25.6,* Focus, *1e: 23.4***)

- Evolution is like tinkering—a process in which new forms arise by the slight modification of existing forms. Even large changes, like the ones that produced the first mammals, can result from the modification of existing structures or existing developmental genes.

Level 1: Knowledge/Comprehension Questions

The Knowledge/Comprehension questions review essential content knowledge but don't represent the type of question you will see on the AP exam. Level 2, Application/Synthesis, questions are similar to those you will see on the exam.

1. The condition in which there are barriers to reproduction between individuals of the same species because they are separated by a portion of a mountain range is referred to as
   (A) mechanical isolation.
   (B) geographic isolation.
   (C) gametic isolation
   (D) temporal isolation

2. Which of the following statements best expresses the concept of punctuated equilibrium?
   (A) Very small changes in the genome of individuals eventually lead to the evolution of a population.
   (B) The five conditions of Hardy-Weinberg equilibrium will prevent populations from evolving quickly.
   (C) Evolution occurs in rapid bursts of change alternating with long periods in which species remain relatively unchanged.
   (D) Profound change over the course of geologic history is the result of an accumulation of slow, continuous processes.

3. In a particular bird species, individuals with average-sized wings survive severe storms more successfully than other birds in the same population with longer or shorter wings. This illustrates which of the following modes of selection?
   (A) directional selection
   (B) stabilizing selection
   (C) diversifying selection
   (D) artificial selection

Directions: Questions 4–7 consist of four lettered choices followed by a list of numbered phrases or sentences. For each numbered phrase or sentence, select the one choice that is most closely related to it. Each choice may be used once, more than once, or not at all.

Questions 4–7

(A) The bottleneck effect
(B) Homologous structures
(C) Analogous structures
(D) The founder effect

4. Results in a new population with a limited gene pool, often seen when a small population is separated from the larger population

5. One result of evolution from a common ancestor; exhibited by certain structures with similar features having different functions

6. A result of drastic reduction in population size due, such as the reduction of the California condor population to nine individuals

7. Features that have evolved independently to serve similar functions but do not represent evolutionary relatedness

8. One general pattern that will maintain two related groups as separate species requires their gametes to not combine. These are called prezygotic barriers. Select the one example here that does not describe this type of isolating mechanism.
   (A) hybrid breakdown
   (B) behavioral isolation
   (C) temporal isolation
   (D) mechanical isolation

9. A marsupial species found only in Australia has evolved to eat tree leaves, be diurnal, and raise its young until they are of reproductive age. A grazing placental mammal species found in Africa has also evolved to eat tree leaves, be diurnal, and raise its young until they are of reproductive age. This is an example of which of the following types of evolution?
   (A) divergent evolution
   (B) species-specific evolution
   (C) convergent evolution
   (D) neutral evolution

10. Banded iron formations in marine sediments provide evidence of
    (A) the crashing of meteorites onto Earth, possibly transporting abiotically produced organic molecules from space.
    (B) oxidized iron layers in terrestrial rocks.
    (C) the accumulation of oxygen in the seas from the photosynthesis of cyanobacteria.
    (D) the evolution of photosynthetic archaea near deep-sea vents.

11. Which of the following constitutes the smallest unit capable of evolution?
    (A) an individual
    (B) a family unit
    (C) a population
    (D) a community

Level 2: Application/Analysis/Synthesis Questions

1. Which of the following would not be important in Darwin's theory of evolution?
   (A) Acquiring genetic change during the life span of the organism contributes to evolution.
   (B) More individuals are born in a population than will survive to reproduce.
   (C) Natural selection occurs as a result of the differing reproductive success of individuals in a population.
   (D) A result of evolution is the adaptation of a population of organisms to their environment.

2. In a certain group of rabbits, the presence of yellow fur is the result of a homozygous recessive condition in the biochemical pathway producing hair pigment. If the frequency of the allele for this condition is 0.09, which of the following is closest to the frequency of the dominant allele in this population? (Assume that the population is in Hardy-Weinberg equilibrium and only two alleles are present for this gene.)
   (A) 0.03
   (B) 0.70
   (C) 0.83
   (D) 0.91

3. In a population of squirrels, the allele that causes bushy tail (B) is dominant, while the allele that causes bald tail is recessive (b). If 64% of the squirrels have a bushy tail, what is the frequency of the dominant allele?
   (A) 0.8
   (B) 0.6
   (C) 0.4
   (D) 0.36

4. Mice that are homozygous for a lethal recessive allele die shortly after birth. In a large breeding colony of mice, you find that a surprising 5% of all newborns die from this trait. In checking lab records, you discover that the same proportion of offspring have been dying from this trait in this colony for the past three years. (Mice breed several times a year and have large litters.) How might you explain the persistence of this lethal allele at such a high frequency?
   (A) Homozygous recessive mice have a reproductive advantage, which helps to maintain the frequency of the allele.
   (B) The rapid reproductive rate is coupled with a large mutation rate, which results in multiple copies of this lethal allele.
   (C) There is some sort of heterozygote advantage and perhaps selection against the homozygous dominant trait.
   (D) Because this is a diploid species, the recessive allele cannot be selected against when it is in the heterozygote.

5. Morphological and genetic comparisons group 30 species of snapping shrimp (genus *Alpheus*) into 15 pairs of closely related species. In sympatric speciation, a new species may arise in the same area as an existing species, where in allopatric speciation, some geographical isolating mechanism is present. What is the best explanation for the fact that one member of each pair lives on the Atlantic side of the Isthmus of Panama, whereas the other member of each pair lives on the Pacific side?
   (A) Different predator pressures in the Atlantic and Pacific selected for differences resulted in the reproductive isolation of these species.
   (B) The pairs of species arose by allopatric speciation when the Isthmus of Panama separated their ancestral species.
   (C) The 30 species of snapping shrimp evolved by sympatric speciation before the Isthmus of Panama formed.

(D) These 15 pairs of species illustrate the pattern of punctuated equilibrium, in which most speciation events take place in a short period of time.

6. Sympatric evolution occurs when a new species arises in the same area as an existing ancestral species. Which of the following can lead to sympatric speciation?
   (A) Migration of a small number of individuals to a geographical remote area.
   (B) A natural disaster cuts off contact between members of a population.
   (C) A flood causes a newly formed river, separating segments of a population.
   (D) A mutation event results in offspring that are fertile tetraploids.

7. Mitochondria and plastids contain DNA and ribosomes and make some, but not all, of their proteins. Some of their proteins are coded for by nuclear DNA and produced in the cytoplasm. What may explain this division of labor?
   (A) Over the course of evolution, some of the original endosymbiont's genes were transferred to the host cell's nucleus.
   (B) The host cell's genome always included genes for making mitochondrial and plastid proteins.
   (C) The smaller prokaryotic ribosomes in these organelles cannot produce the eukaryotic proteins required for their functions.
   (D) Some mitochondria and plastid genes were contributed by early bacterial prokaryotes that shared genes with other primitive cells.

8. Mothers and teachers have often said that they need another pair of eyes on the backs of their heads. And another pair of hands would come in handy in many situations. You can imagine that these traits would have been advantageous to our early hunter-gatherer ancestors as well. According to sound evolutionary reasoning, what is the most likely explanation for why humans do not have these traits?
   (A) Because they actually would not be beneficial to the fitness of individuals who possessed them. Natural selection always produces the most beneficial traits for a particular organism in a particular environment.
   (B) Because every time they have arisen before, the individual mutants bearing these traits have been killed by chance events. Chance and natural selection interact.
   (C) Because these variations have probably never appeared in a healthy human. Tetrapods share a four-limbed, two-eyed body plan; natural selection can only edit existing variations.
   (D) Because humans are a relatively young species. Given enough time for adaptation, it is inevitable that the required adaptations will arise.

9. This phylogenetic tree illustrates the evolutionary relationships of tetrapods and was constructed using both anatomical and DNA sequence data. Using evidence from this tree, which statement most accurately reflects the relationship of mammals to amphibians and birds?

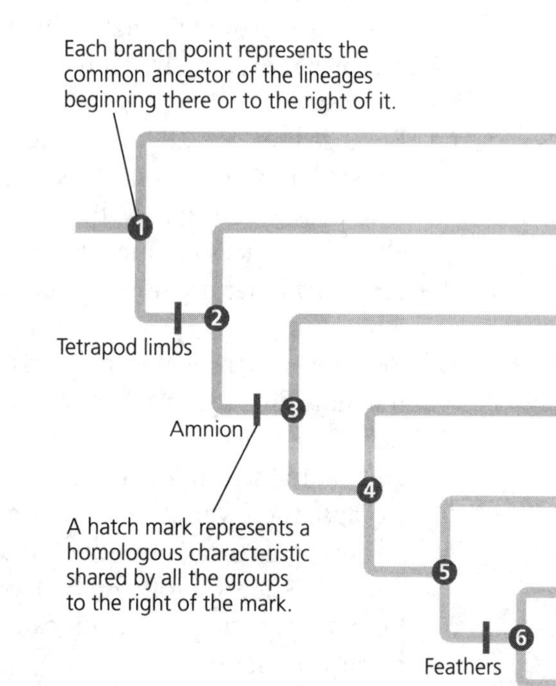

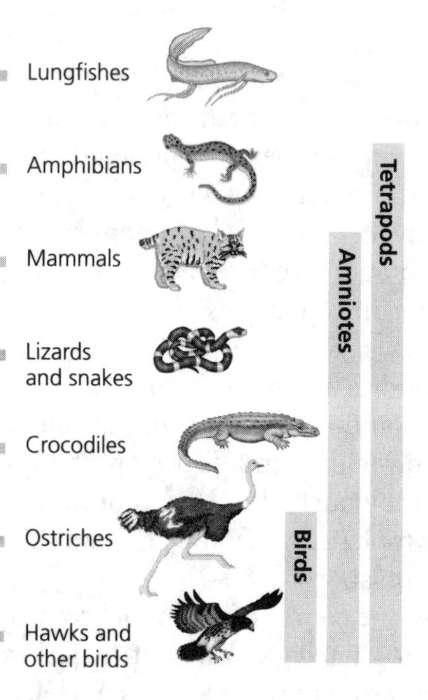

(A) Mammals are more closely related to amphibians and likely share more DNA sequences.
(B) Mammals are more closely related to amphibians because they share anatomical similarities.
(C) Mammals are more closely related to birds because they share a more recent common ancestor.
(D) Mammals, birds, and amphibians are all equally related because they share a common ancestor at the second branch point.

10. According to the figure above, which pair of organisms shares the most recent common ancestor?
(A) lungfish and amphibian
(B) amphibian and lizard
(C) mammal and crocodile
(D) lizard and ostrich

11. According to the figure above, the group most closely related to lizards and snakes are
(A) lungfishes.
(B) amphibians.
(C) mammals.
(D) hawks and other birds.

12. This figure shows the change of a population over time. Which statement best describes the mode of selection depicted in the figure?

13. In plants, speciation can occur when the offspring receive extra sets of chromosomes. This is referred to as being polyploid. Which species of wheat shown is polyploid? The uppercase letters represent not genes but *sets of chromosomes.*

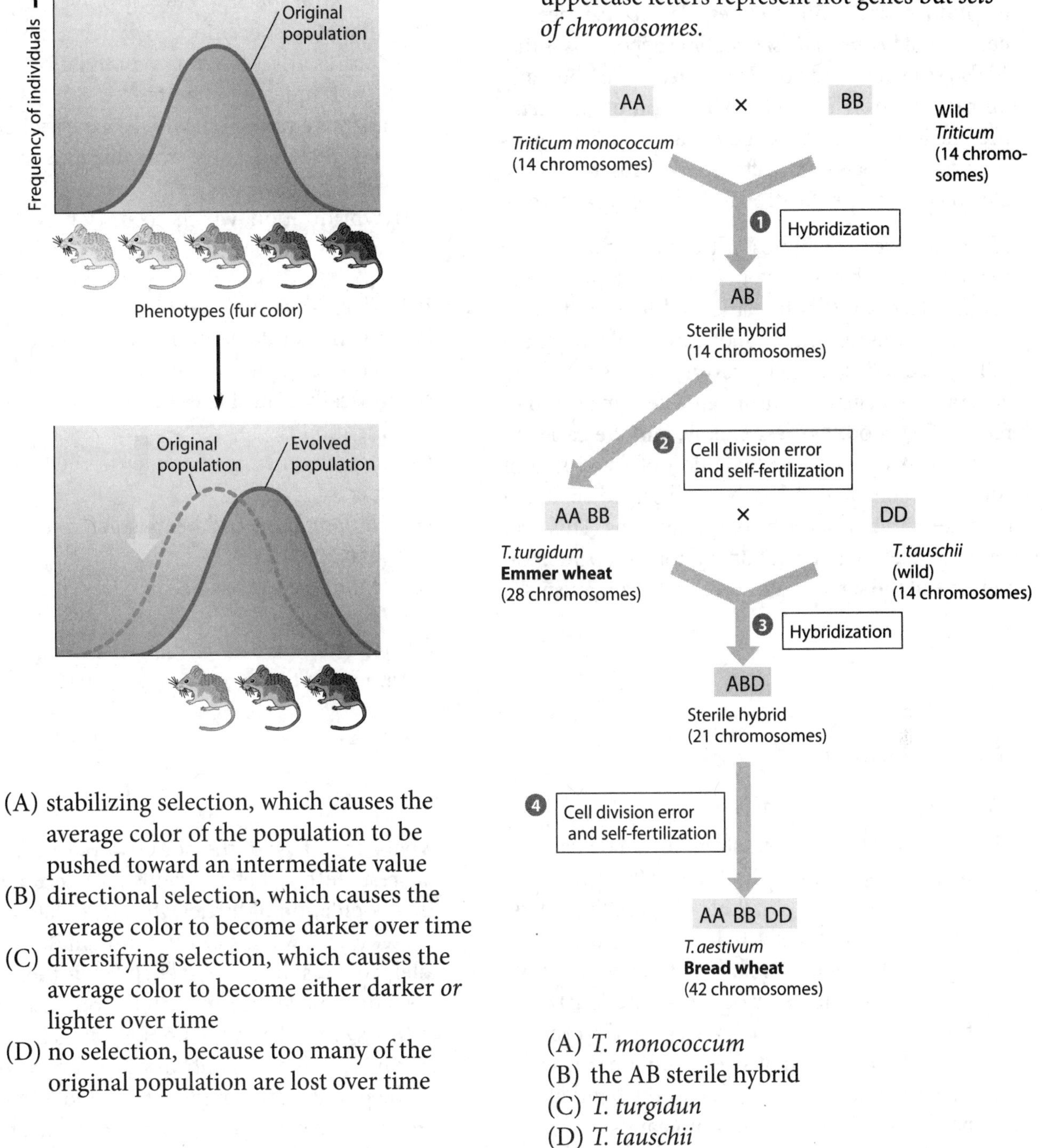

Question 12:

(A) stabilizing selection, which causes the average color of the population to be pushed toward an intermediate value
(B) directional selection, which causes the average color to become darker over time
(C) diversifying selection, which causes the average color to become either darker *or* lighter over time
(D) no selection, because too many of the original population are lost over time

Question 13:

(A) *T. monococcum*
(B) the AB sterile hybrid
(C) *T. turgidun*
(D) *T. tauschii*

After reading the paragraphs, answer the question(s) that follow.

In 2004, scientists announced the discovery of the fossil remains of some extremely short early humans on the Indonesian island of Flores. The new species has been named *Homo floresiensis*. One hypothesis is that *H. floresiensis* evolved from *Homo erectus*, another early human species. How did a population of *H. erectus* become isolated on this remote island? Early humans constructed boats and rafts, so perhaps they were blown far off course by strong winds during a storm.

H. erectus averaged almost 6 feet in height, but the remains show that adults of *H. floresiensis* were only about 3 feet tall. It is hypothesized that limited resources on this hot and humid island (only 31 square miles) exerted selection pressure and succeeding generations began to shrink in size. Small bodies require less food, use less energy, and are easier to cool than larger bodies. Evolution of small size in similar circumstances has been observed in many other species but never before in humans. This find demonstrates that evolutionary forces operate on humans in the same way as on all other organisms.

14. The evolution of *Homo floresiensis* is an example of
    (A) sympatric speciation, which occurs when a new species develops in the midst of an ancestral species, often by polyploidy.
    (B) allopatric speciation, in which a new species arises when two populations are geographically separated.
    (C) adaptive radiation, which occurs when new niches become available and are filled by new species.
    (D) hybridization, which occurs when two species interbreed.

15. If *H. floresiensis* were reunited with *H. erectus* at a much later date but the two populations could no longer interbreed, which would be a scientifically plausible conclusion about their incompatibility?
    (A) *H. floresiensis* is no longer fertile as a species.
    (B) *H. floresiensis* had been isolated for more than 50,000 years.
    (C) *H. floresiensis* has become less fit than *H. erectus*.
    (D) Reproductive barriers had evolved between the two species.

Grid-In Questions

These call for a numerical response.

1. The radioisotope potassium-40 can be used to date past events older than 60,000 years. Potassium-40 has a half life of 1.3 billion years, decaying into Argon-40. If the igneous rock layer that scientists wish to date shows a ratio of Potassium-40 to Argon-40 at one-fourth the current ratio, what is the age of the rock layer? Express your answer in billions of years.
2. In a population of king snakes the banded pattern (*B*) is dominant to no banding (*b*). If 12% of the population shows no banding, what percentage of the population, to the nearest tenth, is heterozygous for banding?
3. In a population of turtles, the allele that causes a yellow shell (*Y*) is dominant to the allele that results in a red shell (*y*). If the dominant allele is present in the population at the 0.72 level and the population is in Hardy-Weinberg equilibrium, what percent of the population would be expected to have a red shell? Express your answer to the nearest tenth of a percent.

Free-Response Question

1. *A number of experimental investigations have provided evidence that the conditions early in the Earth's history provided an environment capable of generating complex organic molecules and simple cell-like structures.*

   (a) Describe one scientific model for the origin of organic molecules on Earth.

   (b) Explain how RNA has the essential features of the earliest genetic material.

   (c) Predict the effect that introduction of free oxygen in the atmosphere had on the prokaryotic species of the time.

TOPIC 7

The Evolutionary History of Life

Phylogeny and the Tree of Life

(Biology, *10e: Chapter 26*, Focus, *1e: Chapter 20*)

YOU MUST KNOW

- The taxonomic categories and how they indicate relatedness.
- How systematics is used to develop phylogenetic trees.
- How to construct a phylogenetic tree that represents processes of biological evolution. (LO 1.13)
- The three domains of life, including their similarities and their differences.
- The significance of widely conserved processes across the three domains.

CONNECT WITH THE CURRICULUM FRAMEWORK

This chapter is a rich source of possible questions. Note these specific objectives, and be sure you can do each of these tasks.

- The student is able to pose scientific questions about a group of organisms whose relatedness is described by a phylogenetic tree or cladogram in order to (1) identify shared characteristics, (2) make inferences about the evolutionary history of the group, and (3) identify character data that could extend or improve the phylogenetic tree. (LO 1.17)
- The student is able to evaluate evidence provided by a data set in conjunction with a phylogenetic tree or a simple cladogram to determine evolutionary history and speciation. (LO 1.18)
- The student is able to create a phylogenetic tree or simple cladogram that correctly represents evolutionary history and speciation from a provided data set. (LO 1.19)

Phylogenies show evolutionary relationships (Biology, *10e: 26.1*, Focus, *1e: 20.1*)

- **Phylogeny** is the evolutionary history of a species or a group of related species. It is constructed by using evidence from **systematics**, a discipline that focuses on classifying organisms and their evolutionary relationships. Its tools include fossils, morphology, genes, and molecular evidence.

- **Taxonomy** is an ordered division of organisms into categories based on a set of characteristics used to assess similarities and differences.
- **Binomial nomenclature** uses a two-part naming system that consists of the **genus** to which the species belongs as well as the organisms' **species** within the genus, such as *Canis familiaris*, the scientific name of the common dog. This system was developed by **Carolus Linnaeus**.
- The hierarchical classification of organisms consists of the following levels, beginning with the most general or inclusive: **domain, kingdom, phylum, class, order, family, genus,** and **species**. Each categorization at any level is called a **taxon**.

> ***STUDY TIP*** Use a mnemonic device, such as "Dear King Phillip climbed over the fence and got shot," to remember the categories in order. Know that the degree of relatedness increases with each successive level down (in other words, the organisms share more common traits).

- Systematists use branching diagrams called **phylogenetic trees** to depict hypotheses about evolutionary relationships. The branches of such trees reflect the hierarchical classifications of groups nested within more inclusive groups. (See Figure 7.1.)
- Notice on Figure 7.1 that two branch points are numbered. Each number represents a *common ancestor* of the two branches. Can you mark the point where all carnivores shared a common ancestor?

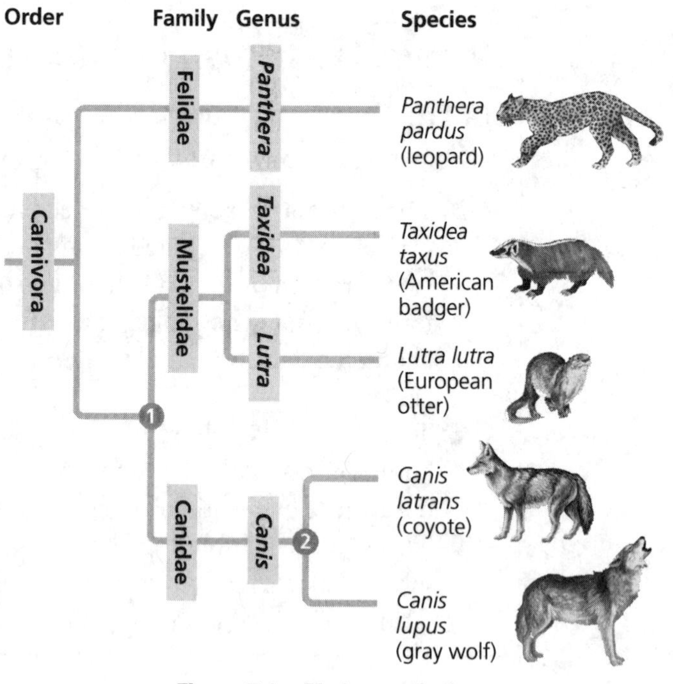

Figure 7.1 Phylogenetic tree

- The basics of how to read a phylogenetic tree are given in Figure 7.2. Carefully work through the explanations by Figure 7.2, then try the following.
  - It is very important for you to be able to work with phylogenetic trees! Here are two more things to do for practice.

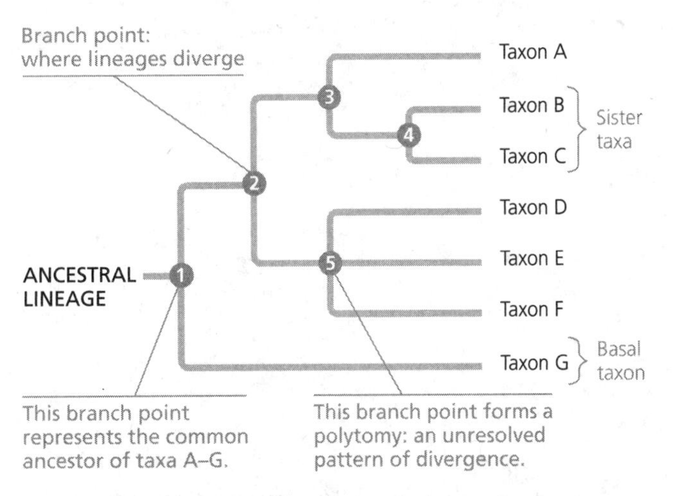

Figure 7.2 How to read a phylogenetic tree

1. Redraw the tree, rotating the branches around branch points 2 and 4. Does this change the evolutionary relationships of the taxa involved? The tree is shown redrawn in the answer section of your textbook (Figure 26.5 in *Biology*, 10e, and Figure 20.5 in *Focus*, 1e). Check your work!
2. Answer Concept Check 26.1, question 2, in *Biology*, 10e, or Concept 20.1, question 3 in *Focus*, 1e. Use the Answer section to check your understanding.

Phylogenies are inferred from morphological and molecular data (Biology, *10e: 26.2*, Focus, *1e: 20.2*)

- **Homologous structures** are similarities due to shared ancestry, such as the bones of a whale's flipper and a tiger's front limb.
- **Convergent evolution** has taken place when two organisms developed similarities as they adapted to similar environmental challenges—not because they evolved from a common ancestor. *Example*: The streamlined bodies of a tuna and a dolphin show convergent evolution.
- The likenesses that result from convergent evolution are considered **analogous** rather than homologous. They do not indicate relatedness but rather similar solutions to similar problems. *Example*: The wing of a butterfly is analogous to the wing of a bat. Both are adaptations for flight.
- **Molecular systematics** uses DNA and other molecular data to determine evolutionary relationships. The more alike the DNA sequences of two organisms, the more closely related they are evolutionarily.

Shared characters are used to construct phylogenetic trees (Biology, *10e: 26.3*, Focus, *1e: 20.3*)

- A **cladogram** depicts patterns of shared characteristics among taxa and forms the basis of a **phylogenetic tree**.
- A **clade**, within a tree, is defined as a group of species that includes an ancestral species and all its descendants. Clades are monophyletic.

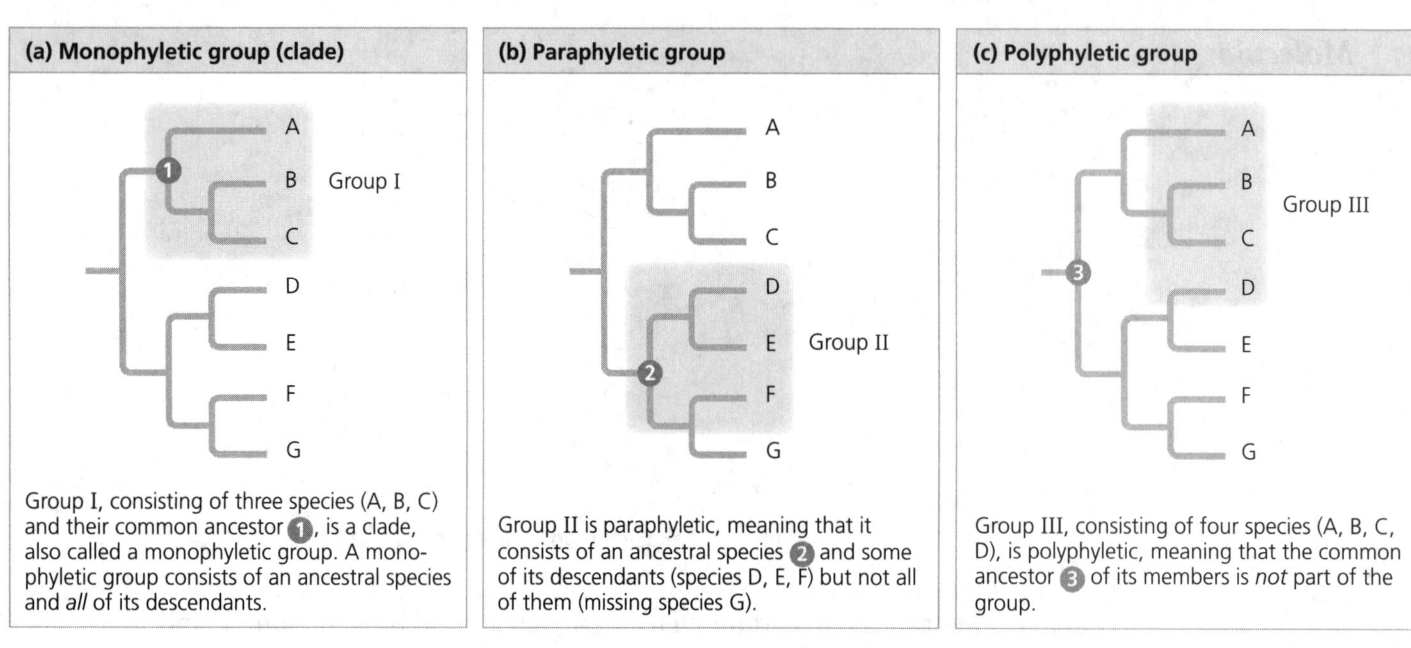

Figure 7.3 Cladograms

- **Shared derived characters** are used to construct cladograms. They are evolutionary novelties unique to a particular clade. For example, hair is a shared derived character of mammals.
- A **shared ancestral characteristic** is one that originated in an ancestor of the taxon. For example, all mammals have backbones, but a backbone does not distinguish a mammal from other vertebrates because all vertebrates have a backbone.
- Study Figure 7.3 to understand why a clade is *monophyletic*, and what is meant by paraphyletic and polyphyletic groupings. Each cladogram is a hypothesis about the evolutionary relatedness of the organisms included. The goal of taxonomy is to understand the lines of descent from ancestral forms well enough to produce monophyletic cladograms. This means the cladogram reflects the common ancestor and all of its descendants. Many times, however, scientists lack the data to make a monophyletic cladogram, resulting in paraphyletic or polyphyletic groupings. As data become available, the cladograms are improved until, hopefully, a monophyletic cladogram can be formed.
- The terms *monophyletic, paraphyletic*, and *polyphyletic* are not in the Curriculum Framework, but questions about these ideas are common. The terms help to organize your learning and will be ones your college professors will expect you to know.

An organism's evolutionary history is documented in its genome (Biology, *10e: 26.4*)

- The rate of evolution of DNA sequences varies from one part of the genome to another; therefore, comparing these different sequences helps us to investigate relationships between groups of organisms that diverged a long time ago.
  - DNA that codes for ribosomal RNA changes relatively *slowly* and is useful for investigating relationships between taxa that diverged hundreds of millions of years ago.
  - DNA that codes for mitochondrial DNA (mtDNA) evolves *rapidly* and can be used to explore recent evolutionary events.

Molecular clocks help track evolutionary time (Biology, *10e: 26.5*, Focus, *1e: 20.4*)

- **Molecular clocks** are methods used to measure the absolute time of evolutionary change based on the observation that some genes and other regions of the genome appear to evolve at constant rates. The underlying assumption for molecular clocks is that the number of nucleotide substitutions in related genes is proportional to the time that has elapsed since the genes branched from their common ancestor.

New information continues to revise our understanding of the tree of life (Biology, *10e: 26.6*, Focus, *1e: 20.5*)

- Taxonomy is in flux! When your authors were in high school, we were taught that there were two kingdoms, plants and animals; then in our college courses, we were introduced to five kingdoms: Monera, Protista, Plantae, Fungi, and Animalia.
- Now biologists have adopted a **three-domain** system, which consists of the domains Bacteria, Archaea, and Eukarya. This system arose from the finding that there are two distinct lineages of prokaryotes.
- The domains Bacteria and Archaea contain *prokaryotic* organisms, and Eukarya contains *eukaryotic* organisms. As we gain more tools for analysis, earlier ideas about evolutionary relatedness are changed, and so taxonomy, too, continues to evolve. That said, let's look at the three domains. Principal differences between the groups are simplified and presented below.

A Comparison of the Three Domains of Life

Characteristic	Bacteria	Archaea	Eukarya
Nuclear envelope	No	No	Yes
Membrane-enclosed organelles	No	No	Yes
Introns	No	Yes	Yes
Histone proteins associated with DNA	No	Yes	Yes
Circular chromosome	Yes	Yes	No

- Did you carefully study the chart? List three ways in which Bacteria and Archaea are similar.
  (Answer: Bacteria and Archaea lack a nuclear membrane, have no membrane bound organelles, and have a circular chromosome.)
- Archaea are more closely related to Eukarya than Bacteria. Provide evidence from the chart to justify this statement.
  (Answer: Archaea and Eukarya have introns as part of their gene complexes and have histones complexed with DNA to form a chromosome. Prokaryotes do not have these characteristics.)

> ***TIP FROM THE READERS***
> We have seen great emphasis placed on the importance of using data from several sources to develop cladograms, or using cladograms to infer relatedness in both multiple-choice and free-response questions. You will be glad you invested your time to work with cladograms and phylogenetic trees!

Bacteria and Archaea

(Biology, *10e: Chapter 27*, Focus, *1e: Chapter 24*)

YOU MUST KNOW

- The key ways in which prokaryotes differ from eukaryotes with respect to genome, membrane-bound organelles, size, and reproduction.
- How horizontal acquisition of genetic information occurs in prokaryotes via transformation, conjugation, and transduction,
- How these mechanisms plus mutation contribute to genetic diversity in prokaryotes. (EK 3.C.2)

Diverse structural and metabolic adaptations have evolved in prokaryotes (Biology, *10e: 27.1*, Focus, *1e: 24.2*)

- Life is divided into three domains: **Archaea, Bacteria, and Eukarya**. Both domain Bacteria and domain Archaea are made up of prokaryotes.
- Prokaryotes are perhaps 1/10 the size of a typical eukaryotic cell.
- As you read the following sections, study Figure 7.4. Prokaryotes have *no true nuclei* or internal compartmentalization. The DNA is concentrated in a *nucleoid region* and has little associated protein. The small genome consists of a single circular chromosome. Prokaryotes reproduce through an asexual process called **binary fission** and have short generation times.

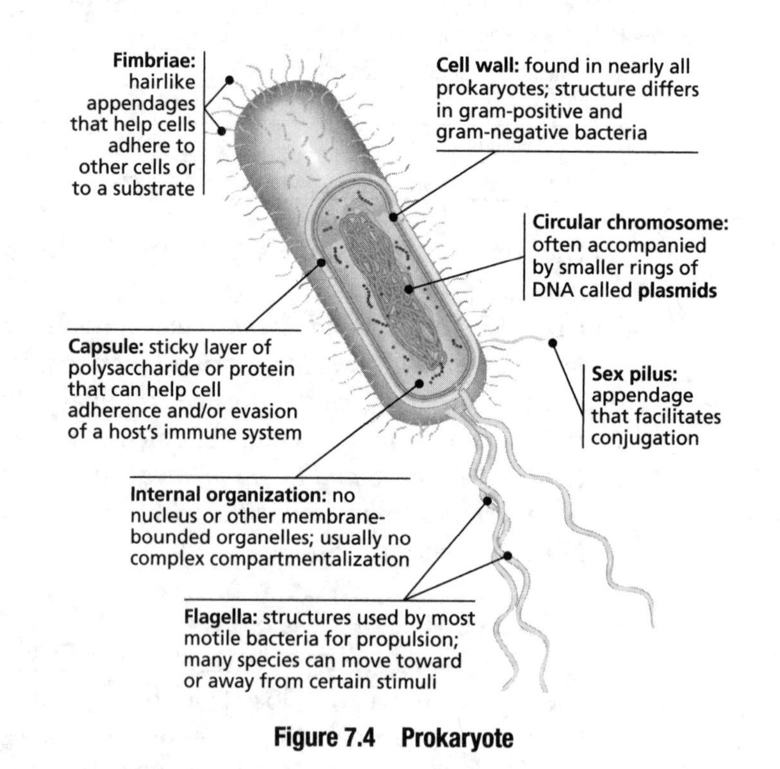

Figure 7.4 Prokaryote

- In addition to their one major chromosome, prokaryotic cells may also possess smaller, circular, self-replicating pieces of DNA called **plasmids**. In biotechnology plasmids are often used to carry the gene of interest into a prokaryotic cell.
- Outside their cell membranes, most prokaryotes possess a cell wall that contains **peptidoglycans**. (It is worth noting that cell walls are found in three kingdoms—plants have cell walls of cellulose, fungi have cell walls of chitin, and bacteria have cell walls of peptidoglycans.)
- Prokaryotes use appendages called **pili** that adhere to each other or to surrounding surfaces. About half of the prokaryotes are **motile**, because they possess whiplike **flagella**.
    - Because the flagellum of a bacterium is structurally different from the eukaryotic flagellum, this is another example of analogous structures.

Rapid reproduction, mutation, and genetic recombination promote genetic diversity in prokaryotes (**Biology,** ***10e: 27.2,*** **Focus,** ***1e: 24.3***)

- It is important for you to know that there are many ways for genetic information to be transferred between bacterial cells. In eukaryotes, the sexual processes of meiosis and fertilization combine DNA from two individuals. Meiosis and fertilization do not occur in prokaryotes. However, genetic recombination—the combining of DNA from two different sources—can occur in prokaryotes through the three mechanisms explained below. When the individuals are different species, this movement is called **horizontal gene transfer**, also referred to as **horizontal acquisition of genetic information**. Note that all three processes result in increased genetic variation without sexual reproduction.
- Three mechanisms by which bacteria can transfer genetic material between each other are:
    - **Transformation**, in which a prokaryote takes up DNA from its environment. The foreign DNA is integrated into the chromosome by the exchange of homologous DNA segments. Investigation 8 utilizes transformation to introduce a gene into *E. coli*. The investigation is reviewed in the lab section of this book on page 321.
    - **Transduction** is a process in which a bacteriophage (virus) transfers genes between one prokaryote and another. A random piece of DNA is accidently packaged into the head of the bacteriophage, which is then introduced to the next host during infection. Recombination may occur when the newly introduced DNA replaces its homologous section in the host cell's chromosome.
    - **Conjugation** occurs when genes are directly transferred from one prokaryote to another when they are temporarily joined by a "mating bridge." Once again recombination may occur when the newly introduced DNA replaces its homologous section in the host celles chromosome.
- Although **mutations** are rare, they are the major source of genetic variation in prokaryotes due to the short generation time and large population sizes of bacteria.

Diverse nutritional and metabolic adaptations have evolved in prokaryotes (Biology, *10e: 27.3*, Focus, *1e: 24.2*)

- Prokaryotes have diverse strategies for taking in carbon and obtaining energy. Some are photosynthetic, and others get their energy from organic and inorganic compounds.
- **Obligate aerobes** cannot grow without oxygen because they need oxygen for cellular respiration.
- **Obligate anaerobes** are poisoned by oxygen. Many obligate anaerobes live exclusively by fermentation.
- **Facultative anaerobes** use oxygen if it is available but can also carry out fermentation.
- In the early period of evolution on Earth, life-forms were anaerobic prokaryotes. Once oxygen began to accumulate in the atmosphere as a result of photosynthesis, many species of anaerobic prokaryotes became extinct.
- Some prokaryotes can use atmospheric nitrogen as a direct source of nitrogen in a process called **nitrogen fixation**. They convert N_2 to NH_4^+. You might want to review the nitrogen cycle, explained in Concept 55.4 in *Biology,* 10e, or Concept 42.4 in *Focus,* 1e.

Prokaryotes have radiated into a diverse set of lineages (Biology, *10e: 27.4*, Focus, *1e: 24.4*)

- Almost since their origin 3.5 billion years ago, prokaryotes have evolved in two separate lineages, the bacteria and the archaea. Although this chapter concentrates on the bacteria, the archaea are also important. The first prokaryotes that were classified in the domain Archaea are known as **extremophiles** and live in extreme environments.
  - **Extreme halophiles** live in saline environments (highly concentrated with salt) such as the Great Salt Lake.
  - **Extreme thermophiles** live in very hot environments, such as geyser pools.
- Other archaea do not live in extreme environments. **Methanogens**, found in deep sea vents as well as swamps, use carbon dioxide to oxidize H_2 and produce methane as a waste product.

Prokaryotes play crucial roles in the biosphere (Biology, *10e: 27.5*, Focus, *1e: 24.5*)

- Many prokaryotes are **decomposers,** breaking down dead organisms and waste products.
- Many prokaryotes are **symbiotic,** meaning that they form relationships with other species:
  - **Mutualism:** Both symbiotic organisms benefit.
  - **Commensalism:** One organism benefits, whereas the other is neither helped nor harmed.
  - **Parasitism:** One organism benefits at the expense of the other.

These ideas are further explored in Concept 54.1 in *Biology,* 10e, or Concept 41.1 in *Focus,* 1e.

Prokaryotes have both beneficial and harmful impacts on humans (Biology, *10e: 27.6*, Focus, *1e: 24.5*)

- Some prokaryotes are **pathogenic** and cause illness by producing poisons.
- **Antibiotics** are chemicals that can kill prokaryotes. They are *not* effective against viruses. Many bacterial plasmids contain resistance genes to different antibiotics.
- Prokaryotes are used by humans in many diverse ways:
  - **Bioremediation**, removing pollutants from soil, air, or water. This includes treating sewage, cleaning up oil spills, and precipitating radioactive materials.
  - Symbionts in the gut, manufacturing vitamins, and digesting foods.
  - Gene cloning and producing transgenic organisms.
  - Production of cheese and yogurt and other products.

Protists

(Biology, *10e: Chapter 28*, Focus, *1e: Chapter 25*)

WHAT'S IMPORTANT TO KNOW?
The Curriculum Framework does not include specific information that you must know about the taxonomy of Protists. However, this chapter contains information that may be used to develop both multiple-choice and essay questions. A general understanding of the Protists will make you more confident in dealing with potential questions.

Most eukaryotes are single-celled organisms (Biology, *10e: 28.1*, Focus, *1e: 25.3 and 25.4*)

- **Protist** is now a term used to refer to eukaryotes that are not plants, animals, or fungi. Biologists no longer consider Protista a kingdom because it is *paraphyletic*. Refer back to Figure 7.3 to review what this means. Some protists are more closely related to plants, fungi, or animals than to other protists.
- Protists vary in structure and function more than any other group of eukaryotes. Here are some general commonalities, but even these are not true for all groups:
  - Most are unicellular.
  - Most use aerobic metabolism and have mitochondria.
- There is evidence that the enormous diversity of protists has its origins in endosymbiosis. According to current theory, mitochondria and chloroplasts evolved through **endosymbiosis**. They were originally unicellular organisms engulfed by other cells that ultimately became organelles in the host cell.
- Protists can be divided into three categories: photosynthetic (plantlike) algae, ingestive (animal-like) protozoans, and absorptive (fungus-like) organisms.

- Most protists are aquatic and are important constituents of plankton. Many other protists live as symbionts in other organisms.
- Protists are such a diverse group that their classification continues to undergo revision. Despite this, you may recognize these protists:
    - *Giardia intestinalis* (causes "hiker's diarrhea"; always treat your water!)
    - *Trichomonas vaginalis* (sexually transmitted infection)
    - *Trypanosoma* species (sleeping sickness and Chagas' disease)
    - *Euglena* (remember seeing the tiny flagellated green cell with a red eyespot in Bio. I?)
    - Dinoflagellates (blooms cause "red tides"; many are bioluminescent)
    - *Plasmodium* (causative agent of malaria)
    - Ciliates (*Paramecium* and *Stentor* are examples; micro- and macronuclei)
    - Amoebas (move by pseudopodia)
    - Diatoms (two-part glass-like wall made of silica)
    - Golden algae (have yellow and brown carotenoids)
    - Brown algae (kelp)
    - Oomycetes (water molds and their relatives; include causative agent of potato blight)
    - Red algae (multicellular; some found at great depths; sushi wraps)
    - Green algae (*Clamydomonas, Ulva, Volvox*; this group is the closest relative of land plants)
    - Slime molds (often seen as flat, yellowish oozes in damp mulch)

How Plants Colonized Land

(Biology, *10e: Chapter 29*, Focus, *1e: Chapter 26*)

WHAT'S IMPORTANT TO KNOW?

Plant evolution is not an explicit topic of the Curriculum Framework, but your teacher may choose to introduce you to this material. This chapter contains information that may be used to develop both multiple-choice and essay questions. A general understanding of plant evolution will make you more confident in dealing with potential questions. Here are what we consider to be the main points of this chapter:

- Evidence that land plants have evolved from green algae.
- Disadvantages and advantages of life on land.
- The major evolutionary characteristics of bryophytes and seedless vascular plants.
- Plants have a unique life cycle termed *alternation of generations* with a gametophyte generation and a sporophyte generation.
- The typical life cycle pattern of plants, as demonstrated by ferns.

Land plants evolved from green algae (Biology, *10e: 29.1*, Focus, *1e: 26.1*)

- Land plants evolved from green algae more than 500 million years ago. Plants have enabled other life-forms to survive on land. Plants supply oxygen and are the ultimate provider of most of the food eaten or absorbed by animals and fungi.
- Movement onto land had both advantages and challenges:
  - Advantages included increased sunlight unfiltered by water, more carbon dioxide in the atmosphere than the water, soils rich in nutrients, and fewer predators.
  - Challenges included a lack of water, desiccation, and a lack of structural support against gravity.
- All land plants have a life cycle that consists of two multicellular stages, called **alternation of generations**.
- The two stages are the haploid **gametophyte,** which produces gametes, and the diploid **sporophyte,** which makes spores by meiosis.
  - During *fertilization*, egg and sperm fuse to form a diploid zygote, which divides by mitosis to form the sporophyte.
  - The zygote develops within the tissues of the female parent, deriving nutrients from it. For this reason land plants are sometimes referred to as embryophytes.

Mosses and other nonvascular plants have life cycles dominated by gametophytes (Biology, *10e: 29.2*, Focus, *1e: 26.3*)

- Bryophytes include three phyla: mosses, liverworts, and hornworts.
- The bryophytes are **nonvascular** plants (no xylem or phloem tissue). The lack of vascular tissue accounts for the small size of bryophytes.
- Unlike vascular plants, in all three bryophyte phyla the haploid gametophytes are the dominant stage of the life cycle. The organs that produce sperm and egg are found on the gametophytes. Bryophytes require water for the sperm to swim to the egg during fertilization.
- Bryophytes have the smallest, simplest sporophytes of all plant groups. Even though the sporophytes are photosynthetic when young, they must absorb water, sugars, and other nutrients from parental gametophytes.

Ferns and other seedless vascular plants were the first plants to grow tall (Biology, *10e: 29.3*, Focus, *1e: 26.3*)

- Besides ferns, this group includes two other plants you might commonly recognize, the club mosses and horsetails.
- The evolution of vascular tissues allowed vascular plants to grow taller than bryophytes, gaining access to sunlight. Ferns and other seedless vascular plants, however, still require a film of water for the sperm to reach the egg.
- In this group—as with all other groups of plants except mosses—the diploid sporophyte is the dominant stage. Meiosis produces haploid spores and occurs within the sporangia. The haploid spores may grow into gametophytes, where swimming sperm fertilize eggs, yielding the diploid zygote, which will grow into the sporophyte generation.

- The details of the fern life cycle in Figure 7.5 are not as important as the patterns demonstrated. In all sexually reproducing organisms meiosis and fertilization are the key events of the life cycle. Note where both of these occur. Also note the alternation from haploid gametophyte to diploid sporophyte—a unique, singular character of the plant life cycle known as alternation of generations.
- Seedless vascular plants formed forests of tall plants in the Carboniferous period, eventually forming deposits of coal that we use for fuel today.

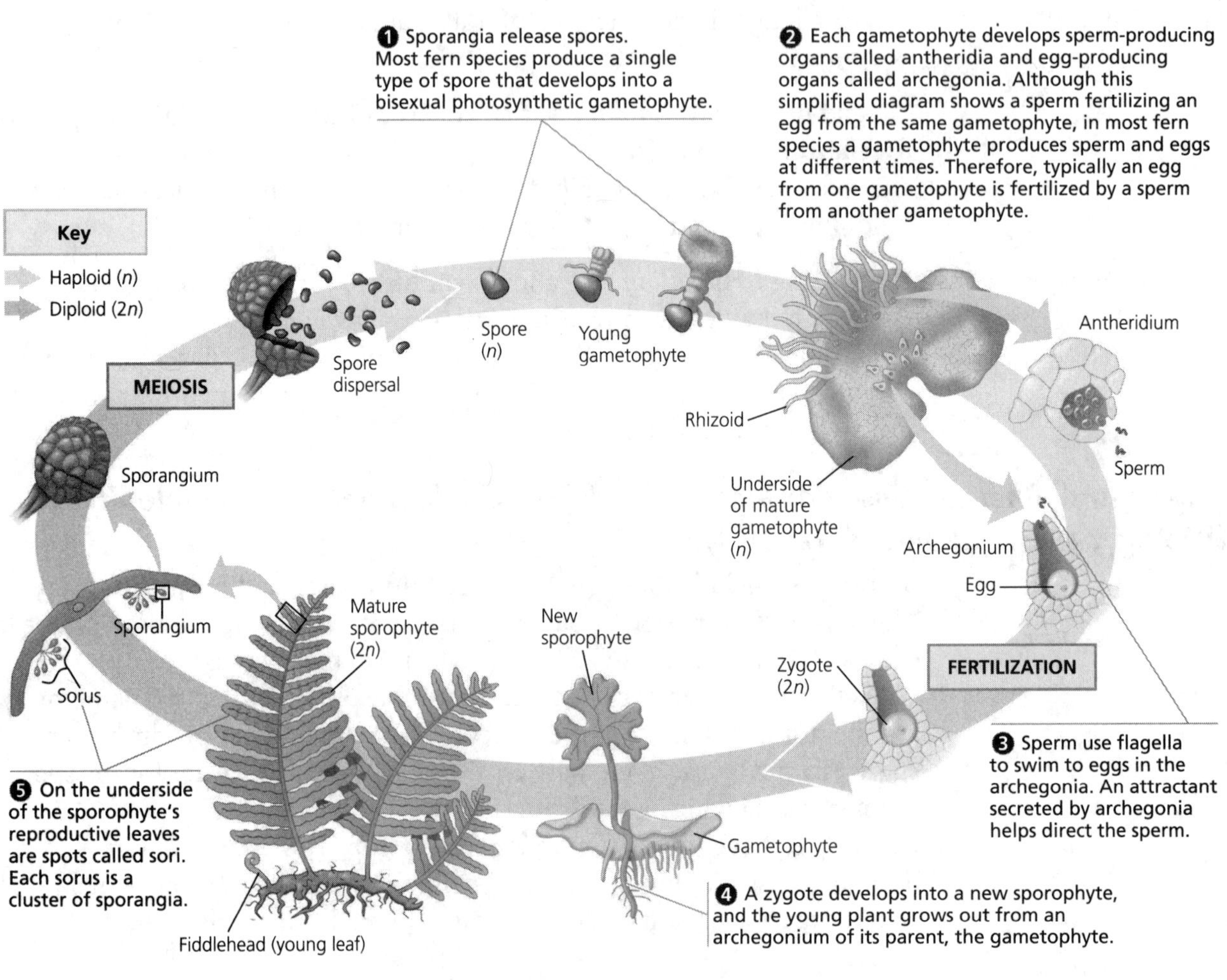

Figure 7.5 Fern life cycle

The Evolution of Seed Plants

(Biology, *10e: Chapter 30*, Focus, *1e: Chapter 26*)

> **WHAT'S IMPORTANT TO KNOW?**
> Plant evolution is not an explicit topic of the Curriculum Framework, but your teacher may choose to introduce you to this material. This chapter contains information that may be used to develop both multiple-choice and essay questions and a general understanding of plant evolution will make you more confident in dealing with potential questions. Here are what we consider to be the main points of this chapter:
>
> - The evolutionary significance of seeds and pollen as key adaptations to life on land.
> - The evolutionary significance of flowers and fruits in angiosperm reproduction.

Seeds and pollen grains are key adaptations for life on land (Biology, *10e: 30.1*, Focus, *1e: 26.4*)

- **Seeds** are plant embryos packaged with a food supply in a protective coat. Mature seeds are dispersed from their parent by wind or other means.
- Seeds have several advantages over spores. Seeds are multicellular, with several layers of protective tissue, safeguarding the embryo. Unlike spores, seeds have a supply of stored energy, which allows the seed to wait for good germination conditions and use stored energy to finance the early growth of the embryo.
- *Pollen and production of sperm.* A pollen grain is a male gametophyte, containing two sperm nuclei. The pollen grain has a waterproof coating, allowing for transfer by the wind. Until pollen, water was required for sperm transfer. *The evolution of pollen was a key adaptation to land.*

Gymnosperms bear "naked" seeds, typically on cones (Biology, *10e: 30.2*, Focus, *1e: 26.4*)

- Gymnosperms are plants that have "naked" seeds that are not enclosed in ovaries. Their seeds are often exposed on modified leaves that form cones. To compare, angiosperms (flowering plants) have seeds enclosed in fruits, which are mature ovaries. Gymnosperms do not have fruits.
- Four phyla of plants are considered gymnosperms, but the most ecologically significant group is the conifers (*Coniferophyta*), which include pines, spruces, firs, and redwoods.

The reproductive adaptations of angiosperms include flowers and fruits (Biology, *10e: 30.3*, Focus, *1e: 26.4*)

- **Angiosperms** are seed plants that produce the reproductive structures called flowers and fruits. Today, angiosperms account for about 90% of all plant species.
- The **flower** is a unique angiosperm structure specialized for sexual reproduction. The flower greatly improved the efficiency of pollination compared to

gymnosperms, which rely entirely on the wind for pollen dispersal. Flowers attract insects and other animals, which transfer pollen from the *anthers* (male structures)of one flower to the ***carpels* (female structures)of another flower, resulting in fertilization. However, not all flowers attract pollinators. Some, like the grasses, produce large quantities of wind-borne pollen.**

- **Fruits** are mature ovaries of the plant. As seeds develop from ovules after fertilization, the wall of the ovary thickens to become the fruit. Fruits help disperse the seeds of angiosperms.
- Angiosperms have traditionally been divided into monocots and eudicots:
    - **Monocots** (about 70,000 species) have one cotyledon in the seed, parallel leaf venation, and flowering parts in multiples of threes. Examples include orchids, lilies, and grasses.
    - **Eudicots** (about 170,000 species) have two cotyledons in the seed, net leaf veination, and flowering parts usually in multiples of fours or fives. Examples include roses, peas, beans, and oaks.

Fungi

(Biology, *10e: Chapter 31*, Focus, *1e: Chapter 26*)

WHAT'S IMPORTANT TO KNOW?
The study of fungi is not an explicit topic of the Curriculum Framework, but your teacher may choose to introduce you to this material. This chapter contains information that may be used to develop both multiple-choice and essay questions and a general understanding of fungi will make you more confident in dealing with potential questions. Here are what we consider to be the main points of this chapter:

- The characteristics of fungi.
- Important ecological roles of fungi including mycorrhizal associations and as decomposers and parasitic plant pathogens.

Fungi are heterotrophs that feed by absorption (Biology, *10e: 31.1*, Focus, *1e: 26.2*)

- Fungi are eukaryotes with these characteristics:
    - **Multicellular heterotrophs** that obtain nutrients by **absorption**. Fungi secrete hydrolytic enzymes, digest food outside their bodies, and absorb the small molecules. Note that this is an example of extra-cellular digestion.
    - The cell walls of fungi are made of *chitin,* a polysaccharide that also makes up the exoskeleton of arthropods.
    - Bodies are composed of filaments called **hyphae** that are entwined to form a mass, the **mycelium**.
    - Most fungi are multicellular with hyphae divided into cells by cross-walls called **septa**. *Coenocytic* fungi lack septa and consist of a continuous cytoplasmic mass containing hundreds of nuclei.

- Fungi reproduce by spores.
- Modes of nutrition include decomposers, parasites, and mutualists.

Fungi play key roles in nutrient cycling, ecological interactions, and human welfare (Biology, *10e: 31.5*, Focus, *1e: 26.2, 26.5*)

- Fungi are important decomposers of organic material, including cellulose and lignin.
- **Mycorrhizal fungi** are found in association with plant roots and may improve delivery of minerals to the plant, while being supplied with organic nutrients. This is a fine example of mutualism.
- **Lichens** are symbiotic associations of photosynthetic microorganisms (algae) embedded in a network of fungal hyphae. They are very hardy organisms that are pioneers on rock and soil surfaces.
- Thirty percent of known species of fungi are parasites. Many are plant pathogens (for example, Dutch elm disease, chestnut blight, dogwood anthracnose, wheat rust, and ergot). Some infect animals (for example, ringworm, athlete's foot, and *Candida*).
- **Yeasts** are unicellular fungi that figure prominently in molecular biology and biotechnology. Current research includes work into the genes involved in Parkinson's disease and Huntington's disease by examining homologous genes in yeast (*Saccharomyces cerevisiae*).

The Rise of Animal Diversity

(Biology, *10e: Chapters 32, 33, 34*, Focus, *1e: Chapter 27*)

WHAT'S IMPORTANT TO KNOW?

Animal diversity is not an explicit topic of the Curriculum Framework, but your teacher may choose to introduce you to this material. This chapter contains information that may be used to develop both multiple-choice and essay questions and a general understanding of animal diversity will make you more confident in dealing with potential questions. Here are what we consider to be some main points of this topic:

- The unique characteristics of animals.
- Important events in evolution of animals with respect to symmetry and the development of a body cavity.
- Adaptations that made life on land possible, including amniote eggs.

Animals are multicellular, heterotrophic eukaryotes with tissues that develop from embryonic layers (Biology, *10e: 32.1*, Focus, *1e:* Overview, *27.1*)

- Animals have the following characteristics:
  - They are multicellular heterotrophs.
  - Most have muscle and nervous tissue.

- Most reproduce sexually, with a flagellated sperm and a large egg uniting to form a diploid *zygote*. The diploid stage dominates the life cycle.

- Some animals have **larvae**, an immature form distinct from the adult stage that will undergo **metamorphosis**.
- Animals share *Hox* genes, a unique homeobox-containing family of genes that plays important roles in development.

Animals can be characterized by "body plans" (Biology, *10e: 32.3*, Focus, *1e: Overview, 27.3*)

- **No symmetry:** the sponges.
- **Radial symmetry** occurs in jellyfish and other organisms, in which any cut through the central axis of the organism would produce mirror images.
- **Bilateral symmetry** occurs in lobsters, humans, and many other organisms. These animals have a right side and a left side, and a single cut would divide the animal into two mirror image halves. There is also a *dorsal* (back) side, a *ventral* (belly) side, and *anterior* (head) and *posterior* (tail) ends.
- **Cephalization** is the concentration of sensory equipment at one end (usually the anterior, or head end) of the organism.
- A **coelom** is a fluid or air-filled space located between the digestive tract and the outer body wall (also referred to as the *body cavity*). Body cavities have several functions, including cushioning suspended organs; acting as a hydrostatic skeleton; and enabling internal organs to grow and move independently.
- Evolutionary relationships among living animals provide a useful framework in studying the rise of animals.

1. All animals share a common ancestor; that is, animals are a monophyletic clade.
2. Sponges are basal animals; that is, sponges branch from the base of the animal tree.
3. Most animals belong to the clade Bilateria. Bilateral symmetry is a shared derived character for this clade, which includes the majority of animal phyla.
4. Most animal phyla are invertebrates, animals that lack a backbone. Only one animal phylum, Chordata, includes vertebrates, such as fish, amphibians, reptiles, and mammals.

The Invertebrates

- **Phylum Porifera** (pore-bearing animals) includes the sponges.

  - Sponges lack true tissues and are *diploblastic* (their body consists of only two layers of cells). They are basal animals forming the base of the animal tree.
  - Sponges have no nerves or muscles. The body of a sponge looks like a sac with holes in it. Water is drawn in through the pores and flows out through the top through the movement of flagellated cells, circulating nutrients and removing wastes.

- **Phylum Cnidaria** includes hydras, jellyfish, and corals. All have radial symmetry, a central digestive compartment known as a gastrovascular cavity, tentacles to capture food, and nerves and muscles in their simplest form.
- **Phylum Platyhelminthes** (flatworms) includes planarians, tapeworms, and flukes. This group also has a gastrovascular cavity and no coelom. Their bodies are thin and flat, providing sufficient surface area for gas exchange and circulation without special organs. There are sensory cells clustered in the anterior end, which is called *cephalization.* This group and all other animal phyla have **bilateral symmetry,** placing them in the clade Bilateria.
- **Phylum Nematoda** (roundworms) includes pinworms, hookworms, and the model organism, *C. elegans.* These small, cylindrical worms are *pseudocoelomates,* which means they have a body cavity but it is lined with mesoderm on only one side.
- **Phylum Mollusca** includes slugs, clams, snails, squids, and octopuses. They have a true coelom (a body cavity lined on both sides with mesoderm) as do all the phyla that follow. They use a muscular "foot" for locomotion.
- **Phylum Annelida** is the segmented worms and includes earthworms and leeches. They show internal and external segmentation and have a digestive system with specialized regions (the crop, gizzard, esophagus, and intestine).
- **Phylum Arthropoda** includes insects, arachnids, millipedes, centipedes, and crustaceans. They all have an *exoskeleton* of *chitin*, which must be *molted* to grow, and jointed appendages.
- Insects undergo metamorphosis during development.
    - **Incomplete metamorphosis:** egg, nymph, adult (for example, grasshoppers)
    - **Complete metamorphosis:** egg, larva, pupa, adult (for example, butterflies)

Clade Deuterostomia

The clade Deuterostomia contains a diverse array of organisms, from sea stars to chordates. All have radial cleavage during embryonic development. You are a chordate . . . so what other animals are deuterostomes and therefore more closely related to you than grasshoppers? Read on.

- **Phylum Echinodermata** includes starfish, brittle stars, sea urchins, sand dollars, sea lilies, and sea cucumbers. Echinoderm larvae have bilateral symmetry but adults radiate from the center, often as five spokes.
- **Phylum Chordata** includes two invertebrate subphyla as well as all vertebrates.
    - There are four shared derived characteristics of all chordates. Note that some of these are present only during embryonic development (Figure 7.6).

        1. A **notochord**—a long, flexible rod that provides skeletal support appears during embryonic development between the digestive tube and the dorsal nerve cord. This is *not* the spinal cord or the vertebral column! Students commonly confuse all of these.

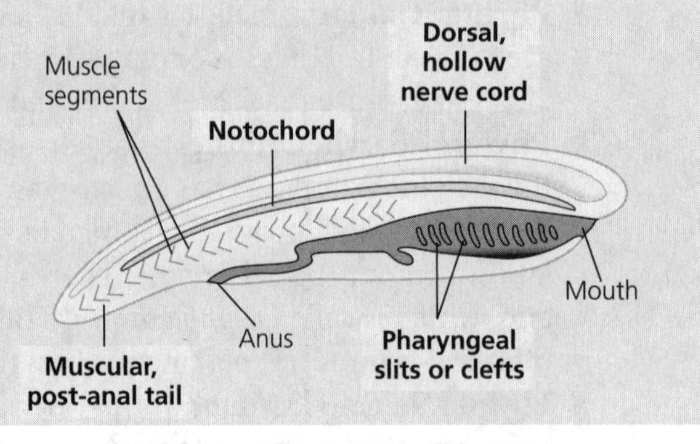

Figure 7.6 Chordate characteristics

2. A **dorsal, hollow nerve cord**—formed from a plate of ectoderm that rolls into a hollow tube and develops into the brain and spinal cord. Other animals have a *ventral* nerve cord.
3. **Pharyngeal slits**—grooves that separate a series of pouches along the sides of the pharynx. In most chordates (but not tetrapods, the critters with four legs), the slits allow water to enter and exit the mouth without going through the digestive tract.
4. A muscular **tail** posterior to the anus.

The Vertebrates

- **Vertebrates** derive their name from the **vertebrae,** a series of bones that make up the backbone. In the majority of vertebrates, vertebrae enclose the spinal cord and have assumed the roles of the notochord.
- Vertebrates are members of the phylum Chordata.
- Let's look at features of several groups of vertebrates.
  - **Chondrichthyes** have flexible endoskeletons composed of cartilage, possess streamlined bodies, are denser than water, and will sink if they stop swimming. Some examples are sharks and rays (such as stingrays).
  - **Osteichthyes** are the bony fishes; these are the most numerous of all vertebrate groups. They have a bony endoskeleton, are covered in scales, and possess a swim bladder for buoyancy control. Some examples are trout and salmon.
  - **Amphibians** include frogs, toads, salamanders, and newts. Some gas exchange occurs across their thin, moist skin, although many members have lungs. Some, such as frogs, have an aquatic larval stage with gills and metamorphosis to an adult stage with lungs. All have external fertilization and external development in an aquatic environment.
  - **Reptiles** include turtles, lizards and snakes, alligators and crocodiles, birds, and the extinct dinosaurs. Modern reptiles have scales containing keratin, which are an adaptation for terrestrial living because they

reduce water loss. They also have *amniote eggs,* which is discussed below. Most reptiles are *ectothermic,* regulating body temperature through behavioral adaptations rather than by metabolism. **Birds** are *endothermic* (warm-blooded) and have *feathers* and a *four-chambered heart.*

- **Mammals** share certain characteristics:
  - All possess **mammary glands** that produce milk and have a body covering of **hair.** These two characteristics are excellent examples of shared, derived characteristics.
  - They have a four-chambered heart and are **endothermic.** (The four-chambered heart of some reptiles, including birds, is analogous—not homologous—to the four-chambered heart of mammals.)
  - Mammals have internal fertilization, and most are born rather than hatched.
- Mammals can be placed into three groups:
  1. **Monotremes** are egg-laying mammals that have hair and produce milk. Examples are platypuses and spiny anteaters.
  2. **Marsupials** are born early in development and complete embryonic development in a marsupium (pouch) while nursing. Examples include kangaroos and opossums.
  3. **Eutherians** (placental mammals) have a longer period of pregnancy; completing their development in the uterus. Examples include mice, dogs, and humans.
- Humans belong to the order **Primates,** along with monkeys and gorillas. Some characteristics common to all **primates** include opposable thumbs, large brains and short jaws, forward-looking eyes, flat nails, well-developed parental care, and complex social behavior.
- One important evolutionary development that allowed extensive colonization of dry habitats was the **amniotic egg,** which has a shell that retains water and so can be laid in a dry environment.
- The clade of **amniotes** consists of mammals and reptiles (including birds).
- Amniotic eggs have four specialized **extraembryonic membranes:**
  - The **amnion** encloses a fluid compartment that bathes the embryo and absorbs shock.
  - The **chorion**, **allantois**, and **yolk sac** function in gas exchange, waste storage, and transfer of stored nutrients.

Level 1: Knowledge/Comprehension Questions

The Knowledge/Comprehension questions review essential content knowledge but don't represent the type of question you will see on the AP exam. Level 2, Application/Synthesis, questions are similar to those you will see on the exam.

1. Which of the following groups is best characterized as being eukaryotic, multicellular, heterotrophic, and without a cell wall?
   (A) Plantae
   (B) Animalia
   (C) Fungi
   (D) Viruses

2. Systematists categorize all living creatures into what three domains?
   (A) Bacteria, Animalia, Eukarya
   (B) Bacteria, Archaea, Eukarya
   (C) Archaea, Plantae, Eukarya
   (D) Protista, Plantae, Eukarya

3. Which of the following is a symbiotic relationship in which both organisms benefit?
   (A) parasitism
   (B) commensalism
   (C) mutualism
   (D) predation

4. Place the following groups of plants in order beginning with those that first appeared on Earth.
   (A) moss, angiosperms, ferns, gymnosperms
   (B) moss, ferns, gymnosperms, angiosperms
   (C) moss, ferns, angiosperms, gymnosperms
   (D) seed plants, cone-bearing plants, bryophytes

5. Evolution of which feature enabled vertebrates to reproduce successfully on land?
   (A) the amniote egg
   (B) origin of limbs with digits
   (C) flagellated sperm
   (D) keratinized body covering

6. Which statement is not supported by the cladogram?

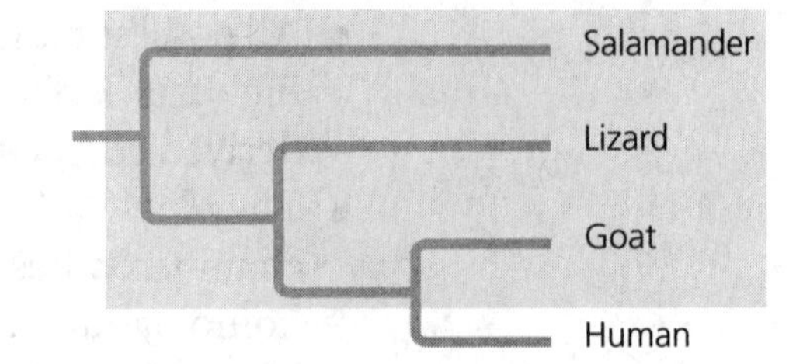

   (A) The lineage leading to salamanders was the first to diverge from the other lineages.
   (B) Humans are more closely related to goats than lizards.
   (C) Salamanders are as closely related to goats as to humans.
   (D) Lizards are more closely related to salamanders than to humans.

7. Genetic variation in bacterial populations is introduced in several ways. Select the one process that will not introduce genetic variation in bacteria.
   (A) transduction
   (B) transformation
   (C) conjugation
   (D) meiosis

8. Shared derived characters are used to develop cladograms. Which is a correct description of them?
   (A) They are features that characterize all the species on a branch of a phylogenetic tree.
   (B) They can be determined through a BLAST comparison of related genes in a group of species.
   (C) They are homologous structures that develop during adaptive radiation.
   (D) They can be used to identify species but not higher taxa.

Level 2: Application/Analysis/Synthesis Questions

1. In a comparison of birds and mammals, having four appendages is
   (A) a shared ancestral character because it originates in an ancestor of the taxa.
   (B) a shared derived character because it is shared by all mammals.
   (C) a character useful for distinguishing birds from mammals.
   (D) an example of analogy rather than homology.

2. An outgroup is a species from an evolutionary lineage that is known to have diverged before the lineage under study. If you were using cladistics to build a phylogenetic tree of cats, which of the following would be the best outgroup?
   (A) lion
   (B) domestic cat
   (C) wolf
   (D) leopard

3. In the 1920s, Frederick Griffith conducted an experiment in which he mixed the dead cells of a bacterial strain that can cause pneumonia with live cells of a bacterial strain that cannot. When he cultured the live cells, some of the daughter colonies proved able to cause pneumonia. Which of the following processes of bacterial DNA transfer does this experiment demonstrate?
   (A) transduction
   (B) conjugation
   (C) transformation
   (D) transposition

4. Molecular data can be used to assess relationships among the major groups of living organisms whose common ancestors lived millions or billions of years ago. Similar techniques can be used to assess relationships among populations within a species. How can molecular techniques be useful for such varied comparisons?
   (A) Studying the relationships among different populations of a single species can be just as effective as studying the relationships of major biological groups if you look at many more genes.
   (B) The same data can be used for any comparison with equal accuracy.
   (C) Faster-evolving gene sequences provide better data for comparisons among close relatives, whereas very slowly evolving sequences work best for distantly related taxa.
   (D) The relationships between very different groups such as bacteria and whales cannot be assessed.

After reading the paragraph, answer the question that follows.

The first fossil of *Archaeopteryx*, which lived about 150 million years ago, was found in the Solnhofen Quarry in Germany. *Archaeopteryx* has an interesting collection of characteristics that led to the hypothesis that it represented an evolutionary transition between modern-day birds and small bipedal dinosaurs. The fossil reveals the imprint of feathers, which connect *Archaeopteryx* to birds, although they don't present direct evidence of flight.

Unlike birds, however, the fossil record reveals that *Archaeopteryx* had teeth, functional claws on the wings that may have been used for climbing trees or holding prey, and a long, bony tail. Birds have a fused collarbone, and this was found in *Archaeopteryx* also. However, there were differences in the structure of the sternum. In birds, the sternum is keeled (raised and slightly concave). The keel of the sternum serves as an attachment site for the flight muscles. *Archaeopteryx* had a flat sternum, similar to that found in reptiles.

5. If you were constructing a phylogenetic tree for the evolution of birds, which characteristics found in *Archaeopteryx* might provide evidence that birds and dinosaurs had a common ancestor?
   (A) feathers, wings, wishbone
   (B) teeth, feathers, keeled sternum
   (C) teeth, flat sternum, claws
   (D) keeled sternum, claws, long forelimbs

6. An experimental forest ecosystem is enclosed in a sealed greenhouse. The entire ecosystem, including the air and soil, is treated with an extremely potent fungicide that kills all fungal life stages including spores. What will probably happen next?
   (A) Tree growth will increase because the dead fungi will act as a fertilizer.
   (B) Plants will enjoy a long-term increase in growth and survival because of the removal of fungal pathogens.
   (C) Dead organic matter will accumulate on the forest floor; plant growth will decline because of a lack of nutrients and the loss of mycorrhizal partners.
   (D) Certain animal populations may decrease due to loss of their fungal food sources, but otherwise the forest will be largely unchanged.

7. The movement of life onto land required numerous adaptations to obtain water, prevent water loss, and reproduce. Plants have a waxy covering, known as cuticle on their leaves that helps them conserve water. Seeds help several plant groups to reproduce effectively on dry land by protecting the embryo from water loss. Which of these pairs correctly shows a trait that prevents water loss and a trait that protects the embryo?
   (A) Reptiles have keratinized scales and produce amniote eggs.
   (B) Mammals have fur or hair and internal fertilization.
   (C) Amphibians have a moist skin and external fertilization.
   (D) Roundworms have a cuticle and are hermaphroditic.

8. This is a figure in which viruses that infect bacteria carry genes from one host cell to another. There is a process in sexually reproducing eukaryotes that uses enzymes similar to the ones that result in the incorporation of the A^+ gene into the host genome shown in the figure. Which of these represents the similar process in eukaryotes?

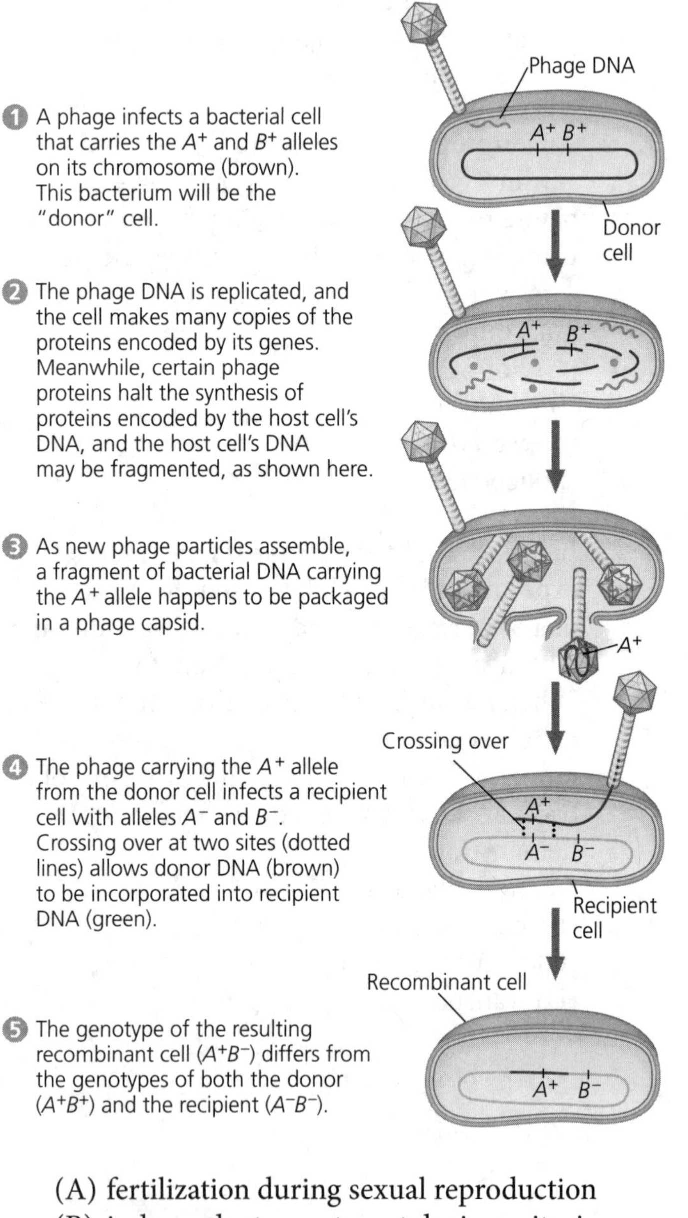

   (A) fertilization during sexual reproduction
   (B) independent assortment during mitosis
   (C) transduction during bacterial recombination
   (D) crossing over during meiosis

9. Monophyletic clades include a common ancestor and all its descendants. According to this figure, which grouping represents a monophyletic clade?

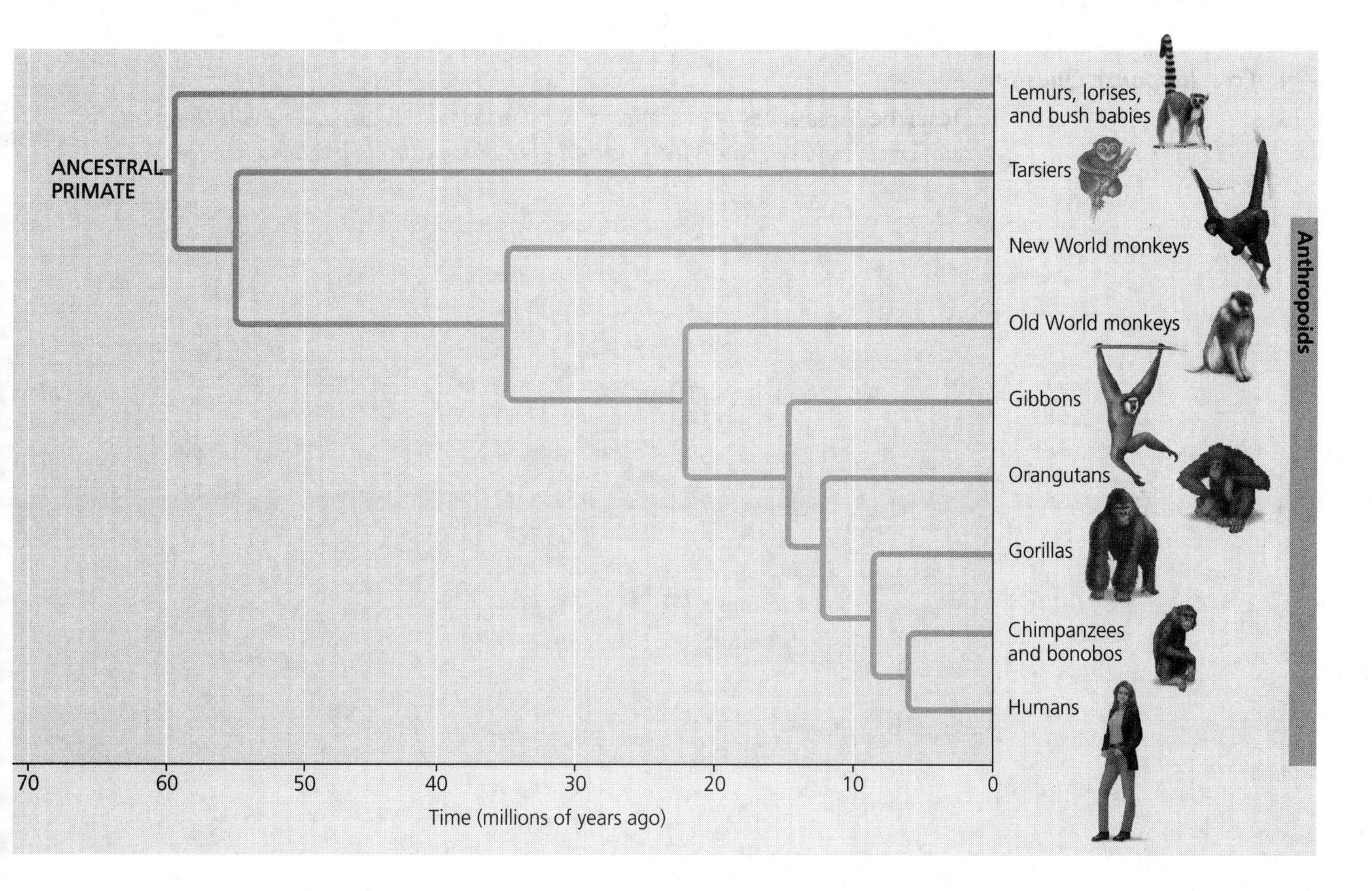

(A) lemurs, tarsiers, New World monkeys, and Old World monkeys
(B) New World monkeys and Old World monkeys
(C) gibbons, orangutans, gorillas, and chimpanzees
(D) gorillas, chimpanzees, and human

10. Which of these relationships is depicted on this cladogram?
(A) Humans have descended from chimpanzees.
(B) Gibbons are the common ancestor of orangutans, gorillas, chimpanzees, and humans.
(C) Old World monkeys and New World monkeys diverged approximately 35 million years ago.
(D) Gorillas are equally related to orangutans and chimps.

Grid-In Question

This calls for a numerical response.

1. Using the figure for question 9, approximately how long ago did the last common ancestor for gibbons, orangutans, gorillas, chimpanzees, and humans diverge?

Free-Response Question

1. **Describe** *three types of evidence that scientists use to discover evolutionary relationships between organisms and give* **one** *example of each.*

TOPIC 8

Plant Form and Function

Plant Structure, Growth, and Development

(Biology, *10e: Chapter 35*, Focus, *1e: Chapter 28*)

WHAT'S IMPORTANT TO KNOW?
Much of plant anatomy is not an explicit topic of the Curriculum Framework, but your teacher may choose to use plant anatomy for illustrative examples. This chapter contains information that may be used to develop both multiple-choice and essay questions, and a general understanding of plant structures will make you more confident in dealing with potential questions. Here are what we consider to be the main points of this chapter:

- Plants grow only at meristems.
- How leaf anatomy relates to photosynthesis.
- The role of root hairs and mycorrhizae in resource acquisition.
- Roots, stems, and leaves interact in essential plant life functions. (EK 4.A.4)

***Plants have a hierarchical organization consisting of organs, tissues, and cells* (Biology, *10e: 35.1*, Focus, *1e: 28.1*)**

- Plants have a **root system** beneath the ground that is a multicellular organ that anchors the plant, absorbs water and minerals, and often stores sugars and starches. Additional structural characteristics of roots include the following:
  - At the tips of the roots vast numbers of tiny *root hairs*, which are extensions of root epidermal cells, increase the surface area enormously, making efficient absorption of water and minerals possible. Plants may also have a symbiotic relationship with fungi at the tips of the roots, termed **mycorrhizae** ("fungus roots"). Mycorrhizae assist in the absorption process and are found in the vast majority of all plants.
- Plants have a **shoot system** above the ground that works as a multicellular organ consisting of stems and leaves.
  - **Stems** function primarily to display the leaves.
  - **Leaves** are the main photosynthetic organ in most plants.

- Plant organs—leaf, stem, and root—are composed of tissue types:

> ***CONNECT WITH THE CURRICULUM FRAMEWORK***
>
> Surface-area-to-volume ratios affect an organism's ability to obtain necessary resources. Root hairs and mycorrhizae each play important roles in greatly enhancing the exposure of the plant to the substrate. Obtaining resources from the soil (water, minerals, organic products) is greatly enhanced through these extensive networks.

  - **Dermal tissue** is a single layer of closely packed cells that cover the entire plant and protect it against water loss (a waxy layer termed the **cuticle** in the leaves) and invasion by pathogens like viruses and bacteria.
  - **Vascular tissue** is continuous through the plant and transports materials between the roots and shoots. Vascular tissue is made up of **xylem**, cells that transport water and minerals up from the roots and are dead at maturity, and **phloem**, which transports sugars and other organic compounds from the leaves to the other parts of the plant.

Meristems generate cells for primary and secondary growth (Biology, *10e: 35.2–35.4*, Focus, *1e: 28.2*)

- **Meristems** are perpetually embryonic tissues. Unlike growth in animals, growth in plants occurs only as a result of cell division in the meristems.
  - **Apical meristems** are located at the tips of roots and in buds of shoots, allowing the plant's stems and roots to extend. This is *primary growth.* Note that this means elongating growth occurs only from the tips of the roots and shoots. If you place a basketball goal at 10 feet on a tree, and the tree grows 6 feet taller over several years, the basketball goal does not move—it stays at 10 feet. Growth only occurs from the tip or apical meristem; this is *primary growth.*
  - **Lateral meristems** result in growth that thickens the shoots and roots. This is termed *secondary growth.*
  - There are two lateral meristem tissues. The **vascular cambium** produces secondary xylem (wood). The **cork cambium** produces a tough covering that replaces epidermis early in secondary growth.
  - **Bark** is all the tissues outside the vascular cambium. Bark includes the living phloem derived from the vascular cambium. Almost all of the wood (xylem) of a large tree is dead when it functions. The living part of the trunk of a large tree is almost exclusively bark. Injuries to the bark may result in the death of the tree when it damages the phloem and interferes with movement of organic compounds through the plant.
- The formation of many cells differing in structure and function all arise from the same plant genome. Cell differentiation depends primarily on the control

of gene expression—just as it does in your body. The regulation of transcription and translation, resulting in different proteins, produces different cell types.

Tissue Organization of Leaves (Biology, *10e: 35.3*, Focus, *1e: 28.3*)

> *TIP FROM THE READERS*
> As you learn plant anatomy, such as the structure of the leaf, focus on *how* the structures enhance functions, such as gas exchange, photosynthesis, and reduction of water loss. The Curriculum Framework asks you to integrate knowledge!

- Refer to Figure 8.1 as you review this. The epidermis of the underside of the leaf is interrupted by **stomata**, which are small pores flanked by *guard cells*, which open and close the stomata. The stomata allow for gas exchange in the leaf: CO_2 necessary for photosynthesis can enter and O_2 and water vapor can escape the leaf. Loss of water from the leaf is part of the Curriculum Framework and is covered in the next chapter under transpiration.
- In Investigation 5, Photosynthesis, the disks float until you use a vacuum to eliminate the gases. Note the air spaces within the leaf where gases can accumulate and diffuse into and out of the cells of the leaf. As oxygen is produced in photosynthesis, these air spaces will refill and float the disks.

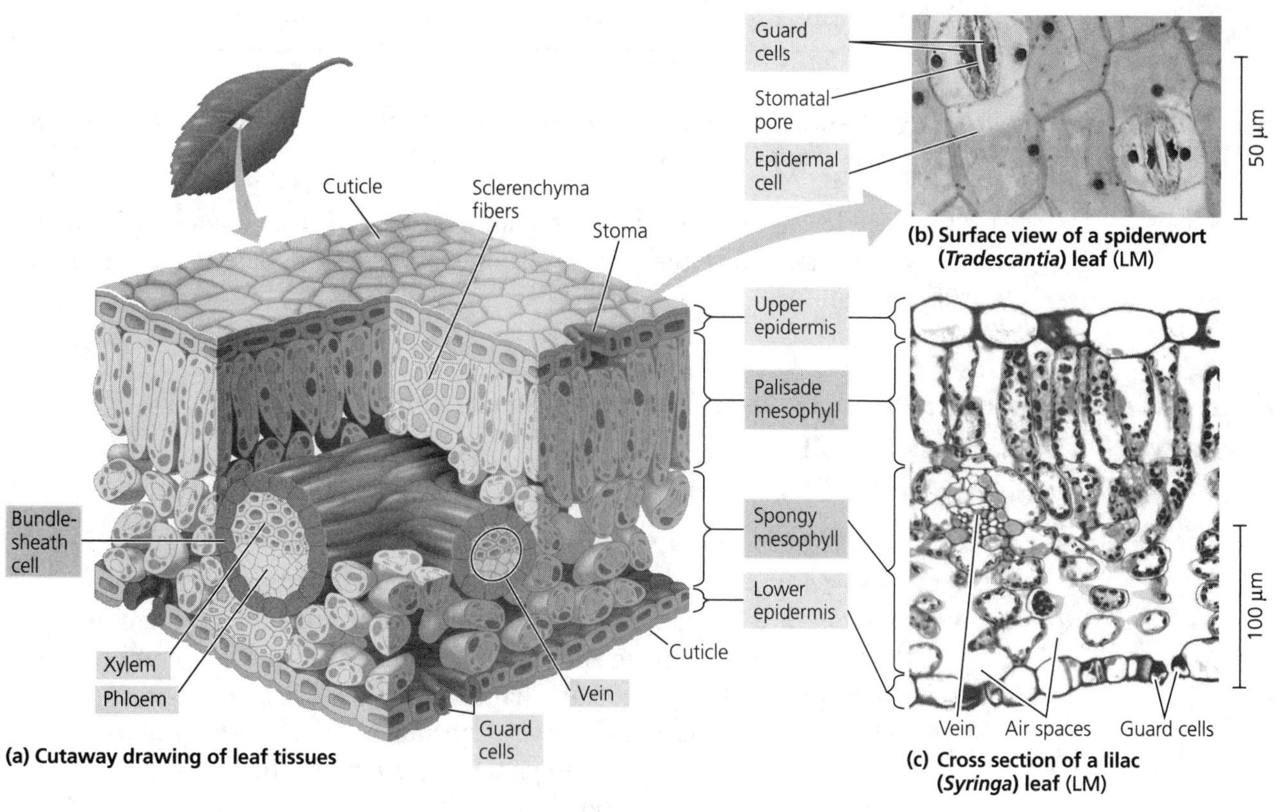

Figure 8.1 Leaf anatomy

Resource Acquisition and Transport in Vascular Plants

(Biology, *10e: Chapter 36*, Focus, *1e: Chapter 29*)

YOU MUST KNOW

- How passive transport, active transport, and cotransport function to move materials across plant cell membranes.
- The role of water potential in predicting movement of water in plants.
- How the transpiration cohesion-tension mechanism explains water movement in plants.
- How bulk flow affects movement of solutes in plants.
- Mechanisms by which plant cells communicate with other distant cells. (EK 3.D.2)

CONNECT WITH THE CURRICULUM FRAMEWORK

This chapter ties together concepts from Big Idea 2 and Big Idea 4. Information that you have learned about transport across membranes will be applied in movement of materials through plants. Focus on integrating your content and knowledge. If your class does AP Lab Investigation 11 on Transpiration, you will find this an important chapter.

Adaptations for acquiring resources were key steps in the evolution of vascular plants **(Biology, *10e: 36.1*, Focus, *1e: 29.1*)**

- Study Figure 8.2 where you will see an overview of resource acquisition and transport. Take your time with the figure because it has much important information.

Different mechanisms transport substances over short or long distances **(Biology, *10e: 36.2*, Focus, *1e: 29.2*)**

- Transport begins with the movement of water and solutes across a cell membrane.
  - Solutes diffuse down their electrochemical gradients. Electrochemical gradients are the combined effects of the concentration gradient of the solute and the voltage or charge differential across the membrane. (Chapter 7 in *Biology*, 10e or Chapter 5 in *Focus*, 1e)
  - If no energy is required to move a substance across the membrane, then the movement is termed *passive transport*. Diffusion is an example of passive transport.

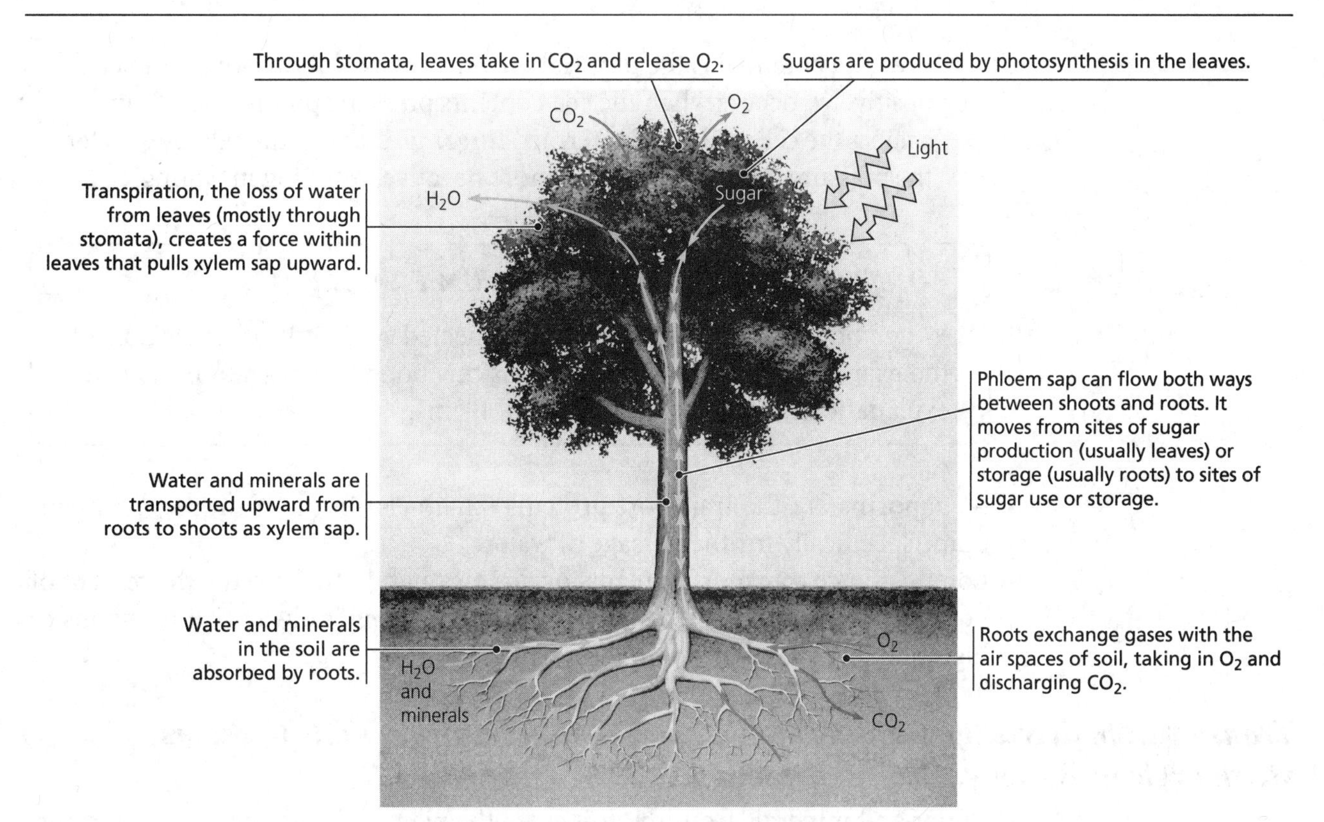

Figure 8.2 Overview of resource acquisition and transport

- If energy is required to move solutes across the membrane, it is termed *active transport*. Because most solutes cannot move across the phospholipid barrier of the membrane, a **transport protein** is required. The most important transport protein in plants is the **proton pump**.
- A proton pump creates an electrochemical gradient by using the energy of ATP to pump hydrogen ions across the membrane. This potential energy can then be used in the process of **cotransport**—the coupling of the steep gradient of one solute (hydrogen in our example) with a solute like sucrose. The drop in potential energy experienced by the hydrogen ion pays for the transport of the sucrose.

- The uptake of water across cell membranes occurs through *osmosis*, the passive transport of water across a membrane.
  - Water moves from areas of high water potential to low water potential. **Water potential** includes the combined effects of solute concentration and physical pressure.
  - The water potential equation is $\psi = \psi_s + \psi_p$, where ψ is water potential, ψ_s is solute potential, and ψ_p is the pressure potential.
  - By definition the ψ_s of pure water is 0. Adding solutes to pure water always lowers water potential. The solute potential of a solution is therefore always negative.

- Pressure potential is the physical pressure on a solution. An example of positive ψ_p occurs when the cell contents press the plasma membrane against the cell wall, a force termed *turgor pressure.* If the cell loses water, the pressure potential becomes more negative, resulting in wilting.

> ***CONNECT WITH THE CURRICULUM FRAMEWORK***
>
> Water potential problems can be expected on the grid-in section of the exam. Refer to the lab section of this book, Investigation 4, for an explanation of how to solve these problems.

- **Aquaporins** are the transport proteins (channels) in the plant plasma membrane specifically for the passage of water.
- Long-distance transport in plants occurs through **bulk flow**, the movement of liquid in response to a pressure gradient. Bulk flow is always from regions of high pressure to regions of low pressure.

Transpiration drives the transport of water and minerals from roots to shoots via the xylem (Biology, *10e: 36.3,* Focus, *1e: 29.5*)

- Water and minerals from the soil enter the plant through the root epidermis, cross the body of the root, and then flow up the xylem.
- Once in the root xylem, water and minerals are transported long distances—to the rest of the plant—by bulk flow. The water and minerals, termed *xylem sap*, flow out of the root and up through the shoot, eventually exiting the plant, primarily through the leaves.
- **Transpiration** is the loss of water vapor from the leaves and other parts of the plant that are in contact with air. It plays a key role in the movement of water from the roots.
  - The **cohesion-tension hypothesis** describes how transpiration provides the pull for the ascent of xylem sap, and the cohesion of water molecules transmits this pull along the entire length of the xylem from shoots to roots.
  - Water is lost through transpiration from the leaves of the plant due to the lower water potential of the air. The *cohesion* of water due to hydrogen bonding plus the *adhesion* of water to the plant cell walls enables the water to form a water column. Water is drawn up through the xylem as water evaporates from the leaves, each evaporating water molecule pulling on the one beneath it through the attraction of *hydrogen bonds.*

> ***TIP FROM THE READERS***
>
> Hydrogen bonding plays a key role in cohesion-tension mechanisms. Be able to explain the importance of cohesion, adhesion (water hydrogen bonded to the xylem walls), and surface tension in this mechanism.

The rate of transpiration is regulated by stomata (Biology, *10e: 36.4,* Focus, *1e: 29.6*)

- Large leaf surface area increases photosynthesis but also increases water loss by the plant through stomata. Guard cells open and close the stomata, controlling the amount of water lost by transpiration, but also the amount of carbon dioxide available from the atmosphere for photosynthesis.
- Guard cells control the size of the stomata opening by changing shape, widening or closing the gap between them. When the guard cells take up K^+ from the surrounding cells, water potential in the guard cells is decreased, causing them to take up water. The guard cells then swell and buckle, increasing the size of the pore between them. When the guard cells lose K^+, the cells then lose water, become less bowed, and the pore closes. Follow this sequence of events as shown in Figure 8.3.
- Guard cells are stimulated to open by the presence of light, loss of carbon dioxide in the leaf, and by normal circadian rhythms. Notice the critical interactions between environmental stimuli and internal molecular changes. This area of study has many essay possibilities!
- Circadian rhythms are part of the plant's internal clock mechanism and cycle with intervals of about 24 hours. Even plants kept in the dark will open their stomata as dawn approaches.

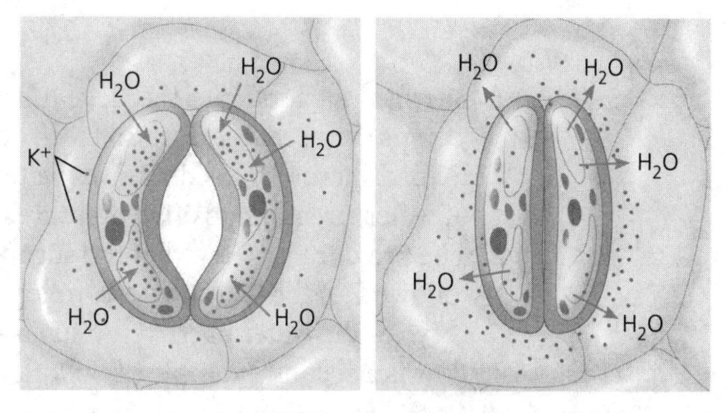

Role of potassium ions (K^+) in stomatal opening and closing.

Figure 8.3 Mechanisms of stomatal opening and closing

Sugars are transported from sources to sinks via the phloem (Biology, *10e: 36.5,* Focus, *1e: 29.7*)

- Phloem transports organic products of photosynthesis from the leaves throughout the plant, a process called **translocation**. The mechanism for translocation is **pressure flow**.
- Phloem always carry sugars from a sugar source to a sugar sink. A **sugar source** is an organ that is a net producer of sugar, such as the leaves. A **sugar sink** is an organ that is a net consumer or storer of sugar, such as a fruit, or roots during the summer. Follow Figure 8.4 while noting the key steps.

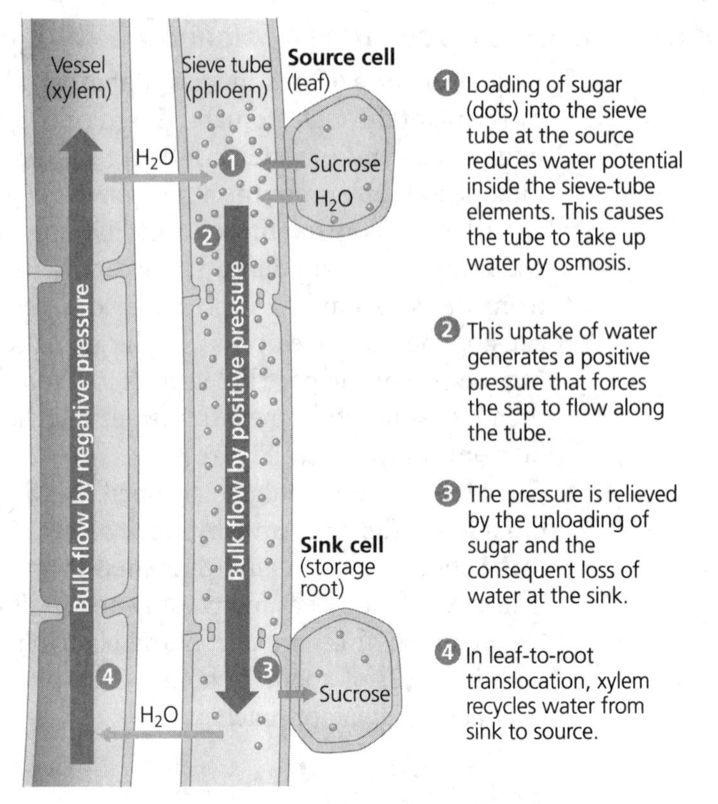

Figure 8.4 Pressure flow in a sieve tube

1. Sucrose is loaded into the sieve tubes at the sugar source. Proton pumps are used to create an electrochemical gradient that is utilized to load sucrose. This decreases water potential and causes the uptake of water, creating positive pressure.
2. The pressure is relieved at the sugar sink by the unloading of sucrose followed by the loss of water. In leaf-to-root translocation, xylem recycles the water back to the sugar source. Translocation via pressure flow is a second example of bulk flow.

The symplast is highly dynamic (Biology, *10e: 36.6*)

- The symplast is a network of living phloem cells that connects all parts of the plant.
  - Plasmodesmata allow for the movement of informational molecules like RNAs and proteins that coordinate development between cells. Plasmodesmata are dynamic and can open or close rapidly in response to changes such as turgor pressure, pH, or calcium levels in the cytoplasm.
  - In some plants the phloem carries out rapid, long-distance electrical signaling. This may lead to changes in gene transcription, respiration, photosynthesis, and other cellular functions in widely spaced organs. This is a nerve-like function, allowing for swift communication.

CONNECT WITH THE CURRICULUM FRAMEWORK

There are several important examples in this section that are within the curriculum.

1. Cells communicate with each other through direct contact with other cells. (EK 3.D.2) The plasmodesmata are an excellent example of this!
2. Timing and coordination of behavior are regulated by various mechanisms and are important in natural selection. (EK 2.E.3) The interactions between environmental stimuli and internal changes, including differential gene expression, fit here.

Soil and Plant Nutrition

(Biology, *10e: Chapter 37*, Focus, *1e: Chapter 29*)

WHAT'S IMPORTANT TO KNOW?

This chapter provides illustrative examples that your teacher might select but are not required content. Here are what we consider to be the main points of this chapter:

- Mutualistic relationships between plant roots and the bacteria and fungi that grow in the rhizosphere help plants acquire important nutrients.
- Plants also form symbiotic relationships that are not mutualistic.
- Interactions between populations (such as competition, predation, mutualism, and commensalism) can influence patterns of species distribution and abundance. (4.B.3)

Plant nutrition often involves relationships with other organisms (Biology, *10e: 37.3*, Focus, *1e: 29.4*)

- A *mutualistic* relationship occurs between nitrogen-fixing bacteria and plants. Nitrogen-fixing bacteria from the genus *Rhizobium* live in the root nodules of the legume plant family, including plants like peas, soybeans, alfalfa, peanuts, and clover. *Rhizobium* bacteria can fix atmospheric nitrogen into a form that can be used by plants. The plant provides food into the root nodule where the bacteria live, hence the designation as a mutualistic relationship.
- Bacteria also play a critical role in the nitrogen cycle (discussed in *Biology*, 10e, Chapter 55, or *Focus*, 1e, Chapter 42). Ammonia is recycled in the soil by bacteria into forms that can be absorbed and used by plants.
- *Mycorrhizae* are another example of mutualistic relationships with roots, this time between the roots and fungi in the soil. In mycorrhizae, the fungus benefits from a steady supply of sugar donated by the host plant.

In return, the fungus increases the surface area for water uptake, selectively absorbs minerals that are taken up by the plant, and secretes substances that stimulate root growth and antibiotics that protect the plant from invading bacteria.

- Plants also form symbiotic relationships that are not mutualistic (See Figure 37.14 in *Biology*, 10e, or Figure 29.15 in *Focus*, 1e).
  - **Parasitic plants**, such as dodder, are not photosynthetic and rely on other plants for their nutrients. Many parasitic plants have completely lost the ability to undergo photosynthesis. Would you consider these plants producers or consumers?
  - **Epiphytes** are not parasitic but just grow on the surfaces of other plants instead of the soil. Many orchids grow as epiphytes.
  - **Carnivorous plants** are photosynthetic, but they get some nitrogen and other minerals by digesting small animals. They are commonly found in nitrogen-poor soil, such as in bogs.

Angiosperm Reproduction and Biotechnology

(Biology, *10e: Chapter 38*, Focus, *1e: Chapter 30*)

WHAT'S IMPORTANT TO KNOW?
Most of this chapter's information is not required knowledge, but below are what we consider to be the main points:

- The biology of pollination and examples of coevolution.
- The relationship between seed and fruit.
- How temperature and moisture determine seed germination. (EK 2.E.1)
- The role of seed dormancy in promoting survival of the species.
- How different modes of plant reproduction affect their genetic diversity.

Flowers, double fertilization, and fruits are unique features of the angiosperm life cycle (Biology, *10e: 38.1*, Focus, *1e: 30.1*)

- **Pollination** is the transfer of pollen from an anther to a stigma. It is accomplished by wind, water, or animals, with animals providing the vast majority of pollination. Many species of flowering plants have evolved with specific animal pollinators in classic examples of *mutualism*. The joint evolution of two interacting species, each in response to selection imposed by the other, is called *coevolution*. Pollination is an excellent and uncomplicated example of mutualism and coevolution!
- The ripe ovary develops into the **fruit**. The ovules within the ovaries develop into seeds. The fruit protects the enclosed seeds and aids in their dispersal by wind or animals.

- As the seed matures, it enters dormancy, in which it has a low metabolic rate and its growth and development are suspended. The seed resumes growth when there are suitable environmental conditions for germination. If you did a laboratory activity on cellular respiration, such as AP Investigation 6, you will know that dry seeds have a very low rate of respiration because they are dormant, not dead.
  - Seed dormancy allows the embryo to wait until a suitable environment exists for the growth and development of the new plant. Seed dormancy can be broken by a number of environmental factors including fire, rain (desert plants), or cold temperatures.
  - Seed germination depends on the uptake of water due to the low water potential of the dry seed. Imbibing water causes the seed to expand and rupture its coat and also triggers metabolic changes in the embryo that enable it to resume growth.
  - Note the importance of timing and coordination with the breaking of seed dormancy and initiation of germination. Environmental cues are received and cell signaling responses turn on genes and stimulate the embryo to start the process of growth.

Flowering plants reproduce sexually or asexually, or both (Biology, *10e: 38.2,* Focus, *1e: 30.2*)

- Asexual reproduction, or **vegetative reproduction**, produces clones. Fragmentation is an example, in which pieces of the parent plant break off to form new individuals that are exact genetic replicas of the parent.
- Whereas some flowers self-fertilize, others have methods to prevent self-fertilization and maximize genetic variation. One of these is **self-incompatibility**, in which a plant rejects its own pollen or that of a closely related plant, thus ensuring cross-pollination.
- Agriculture uses several techniques of artificial vegetative reproduction, such as grafting, cuttings, and test-tube cloning.

People modify crops by breeding and genetic engineering (Biology, *10e: 38.3,* Focus, *1e: 30.3*)

- Humans have intervened in the reproduction and genetic makeup of plants for thousands of years through **artificial selection**.
- The conversion of plant material to sugars, which can be fermented to form alcohols and distilled to yield biofuels, is currently under study.
- **Genetically modified organisms** are engineered to express a gene from another species. *Examples*: "Golden Rice," engineered to include large amounts of vitamin A; and *Bt* corn, engineered to contain a toxin that kills specific crop pests. There is some debate over the creation of these genetically modified (GM) crops due to fear of human allergies and possible effects on nontarget organisms, among other concerns.

Plant Responses to Internal and External Signals

(Biology, *10e: Chapter 39*, Focus, *1e: Chapter 31*)

YOU MUST KNOW

- The three components of a signal transduction pathway and how changes could alter cellular responses.
- The role of auxins in plants.
- How phototropism and photoperiodism use changes in the environment to modify plant growth and behavior. (EK 2.E.2)
- How plants respond to attacks by herbivores and pathogens.

***Signal transduction pathways link signal reception to response* (Biology, *10e: 39.1*, Focus, *1e: 5.6*)**

TIP FROM THE READERS
Essays on the AP Biology Exam often cover ideas in different units in the textbook. This concept brings together the general ideas on cell communication from Chapter 11 with specific examples of cell communication in plants. Focus on mechanisms by which plant cells receive information and the responses that occur from these inputs. Also consider the evolutionary significance of the response or mechanism. This would be an excellent topic for an essay question!

- Let's begin with a review (Concept 11.1 in *Biology*, 10e, or Concept 5.6 in *Focus*, 1e). Signal transduction pathways involve three steps:

  1. **Reception:** Cell signals are detected by receptors that undergo changes in shape in response to a specific stimulus. Two common plasma membrane receptors are *G protein-coupled receptors* and *receptor tyrosine kinase.*
  2. **Transduction:** Transduction is a multistep pathway that amplifies the signal. This allows a small number of signal molecules to produce a large cellular response.
  3. **Response:** Cellular response is primarily accomplished by two mechanisms: (1) turning genes on or off and thereby increasing or decreasing mRNA production, or (2) activating existing enzyme molecules.

***Plant hormones help coordinate growth, development, and responses to stimuli* (Biology, *10e: 39.2*, Focus, *1e: 31.1*)**

- **Hormones** are defined as signaling molecules produced in small amounts in one part of an organism and transported to other parts. Hormones act as chemical messengers that coordinate the different parts of a multicellular organism.

- A **tropism** is a plant growth response that results in the plant growing either toward or away from a stimulus. Tropisms result from hormone production.
- **Phototropism** is the growth of a shoot in a certain direction in response to light. **Positive phototropism** is the growth of a plant organ (the shoot, for example) toward light; **negative phototropism** is growth of a plant organ (like the shoot) away from light. The hormone most directly involved with positive phototropism is auxin.
- The natural auxin in plants is *indoleacetic acid*, usually abbreviated as *IAA*. Auxins have many functions and play key roles in phototropisms and gravitropisms.
- **Auxins** stimulate elongation of cells within young developing shoots. Auxins produced in the apical meristems activate proton pumps in the plasma membrane, which results in a lower pH (acidification of the cell wall). This weakens the cell wall, allowing turgor pressure to expand the cell wall, resulting in cell elongation.
- Phototropism directs shoot growth toward the sunlight that powers photosynthesis. This response results from cells on the darker side elongating faster than the cells on the lighter side. This causes the shoot to grow faster on the side away from the light, bending the plant toward the light source. Can you explain the adaptive value of this response?
- Synthetic auxins are often used as herbicides. Monocots, like grasses, can quickly inactivate synthetic auxins, but eudicots cannot. Thus, the high concentrations of auxins kill broadleaf weeds (dicots), whereas grasses (monocots like turf grass or corn) are not harmed.

> ***STUDY TIP*** You are not required to know the names and functions of the plant hormones. However, you should be able to explain the plant responses of phototropism and photoperiodism. Knowledge of how auxins work will help you understand tropisms.

Responses to light are critical for plant success (Biology, *10e: 39.3*, Focus, *1e: 31.2*)

- Plants can detect not only the presence of light, but also its direction, intensity, and wavelength. Action spectra reveal that red and blue light are the most important colors in plant responses to light.
- *Blue-light photoreceptors* initiate a number of plant responses to light including phototropisms and the light-induced opening of stomata.
- Light receptors termed **phytochromes** absorb mostly red light.
- Phytochromes exist in two isomer forms, P_r and P_{fr}, that can switch back and forth depending on the wavelength of light in greatest supply. P_{fr} is the form of phytochrome that triggers many of a plant's developmental responses to light. The plant produces phytochrome in the P_r form, but upon illumination, the P_{fr} level increases by rapid conversion from P_r. The relative amounts of the two pigments provide a baseline for measuring the amount of sunlight in a day.

- **Circadian rhythms** are physiological cycles that have a frequency of about 24 hours and that are not paced by a known environmental variable. In plants, the surge of P_{fr} at dawn resets the biological clock. The combination of a phytochrome system and a biological clock allows the plant to accurately assess the amount of daylight or darkness and hence the time of year.
- A physiological response to a photoperiod (the relative lengths of night and day), such as flowering, is called **photoperiodism**. Photoperiodism controls when plants will flower. When this was first discovered, scientists thought the critical factor was the length of the day. It was later determined that the length of the night is the actual critical factor. The old terminology, however, is still used, so take note of that in the following categories:
    - Study the response of a short-day plant, such as the chrysanthemum shown in Figure 8.5(a), to day length. Notice that if an amount of *uninterrupted* darkness does not *exceed* the critical period, the plant will not flower. **Short-day plants** flower in early spring or fall. Short-day plants are actually long-night plants; that is, what the plant measures is the length of the night.
    - **Long-day plants** flower only if the night is *shorter* than a critical period. Study the iris in the long-day plant example in Figure 8.5(b). Long-day plants often flower in the late spring or early summer. Long-day plants are actually short-night plants.
    - **Day-neutral plants** can flower in days of any length.

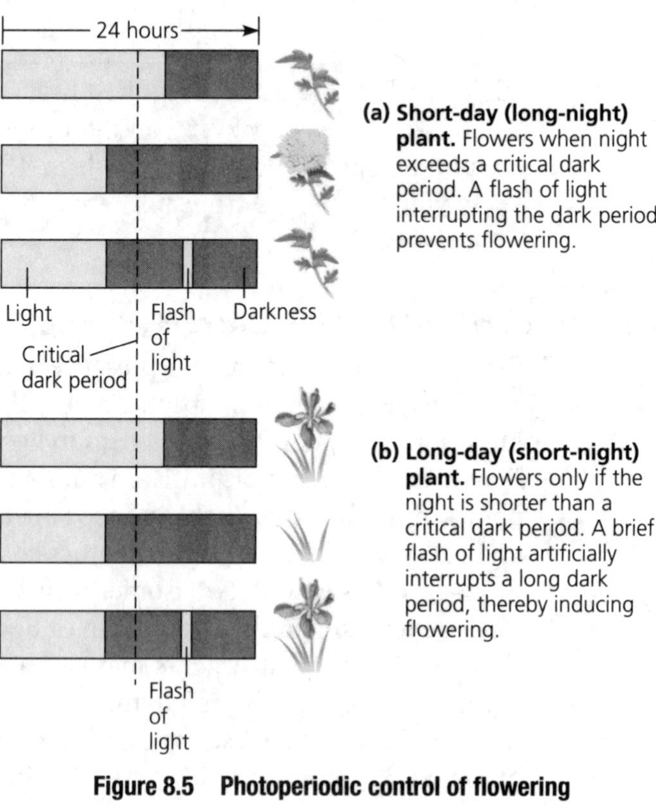

(a) Short-day (long-night) plant. Flowers when night exceeds a critical dark period. A flash of light interrupting the dark period prevents flowering.

(b) Long-day (short-night) plant. Flowers only if the night is shorter than a critical dark period. A brief flash of light artificially interrupts a long dark period, thereby inducing flowering.

Figure 8.5 Photoperiodic control of flowering

> ***TIP FROM THE READERS***
> Phototropism and photoperiodism are both excellent examples of the interaction between environmental stimuli and internal molecular signals leading to specific responses. Because of this, they are included in the framework and are good areas for essay and multiple-choice questions. Be sure you carefully study the receptors and responses involved and consider the evolutionary importance of a plant's ability to respond to light in these ways.

Plants respond to a wide variety of stimuli other than light (Biology, *10e: 39.4,* Focus, *1e: 31.3*)

- What environmental cue causes the shoot of a young seedling to grow up and the root to grow down? **Gravitropism** is a plant's response to gravity. Roots show **positive gravitropism** and grow toward the source of gravity, whereas shoots show **negative gravitropism** and grow away from gravity.
- The hormone **auxin** plays a key role in gravitropism in both roots and stems. In the young root, gravity indirectly causes a high concentration of auxin on the root's lower side. *Higher-than-normal* concentrations of auxins inhibit cell elongation, causing the lower side to grow more slowly, whereas more rapid elongation of cells on the upper side causes the root to curve as it grows.
- **Thigmotropism** is directional growth in a plant as a response to a touch. Vines display thigmotropism when their tendrils coil around supports.
- Plants have various responses to stresses. In times of drought, the guard cells lose turgor. This causes the stomata to close; young leaves will stop growing, and they will roll into a shape that slows transpiration rates. Also, deep roots continue to grow, whereas those near the surface (where there isn't much water) do not grow very quickly.
- Continue to consider why these different plant responses to environmental stimuli are adaptive. What is the value of the behavior in evolutionary terms?

Plants respond to attacks by herbivores and pathogens (Biology, *10e: 39.5,* Focus, *1e: 31.4*)

- Some physical defenses that plants have against predators (herbivores) are thorns and chemicals such as bitter or poisonous compounds. Some plants produce airborne attractants to recruit other animals to kill the herbivores. (Remember reading how parasitoid wasps lay their eggs on caterpillars, and the emerging wasp larva eats the caterpillar from the inside out?)
- The first line of defense against viruses for a plant (as for humans) is the epidermal layer. Plant immune responses involve both localized, specific responses as well as plant-wide responses. The lesions or dead spots you may have seen on leaves can be the result of the plant responding to a pathogen by sealing off the pathogen then killing the cells in the area. This kills the pathogen and prevents the spread of the disease to the rest of the plant. Both plant and animal immune systems depend heavily on signal transduction pathways.

Level 1: Knowledge/Comprehension Questions

The Knowledge/Comprehension questions review essential content knowledge but don't represent the type of question you will see on the AP exam. Level 2, Application/Synthesis questions are similar to those you will see on the exam.

1. Which of the following processes is responsible for the bending of the stem of a plant toward a light source?
   (A) The amount of chlorophyll produced on the side facing the light increases.
   (B) The rate of cell division on the side facing the light increases.
   (C) The rate of cell division on the side away from the light increases.
   (D) The cells on the side of the stem away from the light elongate.

2. Which of the following mechanisms is most important in the movement of water from root to leaves in the xylem?
   (A) the effect of gravity
   (B) root pressure
   (C) transpiration
   (D) osmosis

3. Hydrogen bonding plays a particularly important role in which plant process?
   (A) bulk flow of solutes from source to sink
   (B) photoperiodism resulting in flowering
   (C) attraction of the sperm and egg
   (D) the transpiration-cohesion-tension mechanism

4. In a mesophyll cell of a leaf, the synthesis of ATP takes place in which of the following cell organelles?
   (A) chloroplasts and mitochondria
   (B) mitochondria only
   (C) nucleoli and mitochondria
   (D) ribosomes and mitochondria

5. Which of these does not play a significant role in enhancing the uptake of water by a plant?
   (A) root hairs
   (B) large leaf surface area
   (C) mycorrhizae
   (D) gravitational force

6. How does the sperm of flowering plants reach the egg?
   (A) via a pollen tube that grows from the pollen grain through the carpel tissues to the egg
   (B) via raindrops of a dew film that allows the sperm to swim from the male plant to the female plant
   (C) usually via an insect, which places sperm in the ovary while probing for nectar
   (D) by actively swimming down through the style to the egg

7. Which of the following plant structures will become the seed?
   (A) a flower
   (B) an embryo
   (C) an ovule
   (D) an ovary

Level 2: Application/Analysis/Synthesis Questions

1. In order to flower, short-day plants require a period of
   (A) light greater than a critical period.
   (B) darkness greater than a critical period.
   (C) light less than a critical period.
   (D) darkness less than a critical period.

2. If a long-day plant has a critical length of 9 hours, which 24-hour cycle would prevent flowering?
   (A) 16 hours light/8 hours dark
   (B) 14 hours light/10 hours dark
   (C) 15.5 hours light/8.5 hours dark
   (D) 4 hours light/8 hours dark/4 hours light/ 8 hours dark

3. Select the activity that does not occur in a signal transduction pathway.
   (A) activation of the receptor by a relay molecule
   (B) expression of specific genes
   (C) activation of protein kinases
   (D) phosphorylation of transcription factors

4. The figure shows key elements of Went's experiment. Which plant shows auxin stimulating elongation in the left side of the plant only?

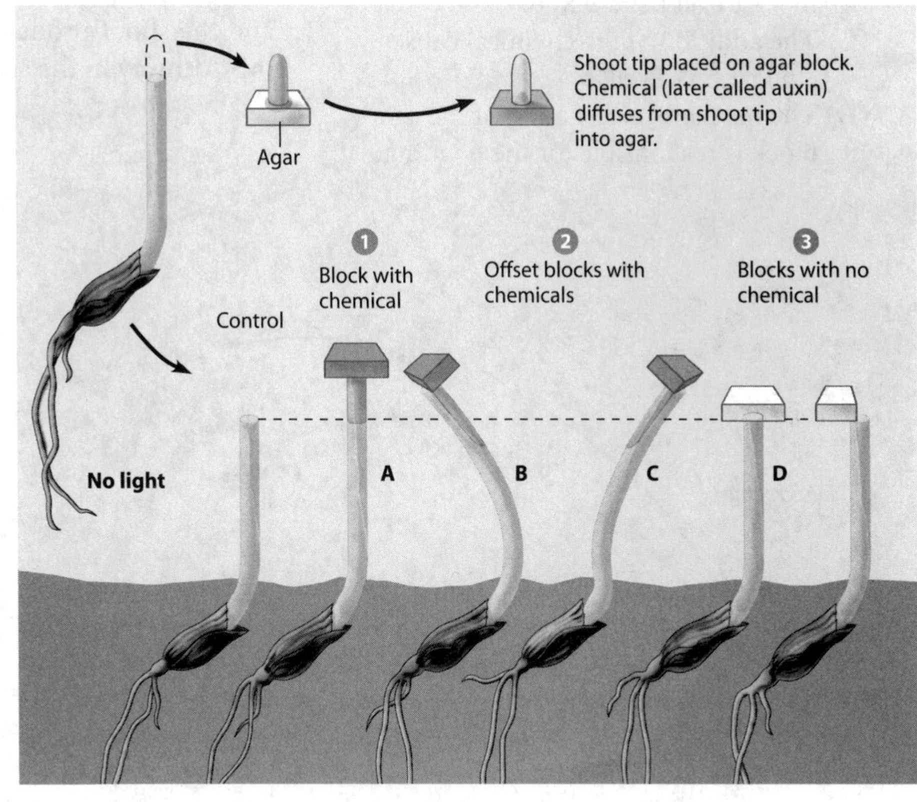

   (A) plant A
   (B) plant B
   (C) plant C
   (D) plant D

5. What prior knowledge must Went have had to design this experiment as shown?
   (A) Plants respond to gravity as well as light.
   (B) Compounds produced in the shoot tip have an effect on phototropism.
   (C) Placing an agar block on a seedling will stimulate its growth.
   (D) Photosynthesis is concentrated in cells of the shoot tip.

6. In Went's experiment, observe what happens to both plants with treatment D, agar blocks with no chemicals. What *conclusion* can be drawn from this?
   (A) These plants serve as the control because they have no chemical.
   (B) It is necessary to add chemical to the block to get bending.
   (C) The addition of the chemical causes elongation of cells.
   (D) The mechanical irritation of the agar block is responsible for the bending.

7. In plants, translocation, the movement of solutes in the phloem, occurs as a result of
   (A) a difference in water potential between a sugar source and a sugar sink.
   (B) evaporation of water through the stomata.
   (C) cohesion of water molecules to each other and adhesion to the transport tubes.
   (D) release of auxin in response to loss of water in the xylem.

Use the following information to answer questions 8–10.

When a seed germinates and the growing plant shoot reaches light, the plant undergoes profound changes known as de-etiolation or greening. Stem elongation slows, leaves expand, roots elongate, and the shoot produces chlorophyll. You can see this pathway in the accompanying figure.

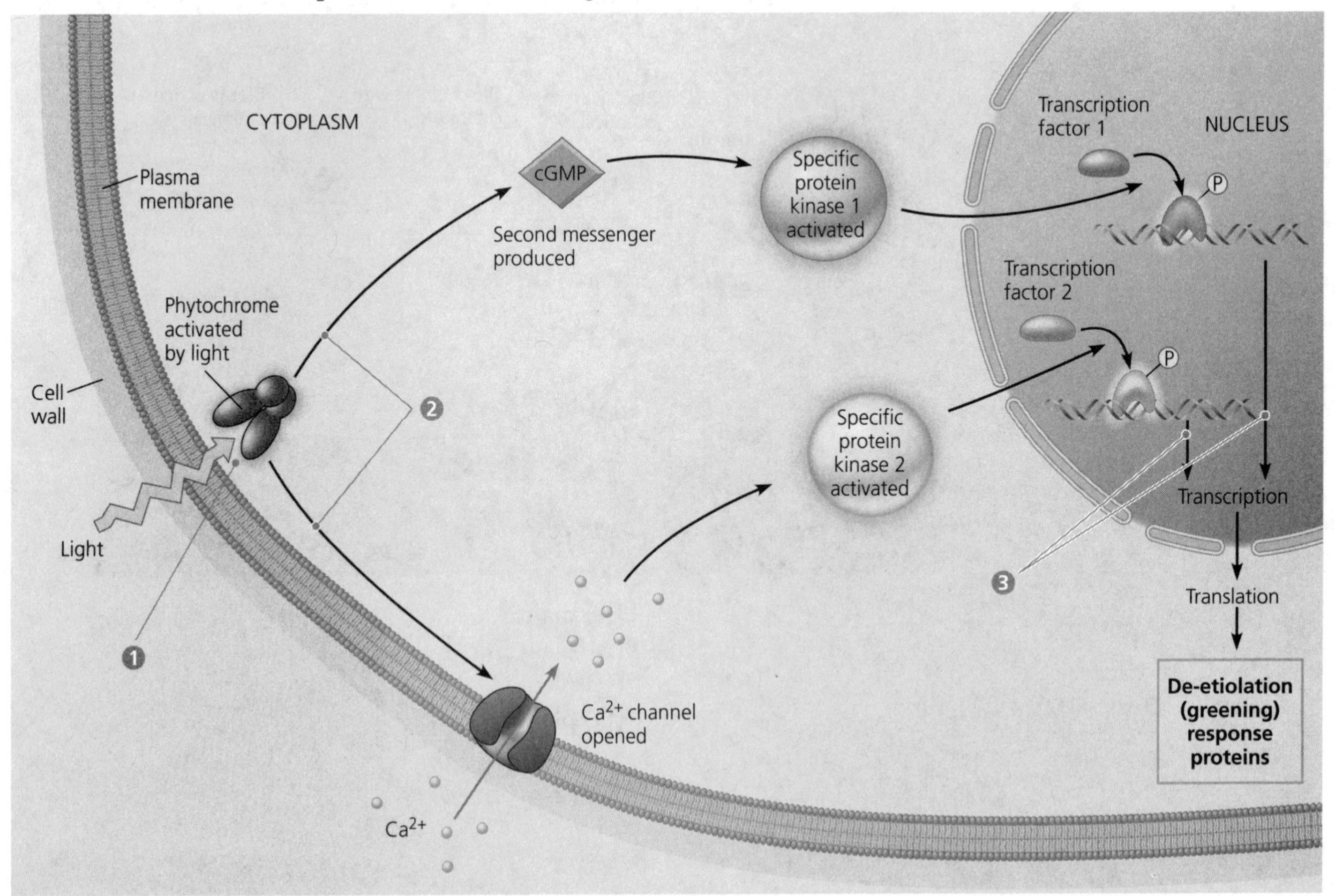

8. What is the signal and its receptor in this pathway?
   (A) Ca^{2+} is the signal, a membrane protein is the receptor.
   (B) Light is the signal, a cytoplasmic protein is the receptor.
   (C) cGMP is the signal, protein kinase is the receptor.
   (D) The protein kinases are the signal, transcription factors are the receptors.

9. Which step in this pathway converts and amplifies the signal resulting in numerous active molecules?
   (A) Step 1
   (B) Step 2
   (c) Step 3
   (D) Both Steps 1 and 3

10. Which of the following would occur if the Ca^2 channels were not functional?
    (A) Reception would be blocked, and phytochrome would not be activated.
    (B) cGMP molecules, the second messengers, would not be produced.
    (C) Transcription and translation of the de-etiolation proteins would not occur.
    (D) Transcription factor 2 would not be produced.

Free-Response Question

1. *Photoperiodism and phototropism are plant behaviors that increase fitness.*

(a) Explain how chrysanthemums, which are short-day plants, can be forced to flower in a greenhouse in the middle of the summer. Include in your discussion the specific signal and its reception.

(b) A plant seedling will respond to a strong directional light source. Explain how this occurs.

(c) Explain how each response increases biological fitness.

TOPIC 9

Animal Form and Function

Chapter 40: Basic Principles of Animal Form and Function

(Biology, *10e: Chapter 40*, Focus, *1e: Chapter 32*)

YOU MUST KNOW

- The importance of homeostasis and examples.
- How feedback systems control homeostasis.
- One example of positive feedback and one example of negative feedback.

Animal form and function are correlated at all levels of organization (Biology, *10e: 40.1*, Focus, *1e: 32.1*)

- **Tissues** are groups of cells that have a common structure and function. Tissues are further organized into functional units called **organs**. Groups of organs that work together make up **organ systems** (for example, the digestive, circulatory, and excretory systems).
- For animal survival tissues, organs, and organ systems must act in a coordinated manner. Two major organ systems specialize in control and coordination:
  - In the **endocrine system**, chemical signals called **hormones** are released into the bloodstream and are broadcast throughout the body. Different hormones cause specific effects but only in cells with specific receptors for the released hormone.
  - In the **nervous system, neurons** transmit information between specific locations. Only three types of cells receive nerve impulses: neurons, muscle cells, and endocrine cells.

Feedback control maintains the internal environment in many animals (Biology, *10e: 40.2*, Focus, *1e: 32.1*)

- In **homeostasis** animals maintain a relatively constant internal environment even when the external environment changes significantly.
- Homeostatic control systems function by having a **set point** (like a body temperature to maintain), **sensors** to detect any stimulus above or below the set point, and a physiological **response** that helps return the body to its set point.

> ***TIP FROM THE READERS***
> Homeostasis is a fundamental concept in biology. You should concentrate on knowing how any change in a system affects it, from the level of cells through an ecosystem. This topic offers many good examples of positive and negative feedback and how small changes can affect a system. You should have a negative- and positive-feedback system ready to discuss before taking the AP Biology Exam.

- In **negative-feedback systems**, the animal responds to the stimulus in a way that reduces the stimulus. For example, in response to exercise, the body temperature rises, which initiates sweating, which cools the body. As the body moves away from the set point, negative-feedback systems return the body to the set point. Homeostasis relies primarily on negative-feedback systems. *More gets you less.*
- In **positive-feedback systems**, a change in some variable triggers mechanisms that amplify rather than reverse the change. For example, during childbirth, pressure of the baby's head against receptors near the opening of the uterus stimulates greater uterine contractions, which cause greater pressure against the uterine opening, which heightens the contractions, and so forth. *More gets you more.*

Homeostatic processes for thermoregulation involve form, function, and behavior (Biology, *10e: 40.3,* Focus, *1e: 32.1*)

- **Thermoregulation** refers to how animals maintain their internal temperature within a tolerable range. It may involve *physiological changes,* such as panting, sweating, shivering, or *behaviors,* such as burrowing and sunning.
- **Endotherms** (such as mammals and birds) are warmed mostly by heat generated by metabolism. **Ectotherms** (such as most invertebrates, fishes, amphibians, and reptiles) generate relatively little metabolic heat, gaining most of their heat from external sources.
- In many birds and mammals, reduction of heat loss relies on **countercurrent exchange**. Heat transfer involves antiparallel arrangement of blood vessels such that warm blood from the core of the animal, en route to the extremities, transfers heat to colder blood returning from the extremities. Heat that would have been lost to the environment is conserved in the blood returning to the core of the animal.

Energy requirements are related to animal size, activity, and environment (Biology, *10e: 40.4,* Focus, *1e: 33.5*)

- An animal's **metabolic rate** is the total amount of energy it uses in a unit of time. Metabolic rates are generally higher for endotherms than for ectotherms.
- Under similar conditions and for animals of the same size, the **basal metabolic rate** of endotherms is substantially higher than the **standard metabolic rate** of ectotherms.
- Metabolic rate is inversely related to body size among similar animals. For example, elephants have a slower metabolic rate, whereas mice have a very fast metabolic rate.

Animal Nutrition

(Biology, *10e: Chapter 41*, Focus, *1e: Chapter 33*)

WHAT'S IMPORTANT TO KNOW?
Digestion is not an explicit topic of the Curriculum Framework, but your teacher may choose to introduce you to this material. This chapter contains information that may be used to develop both multiple-choice and essay questions and a general understanding of digestion will make you more confident in dealing with potential questions.

An animal's diet must supply chemical energy, organic molecules, and essential nutrients (Biology, *10e: 41.1*, Focus, *1e: 33.1*)

- The **essential nutrients** required by an animal include both minerals and preassembled organic molecules that the animal cannot produce from raw materials.

STUDY TIP If your teacher covers this topic, you might refer to Figure 9.1 and consider questions like: How does the structure fit its function? How do the villi and microvilli of the intestine increase digestion and absorption? How does the epithelial lining of the intestine aid in absorption of nutrients?

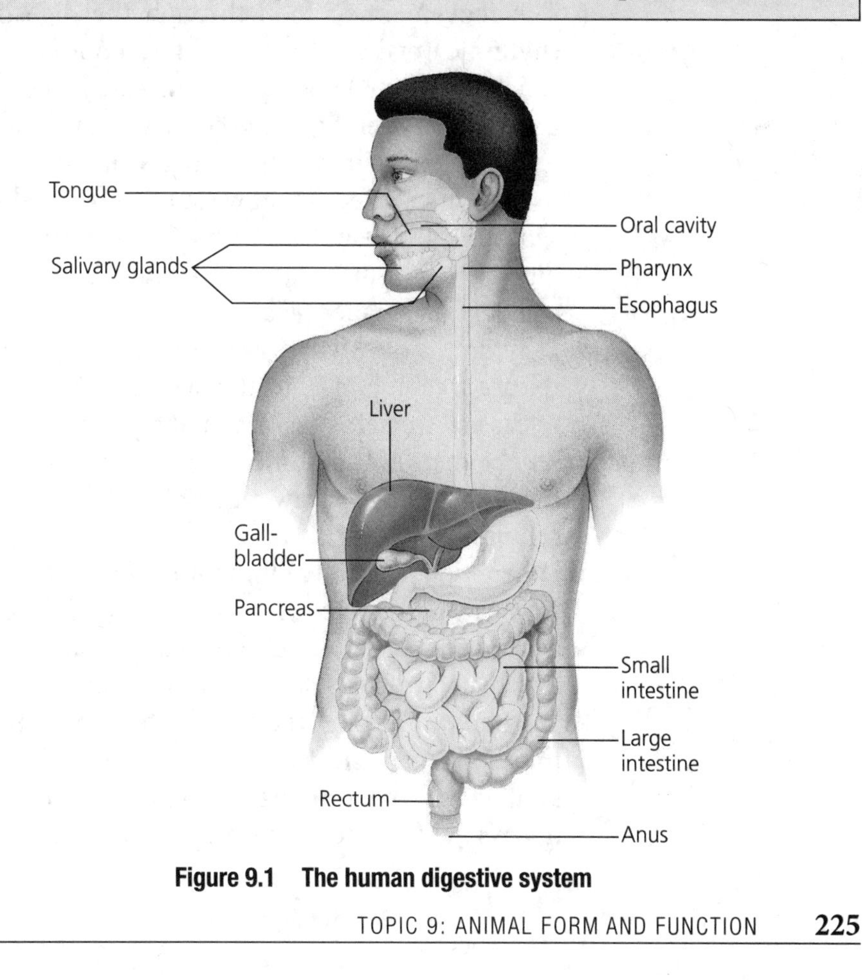

Figure 9.1 The human digestive system

The main stages of food processing are ingestion, digestion, absorption, and elimination (Biology, *10e: 41.2,* Focus, *1e: 33.2*)

- **Intracellular digestion** occurs within a cell enclosed by a protective membrane. Sponges digest their food this way.
- **Extracellular digestion** is carried out by most animals; in this type of digestion, food is broken down outside cells. This process allows the animal to devour much larger sources of food than can be handled using only intracellular digestion.
- More complex animals have **complete digestive tracts (alimentary canals)**, which are *one-way* digestive tubes that begin at the mouth and terminate at the anus.

Organs specialized for sequential stages of food processing form the mammalian digestive system (Biology, *10e: 41.3,* Focus, *1e: 33.3*)

> ***STUDY TIP*** You will not be expected to name the enzymes of digestion or know the anatomy of this system. An important idea here is how the large chunks of food you ingest become reduced to monosaccharides, amino acids, fatty acids, and glycerol and are able to enter your cells. How enzymes work as well as passive and active transport are required knowledge.

- The movement of food through the digestive system is controlled by the rhythmic waves of contraction by smooth muscle in the walls of the alimentary canal (peristalsis) and by muscular, ringlike valves that regulate the passage of material between digestive compartments (sphincters).
- One of the primary functions of the digestive system is to break down polymers to monomers. Recall that we studied how cells make polymers by dehydration synthesis, the removal of water to join two organic molecules chemically. In digestion, enzymes will add the water back, splitting the large compounds into smaller and smaller pieces.
- Dehydration synthesis reactions require energy, while hydrolysis reactions release free energy into the system. The calories from food come from this release of free energy. We intake polymers and through digestive hydrolysis reduce them to monomers. In general, only monomers can be absorbed into the body to serve as raw materials to build new molecules or to serve as fuel for cellular respiration.
- When food is in the mouth, or **oral cavity**, a nervous reflex occurs that causes saliva to be secreted into the mouth. Saliva lubricates the food to facilitate swallowing. It also starts chemical digestion because saliva contains an enzyme that hydrolyzes the polysaccharide starch.
- During swallowing, the food enters the **pharynx**—a junction that opens to the esophagus and the trachea (generally referred to as the throat). During swallowing, the **epiglottis**, a flap made of cartilage, covers the trachea. This diverts the food down the esophagus away from the airway.

- The **esophagus** moves food from the pharynx down to the stomach through peristalsis.
- The stomach's functions include storing food and secreting the digestive fluid termed *gastric juice.* Two components of **gastric juice** carry out chemical digestion:
    - **Hydrochloric acid**, with a pH of about 2, breaks down the extracellular matrix of meat and plant materials, and it also kills most of the bacteria ingested with food.
    - Pepsin is an enzyme in gastric juice that begins to hydrolyze proteins into smaller polypeptides. Pepsin is secreted in an inactive form called *pepsinogen*, which is activated by hydrochloric acid in the stomach. The inactive form protects the cells that produce the protein-digesting enzyme from self-digestion. For further protection, a thick mucus is produced by the lining of the stomach.
- *The small intestine is the major site of digestion and absorption.* The first section of the **small intestine** is known as the duodenum. The duodenum is the specific major site of chemical digestion. In the duodenum the mixture of ingested food and digestive juice mixes with secretions from the pancreas and the liver.
- The **pancreas** releases a complex mix of digestive enzymes and an alkaline fluid rich in bicarbonate, which acts as a buffer against the acidic contents from the stomach. This allows the digestive fluid in the duodenum to be at the correct pH for the enzymes now needed for further digestion. The pancreas produces enzymes for the digestion of all four major organic groups—carbohydrates, fats, proteins, and nucleic acids—as such; it is an essential organ of digestion.
- The **liver** makes bile, which is then stored in the gallbladder. Bile emulsifies fats; that is, bile coats fat droplets, turning large fat droplets into small fat droplets, which are easier to digest.
- Chemical digestion in the duodenum can be summarized as follows:
    - **Carbohydrates:** The breakdown of the polymers starch and glycogen that began in the mouth ends in the small intestines with the production of monosaccharides like glucose. Glucose can now be absorbed into the bloodstream.
    - **Proteins:** Pepsin begins the breakdown of proteins in the stomach. In the duodenum, polypeptides are further broken into their monomers, amino acids. The amino acids are then absorbed into the bloodstream.
    - **Nucleic acids:** The breakdown of nucleic acids starts with the hydrolysis of DNA and RNA to their respective nucleotides. The nucleotides are then broken down to nitrogenous bases, sugars, and phosphate groups, which are absorbed into the bloodstream.
    - **Fats:** Chemical digestion of fats begins in the small intestine, where fats are emulsified by bile secreted by the liver. The enzyme lipase, which is produced in the pancreas, hydrolyzes the small fat droplets into fatty acids and glycerol, which are then absorbed into the bloodstream.

- The epithelial lining of the small intestine has folds called **villi**, and the individual cells have projections called **microvilli**—both of which radically increase the surface area available for absorption. This is an excellent example of structure fitting function.
- In each villus are capillaries for the absorption of monomers, including monosaccharides and amino acids, and a lymph vessel, which absorbs small fatty acids. Passive facilitated diffusion and active transport are used to move monomers across the intestinal membrane and into blood vessels.

> ***STUDY TIP*** Consider questions such as: How are the foods we eat reduced to molecules that can cross into the circulatory system? What adaptations of the digestive system contribute to this? How do passive diffusion and active transport function in absorption?

- Hormones are chemical messengers that travel to target tissues through the bloodstream. Hormones help to coordinate the digestive process.
- The **large intestine**, also called the **colon**, is connected to the small intestine by a sphincter. The point of the connection is the site of the cecum, a small pouch with an extension called the **appendix**.
- The main functions of the large intestine are to compact waste and reabsorb water.
- The large intestine (also called the colon) includes a rich flora of mostly harmless bacteria, including *Escherichia coli*. Some of these bacteria are important symbionts, producing several vitamins including B vitamins and vitamin K. The presence of *E. coli* in lakes and streams is a useful indicator of contamination by untreated sewage.
- At the end of the large intestine is the **rectum**, where feces are stored until they are eliminated.

Evolutionary adaptations of vertebrate digestive systems correlate with diet (Biology, *10e: 41.4*, Focus, *1e: 33.4*)

- A mammal's **dentition** is generally correlated with its diet. In particular, mammals have specialized dentition that best enables them to ingest their food.
- Herbivores generally have longer alimentary canals than carnivores, reflecting the longer time needed to digest vegetation. Much of the chemical energy in herbivore diets comes from the cellulose of plant cell walls. Many vertebrates (as well as termites) house large populations of symbiotic bacteria and protists whose enzymes actually digest the cellulose.

Feedback circuits regulate digestion, energy storage, and appetite (Biology, *10e: 41.5*, Focus, *1e: 33.5*)

- Vertebrates store excess calories as *glycogen* in the liver and muscles and as fat.
- Overnourishment can lead to obesity. Several hormones regulate appetite, including *leptin,* which suppresses appetite in a negative-feedback response.

Circulation and Gas Exchange

(Biology, *10e: Chapter 42*, Focus, *1e: Chapter 34*)

WHAT'S IMPORTANT TO KNOW?
Circulation and gas exchange are not explicit topics of the Curriculum Framework, but your teacher may choose to introduce you to this material. This chapter contains information that may be used to develop both multiple-choice and essay questions and a general understanding of these processes will make you more confident in dealing with potential questions.

Here are what we consider to be the main points of this chapter:

- Red blood cells (RBCs) demonstrate the relationship of structure to function.
- The general characteristics of a respiratory surface.
- How O_2 and CO_2 are transported in the blood.
- The components of blood pressure and how it is measured.
- The roles of diet, blood pressure, and genetics in cardiovascular disease.
- The pathway a molecule of oxygen takes from the air until it is delivered by a red blood cell to the tissues.

Circulatory systems link exchange surfaces with cells throughout the body (Biology, *10e: 42.1*, Focus, *1e: 34.1*)

- Exchange of gases, nutrients, and wastes occurs at the cellular level. Because diffusion is rapid only across small distances, natural selection has led to two general solutions.

    1. Body size and shape that keep many or all cells in direct contact with the environment, such as with sponges. Cnidarians and flatworms possess a **gastrovascular cavity** that serves both in digestion and distribution of substances throughout the body.
    2. Larger animals have a **circulatory system** that moves fluid to the tissues and cells for exchange.

- A *circulatory system* has three components: **blood** (a circulatory fluid), **vessels** (tubes through which blood moves), and a **heart** (a structure that pumps the blood).
- There are three main types of blood vessels:

    - **Arteries** carry blood *away* from the heart and branch into smaller **arterioles**. Their walls are relatively thick and include a significant amount of smooth muscle. The *pulse* is felt in an artery.
    - **Capillaries** are microscopic vessels that are composed of only a single layer of cells. All diffusion occurs here.
    - **Veins** carry the blood back to the heart. They have valves to prevent backflow.

- **Atria** are heart chambers that receive blood and convey it to **ventricles**, which pump blood.

Coordinated cycles of heart contraction drive double circulation in mammals (Biology, *10e: 42.2*, Focus, *1e: 34.2*)

- The **sinoatrial (SA) node** is the pacemaker of the heart. It is located in the upper wall of the right atrium. It generates electrical impulses that set the rate at which cardiac muscle cells contract.
- The **atrioventricular (AV) node**, located in the lower wall of the right atrium, delays the impulses from the SA node to allow the atria to completely empty before the ventricles contract. The AV node generates electrical impulses that cause the ventricles to contract.
- Heart rate is regulated by at least three factors. The sympathetic nerves accelerate heart rate, and the parasympathetic nerves slow it down. Hormones such as epinephrine increase heart rate, as does an increase in body temperature.

Patterns of blood pressure and flow reflect the structure and arrangement of blood vessels (Biology, *10e: 42.3*, Focus, *1e: 34.3*)

- **Blood pressure** is measured with a *sphygmomanometer*.
    - *Short-term regulation of blood pressure* occurs when the smooth muscle of the arterioles contracts or relaxes due to changes in activity or hormonal signals.
    - *Long-term regulation of blood pressure* is accomplished by changes in blood volume due to a negative-feedback system involving the kidney and the endocrine system. (See Chapter 44 on page 237.)
- Blood moves through the veins and venules propelled by rhythmic contractions of smooth muscle in the walls of the blood vessels. Blood flow in veins is also enhanced by the contraction of skeletal muscles during exercise, pushing the blood toward the heart. Valves prevent the backflow of blood in veins and venules.

Blood components function in exchange, transport, and defense (Biology, *10e: 42.4*, Focus, *1e: 34.4*)

- **Plasma** is mostly water, but it also contains ions, electrolytes, and plasma proteins. It transports nutrients, metabolic wastes, gases, and hormones. In addition, blood plasma carries
    - **Red blood cells (erythrocytes or RBCs)**, which transport oxygen via *hemoglobin* (an iron-containing protein).
    - **White blood cells (leukocytes or WBCs)**, which are part of the immune system.
    - **Platelets**, which are fragments of cells responsible for blood clotting.

> ***TIP FROM THE READERS***
> *Structure fits function* is a major theme in AP Biology and has been used as the basis for many essay questions. The red blood cell, as described next, is an excellent example of the correlation of structure and function.

- Red blood cells are *biconcave disks*. This shape increases surface area to enhance oxygen uptake and release. Each RBC contains about 250 million molecules of hemoglobin, and each hemoglobin molecule can bind up to four molecules of oxygen.
- RBCs lack nuclei, which increases space for hemoglobin.
- RBCs lack mitochondria, so the oxygen they carry is not consumed.

Gas exchange occurs across specialized respiratory surfaces (Biology, *10e: 42.5*, Focus, *1e: 34.5*)

- **Gas exchange**, or **respiration**, is the uptake of molecular oxygen (O_2) from the environment and the discharge of carbon dioxide (CO_2) to the environment.
- The **respiratory medium** is the source of the O_2. It is air for terrestrial animals and water for most aquatic animals.
- The **respiratory surface** is the part of an animal's body where gases are exchanged with the surrounding environment. It can be the body wall, the skin, gills, tracheae, or lungs.
    - General characteristics of respiratory surfaces include:
        - Must be moist.
        - Favorable surface-area-volume ratio. Respiratory surfaces are often extensively folded or branched. (Think structure/function.)
        - Closely associated with the vascular system of larger animals.
- **Gills** are respiratory organs in aquatic animals. Water flows through them, and blood flowing through capillaries within the wall of the gill picks up oxygen from the water. Blood flows in a direction opposite to the flow of water. This is called **countercurrent exchange**, and it maximizes the absorption of oxygen.
- Countercurrent exchange mechanisms allow for more diffusion to occur than would otherwise be possible. Follow the countercurrent flow of blood and oxygen-rich water in Figure 9.2. Examine the figure until all the numbers make sense in the context of diffusion.

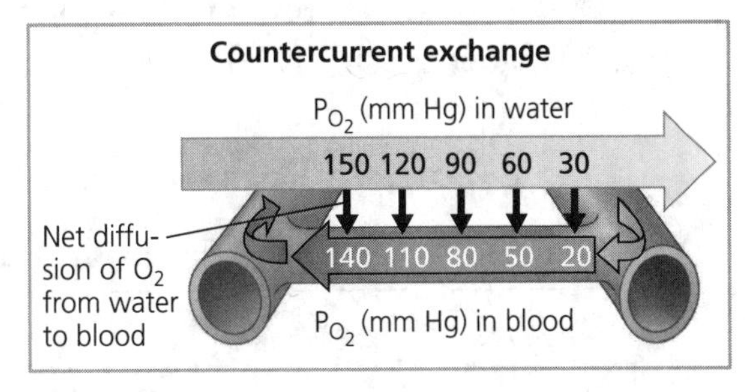

Figure 9.2 Countercurrent exchange

- The **alveoli** are air sacs clustered at the ends of bronchioles in the lungs. They are thin, moist, have a large surface area (in a human, the size of a tennis court if spread out!), and are associated with capillary beds. Here, O_2 diffuses into the blood by passing through the thin membrane of an alveolus and through the thin membrane of a capillary, into the blood, and attaches to a hemoglobin molecule in a red blood cell. Again, think structure/function.

> *STUDY TIP* Refer again to the characteristics of a respiratory surface and note how the mammalian lung fulfills the requirements.

Breathing ventilates the lungs (Biology, *10e: 42.6,* Focus, *1e: 34.6*)

- **Breathing** is the inhalation and exhalation of air that ventilates lungs.
- In mammals, breathing involves movement of the **diaphragm**—a dome-shaped muscle separating the thoracic cavity from the abdominal cavity. Lung volume increases when the rib muscles and diaphragm contract, pressure within the lungs decreases, and air flows into the lungs. During exhalation, lung volume is decreased as the diaphragm relaxes and moves up, and pressure within the lungs increases.
- Breathing is under control of regions located in the brain (the medulla and the pons). It is influenced by pH, which is an indirect indication of blood CO_2 levels.
- Increased metabolic activity lowers pH by increasing the concentration of CO_2 in the blood. CO_2 reacts with water to form carbonic acid, which dissociates into a bicarbonate ion and a hydrogen ion. As blood pH drops, the rate and depth of respiration increase.

Adaptations for gas exchange include pigments that bind and transport gases (Biology, *10e: 42.7,* Focus, *1e: 34.7*)

- **Hemoglobin** is the respiratory pigment found in almost all vertebrates. It consists of four subunits, each of which has a *heme group* with an embedded iron atom. Each iron atom binds O_2, allowing each hemoglobin to carry four oxygen molecules.
- A lowering of the pH in blood (as occurs with exercise) lowers the affinity of hemoglobin for oxygen, and oxygen dissociates from hemoglobin. This is called the **Bohr shift**. This delivers more O_2 to the tissues during times of heavy exercise.
- CO_2 is most commonly carried in the blood in the form of bicarbonate ions (70%). Less commonly, it is transported via hemoglobin (23%) and in solution in the blood plasma (7%).

The Immune System

(Biology, *10e: Chapter 43*, Focus, *1e: Chapter 35*)

YOU MUST KNOW

- Several elements of an innate immune response.
- The differences between B and T cells relative to their activation and actions.
- How antigens are recognized by immune system cells.
- The differences in humoral and cell-mediated immunity.
- Why helper T cells are central to immune responses.

> *STUDY TIP* The immune system is one of three human systems explicitly included in the Curriculum Framework. As you work through this chapter, pay particular attention to the ways cells communicate with each other to maintain defense.

In innate immunity, recognition and response rely on traits common to groups of pathogens (Biology, *10e: 43.1*, Focus, *1e: 35.1*)

- **Innate immune responses** include barrier defenses as well as defenses to combat pathogens that enter the body. Innate immune responses are the same whether or not the pathogen has been encountered previously. All animals have an innate immune response system.
  - **Barrier defenses** include skin and the mucous membranes that cover the surface and line the openings of the animal body. These provide a physical barrier and also produce secretions that result in a skin pH from 3 to 5. The antimicrobial enzyme **lysozyme,** which breaks down bacterial cell walls, is found in saliva, mucous secretions, and tears.
  - **Cellular innate defenses** combat pathogens that get through the skin—for example, in a cut. They include phagocytic white blood cells and antimicrobial proteins.
    - Phagocytic white blood cells recognize microbes using **toll-like receptors**, or **TLRs**. Toll-like receptors recognize fragments of molecules characteristic of a particular type of pathogen. For example, a TLR might recognize a polysaccharide found on the surface of many bacteria. TLRs increase the efficiency of phagocytosis.

Cellular Innate Defenses

- **Neutrophils** are white blood cells that ingest and destroy microbes in a process called **phagocytosis**.
- **Monocytes** are another type of phagocytic leukocyte. They migrate into tissues and develop into **macrophages**, which are even more efficient.
- **Eosinophils** are leukocytes that defend against parasitic invaders such as worms by positioning themselves near the parasite's wall and discharging hydrolytic enzymes.
- **Dendritic cells** populate tissues in contact with the environment, where they capture pathogens, display foreign antigens, and start the primary immune response.
- **Natural killer (NK) cells** help recognize and remove diseased cells. *Examples*: cells containing viruses and cancer cells.

Antimicrobial Proteins

- **Interferon** proteins provide innate defense against viral infections. They cause cells adjacent to infected cells to produce substances to inhibit viral replication.

- The **complement system** consists of roughly 30 proteins with a variety of functions that enhance the immune response. One function is to lyse invading cells.

- A local **inflammatory response** is triggered by damage to tissue by physical injury or the entry of pathogens. It leads to the release of numerous chemical signals. For example, **histamines** are released by *mast cells* in response to injury. Histamines trigger the dilation and permeability of nearby capillaries. This aids in delivering clotting agents and phagocytic cells to the injured area. Systemic inflammatory responses include fever and increased production of white blood cells to fight infection.

In adaptive immunity, receptors provide pathogen-specific recognition (Biology, *10e: 43.2*, Focus, *1e: 35.2*)

- Vertebrates have two types of lymphocytes: **B lymphocytes (B cells)**, which proliferate in the bone marrow, and **T lymphocytes (T cells)**, which mature in the thymus. They circulate through the blood and lymph, and both recognize particular antigens. All blood cells proliferate from stem cells in the bone marrow.
- **Antigens** are foreign molecules that elicit a response by lymphocytes. B and T cells recognize them by specific receptors embedded in their plasma membranes.
- **Antibodies** are soluble proteins secreted by B cells during an immune response.
- **B** or **T cell activation** occurs when an antigen binds to a B or T cell. B cell activation is enhanced by cytokines. B cells form two clones of cells in a process called **clonal selection**. This results in thousands of cells, all specific to this antigen.
  - **Effector cells** combat the antigen by producing antibodies *specific* to that antigen.
  - **Memory cells**, which are long-lived, bear receptors for the same antigen, thus allowing them to quickly mount an immune response in subsequent infections.
- T cell receptors bind antigens that are displayed by **antigen-presenting cells (APCs)** on their **MHCs**.
- **Major histocompatibility complex (MHC) molecules** are proteins that are the product of a group of genes. (Individuals differ in their MHCs. This is a major component of "self.")
- There are two types of MHCs:
  - **Class I MHCs** are found on almost all cells of the body, except RBCs.
  - **Class II MHCs** are made by some cells of the immune system, including dendritic cells, macrophages, and B cells.
- The specificity of B and T cells is a result of the shuffling and recombination of several gene segments and results in more than 1 million *different* B cells and 10 million different T cells.
- Each B or T cell responds to only one antigen.
- A **primary immune response** occurs when the body is first exposed to an antigen and a lymphocyte is activated.

- A **secondary immune response** occurs when the same antigen is encountered at a later time. It is faster and of greater magnitude. (Refer to Figure 9.3.) The secondary immune response starts with memory B cells so it can respond not only faster but with a much higher production of antibodies. Notice how long the primary response takes and the low level of antibody production. No wonder we get sick on first exposures to pathogens!

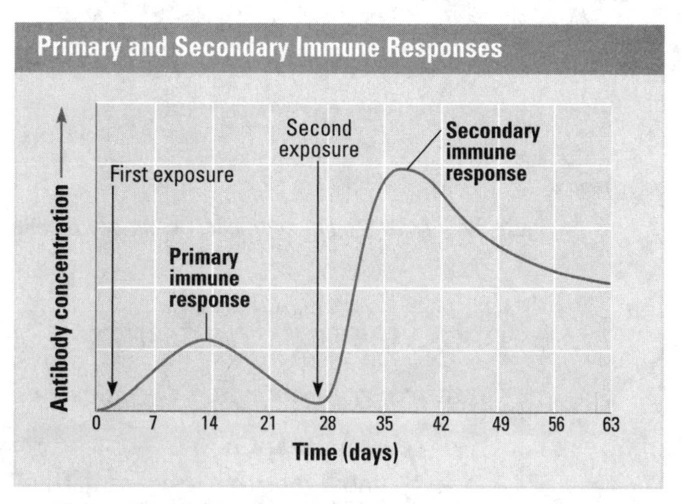

Figure 9.3 Primary and secondary immune responses

Adaptive immunity defends against infection of body fluids and body cells (Biology, *10e: 43.3,* Focus, *1e: 35.3*)

- Adaptive immunity has two branches:
  - **Humoral immune response** involves the activation and clonal selection of effector B cells, which produce antibodies that circulate in the blood.
  - **Cell-mediated immune response** involves the activation and clonal selection of cytotoxic T cells, which identify and destroy infected cells.
- **Helper T cells** aid both responses. When activated by interaction with the class II MHC molecule of an antigen presenting cell (APC), they secrete *cytokines* that stimulate and activate both B cells and cytotoxic T cells. Refer to Figure 9.4 to see the central role of helper T cells.
- **Cytotoxic T cells** bind to class I MHC molecules, displaying antigenic fragments on the surface of infected body cells. Cytotoxic T cells destroy infected body cells.
- Recall that activated B cells produce memory cells as well as plasma cells. Plasma cells secrete antibodies in prodigious numbers. These will circulate in the blood and bind and destroy the antigen.
- **Active immunity** develops naturally in response to an infection; it also develops artificially by immunization (vaccination). In immunization, a nonpathogenic form of a microbe or part of a microbe elicits an immune response, resulting in immunological memory for that microbe.

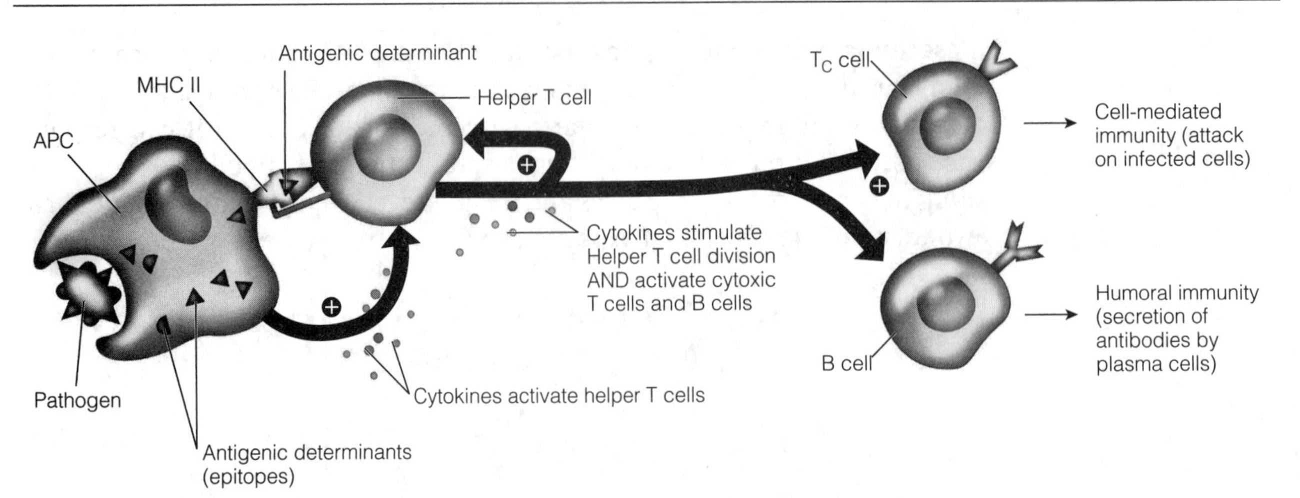

Figure 9.4 Activation of cytotoxic T cells and B cells by helper T cells

> ***ORGANIZE YOUR THOUGHTS***
>
> Confused by all the different cell names?
>
> 1. **B cells** make antibodies, which provide humoral immunity. This helps fight pathogens that are circulating in body fluids.
> 2. **Cytotoxic T cells** destroy body cells that are infected by a pathogen or cancer cells.
> 3. **Helper T cells** activate both B and T cells.

- **Passive immunity** occurs when an individual receives antibodies, such as those passed to the fetus across the placenta and to infants via milk.
- Certain antigens on red blood cells determine whether a person has **type A, B, AB**, or **O blood**. Because antibodies to nonself blood antigens already exist in the body, transfusion with incompatible blood leads to destruction of the transfused cells and a life-threatening situation for the patient.
- **MHC molecules** are responsible for stimulating the rejection of tissue grafts and organ transplants. The chances of successful transplantation are increased if the donor's tissue-bearing MHC molecules closely match the recipient's. The recipient also must take immunosuppressant drugs.

Disruptions in immune system function can elicit or exacerbate disease (Biology, *10e: 43.4,* Focus, *1e: 35.3*)

- In localized **allergies** such as hay fever, IgE antibodies produced after first exposure to an allergen attach to receptors on mast cells. The next time the same allergen enters the body, it bonds to mast cell–associated IgE molecules, inducing the cell to release histamine and other mediators that cause vascular changes and typical allergy symptoms.
- **Lupus, rheumatoid arthritis, type I diabetes mellitus**, and **multiple sclerosis** are examples of autoimmune diseases. In each case, the immune system turns against particular molecules of the body, allowing cytotoxic T cells to attack and damage the body's own healthy cells.

- **HIV** infects helper T cells. Refer again to Figure 9.4. Can you see why people with AIDS are immune suppressed? Note what cell is central in both humoral and cell-mediated immunity.

> ***CONNECT WITH THE CURRICULUM FRAMEWORK***
>
> This section has six specific areas in which you are expected to demonstrate understanding of the immune system and the LO asks you to create a representation or model to describe immune responses. Do not underestimate the importance of mastery in this area! (EK 2.D.4) In addition, EK 3.D.2 looks at cell signaling, and you are expected to know how cells within the immune system communicate.

Osmoregulation and Excretion

(Biology, *10e: Chapter 44*, Focus, *1e: Chapter 32*)

> **WHAT'S IMPORTANT TO KNOW?**
>
> This topic is not an explicit topic of the Curriculum Framework, but the role of a nephron in maintaining solute and water balance makes it an excellent example of a homeostatic mechanism. We consider the role of osmoregulation in homeostasis to be a main point of this chapter.

Osmoregulation balances the uptake and loss of water and solutes (Biology, *10e: 44.1*, Focus, *1e: 32.3*)

- **Osmoregulation** is the process by which animals control solute concentrations and balance water gain and loss.
- Most metabolic wastes must be excreted from the body. One of the most important types is **nitrogenous wastes** from the breakdown of proteins and nucleic acids.
- **Excretion** includes the removal of nitrogenous wastes from the body.

Diverse excretory systems are variations on a tubular theme (Biology, *10e: 44.3*, Focus, *1e: 32.3*)

- Most excretory systems produce urine in a four-step process. Focus on the location of each of these processes in the mammalian kidney, which is introduced in the next two concepts and shown in Figure 9.5.
- **Nephrons** are the functional units of the kidney. Each kidney has approximately 1 million! Within the nephron, liquids are filtered from the blood, wastes removed, and excess ions and toxins moved into the filtrate. Water balance is regulated along the course of passage through the nephron, and then the modified filtrate passes through the collecting tubules of the nephrons and empties into the ureters and passes as urine to the bladder.

The nephron is organized for stepwise processing of blood filtrate (Biology, *10e: 44.4,* Focus, *1e: 32.3*)

- Urine formation in the nephron requires four steps. Study Figure 9.5, and note what occurs in each step.

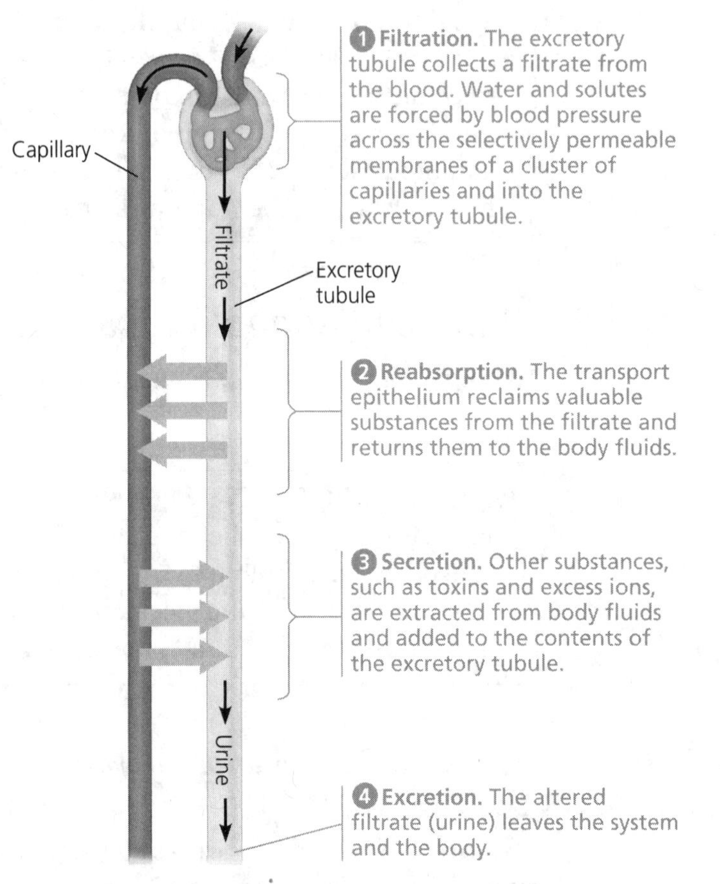

Figure 9.5 Key functions of excretory systems

- **Filtration** occurs when blood enters the kidney and filtrate is formed. Blood cells and proteins do *not* enter the filtrate.
- **Reabsorption** occurs within the tubes of the nephron. Valuable substances are reclaimed as well as water. Aquaporins are water transport proteins that provide channels for the free movement of water.
- **Secretion** adds toxins and excess ions to the filtrate, removing them from the blood to help maintain homeostasis.
- **Excretion** occurs when the filtrate leaves the body as the collecting tubules carry urine toward the ureters, then to the bladder, then out the urethra.

- A feature of the nephron is the loop of Henle, where water is removed from the filtrate. This is an example of a **countercurrent** system and allows the kidney to form concentrated urine with minimal water loss.

Hormonal circuits link kidney function, water balance, and blood pressure (Biology, *10e: 44.5*, Focus, *1e: 32.4*)

- **Antidiuretic hormone** (ADH) is an important hormone in the regulation of water balance. It is produced in the hypothalamus and released from the pituitary gland. It makes the collecting ducts more permeable to water so more water leaves the filtrate, resulting in more concentrated urine and reduced loss of water from the body. (A diuretic helps increase urine volume; *anti*diuretic hormone reduces urine volume.)
- Maintenance of blood volume and blood pressure involves a number of homeostatic mechanisms. What happens if blood pressure or blood volume drops due to dehydration or blood loss? Examine Figure 9.6 while reading the next three steps.

  1. The loss of blood pressure triggers the release of **renin**, an enzyme. Renin activates angiotensin II.
  2. **Angiotensin II** acts as a hormone and causes arterioles to constrict, which will raise blood pressure. It will also cause the adrenal glands to release the hormone **aldosterone**.
  3. **Aldosterone** causes the kidney to reabsorb more Na^+, which increases retention of water and blood volume and pressure.

- In summary, this pathway in the kidney helps in *long-term blood pressure regulation*. Note the two ways in which angiotensin II increases blood pressure.

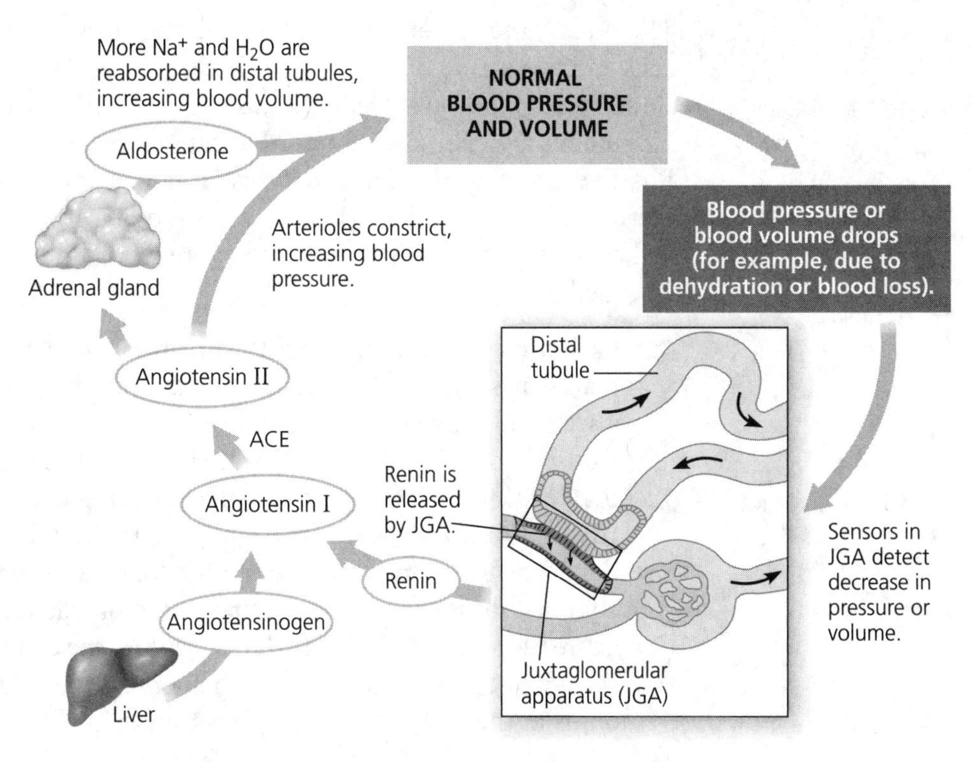

Figure 9.6 Regulation of blood volume and blood pressure

Hormones and the Endocrine System

(Biology, *10e: Chapter 45*, Focus, *1e: Chapter 32*)

YOU MUST KNOW

- How hormones bind to target receptors and trigger specific pathways.
- The secretion, target, action, and regulation of at least two hormones.
- An illustration of both positive and negative feedback in the regulation of homeostasis by hormones.

CONNECT WITH THE CURRICULUM FRAMEWORK

The endocrine system is one of three human systems that are explicitly included in the Curriculum Framework. Feedback systems control the release of hormones so you will want to be familiar with a couple of specific hormones and their regulation.

Hormones and other signaling molecules bind to target receptors, triggering specific response pathways (Biology, *10e: 45.1*, Focus, *1e: 32.2*)

- The **endocrine system** of an animal is the sum of all its hormone-secreting cells and tissues.
- **Endocrine glands** are *ductless* and secrete hormones directly into body fluids.
- **Hormones** are chemical signals that cause a response in *target cells.*
- **Positive and negative feedback** regulate most endocrine secretion.
- Here is a review of the two ways hormones initiate cellular responses. Recall that these mechanisms were covered in some detail earlier in Topic 2, The Cell.
    - **Cell-surface receptors** bind the hormone, and a *signal transduction pathway* is triggered. A signal transduction pathway consists of a series of molecular events that initiate a response to the signal. *Example*: The binding of epinephrine to liver cells causes a cascade that leads to the conversion of glycogen to glucose.
    - **Intracellular receptors** are found within the cell where they are bound by hormones that are lipid-soluble. The receptor-hormone complex then acts as a transcription factor, causing a change in gene expression. *Example*: Testosterone enters the cell and binds to its specific receptor; this complex then enters the nuclei of target cells, binds to the DNA, and stimulates transcription of a specific gene.

Feedback regulation and coordination with the nervous system are common in endocrine signaling (Biology, *10e: 45.2,* Focus, *1e: 32.2*)

- A feedback loop linking the response back to the initial stimulus is characteristic of hormone control pathways.
  - Regulation often involves **negative-feedback** loops, in which the response reduces the initial stimulus. By decreasing hormone signaling, negative-feedback loops prevent excessive pathway activity.
  - **Positive-feedback** loops reinforce a stimulus, leading to an even greater response. Note the examples of both types of feedback at the end of this chapter. You should be ready to go with an example of each type of feedback for the AP exam.
- Hormones in the body can affect one tissue, a few tissues, or most of the tissues in the body (as with the sex hormones), or they may affect other endocrine glands (these last are referred to as **tropic hormones**).
- The **hypothalamus** receives information from nerves throughout the body and from other parts of the brain and then initiates endocrine signals in response.
- The **posterior pituitary** is an extension of the hypothalamus that stores and releases these two hormones (which are produced in the hypothalamus):
  - **Oxytocin** causes contraction of the uterine muscles in childbirth and ejection of milk in nursing.
  - **Antidiuretic hormone (ADH)** makes the collecting tubules of the kidney more permeable to water, increasing water retention.
- The **anterior pituitary** consists of endocrine cells that synthesize and secrete several hormones. Some of these are *tropic hormones*, which stimulate the activity of other endocrine tissues (FSH, LH, TSH, and ACTH).
  - **Follicle-stimulating hormone (FSH)** stimulates development of the ovarian follicles in females and promotes spermatogenesis in males by acting on the cells in the seminiferous tubules.
  - **Luteinizing hormone (LH)** triggers ovulation in females and stimulates the production of testosterone by the interstitial cells of the testes.

Endocrine glands respond to diverse stimuli in regulating homeostasis, development, and behavior (Biology, *10e: 45.3,* Focus, *1e: 32.2*)

- The maintenance of blood calcium level shown in Figure 9.7 is one example of how homeostasis is maintained by *negative feedback*. Remember, in negative feedback, *more gets you less.*
- *Oxytocin regulation* is an example of *positive feedback.* Oxytocin stimulates uterine contractions and also stimulates production of prostaglandins by the placenta. Prostaglandins stimulate release of more oxytocin and more prostaglandins, stimulating more uterine contractions. Remember, in positive feedback, *more gets you more.*

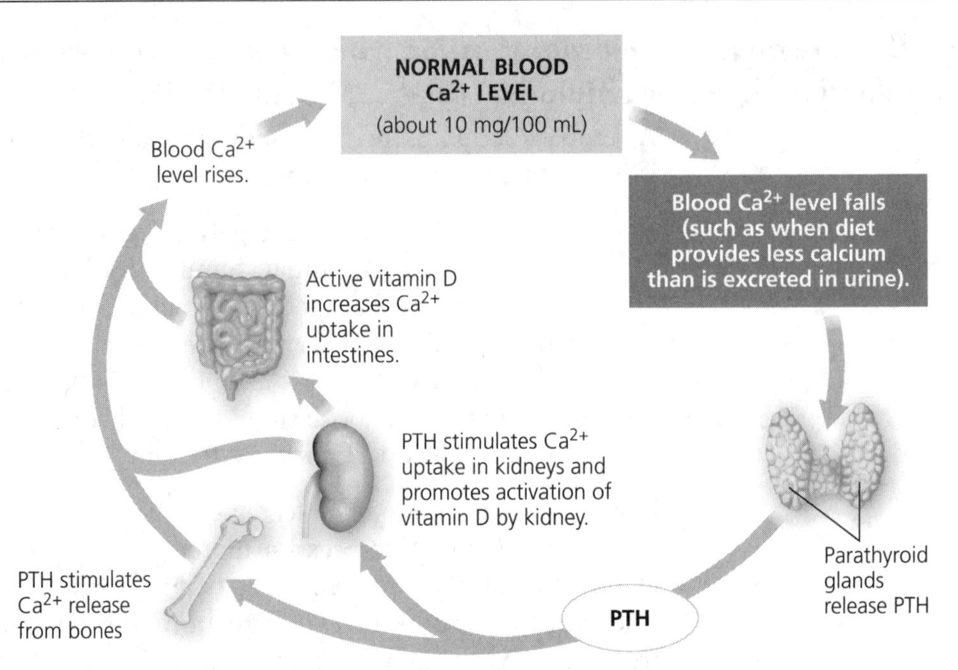

Figure 9.7 Regulation of blood calcium levels

> ***ORGANIZE YOUR THOUGHTS***
>
> Now that you have reviewed the negative-feedback regulation of blood calcium, prepare your own figure to show the regulation of blood glucose by insulin and glucagon.

Animal Reproduction

(Biology, *10e: Chapter 46*, Focus, *1e: Chapter 36*)

> **WHAT'S IMPORTANT TO KNOW?**
>
> Although you are not expected to know the anatomy of the human reproductive system nor specific hormonal controls of reproduction, this chapter comprises material that is included in the framework explicitly and as illustrative examples.

> **YOU MUST KNOW**
>
> - Advantages of asexual vs. sexual reproduction.
> - Various reproductive strategies in response to energy availability.
> - Timing and coordination of reproduction may be triggtered by environmental cues as well as pheromones.

Both asexual and sexual reproduction occur in the animal kingdom (Biology, *10e: 46.1,* Focus, *1e: 36.1*)

- **Sexual reproduction** is the creation of offspring by the fusion of haploid gametes to form a diploid zygote. The female gamete is the *ovum*, and the male gamete is the *sperm*.
- **Asexual reproduction** is reproduction in which all genes come from one parent; there is no fusion of egg and sperm. There are several modes of asexual reproduction.
  - **Fission** is the separation of a parent into two or more individuals of about the same size.
  - **Budding** occurs when new individuals arise from outgrowths of the parent. This is seen in corals and hydras.
  - **Fragmentation** occurs when an individual breaks into several pieces, all of which then may form complete adults. *Regeneration*, the regrowth of body parts, is a necessary part of fragmentation. This mode of reproduction can be seen in starfish, sponges, and cnidarians.
  - **Parthenogenesis** is the process in which a female produces eggs that develop *without* being fertilized. Male bees are produced this way and are always haploid.
- Asexual reproduction has the advantage of production of more offspring identical to the parent. So why sex? It may result in beneficial gene combinations that arise through recombination that speed up adaptation in a changing environment. The shuffling of genes during sexual reproduction might allow a population to rid itself of sets of harmful genes more readily.

Fertilization depends on mechanisms that bring together sperm and eggs of the same species (Biology, *10e: 46.2,* Focus, *1e: 36.1*)

- **Fertilization** is the union of sperm and egg.
- **External fertilization** occurs when eggs are shed by the female and fertilized by the male outside the female's body, usually in water. The release of gametes must be synchronous and often involves environmental cues or courtship behaviors. Species with external fertilization in general produce very large numbers of gametes.
- **Internal fertilization** occurs when sperm are deposited in the female reproductive tract, and fertilization occurs within the tract. It is an adaptation that allows sperm to reach an egg even when the environment is dry. Fewer gametes and fewer zygotes are often produced.
- When fertilized eggs are protected by shells or within the female's body, fewer zygotes are produced.
- *Pheromones* are chemicals produced by members of one species that can influence the physiology and behavior of other members of the species. They are often important as sexual attractants and can trigger reproductive receptiveness.
- The timing of reproductive behaviors is triggered by various cues, such as day length or resource availability. This will allow gametes to be produced by both males and

females at the same time, increasing the probability of fertilization. Offspring will emerge when environmental conditions and food resources are most favorable.

- Global climate change has resulted in many examples of energy resources emerging earlier in the season, such as sprouting green plants and hatching insects. Species that depend on these resources for successful reproduction may not be ready for this early emergence, miss the resource, and suffer negative reproductive consequences due to the lack of energy.

Reproductive organs produce and transport gametes (Biology, *10e: 46.3,* Focus, *1e: 6.2*)

- **Spermatogenesis** is the production of mature sperm cells by meiosis in the testes.
- **Oogenesis** is the development of mature ova by meiosis in the ovaries. In humans, it begins during embryonic development, when the cells progress to prophase I of meiosis I. These egg cells are quiescent until puberty. From puberty onward, hormones periodically stimulate a follicle to grow and its egg cell to complete meiosis I and begin meiosis II.
- Spermatogenesis differs from oogenesis in several ways:
  - Production of four sperm cells versus a single egg.
  - Spermatogenesis continues throughout a mature male's entire life; oogenesis ends with menopause.
  - Spermatogenesis is continuous after puberty, whereas oogenesis has long interruptions. Recall that in humans, eggs are arrested in prophase I prior to a female's birth! Meiosis is not completed until after fertilization.

The interplay of tropic and sex hormones regulates mammalian reproduction (Biology, *10e: 46.4,* Focus, *1e: 36.3*)

> *TIP FROM THE READERS*
> Because the hormonal control of reproduction is rich in examples of feedback and illustrates how the circulation of hormones in the blood affects target tissues, this is a good process to understand.

- In both male and female humans, the coordinated actions of hormones from the hypothalamus, anterior pituitary, and gonads govern reproduction. Study Figure 9.8 to see this interplay in males.
- The hypothalamus produces a "releasing hormone," which stimulates the anterior pituitary to produce FSH and LH. These are both *tropic hormones,* which stimulate other endocrine cells.
- The targets of both FSH and LH are cells in the testes. FSH triggers spermatogenesis and production of inhibin. LH promotes spermatogenesis as well as production of testosterone.
- Note the role of both inhibin and testosterone in this pathway. Why is this considered negative feedback?
- Humans and other primates have **menstrual cycles**. Menstruation occurs when the endometrium is shed from the uterus through the cervix and vagina. Other mammals have **estrous cycles** when the vagina is receptive to mating.

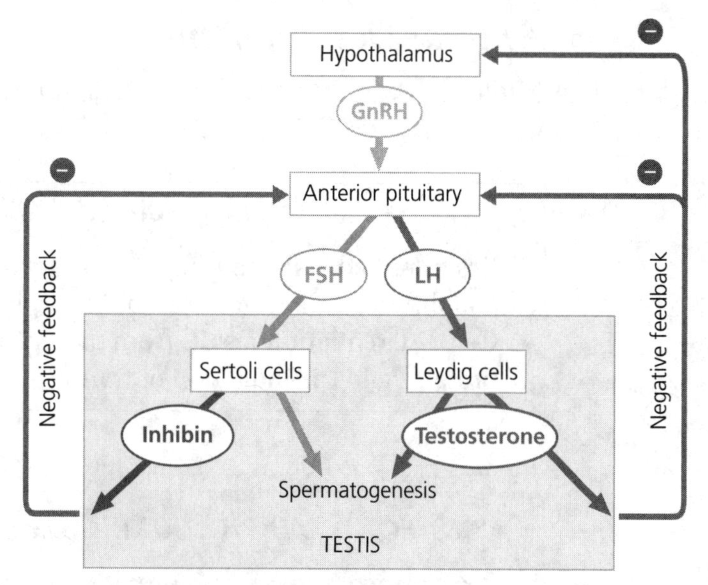

Figure 9.8 Hormonal control of the testes

- This cycling is under hormonal influence, and there is an interplay between pituitary hormones, whose target tissues are in the ovary, and ovarian hormones. This is analogous to what occurs in males as described above.
- As in males, the hypothalamus produces a "releasing hormone," which stimulates the anterior pituitary to produce FSH and LH, which stimulate cells within the ovary.
- The target of FSH is the follicle (where an egg is maturing). As its level increases, the follicle enlarges and matures.
- **LH** induces the final development of the follicle and triggers ovulation.
- The mature follicle will produce two hormones.
    - **Estradiol** is an estrogen secreted by the ovarian follicle. As the follicle develops, the level of estradiol increases and the uterine lining develops.
    - **Progesterone** is produced by the ruptured follicle after ovulation. High progesterone and estradiol levels promote further endometrial development.
- If fertilization of the egg does not occur, there will be a sharp decline in levels of estradiol and progesterone that will lead to menstruation.
- Note that in both males and females, FSH and LH stimulate cells in the ovaries and testes to produce the gametes as well as sex hormones, and that high levels of these sex hormones will inhibit the hypothalamus and anterior pituitary in another example of negative-feedback control.

In placental mammals, an embryo develops fully within the mother's uterus (Biology, *10e: 46.5,* Focus, *1e: 36.4*)

- **Human chorionic gonadotropin** (HCG) is a hormone secreted by the developing embryo that acts to maintain the uterine lining in early pregnancy. Its high level in the urine is used as the basis of a common early pregnancy test.

Animal Development

(Biology, *10e: Chapter 47*, Focus, *1e: Chapter 36*)

YOU MUST KNOW

- Programmed cell death (apoptosis) plays a role in normal development and differentiation (for example, morphogenesis).
- Cell differentiation results from the expression of genes for tissue-specific proteins and the induction of transcription factors.

CONNECT WITH THE CURRICULUM FRAMEWORK

Development involves complex interactions and communication between cells and so is addressed in several parts of the framework. Keep the importance of time and signaling in mind as you study development.

- Timing and coordination of specific events are necessary for the normal development of an organism, and these events are regulated by a variety of mechanisms. (EK 2.E.1) There are four LOs associated with this!
- Interactions between external stimuli and regulated gene expression result in specialization of cells, tissues, and organs. (EK 4.A.3)

Fertilization and cleavage initiate embryonic development (Biology, *10e: 47.1*, Focus, *1e: 36.4*)

- Molecules and events at the egg surface play a crucial role in each step of fertilization.
  - First, sperm dissolve or penetrate any protective layer surrounding the egg to reach the plasma membrane.
  - Next, molecules on the sperm surface bind to receptors on the egg, helping ensure that a sperm of the same species fertilizes the egg.
  - Finally, changes at the surface of the egg prevent polyspermy, the entry of multiple sperm nuclei into the egg. These changes include depolarization, much like what occurs in neuron transmission and removal of the sperm-binding receptors from the egg membrane.
- Across a range of animal species, embryonic development involves common stages that occur in a set order. Study Figure 9.9 as you read the accompanying text.

1. A zygote is formed when the sperm fertilizes the egg. This is followed by the **cleavage** stage, a period of rapid mitotic cell division without accompanying cell growth that convert the embryo to a hollow ball of cells called the **blastula.**

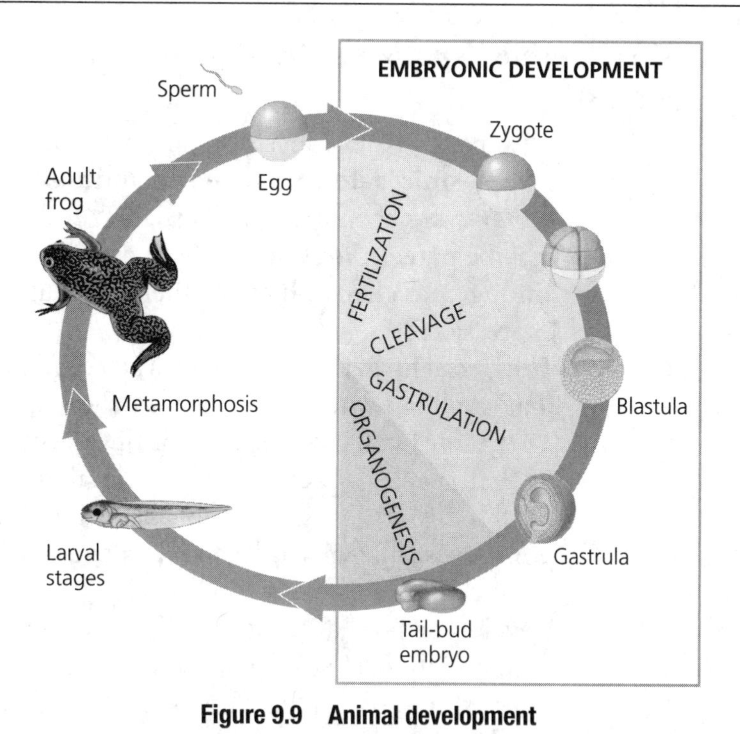

Figure 9.9 Animal development

2. Next, the blastula folds in on itself, rearranging into a three-layered embryo, the **gastrula**, in a process called gastrulation. The three layers are the *ectoderm, endoderm,* and *mesoderm.* The chart that follows gives some of the most important derivatives of each germ layer.

Ectoderm	Mesoderm	Endoderm
• Skin, nails, teeth • Lens of eye • Nervous system	• Skeletal, muscular systems • Excretory, circulatory systems • Reproductive system • Blood, bone, and muscle	• Epithelial *linings* of digestive, respiratory, excretory tracts • Liver, pancreas

3. The last stage, **organogenesis,** is the development of the three germ layers into rudimentary organs.

Morphogenesis in animals involves specific changes in cell shape, position, and survival (Biology, *10e: 47.2*)

- The last two stages of embryonic development are responsible for morphogenesis, the cellular and tissue-based processed by which the animal body takes place.
- Certain cells in the embryo are programmed to change shape or location. Others are programmed to die. A type of *programmed cell death* called **apoptosis** is a common feature of animal development. Familiar examples include the cells in the tail of a tadpole, which undergo apoptosis during frog metamorphosis, or the elimination of webbing between the digits in many birds and mammals.

Cytoplasmic determinants and inductive signals contribute to cell fate specification (Biology, *10e: 47.3*)

- The patterns of development are due to a combination of different **cytoplasmic determinants** and **inductive cell signals**. *Cytoplasmic determinants* are chemical signals such as mRNAs and transcription factors that may be parceled out unevenly in early cleavages. *Induction* is an interaction among cells that influences their fate, usually by causing changes in gene expression.
- The dorsal lip of the blastopore is an "organizer" that induces a series of events that result in formation of the notochord and neural tube.
- **Totipotent cells** are capable of developing into all the different cell types of that species. The cells of mammalian embryos may remain totipotent until the 16-cell stage but have a limited developmental potential from that point onward. Cells at this early stage that are totipotent are referred to as **stem cells**.
- Development is a sequence of events marked by cycles of signaling and differentiation. Cells in a developing embryo receive and respond to different signaling molecules that vary with their location within the embryo, and their developmental potential becomes more limited as embryonic development proceeds.

Neurons, Synapses, and Signaling

(Biology, *10e: Chapter 48*, Focus, *1e: Chapter 37*)

YOU MUST KNOW

- The anatomy of a neuron.
- The role of active transport in establishing the membrane potential of a neuron.
- How long-distance and short-distance signaling is done in neurons.
- The mechanisms of impulse transmission in a neuron.
- The process that leads to release of neurotransmitter, and what happens at the synapse.

CONNECT WITH THE CURRICULUM FRAMEWORK

All of the material in this chapter is covered by EK 3.E.2: Animals have nervous systems that detect external and internal signals, transmit and integrate information, and produce responses.

The nervous system is one of three human systems explicitly included in the framework, so take your time with this material.

Neuron organization and structure reflect function in information transfer (Biology, *10e: 48.1*, Focus, *1e: 37.1*)

- The **neuron** is the functional unit of the nervous system (Figure 9.10). It is composed of a **cell body**, which contains the nucleus and organelles; **dendrites**, which are cell extensions that receive incoming messages from other cells; and **axons**, which transmit messages to other cells.
- Many axons are covered by an insulating fatty **myelin sheath**. This speeds the rate of impulse transmission.
- The **synapse** is a junction between two neurons (or a neuron and a muscle fiber or gland).
- **Neurotransmitters** are chemical messengers released from vesicles in the **synaptic terminals** into the synapse. They will diffuse across the synapse and bind to receptors on the neuron, muscle fiber, or gland across the synapse, effecting a change in the second cell. *Examples*: acetylcholine, dopamine, and serotonin.
- At the synapse, the transmitting neuron is the *presynaptic cell* and the neuron, muscle, or gland that receives the signal is the *postsynaptic cell.*
- The **central nervous system (CNS)** consists of the brain and spinal cord, and the **peripheral nervous system (PNS)** consists of the nerves that communicate motor and sensory signals throughout the rest of the body.

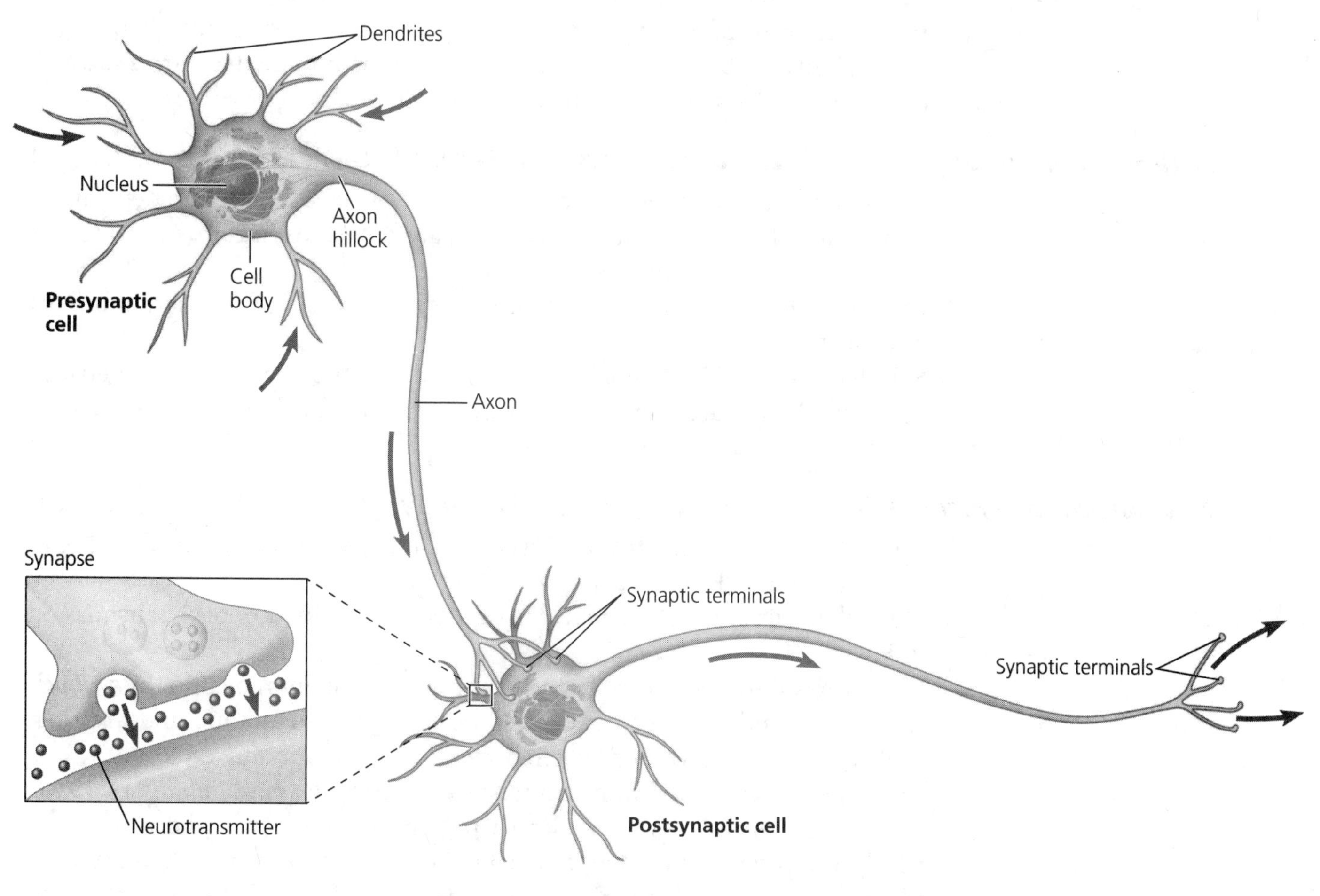

Figure 9.10 Structure of a vertebrate neuron

- **Sensory receptors** collect information about the world outside the body as well as processes inside the body. *Examples*: the rods and cones of the eye; pressure receptors in the skin.
- **Sensory neurons** transmit information from the eyes and other sensors that detect stimuli to the brain or spinal cord for processing.
- **Interneurons** connect sensory and motor neurons or make local connections in the brain and spinal cord.
- **Motor neurons** transmit signals to *effectors*, such as muscle cells and glands.
- **Nerves** are bundles of neurons. A nerve can contain all motor neurons, all sensory neurons, or be mixed.

Ion pumps and ion channels establish the resting potential of a neuron (Biology, *10e: 48.2*, Focus, *1e: 37.2*)

- **Membrane potential** describes the difference in electrical charge across a cell membrane.
- The membrane potential of a nerve cell at rest is called its **resting potential**. It exists because of differences in the ionic composition of the extracellular and intracellular fluids across the plasma membrane.
- The concentration of Na^+ is higher outside the cell, whereas the concentration of K^+ is higher inside the cell. This gradient is maintained by sodium-potassium pumps.
- Changes in the membrane potential of a neuron are what give rise to transmission of nerve impulses (action potentials). A stimulus changes the permeability of the membrane to Na^+, so the membrane depolarizes as Na^+ enter the cell. This may result in an action potential.

Action potentials are the signals conducted by axons (Biology, *10e: 48.3*, Focus, *1e: 37.3*)

- An **action potential** (nerve impulse) is an *all-or-none response* to depolarization of the membrane of the nerve cell. Carefully study Figure 9.11 to follow what happens in an action potential.
- In order to generate an action potential, a certain level of depolarization must be achieved, known as the **threshold**.
- **Saltatory conduction**, which is the jumping of the nerve impulse between *nodes of Ranvier* (areas on the axon not covered by the myelin sheath), speeds up the conduction of the nerve impulse.

Neurons communicate with other cells at synapses (Biology, *10e: 48.4*, Focus, *1e: 37.4*)

- The signal is conducted from the axon of a presynaptic cell to the dendrite of a postsynaptic cell via the **synapse**.
- Study the steps involved in neurotransmitter release and impulse transmission in Figure 9.12.
- Neurotransmitters are released by the presynaptic membrane into the *synaptic cleft*. They bind to receptors on the postsynaptic membrane and are then broken down by enzymes or taken back up into surrounding cells.
- There are two categories of neurotransmitters: *excitatory* and *inhibitory*. Excitatory neurotransmitters cause depolarization of the postsynaptic membrane, which brings the membrane potential closer to the threshold. Inhibitory neurotransmitters cause hyperpolarization of the postsynaptic membrane, which moves the membrane potential further from the threshold.

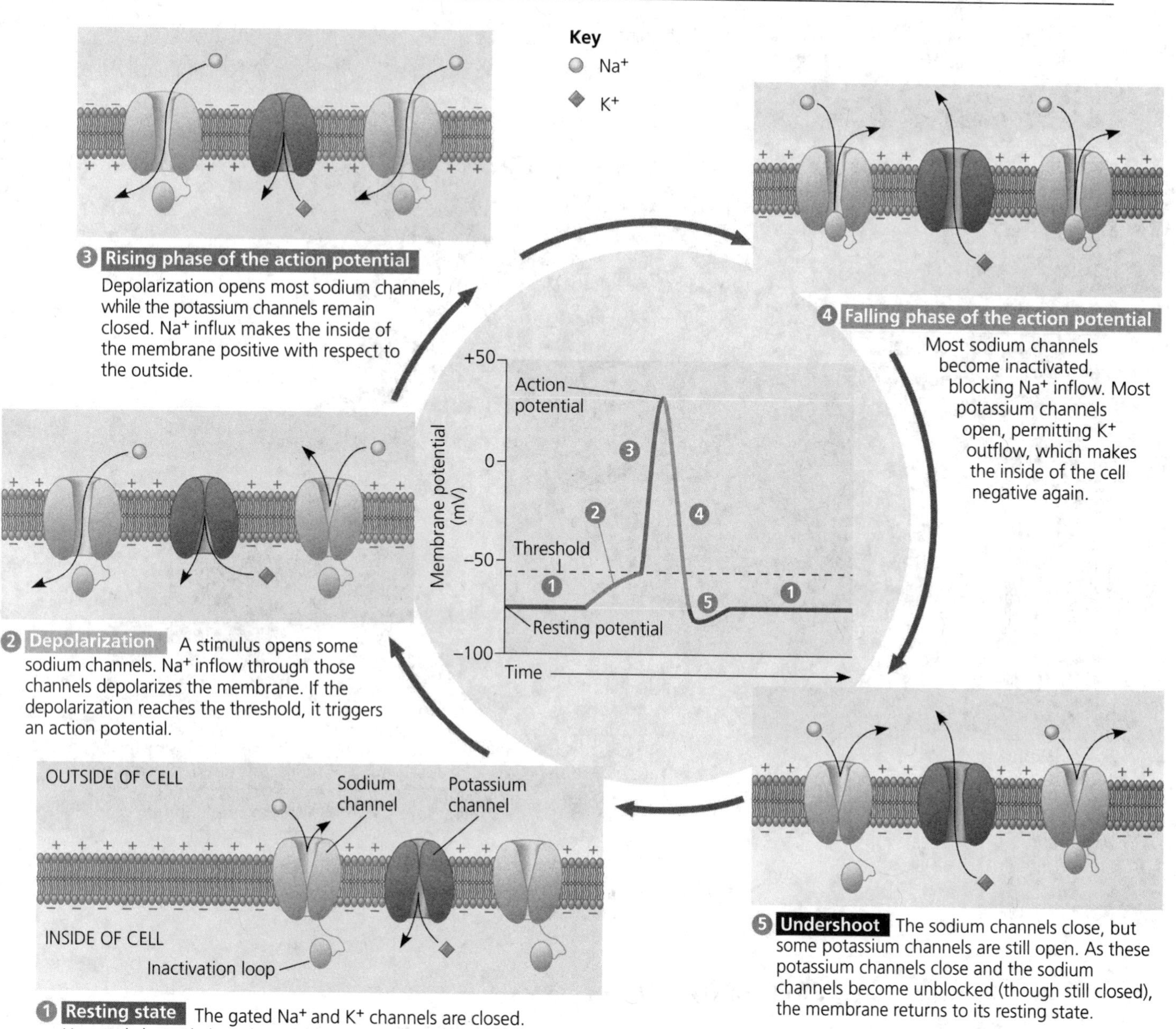

Figure 9.11 Conduction of an action potential

- How are signals terminated? Neurotransmitters are removed from the synapse and signaling ends by one of two mechanisms: enzymes break down the neurotransmitter or the neurotransmitter is recaptured into the presynaptic neuron and repackaged into synaptic vesicles.
- **Acetylcholine** is a very common neurotransmitter; it can be inhibitory or excitatory. It is released by neurons at the neuromuscular junction. Other common **neurotransmitters** are epinephrine, norepinephrine, dopamine, and serotonin.
- Acetylcholine is removed from the synapse by the enzyme acetylcholinesterase. The nerve gas sarin inhibits the enzyme, and the buildup of acetylcholine results in paralysis. On the other hand, botulism toxin prevents the release of acetylcholine, and so muscles cannot contract.

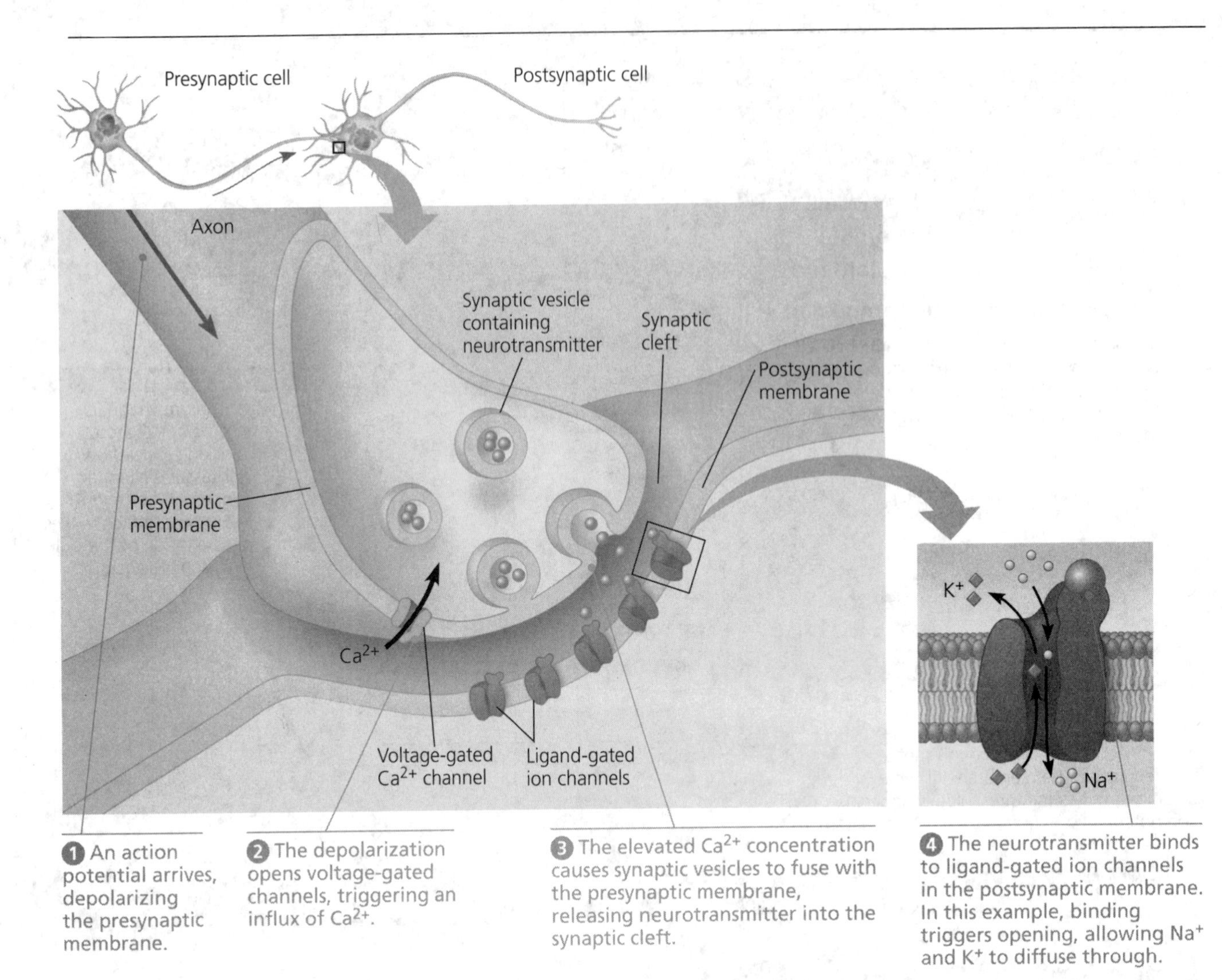

Figure 9.12 A chemical synapse

Nervous Systems

(Biology, *10e: Chapter 49*, Focus, *1e: Chapter 38*)

YOU MUST KNOW

- The brain serves as a master neurological center for processing information and directing responses.
- Different regions of the brain have different functions.
- Structures and associated functions for animal brains are products of evolution, and increasing complexity follows evolutionary lines.
- How the vertebrate brain integrates information, which leads to an appropriate response.

Nervous systems consist of circuits of neurons and supporting cells (Biology, *10e: 49.1*, Focus, *1e: 38.1*)

- A **reflex** is a simple automatic nerve circuit in response to a stimulus. Let's use a protective reflex, such as jerking your finger off a flame, as an example:
    - A *stimulus* is detected by a *receptor* in the skin, conveyed via a *sensory neuron* to an *interneuron* in the spinal cord, which synapses with a *motor neuron*, which will cause the *effector*, a muscle cell, to contract.
- Note that at its simplest level, conscious thought is not required in a reflex.
- **Cerebrospinal fluid** circulates through a central canal in the spinal cord and the ventricles of the brain, bathing cells with nutrients and carrying away wastes. It also cushions the brain and spinal cord.
- **Gray matter** consists of mainly neuron cell bodies and unmyelinated axons.
- **White matter** is white because of the myelin sheaths about the axons.
- **Glia** are cells that support neurons. **Schwann cells**, which form myelin sheaths in the PNS, are glial cells.
- Study Figure 9.13 as you review the organization of the nervous system.

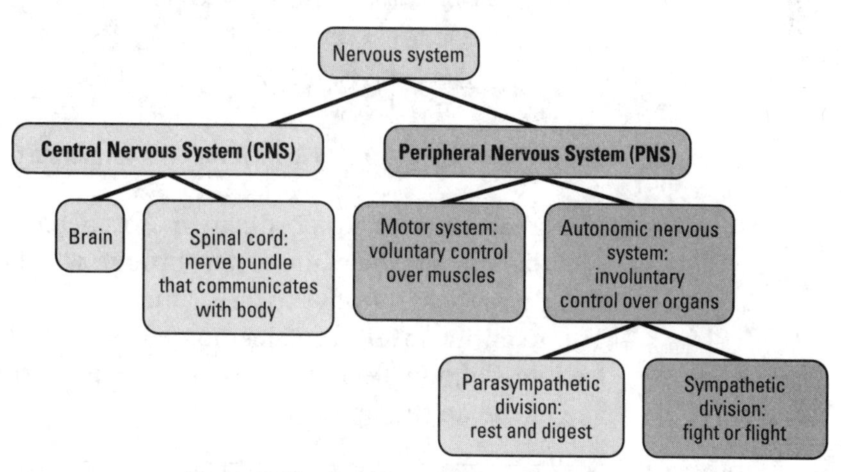

Figure 9.13 Branches of the nervous system

- The **central nervous system (CNS)** is the brain and spinal cord.
- The **peripheral nervous system (PNS)** consists of paired cranial and spinal nerves and associated ganglia. It is divided into the following:
    - The *motor (somatic) nervous system*, which carries signals to skeletal muscles. It is a voluntary system.
    - The *autonomic nervous system*, which regulates the primarily automatic, visceral functions of smooth and cardiac muscles. This is the involuntary system.
- The **autonomic nervous system** transmits signals that regulate the internal environment by controlling smooth and cardiac muscle, including those in

the gastrointestinal, cardiovascular, excretory, and endocrine systems. Its divisions are as follows:

- The **sympathetic division**, which, when activated, causes the heart to beat faster and adrenaline to be secreted (with all its effects).
- The **parasympathetic division**, which has the opposite effect when activated, slowing heartbeat and digestion.

The vertebrate brain is regionally specialized (Biology, *10e: 49.2*, Focus, *1e: 38.2*)

- The **brainstem** is made up of the medulla oblongata, pons, and midbrain. The brainstem controls homeostatic functions, such as breathing, swallowing, and digestion, and conducts sensory and motor signals between the spinal cord and higher brain centers (Figure 9.14).

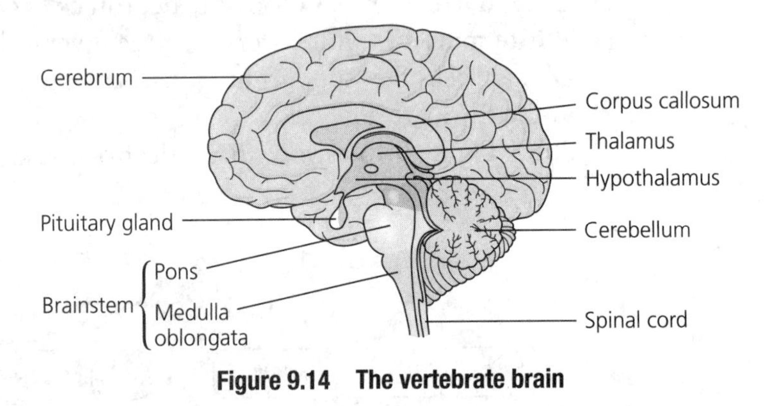

Figure 9.14 The vertebrate brain

- The **cerebellum** helps coordinate motor, perceptual, and cognitive functions.
- The **thalamus** is the main center through which sensory and motor information passes to and from the cerebrum.
- The **hypothalamus** regulates homeostasis; basic survival behaviors such as feeding, fighting, fleeing, and reproducing; thermostat, appestat, thirst center, and circadian rhythms.

> ***STUDY TIP*** Different regions of the brain have different functions. You are not expected to know each region and its function, but you should have a firm grasp of an illustrative example.

- The **cerebrum** has two hemispheres, each with a covering of gray matter over white matter. Information processing is centered here, and this region is particularly extensive in mammals.
- The **cerebral cortex** controls voluntary movement and cognitive functions.
- The **corpus callosum** is a thick band of axons that enables communication between the right and left cortices.

> ***CONNECT WITH THE CURRICULUM FRAMEWORK***
>
> There are eight different objectives (LO 3.43–3.50) specific to the nervous system, so work hard on this area!

Sensory and Motor Mechanisms

(Biology, *10e: Chapter 50*, Focus, *1e: Chapters 38, 39*)

WHAT'S IMPORTANT TO KNOW?

Sensory and motor mechanisms are not an explicit topic of the Curriculum Framework, but your teacher may choose to introduce you to this material. This chapter contains information that may be used to develop both multiple-choice and essay questions and a general understanding of this section will make you more confident in dealing with potential questions. Here are what we consider to be the main points of this chapter:

- Different sensory receptors respond to various types of input.
- Neurons communicate with muscle fibers to stimulate contraction.
- Interaction of cellular organelles leads to muscle contraction.

Sensory receptors transduce stimulus energy and transmit signals to the central nervous system (Biology, *10e: 50.1*, Focus, *1e: 38.4*)

- A sensory receptor is a sense organ or cell that detects stimuli. Different types respond to different inputs. For example, *photoreceptors* in the eye respond to light, *chemoreceptors* function in taste and smell, and *thermoreceptors* in the skin detect heat or cold.
- Other types of receptors include mechanoreceptors, which respond to physical stimuli, such as touch or vibration, and pain receptors.
- **Reception** occurs when a receptor detects a stimulus. **Perception** occurs in the brain as this information is processed. As an example, when you view an optical illusion in which a figure seems to change, what is actually changing is your perception of the object.

The physical interaction of protein filaments is required for muscle function (Biology, *10e: 50.5*, Focus, *1e: 39.1*)

- **Skeletal muscle** is attached to bones and is responsible for their movement. It consists of long fibers, each of which is a single muscle cell.
- Each muscle fiber is a bundle of **myofibrils**, which in turn are composed of two kinds of myofilaments: **thin filaments** and **thick filaments**.
    - The thin filaments are *actin* and a regulatory protein.
    - The thick filaments are *myosin*.
- The **sarcomere** is the basic contractile unit of the muscle. Study the sarcomere in Figure 9.15 as you read about the general action of a sarcomere.
- Notice that during **muscle contraction**, the length of the sarcomere is reduced. The actin filaments are fixed and slide over the myosin in such a way that their degree of overlap increases. This is known as the **sliding-filament model** of muscle contraction.

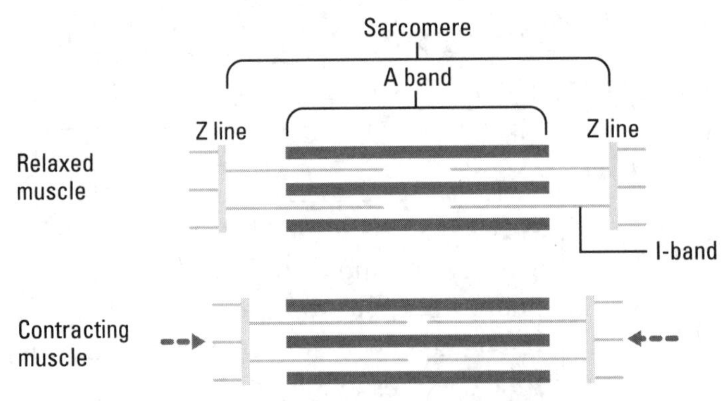

Figure 9.15 Sections of a sarcomere

- Use Figure 9.16 as you review what happens during depolarization of a muscle fiber and what leads to the movement of actin and myosin.
- A **motor neuron** will cause a muscle fiber to contract when its depolarization causes the neurotransmitter *acetylcholine* to be released into the synapse of the *neuromuscular junction.*
- As acetylcholine binds to receptors on the muscle fiber, ion channels open on the muscle cell membrane. This triggers an *action potential* (wave of depolarization) in the muscle cell.
    - The action potential spreads along **T tubules (transverse tubules)** to the **sarcoplasmic reticulum**.
    - This depolarization causes the sarcoplasmic reticulum to release **calcium ions** and now actin and myosin can interact as myosin heads attach to the actin filaments after being phosphorylated by ATP.
    - The muscle contracts, with actin filaments sliding over myosin.

CONNECT WITH THE CURRICULUM FRAMEWORK

How a neuron stimulates a muscle cell and the mechanism of muscle contraction both involve cell signaling and the interaction of many organelles. Because this integrates several areas of your course, we consider this a useful example to understand.

The electrical, chemical, and molecular events regulating skeletal muscle contraction are shown in a cutaway view of a muscle cell and in the enlarged diagram below. Action potentials (horizontal arrows) triggered by the motor neuron sweep across the muscle fiber and into it along the transverse (T) tubules, initiating the movements of calcium that regulate muscle activity.

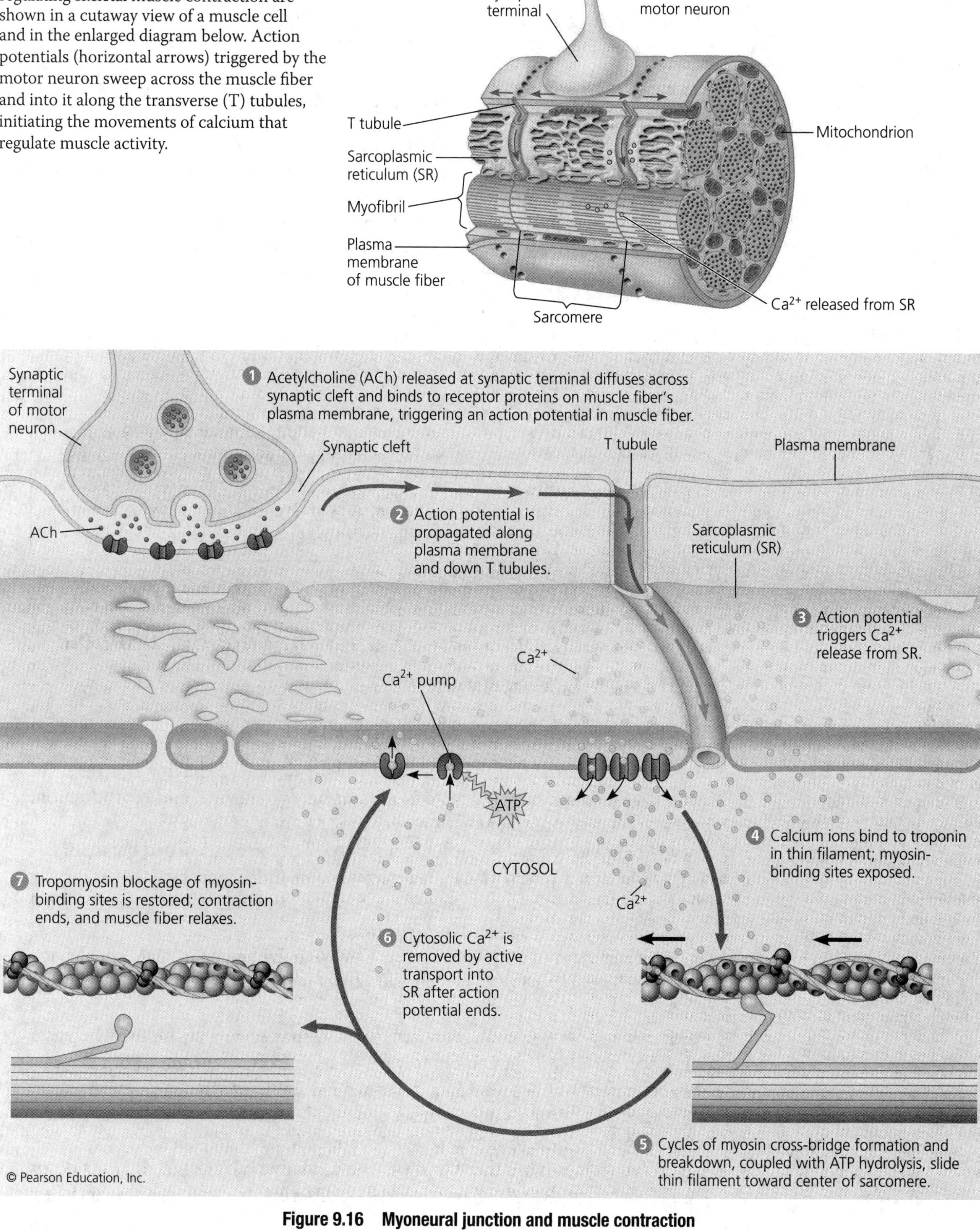

Figure 9.16 Myoneural junction and muscle contraction

Animal Behavior

(Biology, *10e: Chapter 51*, Focus, *1e: Chapter 39*)

YOU MUST KNOW

- How behaviors are the result of natural selection.
- How innate and learned behaviors increase survival and reproductive fitness.
- How organisms use communication to increase fitness.
- The role of altruism and inclusive fitness in kin selection.

CONNECT WITH THE CURRICULUM FRAMEWORK

There are at least six different LOs (2.38–2.40, 3.40–3.42) that specifically address behavior and the common theme for all of them is that organisms are able to communicate with each other in various ways and that behaviors are selected because they serve to increase an individual's fitness while contributing to the overall survival of the population. With each aspect of behavior, keep the evolutionary significance in mind.

Discrete Sensory Inputs Can Stimulate Both Simple and Complex Behaviors

(Biology, *10e: 51.1*, Focus, *1e: 39.3*)

- **Behavior** is what an animal does and how it does it. Behavior is a result of genetic and environmental factors, is essential for survival and reproduction, and is subject to natural selection over time.
- **Innate behaviors** are developmentally fixed. They are unlearned behaviors.
- A **fixed action pattern (FAP)** is a sequence of unlearned acts that is largely unchangeable and usually carried to completion once it is initiated. Fixed action patterns are triggered by *sign stimuli.*
- A classic example of an FAP was noted by *Niko Tinbergen* in male stickleback fish, which attack red objects. The red object is the sign stimulus; the attack is the FAP.
- A **kinesis** is a simple change in activity in response to a stimulus, whereas a taxis is an automatic movement toward or away from a stimulus. For example, the movement of pill bugs toward a moist habitat is a taxis.
- **Migration** is a complex behavior seen in a wide variety of animals. Navigation may be by detection of the earth's magnetic field or visual cues.
- **Circadian rhythms** are those that occur on a daily cycle. Other rhythms occur over longer periods and can be triggered by differing day lengths or lunar cycles.

- A **signal** is a stimulus that causes a change in the behavior of another individual and is the basis for animal communication. Examples of signals follow.
  - **Pheromones,** the chemical signals that are emitted by members of one species that affect other members of the species.
  - **Visual signals,** such as the warning flash of white of a mockingbird's wing.
  - **Auditory signals,** such as the screech of a blue jay or song of a warbler.

Learning establishes specific links between experience and behavior (Biology, *10e: 51.2,* Focus, *1e: 39.4*)

- **Learned behaviors** are behaviors that are modified based on specific experiences. (Recall that innate behaviors are inherited.)
- **Imprinting** is a combination of learned and innate components that are limited to a *sensitive period* in an organism's life and is generally irreversible. Konrad Lorenz demonstrated imprinting in greylag geese. When hatchlings spent their first few hours with him, they followed him as though he were their mother!
- **Associative learning** is the ability of many animals to associate one feature of their environment with another feature, such as associating certain stimuli or behaviors with reward or punishment.
- Twin studies in humans indicate that both environment and genetics contribute significantly to behaviors.
- Behavior can be directed by genes. For example, a single gene appears to control the courtship ritual in fruit flies.

Selection for individual survival and reproductive success can explain most behaviors (Biology, *10e: 51.3,* Focus, *1e: 39.5*)

- Keep in mind that behaviors are selected for the advantages they provide for survival and reproduction. Consider different methods for obtaining food, or foraging. What strategies provide the most calories for the effort to obtain the food?
- **The optimal foraging model** proposes that there is a compromise between the benefits of nutrition and the cost of obtaining food. This is why a lion does not feed on ants.
- **Mating systems** vary between species. The needs of the young are important constraints in the development of these systems. Various systems are:
  - **Promiscuous** with no strong pair-bonds.
  - **Monogamous** with one male/one female.
  - **Polygamous** with one individual mating with several others.
- Continue to keep in mind the evolutionary advantage of a behavior. Which of the mating systems above would be most advantageous to a species that produce helpless young that require food delivered to them? (This would probably be monogamy because both parents would be able to guard the young while one forages.)

Inclusive fitness can account for the evolution of behavior, including altruism (Biology, *10e: 51.4,* Focus, *1e: 39.6*)

- **Altruism** occurs when animals behave in ways that reduce their individual fitness but increase the fitness of other individuals in the population. *Example*: A blue jay giving an alarm call attracts attention to its location.
- **Inclusive fitness** is the total effect an individual has on proliferating its genes by producing its own offspring and by providing aid that enables other close relatives to produce offspring. The natural selection that favors this kind of altruistic behavior by enhancing reproductive success of relatives is called **kin selection.**
- Just checking to see if you are really awake as you read all this! Here is one of our favorite quotes on the topic, from J. B. S. Haldane: "I would lay down my life for two brothers or eight cousins." Can you explain what is meant?

Level 1: Knowledge/Comprehension Questions

The Knowledge/Comprehension questions review essential content knowledge but don't represent the type of question you will see on the AP exam. Level 2, Application/Synthesis questions are similar to those you will see on the exam.

1. Which of the following is required in ALL living things in order for gas exchange to occur?
   (A) lungs
   (B) gills
   (C) moist membranes
   (D) blood

2. During transmission across a typical chemical synapse,
   (A) the binding of neurotransmitters to receptors initiates exocytosis.
   (B) action potentials trigger chemical changes that make the synaptic vesicles fuse with each other.
   (C) vesicles containing neurotransmitters diffuse to the receiving cell's plasma membrane.
   (D) neurotransmitter molecules bind to receptors in the receiving cell's plasma membrane.

3. Epinephrine (adrenalin) raises blood glucose levels, increases metabolic rate, and increases blood flow to skeletal muscles. Which of the following is most likely to result in a release of epinephrine from the adrenal glands?
   (A) falling asleep in front of the TV
   (B) watching a checkers tournament
   (C) doing yoga
   (D) fleeing from a harmful situation

4. Oxygen is transported in human blood by which type of cell?
   (A) erythrocytes (red blood cells)
   (B) leukocytes (white blood cells)
   (C) lymphocytes (type of white blood cell)
   (D) platelets

5. Once the threshold potential is reached after a neuron is stimulated, which of the following occurs?
   (A) The interior of the cell becomes negative with respect to the outside.
   (B) K^+ channels open.
   (C) Na^+ channels close.
   (D) An action potential is generated.

6. Salivary amylase, an enzyme secreted in saliva, begins the breakdown of molecules that will produce monosaccharides. Which category of nutrients does this enzyme digest?
   (A) starches
   (B) proteins
   (C) lipids
   (D) nucleic acids

7. Pepsin in the stomach begins the enzymatic digestion of protein. What molecules are the end products of protein digestion?
   (A) nucleic acids
   (B) amino acids
   (C) fattty acids
   (D) monosaccharides

8. In the mammalian heart, the sinoatrial (SA) node generates electrical impulses and is sometimes called the pacemaker of the heart. Which of the following correctly describes its function?
   (A) It delays the nerve impulse to the walls of the ventricle.
   (B) It controls the action of the atrioventricular valve.
   (C) It regulates the amount of blood that exits with each heartbeat.
   (D) It sets the rate and timing of cardiac muscle contraction.

9. Fluid lost from the capillaries and not reabsorbed is returned to the blood via
   (A) the venous system.
   (B) the arteriole system.
   (C) the lymphatic system.
   (D) capillary beds.

10. In the blood, carbon dioxide is primarily transported in what way?
    (A) by hemoglobin
    (B) as carbon monoxide
    (C) in erythrocytes (RBCs)
    (D) as bicarbonate ions

11. Barrier defenses are an important non-specific arm of the immune system. All of the following are barrier defenses EXCEPT
    (A) skin.
    (B) phagocytes.
    (C) lysozyme in saliva.
    (D) mucous secretions.

12. An immune response to a specific antigen generates the production of which type of cell that launches an attack the next time that same antigen infects the body?
    (A) effector cells
    (B) memory cells
    (C) T cells
    (D) antibodies

13. Muscle cell contraction occurs via the
    (A) contraction of the sarcoplasmic reticulum.
    (B) plasma membrane.
    (C) flow of calcium ions out of the cell.
    (D) sliding of the thin filaments by the thick filaments.

14. The succession of rapid cell division that follows fertilization is called
    (A) gastrulation.
    (B) cleavage.
    (C) morphogenesis.
    (D) organogenesis.

15. Which of the following correctly describes a reflex arc?
    (A) receptor, sensory neuron, effector
    (B) motor neuron, associative neuron, sensory neuron
    (C) receptor, motor neuron, sensory neuron
    (D) effector, sensory neuron, receptor

16. Which of the following is released into the synaptic cleft and acts as a signaling molecule?
    (A) sodium ions
    (B) calcium ions
    (C) neurotransmitter
    (D) receptor proteins

Use this figure to answer the next question.

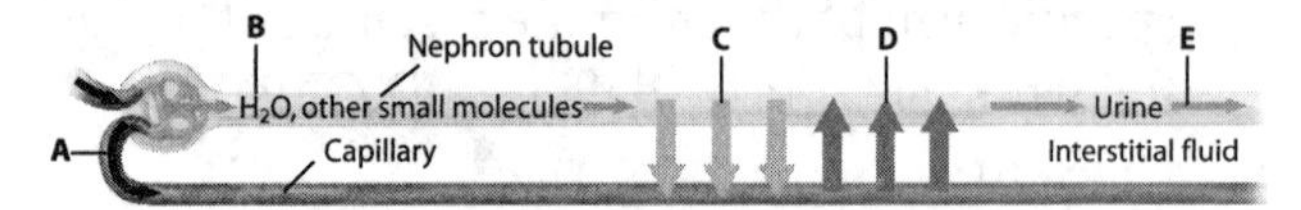

17. Which arrow in this schematic view of the nephron shows reabsorption?
    (A) arrow A
    (B) arrow B
    (C) arrow C
    (D) arrow D

18. What type of cell acts as an intermediary between humoral and cell-mediated immunity?
    (A) plasma cell
    (B) cytotoxic T cell
    (C) B cell
    (D) helper T cell

19. What occurs during gastrulation?
    (A) Individual cells of the embryo divide but do not grow.
    (B) Cytoplasmic determinants signal the development of right and left sides of the embryo.
    (C) A hollow blastula is changed into an embryo that has three tissue layers.
    (D) A neural tube is created by invagination of the ectoderm.

Use this figure to answer the next two questions.

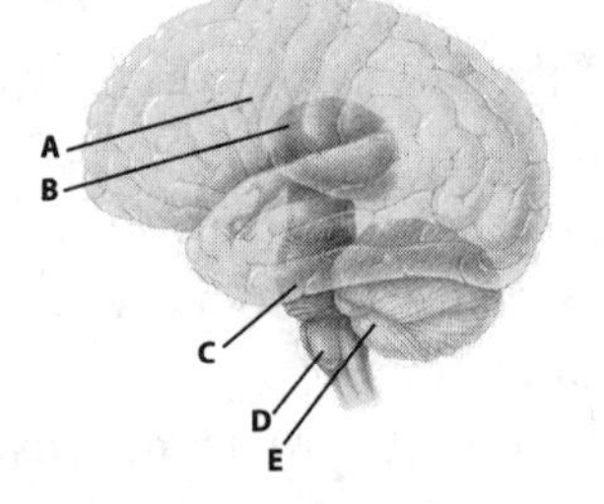

20. Which part of this diagram of the human brain depicts the brain region that coordinates muscular functions such as walking and tying shoelaces?
    (A) part A
    (B) part B
    (C) part C
    (D) part E

21. Which structure in the diagram of the human brain is the center for the control of the involuntary centers for breathing, heartbeat, respiration, and digestion?
    (A) part B
    (B) part C
    (C) part D
    (D) part E

22. Which of these correctly describes the autonomic nervous system?
    (A) It integrates sensory inputs to the brain.
    (B) It carries signals to and from skeletal muscles.
    (C) It regulates the internal environment of the body.
    (D) It is part of the central nervous system.

23. Which part of this figure shows an active plasma B cell?

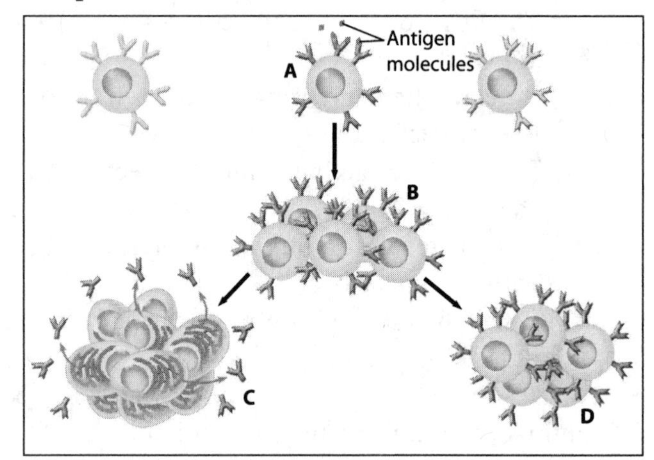

    (A) part A
    (B) part B
    (C) part C
    (D) part D

Level 2: Application/Analysis/Synthesis Questions

After reading the paragraph, answer the question(s) that follow.

Under normal conditions, blood sugar levels are controlled within a narrow range by negative feedback. Two hormones are involved in maintaining blood sugar levels at the set point (about 90 mg of glucose/100 mL of blood). When blood sugar levels rise above the set point, the hormone insulin signals the liver to absorb the excess sugar. When blood sugar levels drop below the set point, the hormone glucagon signals the liver to release its stored glucose to the bloodstream. In type 1 diabetes, the body doesn't produce enough insulin and insulin supplements are required.

1. Based on your understanding of homeostasis, for negative-feedback control of blood glucose levels to function properly
   (A) the control center for glucose must be somewhere in the digestive system.
   (B) there must be sensors that monitor blood glucose levels.
   (C) there must be several other hormones involved (in addition to insulin and glucagon).
   (D) the body must prevent glucose levels from changing even slightly.

2. If you hadn't eaten for several hours, how would your glucose levels be returned to the set point?
   (A) Insulin would be released to move glucose into cells and the liver.
   (B) Glucagon would be released to cause the liver to release glucose.
   (C) Pathways in cellular respiration would switch to using lipids and proteins for ATP production.
   (D) Insulin and glucagon would each be released to quickly return glucose levels to the set point.

3. An animal's inputs of energy and materials would exceed its outputs
   (A) if the animal is an endotherm, which must always take in more energy because of its high metabolic rate.
   (B) if it is actively foraging for food.
   (C) if it is growing and increasing its mass.
   (D) never; to maintain homeostasis, these energy and material budgets always balance.

4. Which statement best describes the difference in responses of effector B cells (plasma cells) and cytotoxic T cells?
   (A) B cells confer active immunity; cytotoxic T cells confer passive immunity.
   (B) B cells kill pathogens directly; cytotoxic T cells kill host cells.
   (C) B cells secrete antibodies against a pathogen; cytotoxic T cells kill pathogen-infected host cells.
   (D) B cells accomplish the cell-mediated response; cytotoxic T cells accomplish the humoral response.

5. Select the one description that is not a feature of the immune system.
   (A) An antibody has more than one antigen-binding site.
   (B) An antigen can cause different antibodies to be generated.
   (C) A pathogen may present more than one antigen.
   (D) A lymphocyte has receptors for multiple antigens.

After reading the paragraph, answer questions 6 and 7.

To protect U.S. soldiers serving overseas, each soldier receives vaccinations against several diseases, including smallpox, before deployment. Following intelligence about an imminent smallpox threat, the Army wants to ensure that soldiers stationed in war zones are fully protected from exposure to the disease, so all the soldiers in the threat zone are given a second vaccination against smallpox.

6. The first vaccination provides immunity because
   (A) a localized inflammatory response is initiated.
   (B) the vaccine contains manufactured antibodies against smallpox.
   (C) antigenic determinants in the vaccine activate B cells, which form plasma cells as well as memory cells.
   (D) the vaccine contains antibiotics and other drugs that kill the smallpox virus.

7. The second vaccination is beneficial because it
   (A) contains plasma cells that survive longer than 4–5 days.
   (B) stimulates production of a higher concentration of antibodies in the bloodstream.
   (C) requires two injections to stimulate antibody formation.
   (D) keeps previously produced plasma cells circulating in the bloodstream.

Figures A and B show two typical cell signaling cascades. Use them to answer questions 8–10.

8. Which figure shows a water-soluble signaling molecule, and what is the evidence for this?
   (A) Figure A because the signaling molecule results in a response within the cytoplasm.
   (B) Figure A because the signaling molecule does not pass through the phospholipid bilayer.
   (C) Figure B because the signaling molecule enters the cell to bind the receptor.
   (D) Figure B because the activated receptor is able to enter the nucleus of the cell.

9. In figure A, which molecule represents the receptor protein and how does it function?
   (A) Molecule A is the receptor protein because it binds to a plasma membrane surface protein.
   (B) Molecule B is the receptor protein, and molecule A is the ligand that binds it.
   (C) Molecule C is the receptor protein, and it is lipid soluble.
   (D) Molecule D is the receptor protein and once activated initiates a response.

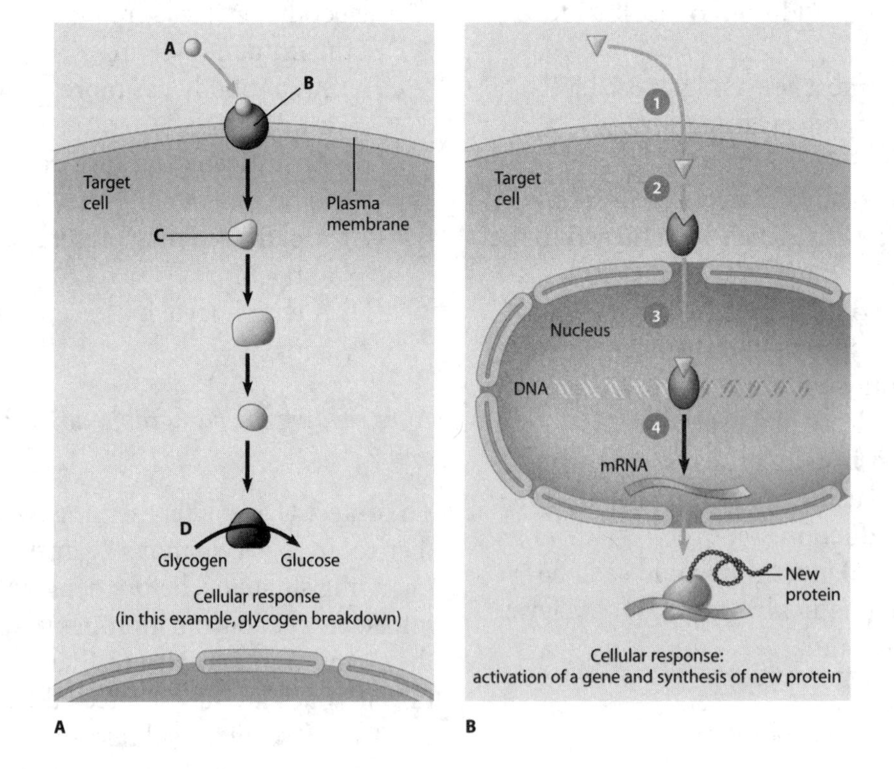

10. Which of the following correctly represents the events in Figure B?
    (A) In step 1, a signaling molecule binds a G protein-coupled receptor.
    (B) In step 2, the G protein-coupled receptor is activated.
    (C) In step 3, the bound receptor passes through the plasma membrane.
    (D) In step 4, the bound receptor functions as a transcription factor.

11. Which statement correctly describes the function of this neuron?

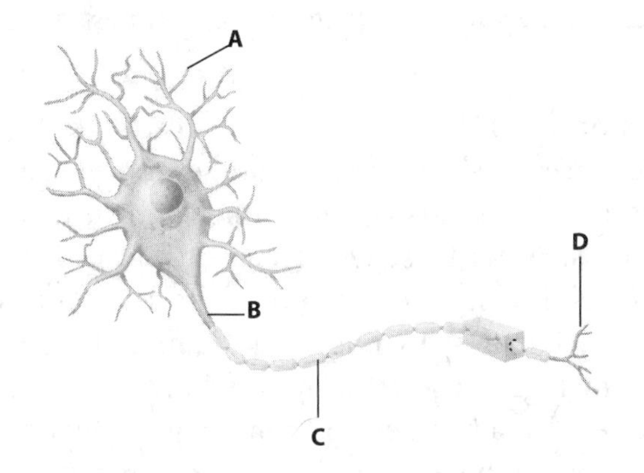

    (A) Structure C is produced by glial cells and increases the rate of impulse transmission.
    (B) Structure A releases a neurotransmitter when an action potential is generated.
    (C) Structure B is depolarized when calcium ions are released in this area.
    (D) Structure D initiates the action potential when depolarized.

12. Altruism is putting the well-being of another ahead of one's own. How could it persist in terms of natural selection?
    (A) Survival of other members of the species will maintain Hardy-Weinberg equilibrium.
    (B) It can result in the survival of the recipients of the behavior.
    (C) This is just one aspect of an important evolutionary principle, survival of the fittest.
    (D) If the altruistic member's behavior is directed toward a close relative, his genes will persist in the population.

After reading the paragraph, answer questions 13 and 14.

One commonly used insecticide contains a permanent acetylcholinesterase inhibitor. Acetylcholine is a neurotransmitter that stimulates skeletal muscle contraction. Acetylcholinesterase removes acetylcholine from the synapse after the signal is received. When insects are exposed to this insecticide, they will twitch for a brief period and then die. Exposure to high pesticide concentrations has a similar effect on humans and can also be caused by exposure to the nerve gas Sarin and other chemical agents.

13. Why does the insecticide cause uncontrollable twitching in insects?
    (A) Acetylcholine is released, but the insecticide prevents it from diffusing across the synapse.
    (B) Acetylcholine is released, but the insecticide prevents it from binding to the receptor sites of the postsynaptic neurons.
    (C) The insecticide causes continuous stimulation of the muscles.
    (D) The insecticide prevents acetylcholinesterase from being removed from the synapse.

14. Because pesticides affect humans in a manner similar to that of roaches, it would be valid to conclude that
    (A) acetylcholinesterase affects the DNA of all animals.
    (B) the mechanism of stimulating skeletal muscle contraction must be similar in humans and roaches.
    (C) pesticides are harmful to roaches but not to humans.
    (D) the terminal end of the axon releases acetylcholine in roaches, but not in humans.

15. Which of these describes a feature that is common to all action potentials (nerve impulses) generated along a neuron?
    (A) An action potential causes the membrane of the neuron to hyperpolarize and then depolarize.
    (B) An action potential can undergo temporal and spatial summation.
    (C) Action potentials are triggered by a depolarization that reaches the threshold.
    (D) All action potentials move at the same speed along all axons.

16. When the action potential reaches the synaptic terminals, which event will occur next?
    (A) Voltage-gated calcium channels in the membrane open.
    (B) Synaptic vesicles fuse with the membrane.
    (C) The postsynaptic cell produces an action potential.
    (D) Ligand-gated ion channels open, allowing neurotransmitters to enter the synaptic cleft.

17. Why are action potentials usually conducted in only one direction?
    (A) The nodes of Ranvier can conduct potentials in only one direction.
    (B) The brief refractory period prevents reopening of voltage-gated Na^+ channels.
    (C) The axon hillock has a higher membrane potential than the terminals of the axon.
    (D) Voltage-gated channels for both Na^+ and K^+ open in only one direction.

18. Which of the following conclusions is supported by this graph?

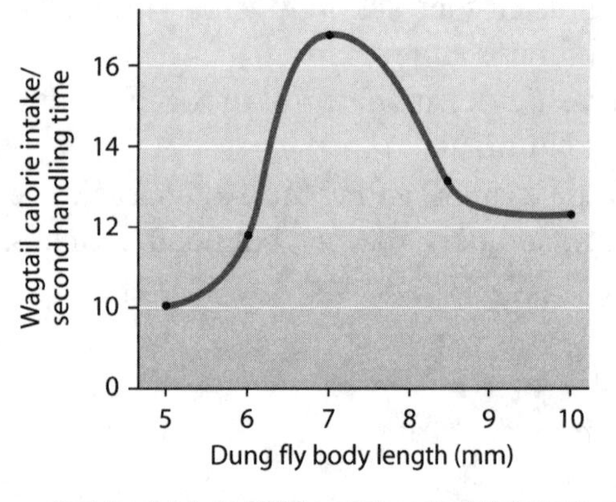

Graph from N. B. Davies. 1977. Prey selection and social behaviour in wagtails (Aves: Motacillidae). *Journal of Animal Ecology* 46: 37-57, fig. 9.

    (A) Prey size does not affect the number of calories gained per second of handling time by wagtails.
    (B) Wagtails get more calories per second of handling time with larger flies than with smaller ones.
    (C) Wagtails get more calories per second of handling time with smaller flies than with larger ones.
    (D) Wagtails get more calories per second of handling with 7-mm flies than with either larger or smaller ones.

After reading the paragraphs, answer questions 19 and 20.

Salmon hatch in freshwater streams, swim to the open ocean where they spend several years and migrate thousands of miles, then return to their home streams to spawn (reproduce) and die. A researcher is investigating how the salmon locate their original spawning area. Salmon from two distant regions, Ketchican and Cold Bay, will be studied. Data from experiments suggest that more than one type of homing mechanism may be involved in this behavior. When salmon arrive at a river mouth from the open sea, they appear to use olfactory cues to find their home streams, but how do they find their way back to the correct spot along the coastline from the open ocean?

19. Which of the following experimental approaches might the researcher propose to test the hypothesis that geomagnetic factors (the influence of Earth's magnetic field) play a key role in the ability of salmon to find the proper location along the coast?
    (A) Trap adult salmon in Ketchikan, Alaska, transport them to a different location on the Alaska Peninsula, Cold Bay, and see if they can return to their home stream.
    (B) Hatch salmon from Ketchikan, Alaska, subject them to the geomagnetic characteristics of Cold Bay, Alaska, then release them and see where they eventually return.

(C) Collect hatchling salmon from Ketchikan, Alaska, release them into the ocean, then see where they eventually return.
(D) Hatch salmon Ketchikan, Alaska, raise them in water from Cold Bay, Alaska, then release them and see where they eventually return.

20. What type of behavior would explain the ability of the salmon to return to their home streams?
(A) imprinting, which occurs when a behavior is learned during a critical period in early life
(B) innate behavior, which is developmentally fixed
(C) social learning, which is passed from one individual to another
(D) habituation, which occurs when the same stimulus is applied numerous times

21. Behaviors are subject to natural selection. Choose the description below that does not support this.
(A) Most essential behaviors are the result of some type of learning.
(B) An individual's reproductive success depends in part on how the behavior is performed.
(C) Some component of the behavior is genetically inherited but may be modified by learning.
(D) An individual's genotype influences its behavioral phenotype.

22. Which of the following describes an example of negative feedback?
(A) the movement of sodium across a membrane through a transport protein and the movement of potassium in the opposite direction through the same transport protein
(B) the pressure of the baby's head against the uterine wall during childbirth that stimulates uterine contractions, causing greater pressure against the uterine wall, which produces still more contractions
(C) a population of bacteria in a petri dish that grows until it has used all its nutrients and then declines
(D) a heating system in which the heat is turned off when the temperature exceeds a certain point and is turned on when the temperature falls below a certain point

23. Pepsinogen is an inactive precursor of pepsin that can be activated by hydrochloric acid. The muscular stomach wall could be digested by the active proteolytic enzyme, pepsin. There are several mechanisms, listed below, that will prevent digestion of the stomach wall from happening. Identify the one that will not help prevent this autodigestion.
(A) Pepsin is stored and secreted in an inactive form as pepsinogen.
(B) Mucus lines the inside surface of the stomach and presents a barrier to digestive enzymes.
(C) Mitosis generates enough new cells to replace the stomach lining every few days.
(D) Pepsinogen is activated by pepsin and hydrochloric acid.

24. Which of the following represents a failure of a homeostatic mechanism?
(A) activation of the thirst center after profuse sweating
(B) production of insulin after eating a jelly donut
(C) hypothermia after falling into a glacier-fed lake
(D) skin flushing and sweating during exercise

Grid-In Questions

These call for a numerical response.

1. At this moment your heart is pumping about 70 mL of blood per heartbeat and your heart is beating at a rate of 72 beats per minute. How many liters of blood will you pump in the next hour? (Answer to the nearest tenth.)

2. The partial pressure of a gas is the pressure exerted by a particular gas in a mixture of gases. For example, at sea level the pressure of the atmosphere, a mixture of gases, is 760 mm Hg. Oxygen makes up 21% of the mixture, so the partial pressure of oxygen is ($760 \times 0.21 = 159.6$ mm Hg) about 160 mm Hg. A mountain climber is about to summit a small peak with an atmospheric pressure of 510 mm Hg. What is the difference in the partial pressure of oxygen at sea level compared to the mountain peak?

Free-Response Question

1. *Natural selection favors behaviors that increase survival and reproductive behaviors. For each of the following types of behaviors, describe an example in nature, and justify how this behavior is adaptive.*

(a) Innate behavior
(b) Learned behavior
(c) Cooperative behavior
(d) Chemical signals

TOPIC 10

Ecology

An Introduction to Ecology and the Biosphere

(Biology, *10e: Chapter 52*, Focus, *1e: Chapter 40*)

YOU MUST KNOW

- The role of abiotic factors in the formation of biomes.
- How biotic and abiotic factors affect the distribution of biomes.
- How changes in these factors may alter ecosystems.

Earth's climate varies by latitude and season and is changing rapidly (Biology, *10e: 52.1,* Focus, *1e: 40.1*)

- **Ecology** is the scientific study of the interactions between organisms and the environment.
- **Climate** is the long-term prevailing weather conditions in a given area. The major components that make up the climate are temperature, precipitation, sunlight, and wind. Climate patterns can be described on two scales: *macroclimate* and *microclimate.*
  - **Macroclimate patterns** work at the global, regional, or local level.
  - The changing angle of the sun over the year, bodies of water, and mountains exert seasonal, regional, and local effects on **macroclimate**.
  - **Microclimate** is determined by fine-scale variations, such as sunlight and temperature under a log.
  - Increasing greenhouse gas concentrations in the air are warming Earth and altering the distributions of many species. Some species will not be able to shift their ranges quickly enough to survive.

The structure and distribution of terrestrial biomes are controlled by climate and disturbance (Biology, *10e: 52.2,* Focus, *1e: 40.1*)

- **Biomes** are the major types of ecosystems that occupy very broad geographic regions.
- The importance of climate, especially precipitation and temperature, are reflected in the climograph for the major biomes of North America featured in Figure 10.1.

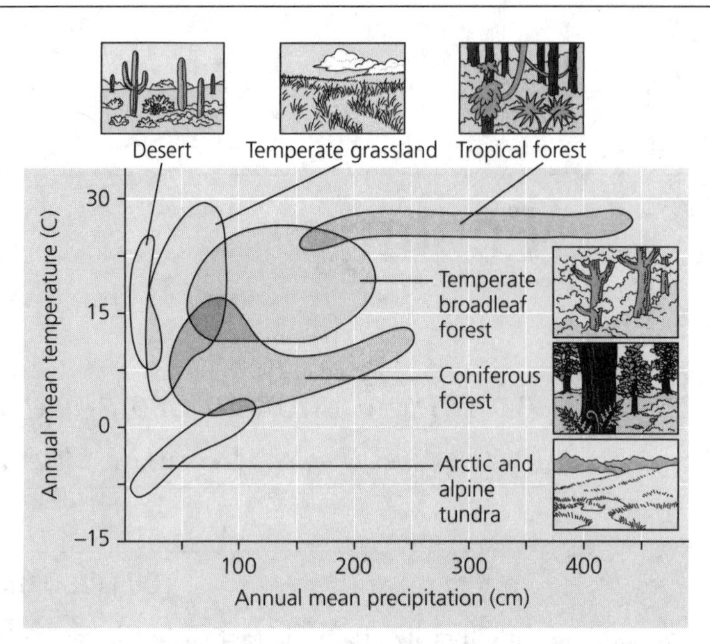

Figure 10.1 Climograph for some major biomes in North America

> ***STUDY TIP*** Why are some biomes rich in species abundance? Why is productivity higher in some ecosystems? You will not be expected to know the names of specific biomes, but you may be expected to predict and justify how alteration of a factor such as moisture or light availability or soil nutrients would impact the system. Always look for the big picture! Think *why? What if?*
> (Species abundance is related to primary productivity, which provides energy resources, and numerous niches. Productivity is higher in ecosystems that have abundant light energy, warmer temperatures, and longer growing seasons.)

- **Desert** is marked by sparse rainfall, and desert plants and animals are adapted to conserve and store water. Deserts contain many CAM plants and plants with adaptations that prevent animals from consuming them, such as the spines on cacti. Temperature (either hot or cold) can be extreme.
- **Chaparral** is dominated by dense, spiny, evergreen shrubs. These are coastal areas with mild rainy winters and long, hot, dry summers. Plants are adapted to fires.
- **Temperate grassland** is marked by seasonal drought with occasional fires and by large grazing mammals. All these factors prevent the significant growth of trees. Grassland soil is rich in nutrients, making these areas good for agriculture.
- **Temperate broadleaf forest** is marked by dense stands of deciduous trees that require sufficient moisture. These forests are more open than (and not as tall as) rain forests. They are stratified—the top layer contains one or two strata of trees; beneath that are shrubs; and under that is an herbaceous stratum. **Canopy** refers to the upper layers of trees in a forest. These trees drop their

leaves in fall, and many mammals enter hibernation. Many birds migrate to warmer climates.

- **Coniferous forest** is dominated by cone-bearing trees such as pine, spruce, and fir. The conical shape of conifers prevents much snowfall from accumulating on—and breaking—these trees' branches.
- **Tundra** is marked by permafrost (permanently frozen layer of soil), very cold temperatures, high winds, and little rainfall. Tundra supports no trees or tall plants. It accounts for about 20% of Earth's terrestrial surface.
- **Tropical rain forest** has pronounced vertical stratification. The canopy is so dense that little light breaks through. These forests are marked by epiphytes, which are plants that grow on other plants instead of the soil. Rainfall varies by season but averages 200–400 cm annually. Biodiversity is greatest of all the terrestrial biomes.

Aquatic biomes are diverse and dynamic systems that cover most of Earth (Biology, *10e: 52.3,* Focus, *1e: 40.2*)

- **Aquatic biomes** make up the largest part of the biosphere because water covers roughly 75% of Earth's surface. These biomes are classified into **freshwater biomes** and **marine biomes**.
- Most aquatic biomes display vertical stratification. For example, the **photic zone** includes the upper layer of water in which there is enough light for photosynthesis to occur, whereas the **aphotic zone** is characterized by very low light penetration.
- The **two types of freshwater biomes** are standing bodies of water, such as lakes and wetlands, and moving bodies of water, such as streams and rivers.
    - In lakes, communities are distributed according to the water's depth. **Oligotrophic lakes** are deep lakes that are nutrient-poor and oxygen-rich and contain sparse phytoplankton. **Eutrophic lakes** are shallower, and they have higher nutrient content and lower oxygen content with a high concentration of phytoplankton.
    - The prominent physical attribute of **streams and rivers** is current. Organisms are distributed in vertical zones and from the headwaters to the mouth.
    - **Estuaries** are areas where freshwater streams or rivers merge with the ocean. Estuaries are among the most productive habitats on Earth.

> *TIP FROM THE READERS*
> What biome do you live in? Focus on it as an illustrative example of a biome. Know the typical species found there and reasons for their distribution.

- ***Marine biomes*** include **coral reefs,** a biome created by a group of cnidarians that secrete hard calcium carbonate shells, which vary in shape and support the growth of other corals, sponges, and algae. Coral reefs are among the most productive ecosystems on Earth.

Interactions between organisms and the environment limit the distribution of species (Biology, *10e: 52.4*, Focus, *1e: 40.3*)

- The ecological study of species involves **biotic** (living) and **abiotic** (nonliving) influences.
  - **Biotic factors** may include behaviors as well as interactions with other species. Population and community ecology (see Chapters 53 and 54) will explore many of the interactions between organisms.
  - The **abiotic components** of an environment are the nonliving, chemical, and physical components. Some important abiotic factors include temperature, water, salinity, sunlight, and soil. Temperature and precipitation are the two most important abiotic factors in determining the distribution of biomes.

Population Ecology

(Biology, *10e: Chapter 53*, Focus, *1e: Chapter 40*)

YOU MUST KNOW

- How density, dispersion, and demographics can describe a population.
- The differences between exponential and logistic models of population growth.
- How density-dependent and density-independent factors can control population growth.
- How a change in matter or energy will affect the population or community. (LO 4.16)
- The effect of age distributions and fecundity on human populations as presented in age-structure pyramids.

Dynamic biological processes influence population density, dispersion, and demographics (Biology, *10e: 53.1*, Focus, *1e: 40.4*)

- A **population** is a group of individuals of a single species living in the same general area. **Population ecology** explores how biotic and abiotic factors influence the density, distribution, size, and age structure of populations.
- Three fundamental characteristics of the organisms in a population follow.
  - **Density** is the number of individuals per unit area or volume. The density of a population increases by births or immigration and decreases by deaths or emigration.
  - **Dispersion** is the pattern of spacing among individuals within the boundaries of the population.
    - The most common pattern of dispersion is *clumped*, with individuals in patches, usually around a required resource. *Example*: cottonwood trees along a stream in the arid Southwest.

- A *uniform* dispersion pattern is often the result of antagonistic interactions. Animals that defend territories often show a uniform pattern. *Example*: red-winged blackbirds during mating season.
- *Random* dispersion shows unpredictable spacing. This is not a common spacing in nature because there is usually a reason for a pattern of spacing.

- **Demography** is the study of vital statistics of a population, especially birth and death rates. A graphic way to show birth and death rates in a population is survivorship curves. Three types are shown in Figure 10.2.
  - **Type I** shows low death rates during early and midlife; then the death rate increases sharply in older age groups. This is a typical pattern for large organisms with long life spans.
  - **Type II** shows a constant death rate over the organism's life span. These are often organisms that are heavily preyed upon, so individuals die before reaching old age.
  - **Type III** shows very high early death rates, then a flat rate for the few surviving to older age groups. Many bird species show a high death rate for the first year, then a slowing for the remainder of their life span.

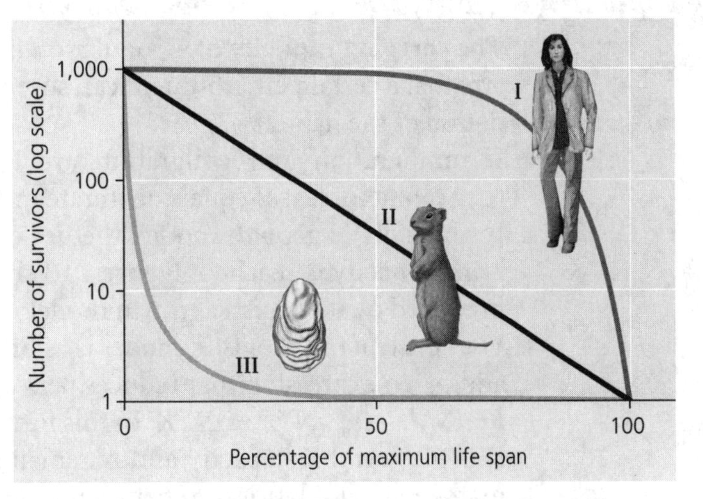

Figure 10.2 Survivorship curves: types I, II, and III

The exponential model describes population growth in an idealized, unlimited environment (Biology, *10e: 53.2,* Focus, *1e: 40.5*)

- **Exponential population** growth refers to population growth under ideal conditions. Figure 10.3 shows a graph of population growth as predicted by the exponential model. Any species, regardless of its life history, is capable of exponential growth if resources are abundant.
- Exponential population growth is shown by the equation $dN/dt = r_{max}N$. In this equation dN is the size of the population at a particular instant in time; dt is the time interval involved in the calculation; r_{max} is the maximum per capita rate of increase for the species under study; N is the population size.

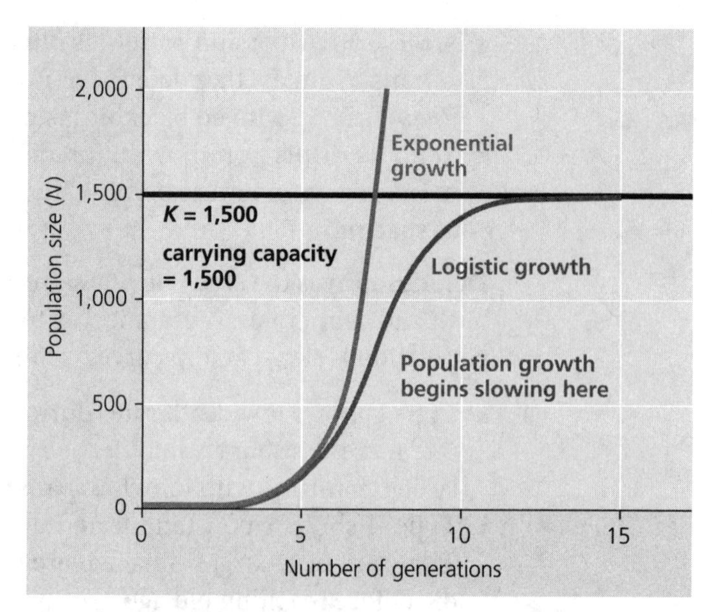

Figure 10.3 Population growth predicted by the exponential and logistic model

The logistic model describes how a population grows more slowly as it nears its carrying capacity (Biology, *10e: 53.3,* Focus, *1e: 40.5*)

- The **carrying capacity** of a population is defined as the maximum population size that a certain environment can support at a particular time with no degradation of the habitat.
- If immigration and emigration are ignored, a population's growth rate (per capita increase) equals birth rate minus death rate: $dN/dt = B - D$.
- In the **logistic growth model**, the per capita rate of increase declines as carrying capacity is reached. Figure 10.3 shows a graph of population growth as predicted by the logistic growth model.
- We construct the logistic model by starting with the exponential model and adding an expression that reduces per capita rate of increase as N approaches K: $dN/dt = r_{max}N\,(K - N)/K$. In this formula the new term is $(K-N)/K$, where K is the carrying capacity and N remains the population estimate. Study the mathematical relationship between K and N in the formula. What would the value of the expression $(K-N)/K$ be if $K=N$ and what growth would you predict for the population? You may check your answer on page 286 at the end of this topic.

> ***TIP FROM THE READERS***
> You will be given a formula sheet to use on the AP exam. The formulas for population growth, exponential growth, and logistic growth rate are included. You will need to practice using each of these! There is a practice problem at the end of this section.

Life history traits are products of natural selection (Biology, *10e: 53.4,* Focus, *1e: 40.6*)

- Traits that affect an organism's schedule of reproduction and survival make up its **life history**. Life histories entail three variables: How early? How often? How many?
  - How early in the life cycle does reproduction begin?
  - How often does the organism reproduce? Some organisms save their resources for one big reproductive event (*big-bang reproduction*), whereas others produce offspring in *repeated reproduction.*
  - How many offspring per reproductive event?
- Life history traits are evolutionary outcomes, *not* conscious decisions by organisms.
- Selection of life history traits that are sensitive to population density and carrying capacity are known as ***K*-selection**. *K*-selection operates in populations living close to the density imposed by the carrying capacity. By contrast, selection for life history traits that maximize reproductive success is called ***r*-selection**.
- The logistic growth model is sometimes associated with *K*-selection, whereas the exponential growth model is often associated with *r*-selection. Both *K*-selection and *r*-selection are two ends of a continuum of life history strategies.

Many factors that regulate population growth are density dependent (Biology, *10e: 53.5,* Focus, *1e: 40.6*)

- A death rate that rises as population density rises and a birth rate that falls as population density rises are **density-dependent factors**. Examples of the major factors that reduce birth rates or increase death rates include the following:
  - *Competition for resources.* As population density increases, competition for resources intensifies. This might include competition for food, space, or essential nutrients.
  - *Territoriality.* Available space for territories or nesting may be limited, thus controlling the population.
  - *Disease.* Increasing densities allow for easier transmission of diseases.
  - *Predation.* As prey populations increase, predators may find the prey more easily.
- When a death rate does not change with increase in population density, it is said to be **density independent**. Natural disasters are examples of density-independent factors.
- All populations exhibit some size fluctuations. Many populations undergo regular boom-and-bust population cycles that are influenced by complex interactions between biotic and abiotic factors.

The human population is no longer growing exponentially but is still increasing rapidly (Biology, *10e: 53.6,* Focus, *1e: 43.5*)

- The exponential growth model in Figure 10.3 approximates the population explosion of humans over the last four centuries. Although the human population is still growing, the rate of growth began to slow in the 1960s. The annual growth rate peaked at 2.2% in 1962 but had dropped to 1.1% by 2011.
- One reason for falling human population growth is demographic transition. **Demographic transition** occurs when a population goes from high birth rates and high death rates to low birth rates and low death rates. Demographic transition may regularly take 150 years to complete. First, death rates fall, usually due to increased medical care and sanitation; however, falling birth rates take much longer, thus delaying transition.
- **Age-structure pyramids** show the relative number of individuals of each age in a population and can be used to predict and explain many demographic patterns. Study Figure 10.4. Why is Afghanistan poised for rapid growth? Why might economic growth in Italy be predicted to slow?

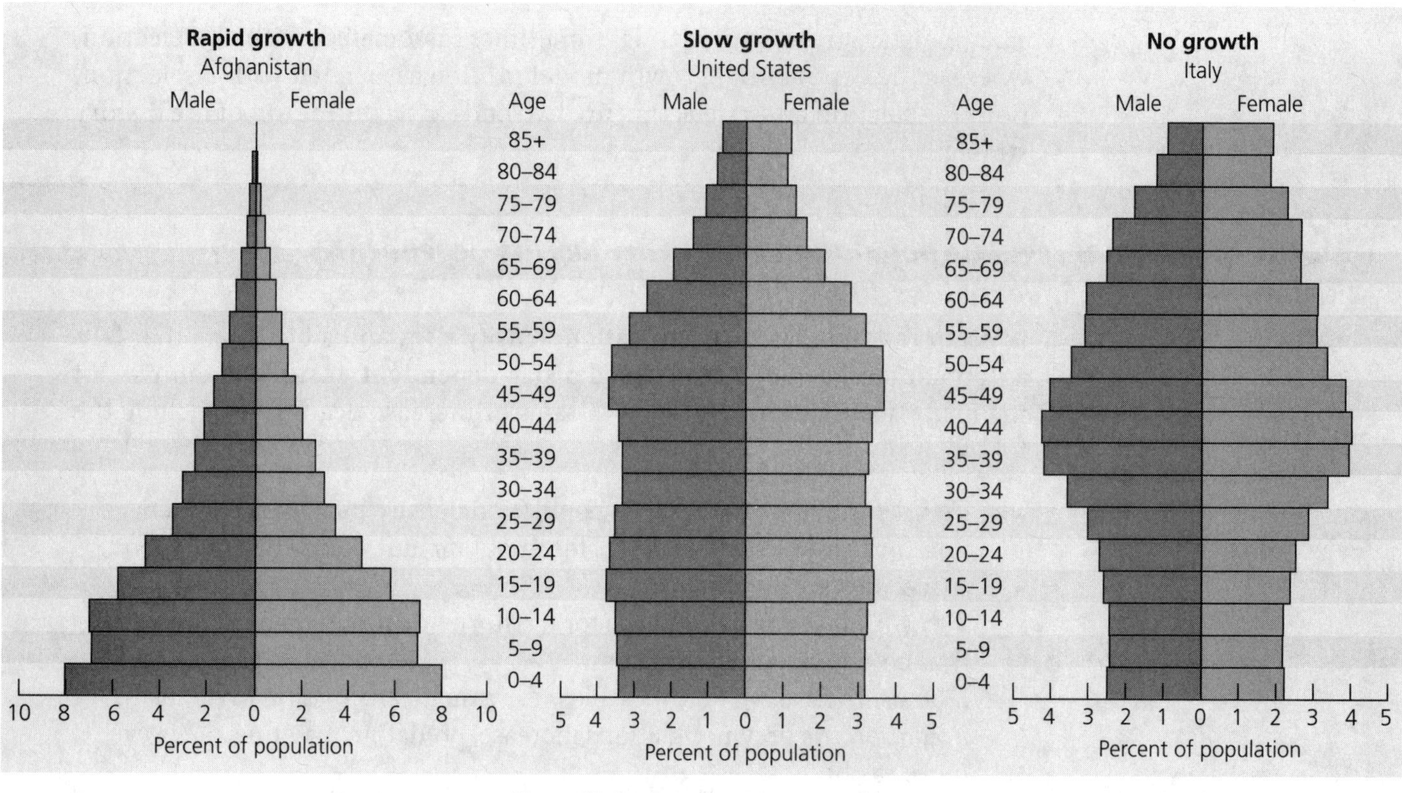

Figure 10.4 Age structure pyramids

- Global carrying capacity for humans is not known. A concept termed the **ecological footprint** examines the total land and water area needed for all the resources a person consumes in a population. Currently, 1.7 hectares per person is considered sustainable. A typical person in the United States has a footprint of 10 hectares.

Practice problem: Can you use and apply a formula to calculate population growth rate? Use the table below to calculate the population growth rate of a hypothetical population where the carrying capacity (K) = 1,500 individuals and r_{max} is 1.0. See page 286 for the solution to the problem.

Population Size (N)	Maximum Rate of Increase (r_{max})	$\frac{K-N}{K}$	Per Capita Rate of Increase $r_{max}\left(\frac{K-N}{K}\right)$	Population Growth Rate $r_{max}N\left(\frac{K-N}{K}\right)$
1,600	1.0			
2,000	1.0			
1,200	1.0			

Community Ecology

(Biology, *10e: Chapter 54*, Focus, *1e: Chapter 41*)

YOU MUST KNOW

- The difference between a fundamental niche and a realized niche.
- The role of competitive exclusion in interspecific competition.
- The symbiotic relationships of parasitism, mutualism, and commensalism.
- The impact of keystone species on community structure.
- The difference between primary and secondary succession.

Community interactions are classified by whether they help, harm, or have no effect on the species involved (Biology, *10e: 54.1*, Focus, *1e: 41.1*)

- A **community** is a group of populations of different species living close enough to interact. **Interspecific interactions** may be positive for one species (+), negative (−), or neutral (0) and include *competition, predation*, and *symbioses*.

> ***STUDY TIP*** The prefix *inter-* means between different groups, whereas *intra-* means within the same group. *Intraspecific competition* is competition within the same species, like two males fighting over a territory. *Interspecific competition* is competition between two different species for resources, like food. Pay attention to the prefix! You could be asked to write about either type of competition in an essay.

- **Interspecific competition** for resources occur when resources are in short supply. Competition is a −/− interaction between the species involved. Central to the idea of competition and community structure are these two concepts:
    - The **competitive exclusion principle** states that when two species are vying for a resource, eventually the one with the slight reproductive advantage will eliminate the other.

    - An organism's **ecological niche** is the sum total of biotic and abiotic resources that the species uses in its environment. A species' **fundamental niche**, the niche potentially occupied by the species, is often different from the **realized niche**, the portion of the fundamental niche the species actually occupies. The classic study of barnacle species in your text (see Inquiry Figure 54.3 in *Biology*, 10e, or Figure 41.3 in *Focus*, 1e) is worth reviewing for both an understanding of niche types and competitive exclusion.

- **Predation** is a +/− interaction between two species in which one species (the **predator**) eats the other species (the **prey**). Defenses for predators include the following:
    - **Cryptic coloration**, in which the animal is camouflaged by its coloring.
    - **Aposematic,** or **warning coloration**, in which a poisonous animal is brightly colored as a warning to other animals.
    - **Batesian mimicry,** referring to a situation in which a harmless species has evolved to mimic the coloration of an unpalatable or harmful species. In **Müllerian mimicry**, two bad-tasting species resemble each other, ostensibly so that predators will learn to avoid them equally.
- **Herbivory** is also a +/− interaction in which an herbivore eats part of a plant or alga. It is advantageous for an animal to be able to distinguish toxic from nontoxic plants. A plant's main protective devices are chemical toxins, spines, and thorns.
- **Symbiosis** occurs when individuals of two or more species live in direct contact with one another.
    - **Parasitism** is a +/− symbiotic interaction in which the parasite derives its nourishment from its host. Parasites may have a significant effect on the survival, reproduction, and density of their host population. Would you support the position that parasitism is a form of predation?
    - **Mutualism** is an interspecific interaction that benefits both species (+/+). Both pollinators and flowering plants benefit from their relationship.
    - **Commensalism** benefits one of the species but neither harms nor helps the other species. A fern growing in the shade of another plant could be a commensal relationship.

CONNECT WITH THE CURRICULUM FRAMEWORK

- How do specific interactions between individuals or populations affect species distribution and abundance? (EK 4.B.3)
- How do human and global natural events impact ecosystem distribution? (EK 4.B.4)

Diversity and trophic structure characterize biological communities (Biology, *10e: 54.2,* Focus, *1e: 41.2*)

- **Species diversity** measures the number of different species in a community (species richness) *and* the relative abundance of each species. A community with an even species abundance is more diverse than one in which one or two species are abundant and the remainder are rare.
- The **trophic structure** of a community refers to the feeding relationships among the organisms. **Trophic levels** are the links in the trophic structure of a community.
- The transfer of food energy from plants through herbivores through carnivores through decomposers (from one trophic level to another) is referred to as a **food chain. Food webs** consist of two or more food chains linked together.
- **Dominant species** in a community have the highest **biomass** (the sum weight of all the members of a population) or are the most abundant.
- **Keystone species** exert control on community structure by their important ecological niche. Notice in Figure 10.5 (and Inquiry Figure 54.18 in *Biology,* 10e, or Figure 41.15 in *Focus,* 1e) the impact of the keystone predator *Pisaster* (a sea star) on the diversity of species present in a tidal pool.

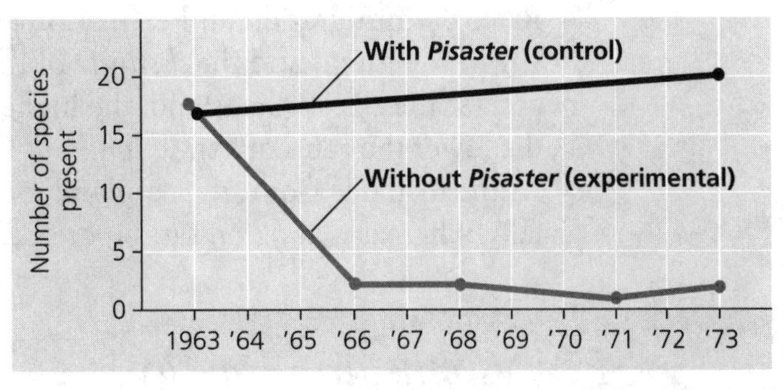

Figure 10.5 Impact of keystone predator on species diversity

Disturbance influences species diversity and composition (Biology, *10e: 54.3,* Focus, *1e: 41.3*)

- A disturbance—storm, fire, flood, drought, or human activity—changes a community by removing **organisms or changing resource availability. Disturbance is not necessarily bad for a community.** The intermediate disturbance hypothesis **states that moderate levels of** disturbance create conditions that foster greater species diversity than low or high levels of disturbance.
- **Ecological succession** refers to transitions in species composition in a certain area over ecological time.
  - In **primary succession**, plants and animals gradually invade a region that was virtually lifeless where soil has not yet formed. The gradual colonization of a newly formed volcanic island would be an example.

- **Secondary succession** occurs when an existing community has been cleared by a disturbance that leaves the soil intact. An abandoned farm will show secondary succession as it starts with the soil intact.

Biogeographic factors affect community diversity (Biology, *10e: 54.4*, Focus, *1e: 41.4*)

- Two biogeographic contributions are especially important in community diversity:
  - *The latitude of the community.* Plant and animal life is generally more abundant and diverse in the tropics, becoming less so moving toward the poles.
  - *The area of the community.* If all other factors are held equal, the larger the geographic area of a community, the more species it has.
- Because of their isolation and limited size, islands are natural laboratories for studying biogeographic factors. In addition to actual islands, this idea also pertains to islands of land, like national parks surrounded by development.
- **Island biogeography** is primarily influenced by two factors:
  - Rates of immigration and extinction are influenced primarily by the *size* of the island and the *distance* of the island from the mainland. The greater the size of the island, the higher *immigration* rates will be and the lower the rates of *extinction*.
  - As the distance from the mainland increases, the rate of *immigration* falls, whereas *extinction* rates increase.

Ecosystems and Energy

(Biology, *10e: Chapter 55*, Focus, *1e: Chapter 42*)

YOU MUST KNOW

- How energy flows through the ecosystem by understanding the terms in bold that relate to food chains and food webs.
- The difference between gross primary productivity and net primary productivity.
- The carbon and nitrogen biogeochemical cycles.
- How biogeochemical cycles impact individual organisms and/or populations and ecosystems.

> ***CONNECT WITH THE CURRICULUM FRAMEWORK***
>
> There are many LOs that relate to ecology. Some emphasize the role of environmental factors on phenotypes and natural selection. A second important theme is the movement of matter and energy through ecosystems. Pay attention to this as you work through this section.
>
> - Be able to use a representation to illustrate the movement of matter or energy through an ecosystem. (LO 4.15) Study Figures 10.6 and 10.7 carefully!
> - Predict the effects of a change of matter or energy availability in an ecosystem. (LO 4.16)
>
> To demonstrate your mastery of an LO, set up a scenario and then explain consequences of a change.

Physical laws govern energy flow and chemical cycling in ecosystems (Biology, *10e: 55.1,* Focus, *1e: 42.1*)

- An **ecosystem** is the sum of all the organisms living within its boundaries (biotic community) and all the abiotic factors with which they interact. Ecosystem ecology involves two unique processes: *energy flow* and *chemical cycling.*
- The flow of energy can be traced through the feeding or trophic levels in food chains and food webs. *Energy cannot be recycled*; therefore, energy must be constantly supplied to an ecosystem—in most cases by the sun.
  - The first law of thermodynamics states that energy cannot be created or destroyed but only transferred and transformed. What role do producers like plants play in an ecosystem? Plants convert solar energy to chemical energy. The amount of solar energy converted (not created) sets the energy budget for the ecosystem.
  - An implication of the second law of thermodynamics is that energy conversions are inefficient. Some energy is always lost as heat. As energy moves through food chains much of the energy is lost as heat to the ecosystem.
  - **Primary producers** in an ecosystem are the **autotrophs** ("self-feeders"). They support all others organisms in the ecosystem.
  - Organisms that are in trophic levels above primary producers cannot make their own food and are therefore consumers or **heterotrophs** ("other-feeders").
  - *Herbivores* eat primary producers and are called **primary consumers.**
  - *Carnivores* that eat herbivores are called **secondary consumers,** whereas carnivores that eat secondary consumers are termed **tertiary consumers.**
  - **Detritivores,** or **decomposers,** are consumers that get their energy from *detritus,* which is nonliving organic material such as the remains

of dead organisms, feces, dead leaves, and wood. Detritivores convert organic materials from all trophic levels to inorganic compounds that can be used by producers. In this way, nutrients *cycle* through ecosystems.

- It is not uncommon for a species to feed at more than one trophic level. An animal's diet might consist of berries and fish or algae and insects. The feeding level may also change as the stage in a species' life cycle changes.

Energy and other limiting factors control primary production in ecosystems (Biology, *10e: 55.2*, Focus, *1e: 42.2*)

- The amount of light energy converted to chemical energy by autotrophs is an ecosystem's **primary production**. The amount of all photosynthetic production sets the spending limit for the energy budget of the entire ecosystem.
  - Total primary production in an ecosystem is known as that system's **gross primary production (GPP)**.
  - GPP is not the amount of energy available to consumers, however. Some of the fuel molecules made by the producers must be used as fuel for their own cellular respiration. **Net primary production (NPP)** is equal to gross primary production minus the energy used by the primary producers for their "autotrophic respiration" (R_a):

$$NPP = GPP - R_a$$

  - *Primary production* in aquatic ecosystems is affected primarily by light availability and nutrient availability. In the photic zone, light—and therefore photosynthesis—decreases with depth. The nutrient most often limiting marine production is either nitrogen or phosphorus. A lake that is nutrient-rich and that supports a vast array of algae is said to be **eutrophic**.
  - Temperature and moisture are the key factors controlling primary production in terrestrial ecosystems. A measure of the amount of water transpired by plants and evaporated from the landscape, termed **evapotranspiration**, combines both key terrestrial factors.
  - This important topic is addressed in *Investigation 10, Energy Dynamics,* which is discussed beginning on page 328. Even if you did not do this investigation, working through the material presented here would help your understanding of productivity.

Energy transfer between trophic levels is typically only 10% efficient (Biology, *10e: 55.3*, Focus, *1e: 42.3*)

- Energy is lost at each level of transfer as heat or for movement or reproduction or any of the many life processes that consume energy.
- If 10% of energy is transferred from primary producer to primary consumer to secondary consumer, only 1% of the net primary production (10% of 10%) is available to secondary consumers. The loss of energy from trophic level to trophic level is one of the factors that keeps food chains so short.

- *Ecological pyramids* can give insight into food chains. Try to sketch and explain each of these: a *biomass pyramid*, an *energy pyramid*, and a *pyramid of numbers*. A biomass pyramid and a number pyramid can be inverted. It should be clear that an energy pyramid can never be inverted.

Biological and geochemical processes cycle nutrients and water in ecosystems (Biology, *10e: 55.4*, Focus, *1e: 42.4*)

- **Biogeochemical cycles** are nutrient cycles that contain both biotic and abiotic components. Understanding these cycles allows scientists to trace how nutrients flow through ecosystems and how humans may have altered the flow.
- The **carbon cycle** has historically been a close balance between the amount of CO_2 removed from ecosystems by photosynthesis and added by cellular respiration. The burning of fossil fuels has added significant amounts of additional CO_2 to the atmosphere. Examine Figure 10.6 to see the generalized flow of carbon while also considering the effects of CO_2 on global warming.

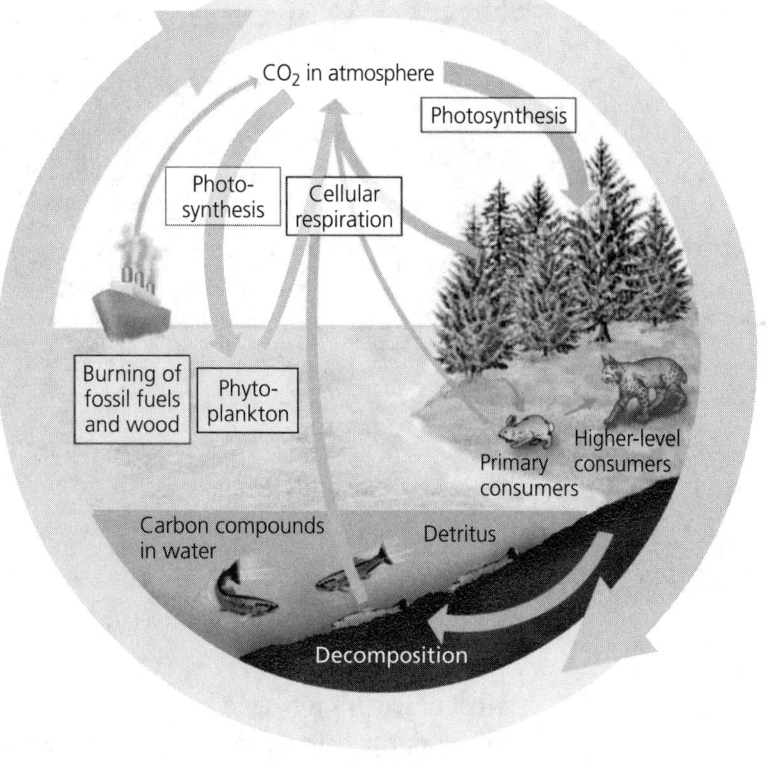

Figure 10.6 The carbon cycle

TIP FROM THE READERS
Remember this:
Matter cycles!
Energy does not cycle!

- The **nitrogen cycle** moves nitrogen from the atmosphere through the living world. Nitrogen is a common limiting factor for plant growth, making its movement through ecosystems especially important. Note the important role of bacteria in the nitrogen cycle while tracing nitrogen flow through Figure 10.7.

> ***TIP FROM THE READERS***
> Work through each figure verbally. You need to be able to explain how a change in the amount of nitrogen or carbon dioxide or another factor would impact an ecosystem. To understand the impact of acid rain or fertilizer runoff or global warming, you need to understand these cycles!

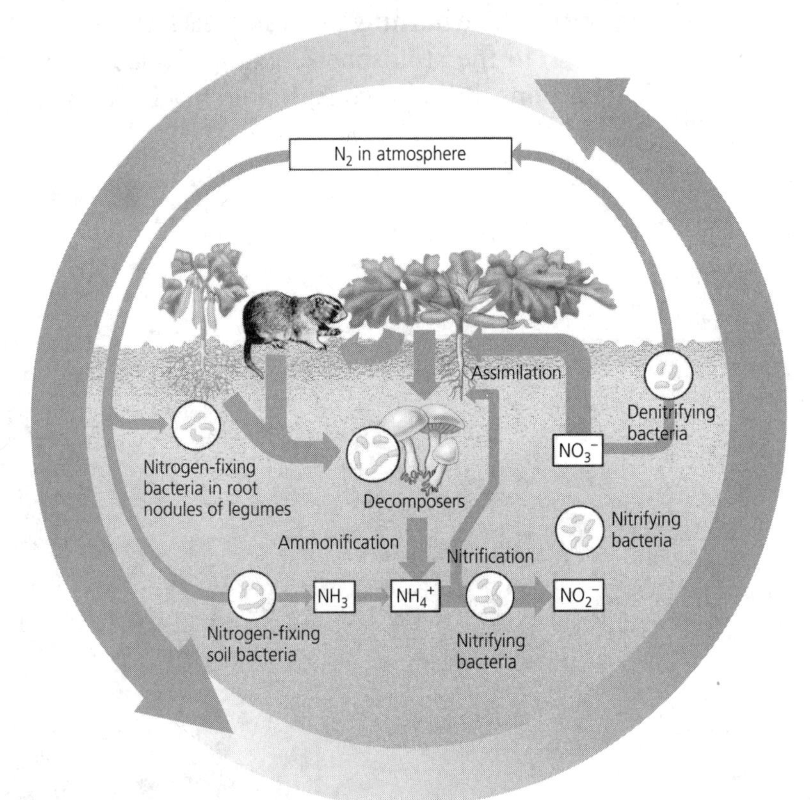

Figure 10.7 The terrestrial nitrogen cycle

- Most of Earth's nitrogen is in the form of N_2, which is unusable by plants. The major pathway for nitrogen to enter an ecosystem is **nitrogen fixation**, the conversion of N_2 by bacteria to forms that can be used by plants. Earlier we noted this relationship between plants that are legumes and the bacterium *Rhizobium* as an example of mutualism.
- **Nitrification** is the process by which ammonium (NH_4^+) is oxidized to nitrite and then nitrate (NO_3^-) by bacteria. Two inorganic nitrogen forms can be absorbed by plants: nitrates and ammonium.
- **Denitrification** by bacteria releases nitrogen to the atmosphere.

- Other important nutrient cycles involve water and phosphorus.

Restoration ecologists help return degraded ecosystems to a more natural state (Biology, *10e: 55.5,* Focus, *1e: 42.5*)

- **Bioremediation** is the use of organisms, usually prokaryotes, fungi, or plants, to detoxify polluted ecosystems. It has been used to restore areas degraded by mining or to remove oil or radioactive elements.
- **Bioaugmentation** is the introduction of desirable species such as nitrogen fixers to add essential nutrients.

Conservation Biology and Global Change

(Biology, *10e: Chapter 56*, Focus, *1e: Chapter 43*)

YOU MUST KNOW

- The value of biodiversity and the major human threats to it.
- How human actions are changing the Earth.
- How to predict consequences on both local and global ecosystems of specific human activities. (LO 45.20)

Human activities threaten Earth's biodiversity (Biology, *10e: 56.1,* Focus, *1e: 43.1*)

- **Biodiversity**—short for biological diversity—can be considered at three main levels: genetic diversity, species diversity, and ecosystem diversity.
- Four major threats to biodiversity are habitat loss, introduced species, overharvesting, and global change.

Population conservation focuses on population size, genetic diversity, and critical habitat (Biology, *10e: 56.2,* Focus, *1e: 43.2*)

- When a population drops below a minimum viable population (MVP) size, its loss of genetic variation due to nonrandom mating and genetic drift can trap it in an *extinction vortex.* One key factor is the loss of genetic variation necessary to enable evolutionary responses to environmental change, such as the appearance of new strains of pathogens.

Landscape and regional conservation help sustain biodiversity (Biology, *10e: 56.3,* Focus, *1e: 43.3*)

- The structure of a landscape can strongly influence biodiversity. As *habitat fragmentation* increases and edges become more extensive, biodiversity tends to decrease. *Movement corridors* can promote dispersal and help sustain populations.
- A **biodiversity hot spot** is a relatively small area with an exceptional concentration of endemic species and a large number of endangered and threatened species. Biodiversity hot spots are also hot spots of extinction and thus prime candidates for protection.

Earth is changing rapidly as a result of human actions (Biology, *10e: 56.4,* Focus, *1e: 43.4*)

- Nutrient cycling is altered by human activities, particularly agriculture. For example, soil nitrogen is often depleted by crops. Excess nitrogen enters aquatic ecosystems as a result of livestock activities and can lead to eutrophication.
- **Acid precipitation** is defined as rain, snow, or fog with a pH less than 5.2. The burning of wood and fossil fuels releases sulfur oxides and nitrogen oxides into the atmosphere. These oxides react with water, forming sulfuric acid and nitric acid.
- In **biological magnification**, toxins become more concentrated in successive trophic levels of a food web. The toxins cannot be broken down biologically by normal chemical means, so they magnify in concentration as they move through the food chain.
- The **greenhouse effect** refers to the absorption of heat the Earth experiences due to certain atmospheric gases. Carbon dioxide and water vapor intercept and absorb much reflected infrared radiation, re-reflecting some back toward Earth.
    - Because of the burning of fossil fuels, CO_2 levels have been steadily increasing. One effect of this increase is that Earth is being warmed significantly (**global warming**).
- The **ozone layer** reduces the amount of UV radiation penetration from the sun through the atmosphere. Chlorine-containing compounds used by humans are eroding the ozone layer, allowing more DNA-damaging UV radiation to penetrate to the surface of the Earth.

Solution to the question on page 274: The value of $(K-N)/K$ would be 0 if $K = N$, and the population would no longer grow. This implies that when a population reaches carrying capacity, there are no longer excess resources to support further growth.

Solution to the practice problem on page 277:

Using a population size of 1,600 as an example,

$$\frac{dN}{dt} = r_{max}N\frac{(K - N)}{K} = \frac{1(1{,}600)(1{,}500 - 1{,}600)}{1{,}500}$$

and the population "growth" rate is -107 individuals per year. The population shrinks even faster when N is farther from the carrying capacity; when N equals 2,000 individuals, the population shrinks by 667 individuals per year. What happens to the population when the population is below carrying capacity as in the last example?

Population Size (N)	Maximum Rate of Increase (r_{max})	$\frac{K-N}{K}$	Per Capita Rate of Increase $r_{max}\left(\frac{K-N}{K}\right)$	Population Growth Rate $r_{max}N\left(\frac{K-N}{K}\right)$
1,600	1.0	(1,500 − 1,600)/1,500 = −0.067	1(−0.067) = (−0.067)	(−0.067)(1,600) = −107
2,000	1.0			
1,200	1.0			

Level 1: Knowledge/Comprehension Questions

The Knowledge/Comprehension questions review essential content knowledge but don't represent the type of question you will see on the AP exam. Level 2, Application/Synthesis questions are similar to those you will see on the exam.

1. Which of the following is a correct description of ozone in the upper atmosphere?
   (A) It is composed of O_2.
   (B) It limits the amount of radiation available for photosynthesis.
   (C) It is thinning, partially as a result of widespread use of certain chlorine-containing compounds.
   (D) It is a result of widespread burning of fossil fuels.

2. The carrying capacity of a population is defined as
   (A) the amount of time the parents in the population spend rearing and nurturing their offspring.
   (B) the maximum population size that a certain environment can support at a particular time.
   (C) the amount of vegetation that a certain geographic area can support.
   (D) the number of different types of species a biome can support.

Directions: Questions 3–6 consist of four lettered choices followed by a list of numbered phrases or sentences. For each numbered phrase or sentence, select the one choice that is most closely related to it. Each choice may be used once, more than once, or not at all.

Questions 3–6

(A) Temperate grassland
(B) Tropical forest
(C) Tundra
(D) Desert

3. Characterized by permafrost and few large plants

4. Characterized by epiphytes, a significant canopy, and abundant rainfall

5. Characterized by occasional fires, nutrient-rich soil, and large grazing animals

6. Characterized by sparse rainfall and extreme daily temperature fluctuations

7. A bacterial colony that exists in an environment displaying ideal conditions will display which of the following growth patterns?
   (A) logistic growth
   (B) intrinsic growth
   (C) demographic growth
   (D) exponential growth

8. Which of the following interspecific interactions represents a form of interaction different from the others?
   (A) a tick on a human
   (B) deer browsing on shrubs
   (C) a honeybee and apple blossoms
   (D) a deadly bacterium and its host

9. When one species was removed from a tide pool, the species richness became significantly reduced. The removed species was probably
   (A) the species with the greatest biomass.
   (B) a potent parasite.
   (C) the species with the highest relative abundance.
   (D) a keystone species.

10. Which of the following would not be a density-dependent factor limiting a population's growth?
    (A) increased predation by a predator
    (B) a limited number of available nesting sites
    (C) a stress syndrome that alters hormone levels
    (D) a very early fall frost

11. There are many reasons the human population is increasing so rapidly. Select the one factor that does not significantly affect this growth rate.
    (A) Technology has increased our carrying capacity.
    (B) The death rate has greatly decreased since the Industrial Revolution.
    (C) The age structure of many countries is highly skewed toward younger ages.
    (D) Fertility rates in many developing countries are above the 2.1 children per female replacement level.

12. A Type I survivorship curve is level at first, with a rapid increase in mortality in old age. This type of curve is
    (A) typical of many invertebrates that produce large numbers of offspring.
    (B) typical of large mammals.
    (C) found most often in *r*-selected populations, which have a rapid rate of reproduction.
    (D) almost never found in nature.

13. The process in which CO_2 in the atmosphere intercepts and absorbs reflected infrared radiation and re-reflects it back to Earth is known as
    (A) global warming.
    (B) ozone depletion.
    (C) the greenhouse effect.
    (D) biological magnification.

14. A species' specific use of the biotic and abiotic factors in an environment is collectively called the species'
    (A) habitat.
    (B) trophic level.
    (C) ecological niche.
    (D) partition.

15. The dominant species in a community is the one that
    (A) has the greatest number of genes per individual.
    (B) is at the top of the food chain.
    (C) has the largest biomass.
    (D) eats all other members of the community.

16. A fire cleared a large area of forest in Yellowstone National Park in the 1980s. When the first plants pioneered this burned area, this was an example of
    (A) primary succession.
    (B) secondary succession.
    (C) biological evolution.
    (D) a keystone species.

Level 2: Application/Analysis/Synthesis Questions

After reading the paragraphs, answer the question(s) that follow.

The largest estuary in the United States is the Chesapeake Bay, which extends through six states. The bay is one of the most productive natural areas in the world. It is home to thousands of plants and animals, including many commercially important species. The water of the bay is relatively shallow. Many areas are no more than 10 feet deep, with an average depth of 30 feet. Light penetrates the shallow water and supports the submerged plants that provide food and shelter for the many species living in the bay ecosystem. However, like many estuaries, the bay receives large amounts of fertilizer runoff from farms, lawns, and wastewater treatment facilities. This runoff introduces large amounts of nutrients.

1. Which of the following is the most probable sequence of events when fertilizer runoff reaches the bay?
   (A) submerged vegetation increases, more food for fish and shellfish, fish and shellfish populations increase
   (B) phytoplankton population increases, more food for fish and shellfish, fish and shellfish populations increase
   (C) phytoplankton population increases, sunlight blocked to submerged vegetation, submerged vegetation dies, fish and shellfish populations decrease
   (D) submerged vegetation decreases, fish and shellfish feed on decaying plants, phytoplankton feed on fish and shellfish, commercial fisheries decline

2. Which of the following pairs of nutrients would have the greatest effect on growth of phytoplankton?
   (A) carbon and hydrogen
   (B) oxygen and carbon dioxide
   (C) nitrogen and phosphorus
   (D) calcium and water

Use the graph below to answer questions 3 and 4.

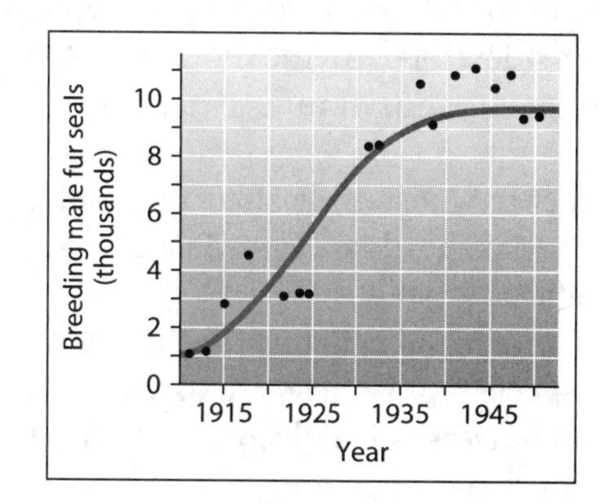

3. According to this graph of the population growth of fur seals, in what year did the population first reach its carrying capacity?
   (A) 1925
   (B) 1930
   (C) 1940
   (D) 1950

4. The formula $dN/dt = r_{max}N(K - N)/K$ describes the pattern of growth for the graph. Which of the following is true concerning the formula and the graph it describes in 1945?
   (A) $K = N$
   (B) $r_{max}N$
   (C) The variable K is constantly changing.
   (D) $(K - N)/K = 9{,}000$

After reading the paragraphs, answer questions 5 and 6.

Introduced species are a problem all over the world, and there are many examples in the United States. Several years ago, a fisherman caught a northern snakehead fish in a pond in Crofton, Maryland (a suburb of Washington, DC). Snakeheads are a favorite food of immigrants from China, and live fish can frequently be found in Asian markets. It's suspected that the fish in the Crofton pond were purchased locally and then intentionally released.

Snakeheads are top predators, and 90% of the northern snakeheads' diet consists of other fishes. The northern snakehead can breathe out of water and travel short distances (about 100 feet) across land. They also reproduce rapidly. Females can lay more than 100,000 eggs per year. Juveniles have also been identified in the Potomac River and other rivers in Pennsylvania.

5. Predict the most likely effect on biodiversity when snakeheads are introduced into an aquatic ecosystem.
   (A) Biodiversity will increase because another species has been added to the ecosystem.
   (B) Biodiversity will decrease because the snakehead will prey on native species.
   (C) Biodiversity will remain the same because local species will prey on the snakeheads and limit their reproduction.
   (D) Biodiversity will decrease because the snakeheads will feed heavily on producer species.

6. Based on the characteristics of the snakehead described, which of the following is most likely to be a productive strategy to reduce the spread of this species?
   (A) Extend the fishing season for prey fishes.
   (B) Introduce a natural predator to feed on juvenile snakeheads.
   (C) Introduce a fungus that prevents fish eggs from hatching.
   (D) Introduce algae and photosynthetic bacteria to reduce nutrient levels in the water.

Use the population pyramid for Nigeria to answer questions 7 and 8.

7. The age-structure data for Nigeria shows that the country has many more individuals under the age of 15 than over the age of 40. What does this imply about the future population of Nigeria?
   (A) The population will probably remain stable.
   (B) The population will probably decrease.
   (C) The population will probably grow rapidly.
   (D) The number of older people will probably increase rapidly.

8. Based on the age structure of the country, which of the following situations would be most likely over the next 20 years?
   (A) strong economic gains stimulated by population growth
   (B) an increased demand for resources based on population growth
   (C) a decreased demand for medical services due to the small number of elderly citizens
   (D) a decline in housing prices based on lack of demand

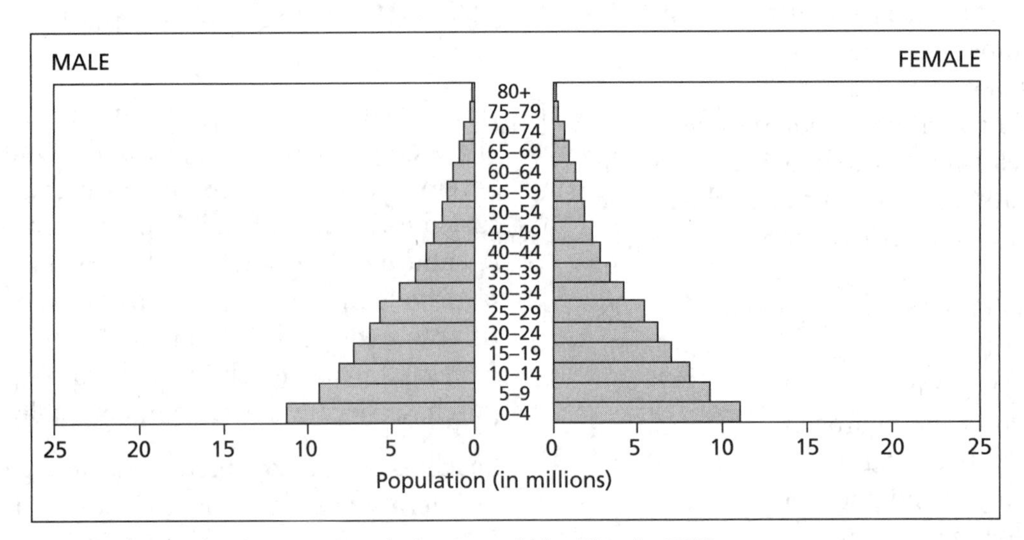

Population Pyramid for Nigeria, 2007

From U.S. Census Bureau International Data Base, 2007

9. Which statement best describes energy transfer in a food web?
   (A) Energy is transferred to consumers, which convert it to nitrogen compounds and use it to synthesize amino acids.
   (B) Energy from producers is converted into oxygen and transferred to consumers.
   (C) Energy from the sun is stored in green plants and transferred to consumers.
   (D) Energy moves from autotrophs to heterotrophs to decomposers, which convert it to a form producers can use again.

10. In ocean kelp forests, kelp perch are preyed upon by kelp bass. In the graph proportional mortality rates were calculated for kelp perch as a function of density. The data from the experiment were plotted as shown on this graph. Which statement is a logical conclusion based on the data?

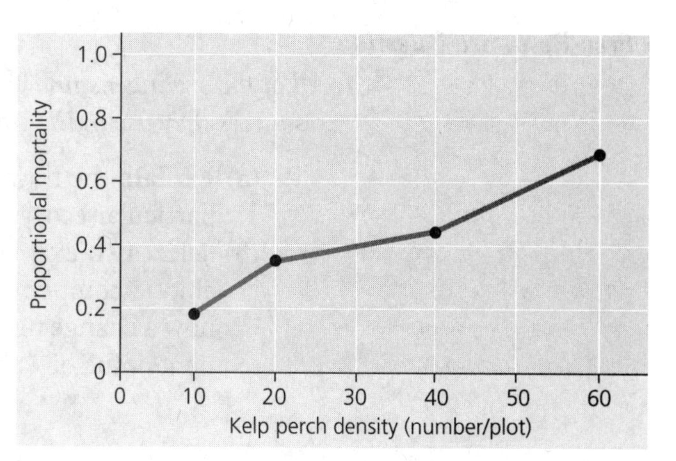

   (A) Fluctuations in kelp perch density are a result of normal population cycles.
   (B) A negative feedback loop regulates the kelp perch population size.
   (C) The equilibrium density for the population of kelp perch can be calculated.
   (D) Higher predation levels lead to healthier kelp perch populations over time.

Grid-In Questions

These call for a numerical response.

1. A population of 20 bobcats was introduced to a barrier island to help control the large rodent population. The bobcat population's birth rate is 0.48 bobcats/year per capita, and the death rate is 0.21 bobcats/year per capita. Given the initial bobcat population, predict the population size after 2 years on the island. (Round to the nearest whole number.)

2. Ecologists often cannot count all the individuals in a population. In such cases, the mark-recapture method is used to estimate population size. An ecologist wished to estimate the population of white-footed deer mice in an old field. On her first trapping 142 mice were caught and marked with a harmless dye on the back of the mouse's head. Two weeks later a second trapping was conducted where 133 mice were caught, of which 41 were marked with the dye. Using the formula

$$N = sn/x$$

where N = population size estimate, s = number of mice marked in the first trapping, n = number of mice caught in the second trapping, and x = the number of mice that were marked in the second trapping. What is the population estimate of the white-footed deer mouse? (Round your answer to the nearest whole number.)

Free-Response Question

1. *All of the organisms in a community are interrelated by the abiotic and biotic resources they use in the course of their lives.*

(a) **Explain** the flow of energy among a hawk, a mouse, and a plant in a particular ecosystem.

(b) Select **two** examples of the impact of human population growth on abiotic components of the environment. For each example, *explain* how a change in the abiotic component will impact the biotic component of the biosphere.

Part III

The Laboratory

AP Biology is designed to be equivalent to a two-semester college course, and laboratory experience must be included in all AP Biology courses. The College Board has published a collection of suggested investigations, and it is expected that all students will complete eight major investigations during the course. Of these, two should come from topics included in each Big Idea. Many of the investigations may be ongoing projects or require several days for completion. Your teacher may select from many possible investigations, so the labs you do may be different from the ones we present here. Your success in the course does not hinge on completion of a particular lab, but rather on your mastery of the underlying Science Practices.

Our students find that a review of the laboratory exercises is an excellent way to review many of the core concepts of their AP Biology course. A great resource for your review is LabBench, available at http://www.phschool.com/science/biology_place/labbench/. Although it was written to accompany the "Classic" lab manual, there is a wealth of information about some labs that your teacher may choose to use as starting points for investigations. At LabBench you will find exercises that take you through 12 basic college labs, along with sample questions to test your understanding.

Inquiry-Based Investigations

In laboratory, your teacher will engage you in inquiry labs in which you are not given step-by-step instruction, but rather are asked to select your own question, develop a hypothesis, and determine your own procedure for data collection. During the course of the year, work in the laboratory will help you gain skills in the science practices. Here are some tasks that you should be able to do:

- Design a plan for collecting data to answer a scientific question using adequate controls.
- Predict expected results.
- Describe variables that were not controlled in the experiment, and how those variables might have affected the outcome of the experiment.
- Analyze data to identify patterns or relationships.
- Construct a graph based on the collected data and use it to formulate conclusions or make predictions.
- Summarize the pattern shown by the data.
- Identify outliers in a data set and propose an explanation for them.
- Propose a rationale for dealing with outliers in a data set.
- Identify possible sources of error in an experimental procedure or data set.
- Use mathematics to solve problems, analyze data, and make predictions.

CONNECT WITH THE CURRICULUM FRAMEWORK

Now would be a good time to turn to page 4 to review the Science Practices. Note that all the tasks above use these practices. These are skills you learn by doing, so keep them in mind as you review each investigation.

BIG IDEA 1: Investigation 1, Artificial Selection

Overview of the Lab

- In this investigation, you will investigate the effect of artificial selection in Fast Plants, although your teacher might select any organism with a fairly short life cycle, such as fruit flies. You and your classmates will observe a range of traits, such as height of plant at first flower, number of trichomes (hairs) on the leaf petioles, or leaf color, and select the individual plants that are at one extreme for this trait. These plants will serve as the parents for the next generation of plants, while the others are discarded.

YOU MUST KNOW

- A technique to investigate selection as a mechanism of evolution.
- How to apply mathematical methods to data to predict what will happen to the population in the future.
- Quantitative methods to determine whether the two populations are significantly different, and appropriate use of error bars, graph type, and statistical tests.

SCIENCE PRACTICES: CAN YOU . . .

- Design a plan for collecting data that will investigate the effect of selection on a population?
- Analyze your data to identify patterns and relationships?
- Use data to predict what will happen to a population in the future, based on models of types of selection? (LO 1.22)

Hints and Review

- You will measure or quantify the trait for the first generation, and then do the same for their offspring. If enough measurements are made, data often show a normal or bell-shaped distribution for the first generation.
- The data collected for each generation can be used to generate a histogram. Future generations may show the effect of directional selection.
- Go back to page 163 (see Topic 6) to review the modes of selection.

Questions

The following data were obtained in an artificial selection experiment with Fast Plants.

Height (in cm) at First **Flower Generation 1**	Height (in cm) at First **Flower Generation 2**
8.2	9.6
7.3	9.3
9.2	8.8
8.1	6.4
8.0	8.9
7.8	8.8
7.9	8.2
9.1	7.9
6.2	8.7
8.9	8.8

1. What is the mean height of the plants in the first generation to the nearest tenth?
2. What is the mode of the plants in the second generation to the nearest tenth?
3. To the nearest tenth, what is the percentage increase in height of plants in the second generation?
4. What procedure would be the most useful to test the effect of selection for height at first flowering on the mean height of the plants?
   (A) Select plants for breeding that are closest to the average height of the population.
   (B) Use the entire population for breeding, but increase the amount of fertilizer and light.
   (C) Select seeds from only those plants whose height exceeds the mean for the generation.
   (D) Randomly select five plants for breeding.

BIG IDEA 1: Investigation 2, Mathematical Modeling: Hardy-Weinberg

Overview of the Lab

- This investigation has you manipulate a computer spreadsheet to build a mathematical model to investigate the relationship between changing allelic frequencies in a population and evolution. You will develop an understanding of the Hardy-Weinberg equation, gain expertise with a spreadsheet program, and use your model to answer a question you pose. The skills you develop in creating the spreadsheet model and using it will be invaluable throughout many of your college courses.

YOU MUST KNOW

- The Hardy-Weinberg equation and be able to use it to determine the frequency of alleles in a population.
- Conditions for maintaining Hardy-Weinberg equilibrium.
- How genetic drift, natural selection, and the heterozygote advantage affect Hardy-Weinberg equilibrium.

SCIENCE PRACTICES: CAN YOU . . .

- Use a data set to reflect a change in the genetic makeup of a population over time and apply mathematical methods to investigate the cause(s) and effect(s) of this change?
- Apply mathematical methods to data from a real or simulated population to predict how the genetic composition of a population may change due to natural selection, genetic drift, or gene flow or other factors?
- Evaluate data-based evidence that describes evolutionary changes in the genetic makeup of a population over time?
- Use data from mathematical models based on the Hardy-Weinberg equilibrium to analyze genetic drift and the effect of selection in the evolution of specific populations?
- Justify how data from mathematical models based on the Hardy-Weinberg equation can be used to analyze genetic drift and the effects of selection in the evolution of specific populations?
- Describe a model that represents evolution within a population?
- Evaluate data sets that illustrate evolution as an ongoing process?

Hints and Review

- The **Hardy-Weinberg Law of Genetic Equilibrium** provides a mathematical model for studying evolutionary changes in allelic frequency within a population. It predicts that the frequency of alleles and genotypes in a population will remain constant from generation to generation if the population is stable and in genetic equilibrium.

Five conditions are required in order for a population to remain at Hardy-Weinberg equilibrium:

1. No change in allelic frequency due to mutation.
2. Random mating.
3. No natural selection.
4. The population size must be extremely large (no genetic drift).
5. No gene flow (emigration, immigration, transfer of pollen, etc.).

- **No change in allelic frequency due to mutation**—Any mutation in a particular gene could result in a change in the balance of alleles in the gene pool. Mutations alone never change allelic frequency, but natural selection may make a mutation more common in a population over time. This is evolution.
- **Random mating**—In a population at equilibrium, mating must be random. In assortative mating, individuals tend to choose mates similar to themselves; for example, large blister beetles tend to choose mates of large size and small blister beetles tend to choose small mates. Although this does not alter allelic frequencies, it results in fewer heterozygote individuals than you would expect in a population where mating is random.
- **No natural selection**—No alleles are selected over other alleles. If selection occurs, those alleles that are selected will become more common. For example, if resistance to a particular herbicide allows weeds to live in an environment that has been sprayed with that herbicide, the allele for resistance may become more frequent in the population.
- **Extremely large population size**—A large breeding population helps to ensure that chance alone does not disrupt genetic equilibrium. In a small population, only a few copies of a certain allele may exist. If for some chance reason the organisms with that allele do not reproduce successfully, the allelic frequency will change. This random, nonselective change is what happens in **genetic drift**.
- **No gene flow**—No new alleles can come into the population, and no alleles can be lost. Both immigration and emigration can alter allelic frequency. To estimate the frequency of alleles in a population, we can use the Hardy-Weinberg equation.
- Turn back to page 161 where use of the Hardy-Weinberg equation is in a box. All the information you need to calculate allelic frequencies when there are two different alleles is explained there.

  Sample Problem #1:

  Consider a population of pigs where B = tan coat color and b = black coat color. Use the Hardy-Weinberg equation to determine the percent of the pig population that is heterozygous for tan coat (All the steps necessary to calculate this are described on the next page. See if you can work this problem before reading through our solution.)

1. **Calculate q^2.**

   Count the individuals that are homozygous recessive in the illustration above. Calculate the percent of the total population they represent. This is q^2.

   Answer: Four of the sixteen individuals show the recessive phenotype, so the correct answer is 25% or 0.25.

2. **Find q.**

   Take the square root of q^2 to obtain q, the frequency of the recessive allele.

   Answer: $q = 0.5$

3. **Find p.**

   The sum of the frequencies of both alleles = 100%, $p + q = 1$. You know q, so what is p, the frequency of the dominant allele?

   Answer: $p = 1 - q$, so $p = 0.5$

4. **Find $2pq$.**

   The frequency of the heterozygotes is represented by $2pq$. This gives you the percent of the population that is heterozygous.

   Answer: $2pq = 2(0.5)(0.5) = 0.5$, so 50% of the population is heterozygous.

- Let's go back to the idea of modeling to make predictions about what will happen to allelic frequencies in a population. Here are two questions that should be considered:

  1. *Why aren't negative recessive alleles eliminated?* Negative recessive alleles are seldom eliminated because selection acts on phenotypes and the recessive alleles are "hidden" in the heterozygotes.
  2. *Why do certain negative alleles, such as that for sickle-cell anemia, persist at relatively high levels in certain populations?* The allele for sickle-cell anemia (*s*) conveys protection against malaria in individuals who are heterozygous (*Ss*) called the *heterozygote advantage*. Individuals who lack this allele (*SS*) have normal hemoglobin but are susceptible to malaria. Although individuals who are *ss* may die from the consequences of the anemia, the negative allele persists in populations under pressure from malarial infection because the heterozygotes are less likely to die of malaria.

- You may want to go back to Topic 6, page 161, where we also discuss the Hardy-Weinberg equation.

TIP FROM THE READERS

There are a couple of ways you can get confused when doing these problems. Here's what to watch for:

1. If you are given the number of individuals showing the dominant trait, remember that this is *not* p^2 because it includes heterozygotes. However, you can use it to determine q^2 and from that get a value for q.
2. You also may be given a problem where you are told the frequency of an allele. You are being directly given p or q! Genotypic frequencies, like the 25% of the pigs that were black, represent q^2; to find the allelic frequency of q requires taking the square root. Be sure to read carefully to determine if you are given allelic or genotypic frequencies in problems!

Sample Problem #2:

In a certain population of 1,000 fruit flies, 360 have red eyes, whereas the remainder have sepia eyes. The sepia eye trait is recessive to red eyes. How many individuals would you expect to be homozygous for red eye color?

Hint: The first step is always to calculate q^2! Start by determining the number of fruit flies that are homozygous recessive.

Answer: You should expect 40 to be homozygous dominant.

Calculations:

q^2 for this population is $640/1{,}000 = 0.64$

$q = \sqrt{0.64} = 0.8$

$p = 1 - q = 1 - 0.8 = 0.2$

The homozygous dominant frequency $= p^2 = (0.2)(0.2) = 0.04$.

Therefore, you can expect 4% of 1,000, or 40 individuals, to be homozygous dominant.

STUDY TIP The focus of this lab will not be your ability to do Hardy-Weinberg problems, but to create and use models and mathematical applications. However, we know you will need to be able to do problems like this, so work through them all!

Questions

1. If the frequency of two alleles in a gene pool is 90% *A* and 10% *a*, what is the frequency of individuals in the population with the genotype *Aa*?
   (A) 0.81
   (B) 0.09
   (C) 0.18
   (D) 0.01

2. If a population experiences no migration, is very large, has no mutations, has random mating, and there is no selection, which of the following would you predict?
   (A) The population will evolve but much more slowly than normal.
   (B) The makeup of the population's gene pool will remain virtually the same as long as these conditions hold.
   (C) The composition of the population's gene pool will change slowly in a predictable manner.
   (D) Dominant alleles in the population's gene pool will slowly increase in frequency, whereas recessive alleles will decrease.

3. In a population that is in Hardy-Weinberg equilibrium, the frequency of the homozygous recessive genotype is 0.09. What is the frequency of individuals that are homozygous for the dominant allele?
   (A) 0.7
   (B) 0.21
   (C) 0.42
   (D) 0.49

BIG IDEA 1: Investigation 3, Comparing DNA Sequences to Understand Evolutionary Relationships with BLAST

Overview of the Lab

In this investigation you will be introduced to one of the tools of bioinformatics, BLAST, which stands for Basic Local Alignment Search Tool. When you input a DNA sequence to BLAST, entire genomic libraries are searched for identical or similar sequences in a matter of seconds. The purpose of this lab is to have you use some of the tools of bioinformatics while becoming adept in making and interpreting cladograms or phylogenetic trees. First, you place a newly discovered fossil species onto an existing cladogram using information from a photograph. Then to confirm your placement, you will use BLAST to compare several genes, and use the information to refine your cladogram.

YOU MUST KNOW

- Computer programs have sophisticated ways of measuring and representing relatedness among organisms.
- Similarities in gene or amino acid sequences can be used to determine evolutionary relationships.
- Phylogenetic trees graphically represent ancestral groups and their descendants and can be drawn by using many types of evidence, including morphology, DNA, and protein sequences.

SCIENCE PRACTICES: CAN YOU . . .

- Create a phylogenetic tree that correctly represents evolutionary history and speciation from a provided data set? (LO 1.19)

Hints and Review

- The BLAST website allows you to input either DNA or amino acid/protein sequences to search for species that show similarities. You are able to set the parameters for a search and can compare species.
- Cladograms can represent the same information in many ways. Think of them as mobiles that may be hung from a ceiling and rotated and you will begin to see how the same relationship may be shown in different ways.
- All cladograms show *nodes* that represent a point of divergence from a *common ancestor* of each branch.
- The more branches that occur after a common ancestor, the more distantly related the groups.

Questions

1. Suppose that species 1 and species 2 have similar appearances but very divergent gene sequences, and that species 2 and species 3 have very different appearances but similar gene sequences. Which statement best reflects their relationships?
   (A) Species 1, 2, and 3 are all closely related because they either share similar appearance or similar gene sequences.
   (B) Species 1 and 2 are most closely related because their similar appearance indicates convergent evolution.
   (C) Species 2 and 3 are most closely related because of their genetic similarity.
   (D) Species 1 and 2 are most closely related because they share a common ancestor with species 3.

2. Four of the following trees describe the same phylogenetic relationships among taxa A, B, C, D, E, and F. Which tree shows a different phylogeny?

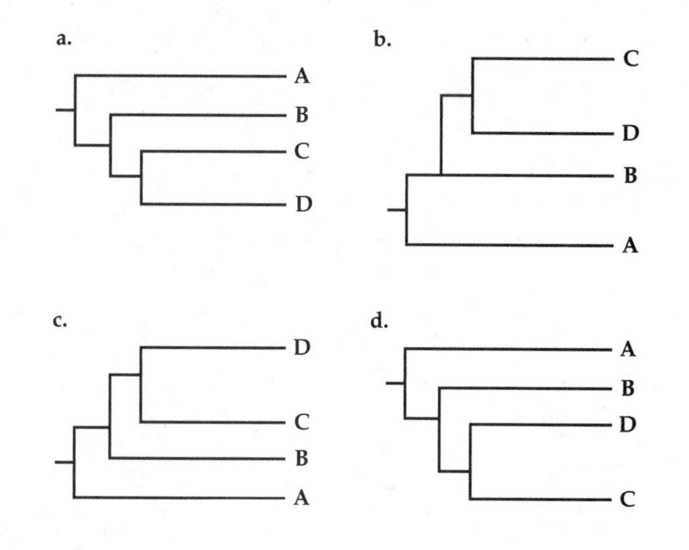

3. Which of the following approaches would allow a biologist studying the evolution of four similar species of birds to choose the best phylogenetic tree from all possible phylogenies?
   (A) Choose the simplest tree that is based on physical appearances.
   (B) From a comparison of DNA sequences, choose the tree that requires the smallest number of evolutionary events.
   (C) Choose the tree that has the most evolutionary changes because this would be the most likely explanation for how these very similar birds evolved into four distinct species.
   (D) Determine which species can interbreed; those that can interbreed evolved from a common ancestor most recently.

4. When cytochrome *c* molecules are compared, yeasts and molds are found to differ by approximately 46 amino acids per 100 residues (amino acids in the protein); insects and vertebrates are found to differ by 29 amino acids per 100 residues. What can one conclude from these data?
   (A) Very little, unless the DNA sequences for the cytochrome *c* genes are compared.
   (B) Insects and vertebrates diverged from a common ancestor more recently than did yeasts and molds.
   (C) Yeasts and molds diverged from a common ancestor more recently than did insects and vertebrates.
   (D) The evolution of cytochrome *c* occurred more rapidly in yeasts and molds than in insects and vertebrates.

BIG IDEA 2: Investigation 4, Diffusion and Osmosis

Overview of the Lab

In Part I you will create cell models with agar cubes and use them to calculate surface-area-to-volume ratios and make predictions about the rate of diffusion. In Part II, you will use dialysis tubing, which is selectively permeable, to create cell models to investigate questions about movement of molecules across the membrane. In Part III you will use a living tissue to observe and understand osmosis in cells.

YOU MUST KNOW

- Factors that affect diffusion across the membrane.
- How water potential is measured and its relationship to solute concentration and pressure potential of a solution.
- Water moves from a region where water potential is high to a region where water potential is low.
- The relationship of molarity to osmotic concentration.
- How to determine osmotic concentration of a solution from experimental data.

SCIENCE PRACTICES: CAN YOU . . .

- Design an experiment to measure the rate of osmosis in a model system?
- Analyze data and make predictions about molecular movement through cellular membranes?
- Connect the concepts of diffusion and osmosis to the structure of the membrane and molarity?
- Use the principles of water potential to predict and justify the movement of water into plant tissue?

Hints and Review

- In the first part of this investigation, you will calculate both the surface area and volume of several sizes of agar cubes, which serve as models of cells. They are prepared in such a way that you can observe a color change in the cubes over a period. This will allow you to see that the distance material diffuses into the cubes is the same regardless of their size, and that the percentage of the total cube volume penetrated goes down drastically as the cubes get larger.
- Because cells exchange materials with their environment by diffusion, the conclusion you can draw from this is that smaller cells have more favorable surface-area-to-volume ratios.
- Refer to the AP Biology Equations and Formulas page to be sure you can calculate surface area and volumes. (See pages 346–347.) We will give you a practice problem on the sample test.
- Get a firm fix on the terminology! This is one topic where you will not get any credit if you understand the concept but garble the vocabulary. Let's add some more terms.

- **Osmosis** is the movement of water from a region of high concentration to a region of low concentration through a selectively permeable membrane.
- In **dynamic equilibrium** molecules are in motion, but there is no net change in concentration.
- Study Figure 4.1 to review some important terms.
- In Figure 4.1a the two solutions are equal in their solute concentrations. We say that they are **isotonic** to each other.

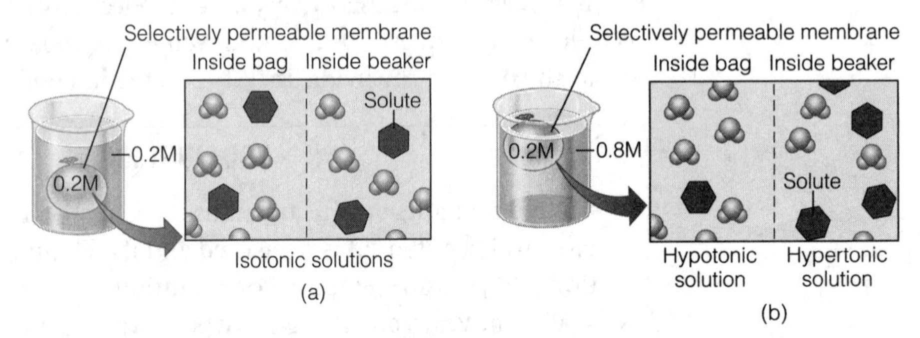

Figure 4.1 Diffusion across a selectively permeable membrane

- In Figure 4.1b, the solution in the bag contains less solute than the solution in the beaker. The solution in the bag is **hypotonic** (lower solute concentration) to the solution in the beaker. The solution in the beaker is **hypertonic** (higher solute concentration) to the one in the bag. Water will move from the hypotonic solution into the hypertonic solution.
- You must have a solid knowledge of the principles of diffusion to understand this lab, but what many students find most difficult in this laboratory is the concept of water potential. Here's a quick review.
- **Remember this:** *Water moves from a region where water potential is high to a region where water potential is low.* Think of water potential as potential energy, and you will understand why this is so.
- **Water potential** (Ψ) involves *two* components: solute potential (Ψ_S) and pressure potential (Ψ_P).
- In this laboratory we use bars as the unit of measure for water potential; 1 bar = approximately 1 atmosphere.
- **Solute potential (Ψ_S)** results from the presence of solutes. An increase in solute concentration will cause solute potential to decrease. (Ψ_S becomes more negative.) Adding solute therefore lowers the water potential.
- **Pressure potential (Ψ_P)** is zero in an open container. When a solution is enclosed by a rigid cell wall, the movement of water into the cell will exert pressure on the cell wall and the pressure potential (Ψ_P) will increase. This increase in pressure within the cell will raise the water potential.
- The water potential of pure water in an open container is zero because there is no solute and the pressure in the container is zero.
- A dehydrated potato slice does *not* have high water potential. Its water potential is low, and if placed in distilled water (which has high water potential), water will move into the potato cells.

Figure 4.2 will help you understand the relationships of water potential, solute potential, and pressure potential.

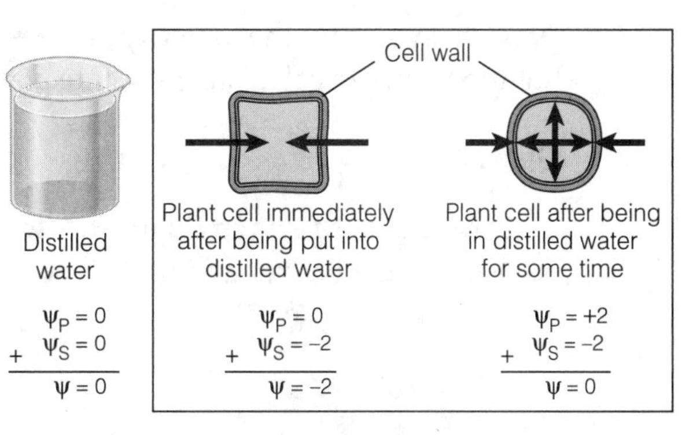

Figure 4.2 Water potential in a plant cell

Calculating Water Potential and Solute Potential

- Water potential is calculated using the following formula:

$$\Psi = \Psi_P + \Psi_S$$

Water potential (Ψ) = pressure potential (Ψ_P) + solute potential (Ψ_S)

- If you know the solute concentration, you can calculate solute potential (Ψ_S) using the following formula: $\Psi_S = -iCRT$

where

i = The number of particles the molecule will make in water; for NaCl this would be 2; for sucrose or glucose, this number is 1

C = Molar concentration (from your experimental data)

R = Pressure constant = 0.0831 liter bar/mole K

T = Temperature in degrees Kelvin = 273 + °C of solution

- The formulas and values for T and R given above will be provided on the AP Biology Equations and Formulas page. (See pages 346–347.)
- What if you don't know the molarity? One way it can be determined is to take a sample of the cells and drop them into solutions of known molarity. In this lab, we use cores taken from potatoes. If the following results are obtained, what is the molarity of the potato cores? (See Figure 4.3)

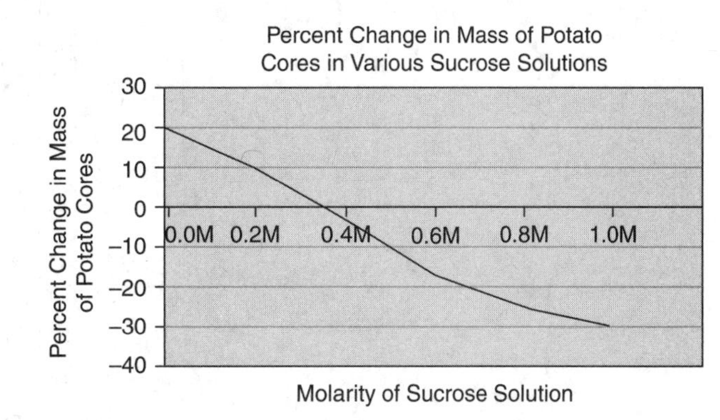

Figure 4.3 Percent change in mass of potato cores in various sucrose solutions

- The correct answer is approximately 0.35 M. This is determined by seeing at what molarity on the graph the cores neither gain nor lose mass. This is where the line crosses the X axis.
- Now that you have the molarity of the potato, if you know the temperature of the solution you can calculate the solute potential using the formula $\Psi_S = -iCRT$.

Questions

1. In beaker b, shown in Figure 4.4, what is the water potential of the distilled water in the beaker and of the beet core?

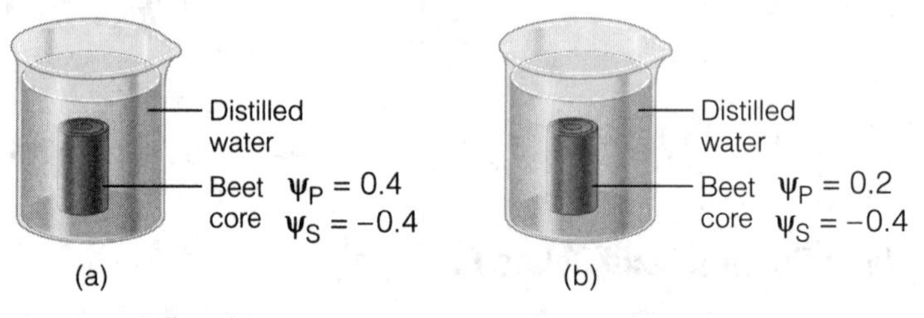

Figure 4.4 Water potential practice problem

(A) water potential in the beaker = 0; water potential in the beet core = 0
(B) water potential in the beaker = 0; water potential in the beet core = −0.2
(C) water potential in the beaker = 0; water potential in the beet core = 0.2
(D) water potential in the beaker cannot be calculated; water potential in the beet core = 0.2

2. Which of the following statements is true for the diagrams in Figure 4.4?
(A) The beet core in beaker a is at equilibrium with the surrounding water.
(B) The beet core in beaker b will lose water to the surrounding environment.
(C) The beet core in beaker b would be more turgid than the beet core in beaker a.
(D) The beet core in beaker a is likely to gain so much water that its cells will rupture.

3. Calculate the solute potential of the potato cores in Figure 4.3. The temperature is 21° C.

Express your answer in bars, rounded to the nearest one-hundredth.

BIG IDEA 2: Investigation 5, Photosynthesis

Overview of the Lab

- What factors affect the rate of photosynthesis in living leaves? In the first part of this lab, you will learn a procedure to measure the rate of photosynthesis. There are several methods for this. Your teacher may elect to use the floating disk procedure to indirectly measure the rate of oxygen production or a DPIP reduction method or probes interfaced with computers.
- In the second part you will design and conduct your own investigation of one factor that affects the rate of photosynthesis.

YOU MUST KNOW

- The equation for photosynthesis and understand the process of photosynthesis.
- The relationship between light wavelength or intensity and photosynthetic rate.
- The anatomy of a typical leaf and how the structures interact in photosynthesis.
- How to determine the rate of photosynthesis.

SCIENCE PRACTICES: CAN YOU . . .

- Measure the rate of photosynthesis using a technique that gives consistent results?
- Apply mathematical routines to calculate the rate of photosynthesis?
- Apply the concepts you have learned in your investigation to describe relationships of cell structure and function?
- Describe strategies for capture, storage, and use of free energy by plants?

Hints and Review

- In photosynthesis, plant cells convert light energy into chemical energy that is stored in sugars and other organic compounds.

The equation for photosynthesis is

$$6\,H_2O + 6\,CO_2 \rightarrow C_6H_{12}O_6 + 6\,O_2$$

- There are several ways you could measure the rate of photosynthesis. One would be to measure the rate of oxygen production. In the floating disk procedure, this is done by submerging leaf disks and then measuring how long it takes for enough oxygen to be produced to float the disks. Another method uses the reduction of DPIP by electrons from chlorophyll that are excited when exposed to light.
- You should be able to explain the chemical and physiological basis for the technique you use to measure the rate of photosynthesis.
  1. In the *floating disk procedure,* a vacuum is created to remove accumulated gases from the air spaces in the leaf, which will cause the leaf disk to sink when placed in water. Because oxygen is produced in photosynthesis, most of these gas molecules are oxygen, and when the disk is exposed to bright light, it will float again as oxygen is generated.
  2. In the *DPIP reduction technique,* you disrupt the chloroplast membrane by tossing the plant material in a blender. When exposed to light, the electrons in chlorophyll will be boosted to higher energy levels, but the electron acceptors embedded in the thylakoid membranes are not able to function. These high-energy electrons from chlorophyll will be picked up by DPIP, a chemical that is readily reduced. When oxidized, DPIP is deep blue. When reduced, it will become colorless. The rate of color change can be measured with a spectrophotometer.
- With a technique that will measure the rate of photosynthesis, such as DPIP reduction or oxygen accumulation, you can design an experiment to test a variable such as exposure to different light intensities (vary the distance from a bulb or vary the wattage of the bulb) or exposure to different wavelengths of light (use colored cellophanes or filters).
- If you removed the gases in leaf disks and placed the disks in water near a bright light source, they would never rise! Can you explain this based on the equation for photosynthesis? (Answer: There is so little CO_2 in water that photosynthesis can scarcely proceed.) You should now be able to explain why the floating disk procedure requires the addition of baking soda to the water.

Questions

1. If a student uses the DPIP technique described above, which of the following statements best describes the role of DPIP?
   (A) It mimics the action of chlorophyll by absorbing light energy.
   (B) It serves as an electron donor and blocks the formation of NADPH.
   (C) It is an electron acceptor and is reduced by electrons from chlorophyll.
   (D) It is bleached in the presence of light and can be used to measure light levels.

2. Some students were not able to get many data points when using the DPIP technique because the solution went from blue to colorless in only 5 minutes when they used chloroplasts exposed to light. What modification to the experiment do you think would be most likely to provide better results?
   (A) Double the volume of chloroplasts used.
   (B) Double the volume of DPIP so that the solution has a lower initial transmittance.
   (C) Boil the chloroplast in order to further disrupt the thylakoid membrane.
   (D) Select a different plant material and blend it more thoroughly.

3. If a student performed this experiment using DPIP that was initially deep blue and got a flat line when the data were graphed (showing very little change in color over time), which of these would be a plausible explanation?
   (A) The rate of photosynthesis was very high.
   (B) The intensity of light may have been too great for a reaction to occur.
   (C) The chlorophyll was damaged and could not respond to light.
   (D) The DPIP was already reduced when the experiment began.

4. A student used the floating disk technique to measure the rate of photosynthesis. After 20 minutes under a bright light, none of the disks had floated. Based on your understanding of photosynthesis, which of the following might be a reasonable explanation for this?
   (A) Not all green plant material does photosynthesis when exposed to light.
   (B) A source of carbon dioxide was not provided.
   (C) A source of oxygen was not available.
   (D) There was no water available for photosynthesis.

BIG IDEA 2: Investigation 6, Cellular Respiration

Overview of the Lab

In this experiment you will learn how to use a respirometer to measure the rate of cellular respiration. You will select a question that interests you and then design an experiment to test the effect of a single variable on the rate of respiration.

YOU MUST KNOW

- The equation for cellular respiration.
- The components of a respirometer and how it works.
- The gas laws that affect volume changes within the respirometer.
- The relationship between movement of water in a respirometer and cellular respiration.
- The effect of temperature or increased metabolic activity on respiration.
- How to calculate the rate of respiration.

SCIENCE PRACTICES: CAN YOU . . .

- Design an experiment to answer a question about cellular respiration?
- Analyze your data and use appropriate mathematical routines to describe your results?
- Justify a claim such as "dormant seeds respire at a low rate" with evidence?

The equation for cellular respiration is

$$\mathbf{C_6H_{12}O_6 + 6\,O_2 \rightarrow 6\,H_2O + 6\,CO_2 + Energy}$$

Hints and Review

- How can the rate of cellular respiration be measured? When you study the equation for cellular respiration, you will see that there are at least three ways:
  1. Measure the amount of glucose consumed.
  2. Measure the amount of oxygen consumed.
  3. Measure the amount of carbon dioxide produced.
- In this experiment, you will use either gas probes or *respirometers* to measure the amount of oxygen consumed.
- A **respirometer** is an air-tight chamber except for one opening for gases to enter or leave. Potassium hydroxide (KOH) is used to soak a cotton ball and

will combine with the CO_2 produced by the organism you are using. A solid precipitate forms in the following reaction:

$$CO_2 + 2KOH \rightarrow K_2CO_3 + H_2O$$

- Because CO_2 and O_2 are produced and consumed in equal amounts, any changes in the sealed container (assuming temperature and pressure remain constant) will be caused by a change in gas volume due to cellular respiration. As O_2 is consumed in respiration and CO_2 removed as a solid precipitate, pressure within the respirometer decreases and water enters the pipette. Study Figure 6.1 to see the components of a respirometer.

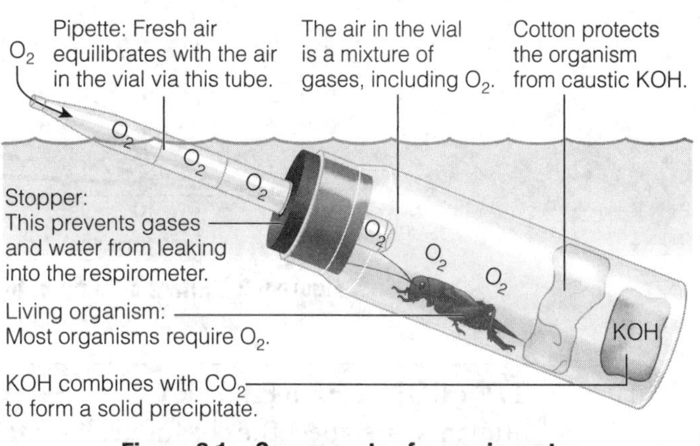

Figure 6.1 Components of a respirometer

- Although no single experimental organism is prescribed in this lab, many students will use some sort of seeds or small invertebrates such as mealworms. We will describe ideas related to seeds or invertebrates, although your experimental organism may differ.
- A seed contains an embryo plant and a food supply surrounded by a seed coat.
- *Germinating* (sprouting) seeds will show a higher rate of respiration than the *nongerminating* seeds because metabolic activity is increased. However, these dry, nongerminating seeds are not dead, but **dormant**. They can be stored for years and, when soaked in water, will germinate.
- As you investigate, consider the size and metabolic activity of the organism you have selected and its rate of cellular respiration. Your knowledge of metabolism should lead you to see that larger or more active or endothermic organisms require more energy and therefore have higher rates of respiration.
- In most cases, chemical reactions occur more slowly at lower temperatures, so seeds or ectotherms that are chilled will show a lower rate of O_2 consumption.
- A vial of glass beads alone might act as a *control* for this experiment. The control will compensate for any change in pressure or temperature.
- Students are often confused by the difference between a *control* (the glass beads in this experiment) and *factors that are held constant.* In this experiment, you will hold constant factors such as the volumes within the containers, the number of peas or mealworms, and the temperature of the water for each experimental group.

- Figure 6.2 shows a graph of typical results from an experiment. What is the question being tested?
- You need to be able to calculate the rate of these reactions. If you need to review how to do this, you will find a lesson on calculating rate in Investigation 13, Enzyme Activity.

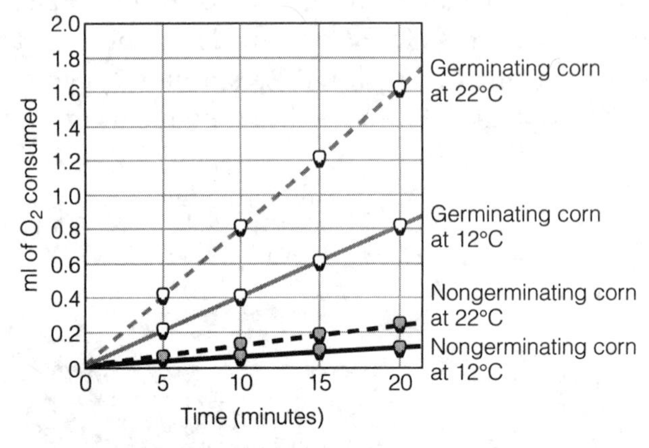

Figure 6.2 Effect of temperature on respiration rate

> ***TIP FROM THE READERS***
> Students often become confused about the effect of cold on the rate of respiration. You may know that there is more O_2 in colder water, but that is *not* the reason organisms respire more slowly when chilled! In general, within a normal physiological range for the organism, **the rate of metabolism is directly related to the temperature of the organism,** whether it is a pea seed, a cricket, or a goldfish.

Questions

1. What is the rate of oxygen consumption in germinating corn at 12°C, as shown in Figure 6.2?
   (A) 0.08 mL/min
   (B) 0.04 mL/min
   (C) 0.8 mL/min
   (D) 0.6 mL/min

2. Using Figure 6.2, which of the following is a true statement based on the data?
   (A) The amount of oxygen consumed by germinating corn at 22°C is approximately twice the amount of oxygen consumed by germinating corn at 12°C.
   (B) The rate of oxygen consumption is the same in both germinating and nongerminating corn during the initial time period from 0 to 5 minutes.
   (C) The rate of oxygen consumption in the germinating corn at 12°C at 10 minutes is 0.4 mL O_2/minute.
   (D) The rate of oxygen consumption is higher for nongerminating corn at 12°C than at 22°C.

3. Which of the following conclusions is supported by the data shown on the graph in Figure 6.2?
   (A) The rate of respiration is higher in nongerminating seeds than in germinating seeds.
   (B) Nongerminating seeds are not alive and show no difference in rate of respiration at different temperatures.
   (C) The rate of respiration in the germinating seeds would have been higher if the experiment were conducted in sunlight.
   (D) The rate of respiration increases as the temperature increases in both germinating and nongerminating seeds.

4. What is the role of KOH in a respirometer?
   (A) It serves as an electron donor to promote cellular respiration.
   (B) As KOH breaks down, the oxygen needed for cellular respiration is released.
   (C) It serves as a temporary energy source for the respiring organism.
   (D) It binds with carbon dioxide to form a solid, removing CO_2 from the respirometer and allowing water to move into the respirometer.

BIG IDEA 3: Investigation 7, Cell Division: Mitosis and Meiosis

Overview of the Lab

This laboratory involves five parts:

1. **Modeling Mitosis**
You will use beads or pipe cleaners or other materials to model the events of the cell cycle, including chromosome duplication and movement.
2. **Effects of Environment on Mitosis**
You will set up and analyze an experiment using root squashes made from onion roots to investigate the effect of a protein (lectin) known to increase the rate of mitosis in the roots. After you collect data, you will use Chi-square analysis to statistically analyze the results.
3. **Loss of Cell Cycle Control in Cancer**
For this part, you will consider HeLa cells and the Philadelphia chromosome and how the genetic changes they show lead to loss of cell cycle control and cancer.
4. **Modeling Meiosis**
You will use the same materials as in Part I to model the events of meiosis, and show how meiosis and crossing-over events increase genetic variation. You will also demonstrate nondisjunction and explain its relationship to genetic disorders.
5. **Meiosis and Crossing Over in *Sordaria***
In this part, you will observe crossover events in meiosis of a fungus and calculate map distance.

YOU MUST KNOW

- The events of mitosis and meiosis in plant and animal cells.
- How mitosis and meiosis differ.
- How normal cells and cancer cells differ from each other.
- What may go wrong during the cell cycle in cancer cells.
- The roles of segregation, independent assortment, and crossing over in generating genetic variation.
- How to calculate map distance from experimental data.
- How to evaluate experimental results using Chi-square analysis.

SCIENCE PRACTICES: CAN YOU . . .

- Make predictions about natural phenomena occurring in the cell cycle? (LO 3.7)
- Describe the events that occur in the cell cycle? (LO 3.8)
- Represent the connection between meiosis and increased genetic diversity? (LO 3.10)
- Use the mathematical routine of Chi-square analysis appropriately?

Hints and Review

Part 1 and Part 5. Modeling Mitosis and Meiosis

Go back to Topic 2 to review mitosis and the cell cycle. Go to Topic 4 to review the important elements of meiosis. Although you have had this process described in class numerous times by this point in your high school career, we find that students struggle to model the process. Buy a cheap sack of colored pipe cleaners and try this:

1. Assemble two pairs of homologous chromosomes. (How did you represent maternal vs. paternal chromosome?)
2. Use them to model a cell in metaphase of mitosis.
3. Model metaphase I of meiosis.
4. Model metaphase II of meiosis.
5. When does crossing over occur? Model it. When do the cells become haploid? Show it with your model.

This activity is deceptive—it seems easy but is surprisingly difficult! Work with a study partner and check with your teacher.

Part 2: Effects of Environment on Mitosis

- To investigate the effect of a chemical believed to affect the rate of cell division, you would need to grow some of the plants in a medium with this chemical and another group in water as a control. You could then determine the effect on cell division by comparing the number of cells actively dividing in each treatment.
- Count the cells that are in any stage of mitosis (prophase, metaphase, anaphase, or telophase) in both the control and experimental groups as well as the cells in interphase. Sample data are shown in Table 7.1.

	Cells in Interphase	Cells in Mitosis	Total # Cells
Control group	176	24	200
Experimental group	186	64	250

Table 7.1 Onion root tip cell phase data

- Actual data will not have such even numbers, but this will make it easier to illustrate what needs to be done. Based on the data, can you now conclude that your treatment increased the rate of cell division? Not yet! In science, you will need to use statistical methods to support such a conclusion. Chi-square analysis would be an appropriate test to apply here. If this is new information for you, proceed to the lesson that begins on page 342, *Statistical Analysis,* and work through the information provided there, then come back to this.
- The observed values are what you actually count in the experimental group. To determine the expected values, calculate the percentage of cells in interphase and mitosis in the control group, and multiply these by the observed cells in the treated groups.

- To illustrate, 12% of the control cells are in mitosis. If the treatment has no effect, you would expect 12% of the experimental cells to be in mitosis. Because there are 250 cells in the experimental group, 12% = $0.12 \times 250 = 30$ cells expected to be in mitosis. Note that 64 cells are actually observed in mitosis.

> ***TIP FROM THE READERS***
> Be sure you practice using Chi-Square analysis of data. This task has been given to students on numerous AP exams, most recently in 2013. Do the problem that follows and check your results!

- Using the data from the Table 7.1 do a Chi-square analysis to complete the following table. State the null hypothesis being tested. The solution to this chart is in the Answers and Explanations for this investigation.

Null Hypothesis:

	Observed (o)	**Expected (e)**	**$(o-e)^2 / e$**
Interphase cells			
Mitosis cells			
Total			
		X^2 =	
Degrees of Freedom =	p value =	critical value =	
Accept or reject the null hypothesis? Explanation:			

- The task above is equivalent to Free Response Question 1 on the 2013 AP exam.

***Part 5: Meiosis and Crossing Over in* Sordaria**

- When the growing filaments of two haploid strains of *Sordaria* that produce spores of different colors meet, fertilization occurs and zygotes form. Figure 7.1 shows spore formation in *Sordaria.*
- Meiosis occurs within fruiting bodies to form four haploid *ascospores,* spores contained in *asci* (special sacs).
- One mitotic division then doubles the number of ascospores to eight.
- The number of **map units** between two genes is calculated by determining the percentage of recombinants that result from crossing over. The greater the frequency of crossing over, the greater the map distance.

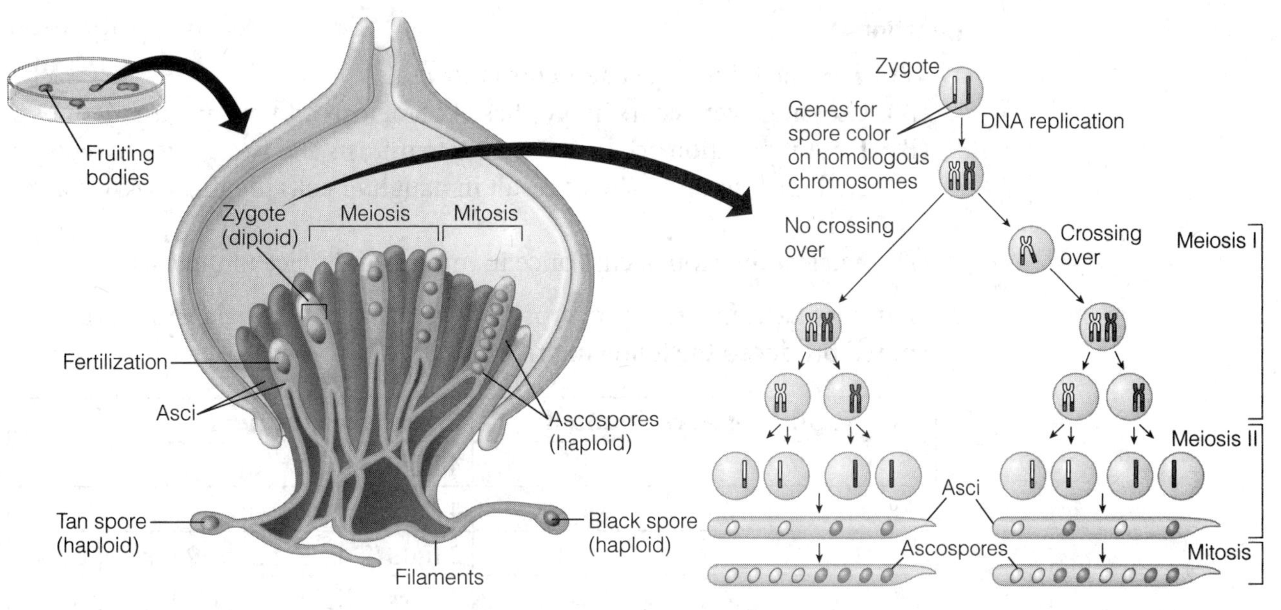

Figure 7.1 Crossing over in *Sordaria*

- Calculate the percent of crossovers by dividing the number of crossover asci (these are the ones with spores arranged 2:2:2:2 or 2:4:2 as you see in Figure 7.2) by the total number of asci × 100.

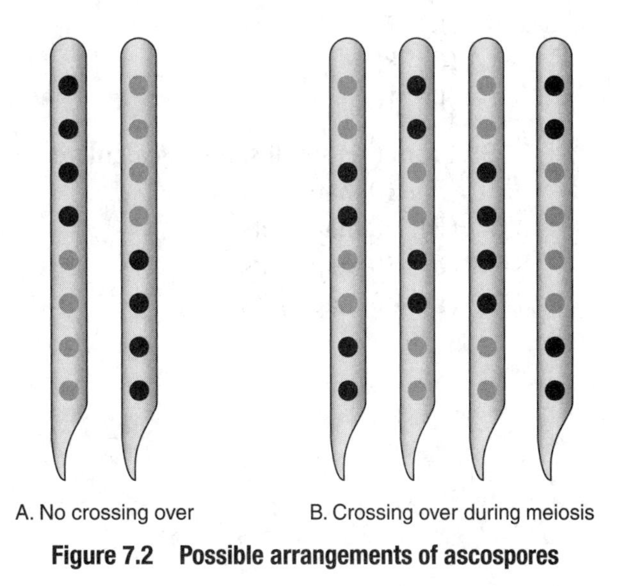

Figure 7.2 Possible arrangements of ascospores

- To calculate the map distance, divide the percent of crossover asci by 2. The percent of crossover asci is divided by 2 because only half of the spores in each ascus are the result of crossing over.

Questions

1. Which of the following statements is correct?
   (A) Crossing over occurs in prophase I of meiosis and metaphase of mitosis.
   (B) DNA replication occurs once prior to mitosis and twice prior to meiosis.
   (C) Both mitosis and meiosis result in daughter cells identical to the parent cells.
   (D) Nuclear division occurs once in mitosis and twice in meiosis.

2. A group of asci formed from crossing light-spored *Sordaria* with dark-spored produced the following results:

Number of Asci Counted	Spore Arrangement
7	4 light/4 dark spores
8	4 dark/4 light spores
3	2 light/2 dark/2 light/2 dark spores
4	2 dark/2 light/2 dark/2 light spores
1	2 dark/4 light/2 dark spores
2	2 light/4 dark/2 light spores

   How many of these asci contain a spore arrangement that resulted from crossing over?
   (A) 3
   (B) 7
   (C) 8
   (D) 10

3. From this small sample, calculate the map distance between the genes.
   (A) 10 map units
   (B) 20 map units
   (C) 30 map units
   (D) 40 map units

BIG IDEA 3: Investigation 8, Biotechnology: Bacterial Transformation

Overview of the Lab

You will use antibiotic-resistance plasmids to transform *Escherichia coli.* A plasmid containing a gene for resistance to the antibiotic ampicillin is introduced into a strain of *E. coli* that is killed by ampicillin. If the susceptible bacteria incorporate the foreign DNA, they will become ampicillin resistant. You will apply mathematical routines to calculate transformation efficiency. Then you may design and conduct an investigation to study transformation in more depth.

YOU MUST KNOW

- The principles of bacterial transformation, including how plasmids are engineered and taken up by cells.
- Factors that affect transformation efficiency.
- How to verify and screen for transformed cells.
- Bacterial transformation is a type of horizontal gene transfer and increases genetic variation.

SCIENCE PRACTICES: CAN YOU . . .

- Calculate transformation efficiency and express the results in scientific notation?
- Predict and justify how a change in the basic protocol for bacterial transformation would affect transformation efficiency?

Key Concepts of Bacterial Transformation

- Genetic **transformation** occurs when a host organism takes in foreign DNA and expresses the foreign gene.
- Bacterial cells have a single main chromosome and circular DNA molecules called **plasmids**, which carry genetic information. All of the genes required for basic survival and reproduction are found in the single chromosome.
- **Plasmids** are circular pieces of DNA that exist outside the main bacterial chromosome and carry their own genes for specialized functions, including resistance to specific drugs. In genetic engineering, plasmids are one means used to introduce foreign genes into a bacterial cell. To understand how this might work, consider the plasmid in Figure 8.1.

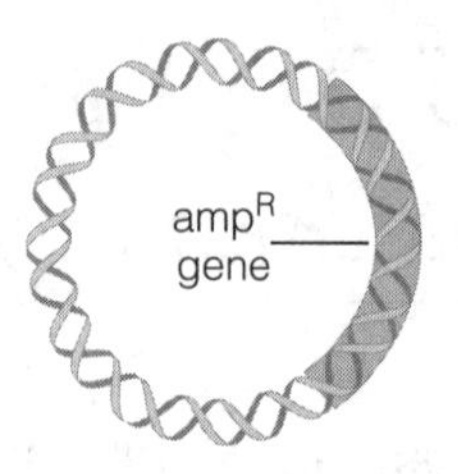

Figure 8.1 Plasmid with ampicillin-resistant gene

- Plasmids with the *amp*^{R} gene, are resistant to the antibiotic ampicillin. *E. coli* cells containing this plasmid, termed **"+amp^R"** cells, can survive and form colonies on LB agar that has been supplemented with ampicillin.
- In contrast, cells lacking the amp^R plasmid, termed **"−amp^R"** cells, are sensitive to the antibiotic, which kills them. An ampicillin-sensitive cell (−amp^R) can be transformed to an ampicillin-resistant (+amp^R) cell by its uptake of a foreign plasmid containing the *amp*^{R} gene.
- **Competent cells** are cells that are most likely to take up extracellular DNA. Competent cells are in logarithmic growth, and chemical conditions are modified to induce the uptake of DNA. Study Figure 8.2 below to review the lab procedure used to prepare competent cells and get them to take up the amp^R plasmids.

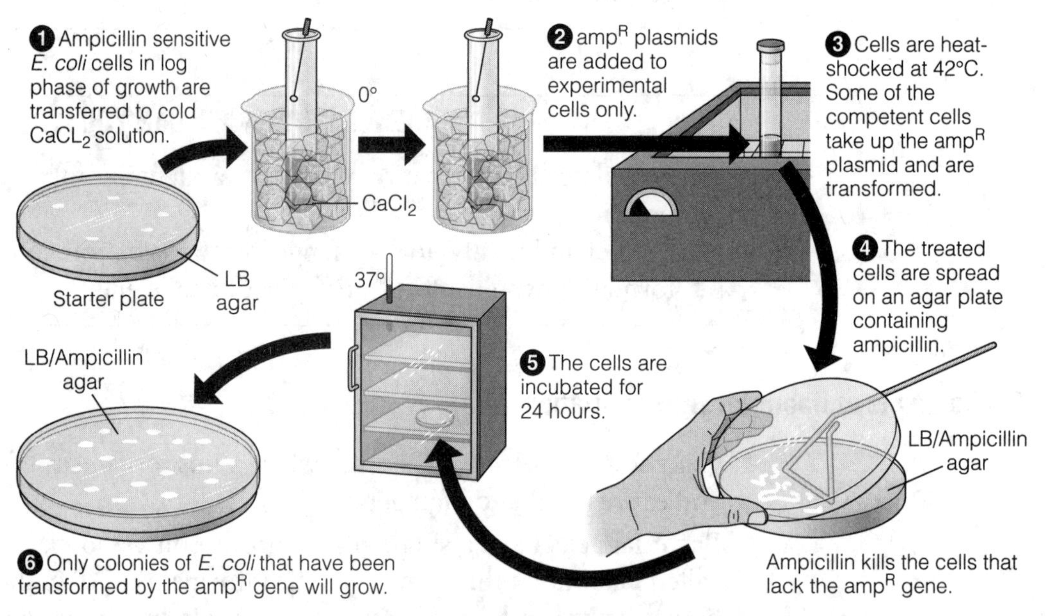

Figure 8.2 Procedure for bacterial transformation with amp plasmids

- Study Figure 8.3. It shows the expected results in this experiment.
- If there is no ampicillin in the agar, *E. coli* will cover the plate with so many cells it is called a "lawn" of cells (Plates A and C).
- Only transformed cells can grow on agar with ampicillin. Because only some of the cells exposed to the amp^R plasmids will actually take them in, only some

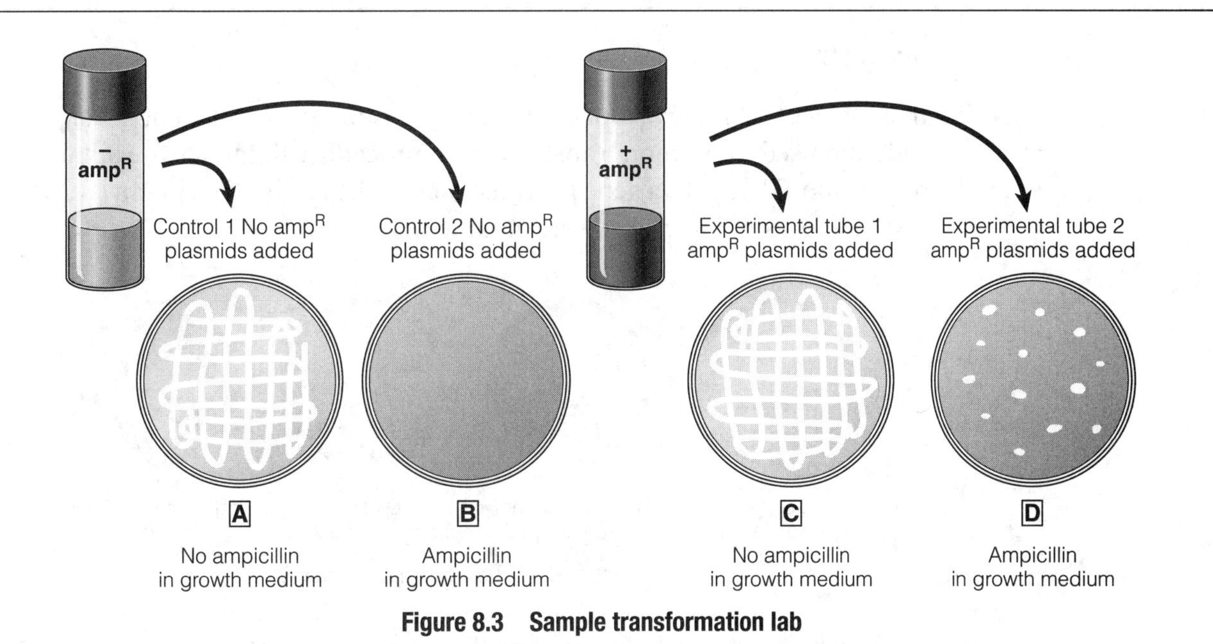

Figure 8.3 Sample transformation lab

cells will be transformed. Thus, you will see only individual colonies on the plate (Plate D).

- If none of the sensitive *E. coli* cells has been transformed, nothing will grow on the agar with ampicillin (Plate B).
- **Restriction enzymes** or endonucleases are bacterial enzymes that will cut DNA at specific DNA sequences known as **recognition sites**. Often the enzymes cut the DNA so that the ends are single-stranded "sticky ends." A **gene of interest** (such as antibiotic resistance) can be introduced into a plasmid using restriction enzymes as described below.
- Here are the general steps used to introduce a gene of interest into bacteria:
  1. Both the gene of interest and the plasmid are cut with the *same* restriction enzyme, so they have the same sticky ends.
  2. DNA ligase is used to anneal and seal the sticky ends.
  3. The recipient cells are transformed with the engineered plasmid.
  4. Colonies carrying the plasmid are isolated.
- How do we know that transformation has been successful?
  1. Use a **selection gene**, such as for antibiotic resistance. Only those cells that have incorporated the plasmid will have antibiotic resistance.
  2. Use a **reporter gene** such as GFP (green fluorescent protein). Transformed cells will glow!
- **Transformation efficiency** is the number of transformed cells per microgram of the plasmid. High transformation efficiencies require cells that are in log phase of growth, suspended in ice-cold calcium chloride, have a rapid heat shock (this makes the membrane permeable to the plasmid), and plasmids that are not too large.

Questions

In a molecular biology laboratory, a student obtained competent *E. coli* cells and used a common transformation procedure to induce the uptake of plasmid DNA with a gene for resistance to the antibiotic kanamycin. The results shown in Figure 8.4 were obtained.

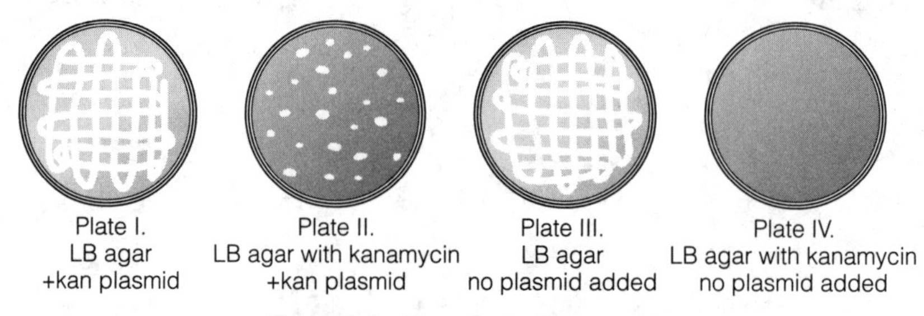

Figure 8.4 Transformation results

1. On which petri dish do only transformed cells grow?
   (A) Plate I
   (B) Plate II
   (C) Plate III
   (D) Plate IV

2. Which of the plates is used as a control to show that nontransformed *E. coli* will not grow in the presence of kanamycin?
   (A) Plate I
   (B) Plate II
   (C) Plate III
   (D) Plate IV

3. If a student wants to verify that transformation has occurred, which of the following procedures should she use?
   (A) Spread cells from Plate I onto a plate with LB agar; incubate.
   (B) Spread cells from Plate II onto a plate with LB agar; incubate.
   (C) Repeat the initial spread of $-\text{kan}^R$ cells onto Plate IV to eliminate possible experimental error.
   (D) Spread cells from Plate II onto a plate with LB agar with kanamycin; incubate.

4. During the course of an *E. coli* transformation laboratory, a student forgot to mark the culture tube that received the kanamycin-resistant plasmids. The student proceeds with the laboratory because he thinks that he will be able to determine from his results which culture tube contained cells that may have undergone transformation. Which plate would be most likely to indicate transformed cells?
   (A) a plate with a lawn of cells growing on LB agar with kanamycin
   (B) a plate with a lawn of cells growing on LB agar without kanamycin
   (C) a plate with 100 colonies growing on LB agar with kanamycin
   (D) a plate with 100 colonies growing on LB agar without kanamycin

BIG IDEA 3: Investigation 9, Biotechnology: Restriction Enzyme Analysis of DNA

Overview of the Lab

You will use restriction endonucleases and gel electrophoresis to create and analyze genetic fingerprints. After electrophoresis, you will use your results to prepare a standard curve and estimate fragment sizes of an unknown sample.

YOU MUST KNOW

- The function of restriction enzymes and their role in genetic engineering.
- How gel electrophoresis separates DNA fragments.
- How to use a standard curve to determine the size of unknown DNA fragments.

SCIENCE PRACTICES: CAN YOU . . .

- Apply mathematical routines to construct a graph of DNA fragments of known size?
- Use a standard curve to determine the size of unknown DNA fragments?
- Use the results of gel electrophoresis to map the restriction sites of a bacterial plasmid?

Key Concepts of Restriction Enzyme Cleavage of DNA and Gel Electrophoresis

- **Gel electrophoresis** is a procedure that separates molecules on the basis of their rate of movement through a gel under the influence of an electrical field.
- The direction of movement is affected by the charge of the molecules, and the rate of movement is affected by their size and shape, the density of the gel, and the strength of the electrical field.
- DNA is a negatively charged molecule, so it will move toward the positive pole of the gel when a current is applied. When DNA has been cut by restriction enzymes, the different-sized fragments will migrate at different rates. Because the smallest fragments move the most quickly, they will migrate the farthest during the time the current is on. Keep in mind that the length of each fragment is measured in number of DNA base pairs.
- You may use restriction enzymes to create your own DNA fragments or use fragments that are commercially prepared. In one lab we often purchase, students are given three samples of DNA obtained from the bacteriophage lambda. One sample is uncut DNA, one is incubated with the restriction enzyme *Hind*III, and one is incubated with *Eco*RI. The fragments of DNA are separated by electrophoresis, stained for visualization, and then analyzed.

- After the DNA samples are loaded into wells in the gel, electricity is applied. The DNA fragments will migrate.
  1. *DNA is negatively charged and will migrate toward the positive pole.*
  2. *Smaller fragments of DNA will migrate faster than larger fragments.*
- DNA is not visible to the naked eye. In order to visualize it, a dye, such as methylene blue, must be added, which will bind to the DNA.
- Each fragment of DNA is a particular number of nucleotides, or base pairs, long. When researchers want to determine the size of DNA fragments produced with particular restriction enzymes, they run the unknown DNA alongside DNA with known fragment sizes. The known DNA acts as a **marker** and is used to help determine the unknown fragment sizes.
- Figure 9.1 shows the results of electrophoresis. In this case semilog paper has been used to plot the results of the *Hind*III digest. Because its fragment sizes are known, this is the *standard curve*. It can now be used to determine the other fragment sizes from the DNA I and DNA II samples by interpolation.
- In the commercial lab described above, *Hind*III is the marker and used to prepare the standard curve. The standard curve is used to determine the fragment sizes in the *Eco*RI digest.

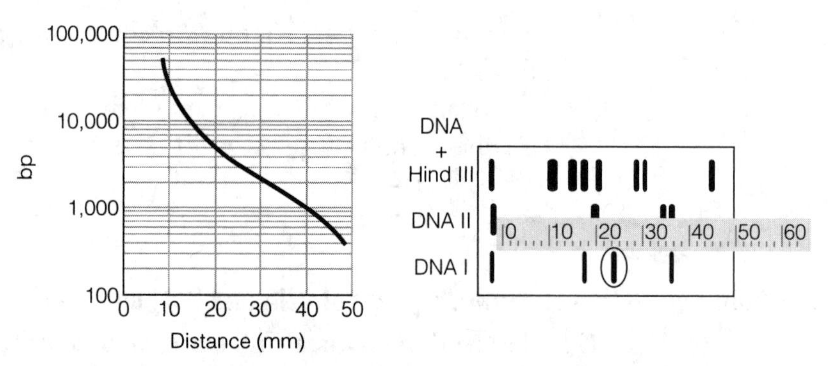

Figure 9.1 Using results of electrophoresis to determine fragment sizes for an unknown sample

Questions

1. How many base pairs is the fragment circled in Figure 9.1?
   (A) 350
   (B) 22
   (C) 2,200
   (D) 3,500

2. Which of the following statements is correct?
   (A) Longer DNA fragments migrate farther than shorter fragments.
   (B) Migration distance is inversely proportional to the fragment size.
   (C) Positively charged DNA migrates more rapidly than negatively charged DNA.
   (D) Uncut DNA migrates farther than DNA cut with restriction enzymes.

Here is a plasmid with restriction sites for *Bam*HI and *Eco*RI. Several restriction digests were done using these two enzymes either alone or in combination. Use Figure 9.2 to answer questions 3 and 4. **Hint:** Begin by determining the number and size of the fragments produced with each enzyme; "kb" stands for kilobases, or thousands of base pairs.

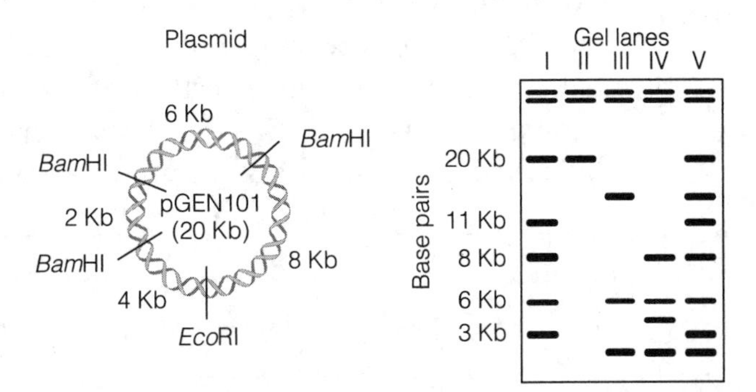

Figure 9.2 Plasmid with restriction sites for *Bam*HI and *Eco*RI

3. Which lane shows a digest with *Bam*HI only?
   (A) I
   (B) II
   (C) III
   (D) IV

4. Which lane shows a digest with both *Bam*HI and *Eco*RI?
   (A) II
   (B) III
   (C) IV
   (D) V

BIG IDEA 4: Investigation 10, Energy Dynamics

Overview of the Lab

Primary productivity is a measure of the amount of light energy converted to chemical energy in the form of organic compounds during a given period and sets the energy budget of an ecosystem. Energy flows constantly through an ecosystem, and this investigation has you look at energy flow from plants (the producers) to the insects (consumers) that feed on them. You do this by measuring the biomass of the producers and then of the consumers.

YOU MUST KNOW

- The difference between gross and net productivity and how it can be measured.
- Energy does not cycle; only matter cycles.
- The relationship between photosynthesis and respiration and how these processes relate to energy flow, NPP, and ecosystem energy dynamics.
- How to measure productivity to investigate a question about energy capture and flow in an ecosystem.

SCIENCE PRACTICES: CAN YOU . . .

- Plan and implement data collection strategies to answer a question about energy flow in an ecosystem?
- Analyze data to identify patterns or relationships?
- Refine your observations and measurements and evaluate the evidence provided by the data?

Hints and Review

- You must have a firm grip on the summary equations for both cellular respiration and photosynthesis to understand energy flow. So here are the summary equations for both:

Photosynthesis

$$6\ CO_2 + 6\ H_2O + \text{Light energy} \rightarrow C_6H_{12}O_6 + 6\ O_2$$

Cellular Respiration

$$C_6H_{12}O_6 + 6\ O_2 \rightarrow 6\ CO_2 + 6\ H_2O + \text{Energy}$$

- **Primary productivity** is a term used to describe the rate at which plants and other photosynthetic organisms produce organic compounds in an ecosystem. There are two aspects of primary productivity:
  - **Gross Primary Productivity (GPP)** = the entire photosynthetic production of organic compounds in an ecosystem over a unit of time.
  - **Net Primary Productivity (NPP)** = the organic materials that remain after photosynthetic organisms in the ecosystem have used some of these compounds for their cellular energy needs (cellular respiration).
  - Net primary productivity is determined by subtracting the energy lost by cellular respiration from gross primary productivity, or NPP=GPP-Cellular respiration.
- Refer to the equation for photosynthesis. Because GPP is the result of photosynthesis, it can be measured in three ways:
  1. The amount of carbon dioxide used
  2. The rate of oxygen production
  3. The rate of sugar formation = amount of biomass
- In an aquatic ecosystem, you could measure the amount of dissolved oxygen to determine the rate of photosynthesis and determine productivity.
- This investigation has two parts before you design your own experiment. Each is described below.

Part I : Estimating Net Primary Productivity of Fast Plants

- In a terrestrial ecosystem, you will use the third measure listed above: the rate of carbon fixation in sugar formation, which is measured by determining the amount of biomass.
- **Biomass** is the weight of organic material produced. In order to determine biomass, the material must be dried so that water is not being included in the measure.
- The plant material will need to be dried and references can give you a reasonable approximation of the conversion factor to use to determine kcal of energy. Because the plant material you measure has been doing both photosynthesis as well as respiration, you are determining *net* productivity, not gross.

Part II: Estimating Energy Flow between Producers and Consumers

- In this part of the lab, you will use a consumer (cabbage butterfly larvae) and a producer (brussels sprouts or another food source for the consumers).
- You will determine the initial wet mass of the larva, let them feed for 3 days, and mass them again. Their change in mass reflects energy gained from the food they ate.
- You will determine the wet mass of the larvae and use published data to approximate the biomass. But *wait* . . . these little guys are pushing out the waste (called *frass*) as fast as they eat, so you must also collect and weigh it! When you combine all this and consult published data, you should be able to calculate the energy represented by the increased size of the larvae and contained in their frass.

- Compare the energy estimate for the plant material (NPP) and for the consumers and you should be able to draw conclusions about energy transfer between trophic levels and its efficiency.
- According to the second law of thermodynamics, energy transfer is never 100% efficient. Most textbooks cite an energy transfer of only 10% between trophic levels. What other sources of energy consumption are not accounted for using this artificial ecosystem?

Questions

1. Figure 10.1 shows productivity in an aquatic environment, determined by the amount of oxygen produced.

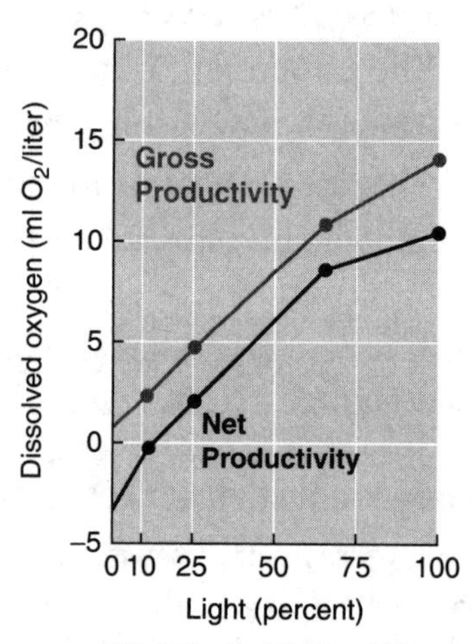

Figure 10.1 Gross and Net Productivity at Varying Light Intensities

At what light intensity do you expect there to be no net productivity?
(A) any intensity below 100%
(B) only at intensities of 0% and 2%
(C) any intensity below 10%
(D) any intensity above 25%

2. What is meant by "net productivity" and how is it calculated?
(A) It is a measure of the organic products of photosynthesis that accumulate after cellular respiration. It is calculated by subtracting the energy consumed in respiration from the total energy captured in photosynthesis.
(B) It is a measure of the amount of respiration in a test area, and it is calculated by subtracting the amount of waste produced by the consumers from their biomass.
(C) It is the total amount of carbon fixed and it is calculated by determining the biomass of all producers in the test area over a unit of time.
(D) It is the amount of energy available to the secondary consumers and it is calculated by determining the biomass of the primary consumers.

3. Consider the following two ecosystems located at the same latitude.

Ecosystem A	Ecosystem B
266 sunny days/year	183 sunny days/year
282 frost-free days/year	125 frost-free days/year
25" rain/year	36" rain/year

(A) Ecosystem A would be expected to have less species diversity because of abundant light and a long growing season.
(B) Ecosystem B would have greater species diversity because it receives more water annually.
(C) Ecosystem A would be expected to have the highest gross primary productivity because of more sunny days and a longer growing season.
(D) Ecosystem B would be expected to have the highest net primary productivity because of longer winters and fewer sunny days.

BIG IDEA 4: Laboratory 11, Transpiration

Overview of the Lab

Transpiration is the major mechanism that drives the movement of water through a plant. The first section of this laboratory begins by calculating leaf surface area and uses this to determine the average number of stomata per square millimeter of leaf. Then you will learn a technique to measure the rate of transpiration, such as a potometer or whole-plant method. This will allow for the design of your own experiment to answer a question about a factor that influences the rate of transpiration.

YOU MUST KNOW

- The function of stomata in gas exchange in plants. What enters? What leaves?
- The role of water potential and transpiration in the movement of water from roots to leaves.
- The effects of various environmental conditions on the rate of transpiration.

SCIENCE PRACTICES: CAN YOU . . .

- Predict and justify whether a plant cell will give or lose water based on water potential?
- Create and annotate a diagram to show what would happen to grass planted near a road that has been salted in winter? Include water potential in your representation.

Hints and Review

- Review **hydrogen bonding** (see Topic 1, Chapter 3). In water, a hydrogen bond is a weak bond between the hydrogen of one water molecule and the oxygen of another, and it accounts for the unique properties of water, including adhesion and cohesion.
- Water enters a plant through the root hairs, passes through the tissues of the root into the xylem, and travels up through the xylem vessels into the leaves.
- **Transpiration** is the evaporation of water from the leaves through the stomata. It is the major factor that pulls the water up through the plant.
- Study Figure 11.1 to see this process. When water enters the roots, hydrogen bonds link each water molecule to the next (*cohesion*) so the molecules of water are pulled up the thin xylem vessels like beads on a string. The water molecules also cling to the thin walls of the xylem cells (*adhesion*). The water moves up the plant, enters the leaves, moves into air spaces in the leaf, and then evaporates (transpires) through the *stomata* (singular, *stoma*).

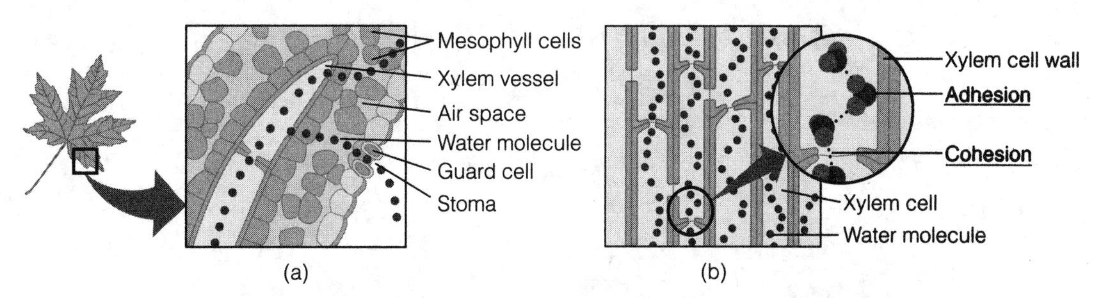

Figure 11.1 Leaf anatomy showing movement of water molecules

- **Stomata** are the pores in the epidermis of a leaf. There are hundreds of stomata in the epidermis of a leaf. Most are located in the lower epidermis. This reduces water loss because the lower surface receives less solar radiation than the upper surface. Each stoma allows the carbon dioxide necessary for photosynthesis to enter, while water evaporates through each one in transpiration.
- **Guard cells** are cells surrounding each stoma. They help to regulate the rate of transpiration by opening and closing the stomata. To understand how they function, study the following figures. As you look at the figures, keep in mind that an increase in solute concentration lowers the water potential of the solution and that water moves from a region with higher water potential to a region of lower water potential.
- Notice that in Figure 11.2a the guard cells are turgid, or swollen, and the stomatal opening is large. This turgidity is caused by the accumulation of K^+ (potassium ions) in the guard cells. As K^+ levels increase in the guard cells, the water potential of the guard cells drops and water enters the guard cells.

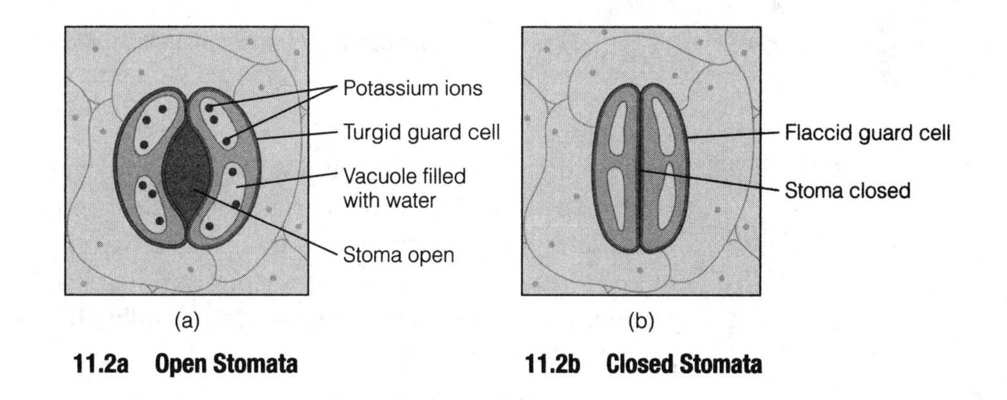

11.2a Open Stomata **11.2b Closed Stomata**

- In Figure 11.2b, the guard cells have lost water, which causes the cells to become flaccid and the stomatal opening to close. This may occur when the plant has lost an excessive amount of water. In addition, it generally occurs daily as light levels drop and the use of CO_2 in photosynthesis decreases.

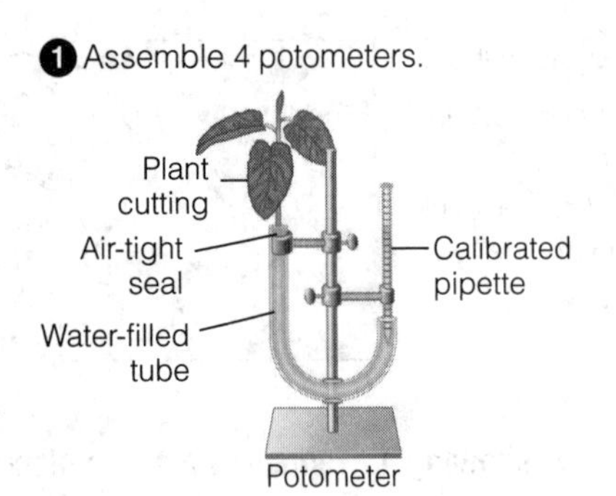

2 Place each potometer in a different environment: room conditions, mist, wind, and bright light.

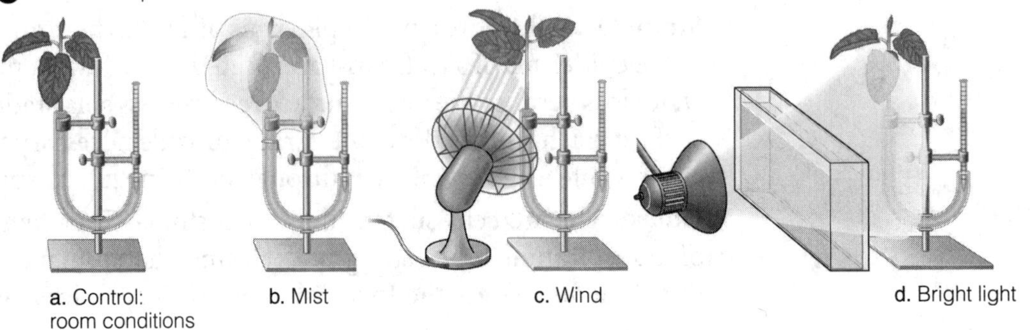

3 Measure water loss in each potometer every 3 minutes for 30 minutes.

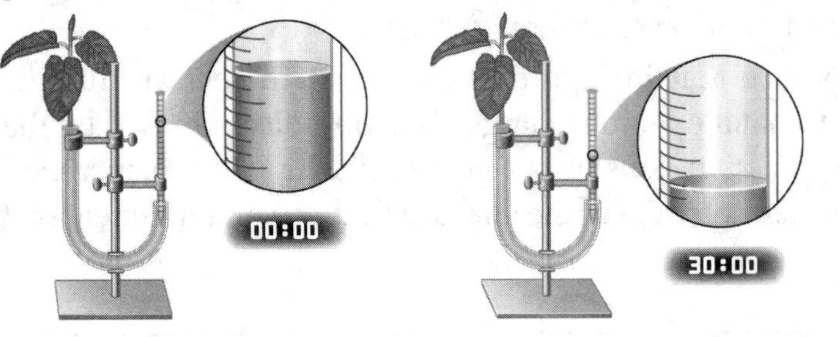

Figure 11.3 Procedure for transpiration lab (potometer method)

- A leaf needs carbon dioxide and water for photosynthesis. For carbon dioxide to enter, the stomata on the surface of the leaf must be open. Transpiration draws water from the roots into the leaf mesophyll. However, the plant must not lose so much water during transpiration that it wilts. The plant must strike a balance between conserving water and bringing in sufficient amounts of CO_2 for photosynthesis.
- One way to measure water loss from a plant is to use a *potometer*, a device that measures the rate at which a plant draws up water. Because the plant draws up water as it loses it by transpiration, you are able to measure the rate of transpiration. The basic elements of a potometer are shown above in Figure 11.3, step 1. They are
  - a plant cutting.
  - a calibrated pipette to measure water loss.
  - a length of clear plastic tubing.
  - an air-tight seal between the plant and the water-filled tubing.

- Figure 11.3 shows a typical setup to investigate a variety of environmental effects on the rate of transpiration. Based on your knowledge of transpiration developed in this investigation, can you answer the following questions? (Answers will be found at the end of this investigation.)

1. What are some factors that you would need to hold constant for a valid *control*?
2. Predict the rate of transpiration in the plant that has been misted and placed in a plastic bag compared to the control, and justify your prediction.
3. Predict the rate of transpiration in the plant in front of the fan compared to the control, and justify your prediction.
4. Predict the rate of transpiration in bright light compared to the control, and justify your prediction.
5. Why is the bright light being shone through a tub filled with water?

Questions

1. All of the following enhance water transport in terrestrial plants EXCEPT
   (A) hydrogen bonds linking water molecules.
   (B) capillary action due to adhesion of water molecules to the walls of xylem.
   (C) evaporation of water from the leaves.
   (D) K^+ being transported out of the guard cells.

2. Under conditions of bright light, in which part of a transpiring plant would water potential be lowest?
   (A) xylem vessels in the leaves
   (B) xylem vessels in the roots
   (C) root hairs
   (D) spongy mesophyll of the leaves

Answers for Questions about Figure 11.3

1. In order for the *control* to be valid, you should consider using the same type of plant, the same relative amount of leaf surface, have the water in all potometers at the same temperature, and place the control and experimental in virtually identical conditions with the exception of the single factor you are testing.
2. The plant cutting inside the plastic bag is in a situation of *high humidity.* Because of high water potential in the area surrounding the leaves, the rate of transpiration will be low.
3. The potometer in the *fan* shows increased water loss compared to the control. The reason is that the air movement results in greater evaporation, lowering water potential outside the plant surface, resulting in a higher rate of water loss or transpiration.
4. The potometer in *bright light* generally shows a higher rate of water loss, indicating more photosynthesis than the control.
5. A bright light produces heat, which operates as another variable. The water-filled tub is a heat sink and will minimize the effect of heat on the plant.

BIG IDEA 4: Investigation 12, Fruit Fly Behavior

Overview of the Lab

- In this lab you will explore behaviors in an invertebrate and design an experiment to answer a question about behavior. Animals exhibit a variety of behaviors, both learned and innate, that promote their survival and reproductive success in a variety of ways. In this investigation, you will make detailed observations of an organism's behavior and design a controlled experiment to test a hypothesis about a specific case of animal behavior.
- If you use the 2012 "AP Investigations" manual, your experimental organisms may be fruit flies (as in the title of the investigation). It is equally possible that your teacher may select another species to use for similar studies such as mealworms or pill bugs.

YOU MUST KNOW

- Descriptions of various animal behaviors, such as orientation behavior, geotaxis, phototaxis, chemotaxis, and how the behavior is adaptive.

SCIENCE PRACTICES: CAN YOU . . .

- Design a plan for collecting data to show how a particular species is affected by biotic or abiotic interactions?
- Analyze data collected to identify possible patterns and relationships between an organism or species and a biotic or an abiotic factor?
- Apply mathematical routines, such as statistical analysis, to evaluate data?

Hints and Review

- This lab is an opportunity to make detailed observations and learn about some interesting animal behaviors. Because the topic is so broad and there are so many local organisms and possibilities for your teacher to choose, we will only make a few comments about animal behavior here. This is a wonderful opportunity to teach experimental design, so this is where we focus our discussion.
- Refer to the Introduction of this book, where we have given some hints about writing the lab essay. At the conclusion of this course, you should have conducted a number of investigations that you developed. You should now be able to design a controlled experiment to test a single variable, record data in a logical manner, and present your conclusions.

- There are hundreds of species of fruit flies—so how do members of the same species find each other and signal willingness to mate? Each species has evolved a complex series of behaviors that appear to be genetically programmed. Students who do Lab 11 in the 2001 AP Biology Lab Manual investigate this behavior in *Drosophila melanogaster*.
- How do fruit flies find their food sources? Orient to gravity? To light? Investigation 12: Fruit Fly Behavior encourages a look at a number of behaviors in fruit flies and directs you in the preparation of a choice chamber.
- A behavior experiment that is frequently done observes how pill bugs respond to their environment. Pill bugs are placed in a choice chamber, half in the side lined with dry filter paper and the other half in the side lined with wet filter paper. Pill bugs are crustaceans so they respire through gills and are generally found in a moist habitat. Because of this, most students hypothesize that more pill bugs will be found in the moist chamber. This is often what occurs but not always!
- This brings us to an important consideration: If a student prepares a single-choice chamber, the exercise is not a controlled experiment. Could there be more light at one end of the choice chamber? More activity and vibration? A chemical residue on one side? Any of these conditions and more could possibly influence the organism's behavior. Without a control, it is very risky to state a conclusion.
- What's needed is a **controlled experiment**. A controlled experiment begins with a ***hypothesis***, a proposed solution for the problem being investigated. A hypothesis is often written as an IF, THEN statement that predicts the outcome we should expect if the hypothesis is correct. A hypothesis should not only predict results, but must also be testable.
- In a controlled experiment, *all variables are held constant* except the one being tested or manipulated. For instance, if the goal is to test response to wet versus dry conditions, the light, temperature, chemicals in the filter paper or on the dish surface, and movement of the table must all remain constant. In addition, all the experimental organisms must be of the same approximate age, size, and state of health. It is not enough to say that you will hold all variables constant; you must be explicit in your explanation of how you will do this.
- To be meaningful, the experiment must include a *large sample size* to be representative of a general condition.
- The *results must be measurable!* Are you going to count, measure, find the mass? Some way to quantify the results must be devised.
- Several *repetitions (or replicates)* of the experiment must be done. Like a large sample size, this lets you verify your result.
- Before you design an experiment, it may be useful to *search the literature* to learn what has already been done and to help develop ideas for a reasonable study.
- Finally, *statistical analysis of your data* (such as the Chi-square test found at the end of this topic) should be done to validate experimental results.

Questions

1. A student wanted to study the effect of nitrogen fertilizer on plant growth, so she took two similar plants and set them on a window sill for a two-week observation period. She watered each plant the same amount, but she gave one a small dose of fertilizer with each watering. She collected data by counting the total number of new leaves on each plant and also measured the height of each plant in centimeters. Which of the following is a significant flaw in this experimental setup?
   (A) There is no variable factor.
   (B) There is no control.
   (C) There is no repetition.
   (D) Measurable results cannot be expected.

2. Students placed five pill bugs on the dry side of a choice chamber and five pill bugs on the wet side. They collected data as to the number on each side every 30 seconds for 10 minutes. After 6 minutes, eight or nine pill bugs were continually on the wet side of the chamber, and several were under the filter paper. Which of the following is *not* a reasonable conclusion from these results?
   (A) It takes the pill bugs several minutes to explore their surroundings and select a preferred habitat.
   (B) Pill bugs prefer a moist environment.
   (C) Pill bugs may find chemicals in dry filter paper irritating.
   (D) Pill bugs demonstrate no significant habitat preference.

3. If a student wanted to determine whether pill bugs prefer a moist or a dry environment, what would be a good first step in looking at the data?
   (A) Total the number of pill bugs on the dry side throughout the entire experiment and compare this with the number on the wet side throughout the experiment.
   (B) After waiting 5 minutes for the pill bugs to acclimate, count the number of pill bugs on the dry side every 30 seconds for 5 minutes and determine the total number on the dry side. Do the same for the wet side and compare the data.
   (C) Compare the number of pill bugs on the dry side at the end of 10 minutes with the number of pill bugs on the wet side at the end of 10 minutes.
   (D) Divide the number of pill bugs on the dry side throughout the experiment by the number on the wet side throughout the experiment.

4. Which of the following hypotheses is stated best?
   (A) If pill bugs are allowed free movement, then more will be found in a moist environment than in a dry environment.
   (B) If pill bugs like a moist environment, then they will move to the wet side of a choice chamber.
   (C) If an experiment with pill bugs is run for 10 minutes, then more pill bugs will be found in the most favorable environment.
   (D) Pill bugs are found in moist habitats, so I predict that more will be found where it is wet.

BIG IDEA 4: Investigation 13, Enzyme Activity

Overview of the Lab

- This experiment investigates enzymatic activity of peroxidase. In Procedure 1 you will learn how to measure the activity of peroxidase, and in Procedure 2 you will investigate the effect of varying pH on enzyme activity and then select your own question about factors that would affect enzyme activity and design an experiment to answer it.
- Your teacher might choose to use another enzyme-substrate system and so the tips that follow will be generalized for any enzyme.

YOU MUST KNOW

- The factors that affect the rate of an enzyme reaction such as temperature, pH, and enzyme concentration.
- How the structure of an enzyme can be altered and how pH and temperature affect enzyme function.

SCIENCE PRACTICES: CAN YOU . . .

- Design a controlled experiment to measure the activity of a specific enzyme under varying conditions?
- Use mathematical routines to calculate the rate of a reaction from a graph or data chart?
- Predict and justify how changing an environmental factor such as temperature or pH would alter an enzyme's activity?

Hints and Review

- Enzymes are large globular proteins. Much of their three-dimensional shape is the result of interactions between the R (variable) groups of their amino acids. Anything that changes these interactions will change the shape of the enzyme and therefore alter the rate of reaction. The *active site* is the portion of the enzyme that will interact with the substrate.
- Remember, *change the shape, change the function!*
- Enzyme activity is affected by pH and temperature because these affect the 3-D shape. Extremes of pH and temperature result in *denaturation* when the 3-D shape is so altered that the enzyme can no longer function.
- Enzymes are not denatured by cold, but the rate of reaction is decreased as temperature decreases.
- Be able to calculate rate from graphed data using Figure 13.1.

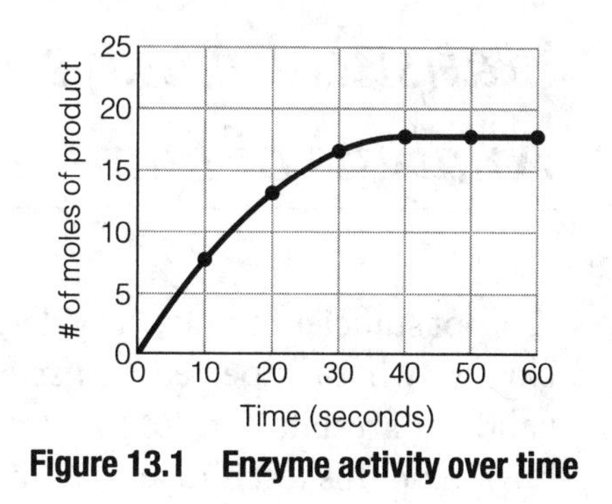

Figure 13.1 Enzyme activity over time

Enzyme Action Over Time

We can calculate the rate of a reaction by measuring, over time, either the disappearance of substrate (as in our catalase example) or the appearance of product (as in Figure 13.1). For example, on the graph above, what is the rate, in moles/second, over the interval from 0 to 10 seconds?

$$\text{Rate} = \frac{\Delta y}{\Delta x}$$

for this example, the rate would be

$$\frac{7 \text{ moles} - 0 \text{ moles}}{(10 \text{ seconds} - 0 \text{ seconds})} = \frac{7}{10}$$

$$= 0.7 \text{ moles/second}$$

- Note that the slope of the graph is steepest during the *initial* time period; this is when the rate of a reaction is greatest and occurs because the substrate is most abundant, allowing for more enzyme–substrate collisions. The rate of the reaction decreases as substrate is consumed and the slope of the graph flattens.
- What is the rate for this same reaction between 40 and 60 seconds? (It is 0.) You should be able to explain this. It is because the substrate has been consumed.

Questions

1. In order to keep the rate of reaction constant over the entire time course, which of the following should be done?
   (A) Add more enzyme.
   (B) Gradually increase the temperature after 60 seconds.
   (C) Add more substrate.
   (D) Add H_2SO_4 after 60 seconds.
2. To determine the rate of enzyme activity, an experiment is done that mixes enzyme and substrate together for 30 seconds, 60 seconds, 90 seconds, and 120 seconds. After the specified times, H_2SO_4 is added to the reaction chamber. What is the role of sulfuric acid (H_2SO_4) in this experiment?
   (A) It is the substrate on which the enzyme acts.
   (B) It denatures the enzyme by altering the active site.
   (C) It accelerates the reaction between enzyme and substrate.
   (D) It blocks the active site of the enzyme.

Statistical Analysis: Chi-Square Analysis of Data

Overview of Chi-Square

It is not sufficient to say, "My data looks really good!" or "The results were very close to what I expected." In science, we impose rigorous tests to support the validity of results. One of these is Chi-square analysis. Here is a short review of how to do this test and at the same time a review of your knowledge of genetic ratios expected with different crosses.

YOU MUST KNOW

- What is meant by degrees of freedom, critical value, probability value, the null hypothesis, and how to do Chi-square analysis of data.

Chi-Square Analysis of Data

Assume you obtained the results shown in Figure 1 for the F_1 generation.

F_1 RESULTS	OBSERVED PHENOTYPE AND NUMBERS
	Red eyes
♂ MALES	12
♀ FEMALES	8

(a)

F_2 RESULTS	OBSERVED PHENOTYPE AND NUMBERS	
	Red eyes	Sepia eyes
♂ MALES	19	4
♀ FEMALES	12	9

(b)

Figure 1 Data table of *Drosophila* (1)

- From the data presented, you can deduce that the F_1 cross was between individuals heterozygous for eye color: se^+ *se* X se^+ *se* (se^+ = red eyes, *se* = sepia)
- The student kept data showing both males and females with the trait because the unknown trait might be sex-linked. The data do not support a sex-linked trait but do support an autosomal trait; thus the data are merged so that only the trait is considered in the cross.
- From this conclusion, you could write the following hypothesis concerning the relationship between the dominant red eye color to the recessive sepia eye color: *If the parents are heterozygous for eye color, there will be a 3:1 ratio of red eyes to sepia eyes in the offspring.* Do your results support this hypothesis?

CALCULATING CHI-SQUARE

The formula for Chi-square is

$$\chi^2 = \Sigma\frac{(o - e)^2}{e}$$

where:
o = observed number of individuals
e = expected number of individuals

- The actual results of an experiment are unlikely to match the expected results precisely. But how great a variance is significant? One way to decide is to use the Chi-square (χ^2) test. This analytical tool tests the validity of a **null hypothesis**, which states that there is no statistically significant difference between the observed results of your experiment and the expected results. When there is little difference between the observed results and the expected results, you obtain a very low Chi-square value; your hypothesis is supported.

Using the Chi-Square Critical Values Table

The Chi-square critical values table provides two values that you need to calculate Chi-square:

- **Degrees of freedom.** This number is one less than the total number of classes of offspring in a cross. In a monohybrid cross, such as our Case 1, there are two classes of offspring (red eyes and sepia eyes). Therefore, there is just one degree of freedom. In a heterozygous dihybrid cross, there are four possible classes of offspring so there are three degrees of freedom.
- **Probability.** The probability value (p) is the probability that a deviation as great as or greater than each Chi-square value would occur simply by chance. Many biologists agree that deviations having a chance probability greater than 0.05 (5%) do not support the null hypothesis. Therefore, when you calculate Chi-square, you should consult the table for the p value in the 0.05 row.

Critical Values Table

	Degrees of Freedom (df)				
Probability (p)	**1**	**2**	**3**	**4**	**5**
0.05	3.84	5.99	7.82	9.49	11.1
0.01	6.64	9.21	11.3	13.2	15.1
0.001	10.8	13.8	16.3	18.5	20.5

Steps to Determining Chi-Square

1. **Set up a data chart as shown.** Because your hypothesis predicted a 3:1 ratio in the offspring, you would expect 3/4 of the total offspring (44) to have red eyes.

Phenotypes	Observed (*o*)	Expected (*e*)	$(o - e)$	$(o - e)^2$	$\frac{(o-e)^2}{e}$
Red Eyes	31	33	2	4	4/33 = 0.12
Sepia Eyes	13	11	2	4	4/11 = 0.36
	44	44			Total = χ^2 = 0.48

2. **Determine the degrees of freedom.** This is the number of categories (red eyes or sepia eyes) minus one. For these data, the number of degrees of freedom is 1.
3. **Find the probability (*p*) value for 1 degree of freedom in the 0.05 row.** This is the **critical value**. For these data, the critical value = 3.84.
4. **Accept or reject the null hypothesis.** The null hypothesis states that there is no statistically significant difference between the observed and expected data. Because the χ^2 value for this data is less than the critical value, you will accept the null hypothesis. This then supports your working hypothesis, *If the parents are heterozygous for eye color, there will be a 3:1 ratio of red eyes to sepia eyes in the offspring.*

- Don't forget this little nugget: **If the Chi-square value is greater than the critical value, the null hypothesis is rejected,** and you must consider reasons for this variation, such as errors in sample size or data collection.

TIP FROM THE READERS
There was a Chi-square problem on the 2013 AP Biology Exam, and we found that a common error was that some students took the square root of the value they got for χ^2. Don't make this mistake! χ^2 is just the shorthand for the name of this mathematical technique, *Chi-square.*

Questions

1. You have been given a vial containing a red-eyed male with normal wings and a red-eyed female with normal wings. These are the F_1 generation. After two weeks, you collect the offspring from this pair and obtain the results shown in Figure 2. On the basis of the results shown in Figure 2, which statement is most likely true?

F_2 RESULTS	OBSERVED PHENOTYPE AND NUMBERS			
	Red eyes normal wings	Red eyes no wings	Sepia eyes normal wings	Sepia eyes no wings
♂ MALES	48	13	16	4
♀ FEMALES	50	9	10	10

Figure 2 Data table of *Drosophila* (2)

(A) The genes for red eyes and normal wings are linked.
(B) The gene for no wings is sex-linked.
(C) The F_1 mates were both homozygous for both eye color and wings.
(D) The gene for eye color is inherited independently of the gene for wings.

2. Based on the hypothesis that this is a dihybrid cross, with the two genes unlinked, calculate χ^2 using the data in the table of observed phenotypes.
   (A) 6.043
   (B) 7.815
   (C) 4.977
   (D) 24.038

3. Compare the Chi-square value obtained in question 2 with the Critical Values Table on page 343 for $p = 0.05$. Which of the following statements would be true?
   (A) Because the calculated value for Chi-square is less than 7.82, the results support the hypothesis that the parents are heterozygous for two unlinked traits.
   (B) Because the calculated value for Chi-square is less than 7.82, the results support the hypothesis that eye color and wings are linked.
   (C) Because the calculated value for Chi-square is less than 7.82, the results are inconclusive. The experiment should be repeated.
   (D) Because the Chi-square value is less than the critical value of 7.82, the null hypothesis is rejected for the hypothesis that the parents are heterozygous for two unlinked traits.

AP Biology Equations and Formulas

The following is the formula list that you will receive as part of your testing materials. *Source:* AP Biology—Course and Exam Description. © 2012. The College Board. www.collegeboard.org. Reproduced with permission.

Statistical Analysis and Probability

Standard Error

$$SE_{\bar{x}} = \frac{s}{\sqrt{n}}$$

Mean

$$\bar{x} = \frac{1}{n}\sum_{i=1}^{n} x_i$$

Standard Deviation

$$s = \sqrt{\frac{\sum (x_i - \bar{x})^2}{n-1}}$$

Chi-Square

$$\chi^2 = \sum \frac{(o-e)^2}{e}$$

Chi-Square Table

	Degrees of Freedom							
p	1	2	3	4	5	6	7	8
0.05	3.84	5.99	7.82	9.49	11.07	12.59	14.07	15.51
0.01	6.64	9.32	11.34	13.28	15.09	16.81	18.48	20.09

s = sample standard deviation (i.e., the sample based estimate of the standard deviation of the population)

$\bar{x}$ = mean

n = size of the sample

o = observed individuals with observed genotype

e = expected individuals with observed genotype

Degrees of freedom equals the number of distinct possible outcomes minus one.

Laws of Probability

If A and B are mutually exclusive, then P (A or B) = P(A) + P(B)

If A and B are independent, then P (A and B) = P(A) x P(B)

Hardy-Weinberg Equations

$p^2 + 2pq + q^2 = 1$

$p + q = 1$

p = frequency of the dominant allele in a population

q = frequency of the recessive allele in a population

Metric Prefixes

Factor	Prefix	Symbol
10^9	giga	G
10^6	mega	M
10^3	kilo	k
10^{-2}	centi	c
10^{-3}	milli	m
10^{-6}	micro	μ
10^{-9}	nano	n
10^{-12}	pico	p

Mode = value that occurs most frequently in a data set

Median = middle value that separates the greater and lesser halves of a data set

Mean = sum of all data points divided by number of data points

Range = value obtained by subtracting the smallest observation (sample minimum) from the greatest (sample maximum)

Rate and Growth

Rate
dY/dt

Population Growth
dN/dt=B-D

Exponential Growth
$$\frac{dN}{dt} = r_{max}N$$

Logistic Growth
$$\frac{dN}{dt} = r_{max}N\left(\frac{K-N}{K}\right)$$

dY= amount of change

t = time

B = birth rate

D = death rate

N = population size

K = carrying capacity

r_{max} = maximum per capita growth rate of population

Water Potential (Ψ)

$\Psi = \Psi p + \Psi s$

Ψp = pressure potential

Ψs = solute potential

The water potential will be equal to the solute potential of a solution in an open container, since the pressure potential of the solution in an open container is zero.

The Solute Potential of the Solution

$\Psi_s = -iCRT$

i = ionization constant (For sucrose this is 1.0 because sucrose does not ionize in water)

C = molar concentration

R = pressure constant (R = 0.0831 liter bars/mole K)

T = temperature in Kelvin (273 + °C)

Temperature Coefficient Q_{10}

$$Q_{10} = \left(\frac{k_2}{k_1}\right)^{\frac{10}{t_2 - t_1}}$$

Primary Productivity Calculation

mg O_2/L x 0.698 = mL O_2/L

mL O_2/L x 0.536 = mg carbon fixed/L

t_2 = higher temperature

t_1 = lower temperature

k_2 = metabolic rate at t_2

k_1 = metabolic rate at t_1

Q_{10} = the *factor* by which the reaction rate increases when the temperature is raised by ten degrees

Surface Area and Volume

Volume of Sphere
$V = 4/3 \pi r^3$

Volume of a cube (or square column)
V = l w h

Volume of a column
$V = \pi r^2 h$

Surface area of a sphere
$A = 4 \pi r^2$

Surface area of a cube
A = 6 a

Surface area of a rectangular solid
A = Σ (surface area of each side)

r = radius

l = length

h = height

w = width

A = surface area

V = volume

Σ = Sum of all

a = surface area of one side of the cube

Dilution - used to create a dilute solution from a concentrated stock solution

$C_iV_i = C_fV_f$

i = initial (starting)   C = concentration of solute
f = final (desired)   V = volume of solution

Gibbs Free Energy

$\Delta G = \Delta H - T\Delta S$

ΔG = change in Gibbs free energy

ΔS = change in entropy

ΔH = change in enthalpy

T = absolute temperature (in Kelvin)

pH = $-\log [H^+]$

Part IV

Sample Test

On the following pages is a sample examination that approximates the actual AP Biology Examination in format, types of questions, and content. Set aside 3 hours to take the test. To best prepare yourself for actual AP exam conditions, use only the allowed time for Section I and Section II.

You may want to review our test-taking hints on page 13! Before you begin, get a copy of the Equations and Formulas (pages 346–347) and a four-function calculator. Remember, answer every question because there is no penalty for guessing! You have approximately 90 seconds for each question, so use your time wisely.

Sample Test

Biology
Section I

60 Multiple-Choice Questions
6 Grid-In Questions
Time—90 minutes

Part A Directions: Each of the questions or incomplete statements below is followed by four suggested answers or completions. Select the one that is best in each case, and then fill in the corresponding circle on the answer sheet. When you have completed Part A, you should continue on to Part B.

Questions 1 and 2 refer to the graph below.

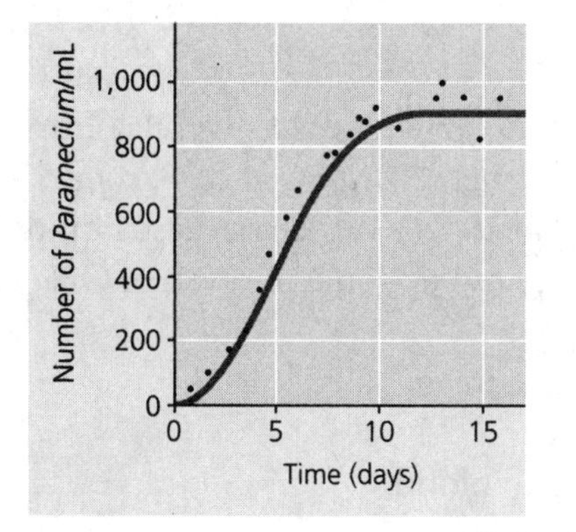

1. From the data in the graph, determine the growth rate over 10 days as well as the carrying capacity of the environment for the population shown.
   (A) Growth Rate: 900; Carrying Capacity: Not Reached
   (B) Growth Rate: 90; Carrying Capacity: 500
   (C) Growth Rate: 900; Carrying Capacity: 900
   (D) Growth Rate: 90; Carrying Capacity: 900

2. The formula $dN/dt = r_{max}N(K - N)/K$ describes the pattern of growth for the *Paramecium* population graph. Which of the following is true concerning the formula and the graph it describes at carrying capacity?
   (A) $K = N$
   (B) $r_{max}N =$ *carrying capacity*
   (C) The variable K is constantly changing.
   (D) $(K - N)/K = 9{,}000$

Gene	Probability of Appearing in Gamete
P	1/4
Q	1/4
R	1/4

3. Three genes—*P*, *Q*, and *R*—are not linked. The probability of each gene appearing in a gamete is shown in the table above. Which of the calculations below represents the probability that all three genes will appear in the same gamete?
   (A) $1/4 \times 1/4 \times 1/4$
   (B) $1/4 + 1/4 + 1/4$
   (C) $1/4 \div 1/4 \times 1/4$
   (D) $(1/4)1/3$

4. The value for water potential in root tissue was found to be Ψ -0.15 MPa. If you take the root tissue and place it in a 0.1 *M* solution of sucrose ($\Psi = -0.23$ MPa), which of these would describe the movement of water?
   (A) The net flow of water would be from the tissue into the sucrose solution.
   (B) The net flow of water would be from the sucrose solution into the soil.
   (C) Water would flow in both directions with no net change.
   (D) Water would flow from the soil into the root tissue where water potential is higher.

5. In deer, fur length is controlled by a single gene with two alleles. When a deer homozygous for long fur is crossed with a deer homozygous for short fur, the offspring all have fur of medium length. If these offspring with medium-length fur mate, what percentage of their offspring will have long fur?
   (A) 100%
   (B) 75%
   (C) 50%
   (D) 25%

6. Which of the following statements best supports the idea that certain cell organelles are evolutionarily derived from symbiotic prokaryotes living in host cells?
   (A) The process of cellular respiration in certain prokaryotes is similar to that occurring in mitochondria and chloroplasts.
   (B) Mitochondria and eukaryotes have similar cell wall structures.
   (C) Like prokaryotes, mitochondria have a double membrane.
   (D) Mitochondria and chloroplasts have DNA and ribosomes that are similar to those of prokaryotes.

7.

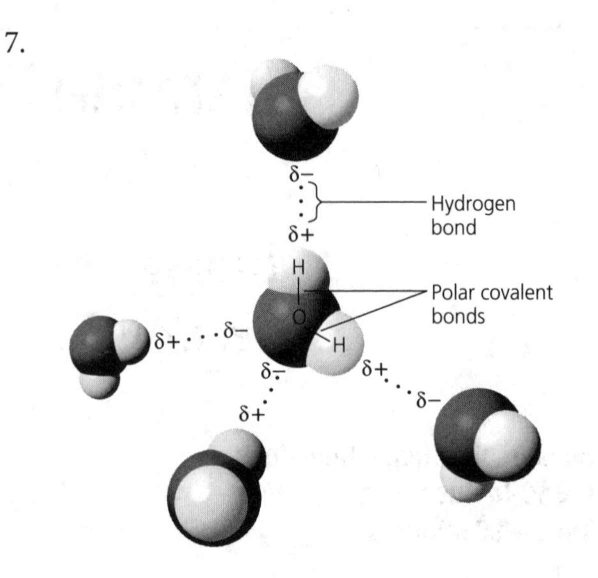

If each of the water molecules in the diagram had all of its potential hydrogen bonding sites filled, like the water molecule in the center of the diagram, what would be the total number of water molecules represented?
   (A) 5
   (B) 8
   (C) 17
   (D) 20

8. Here are the pH values of several common materials: *cola, 2; orange juice, 3; coffee, 5; human blood, 7.4.* Which of these liquids has the *highest* molar concentration of OH^-?
   (A) cola
   (B) orange juice
   (C) coffee
   (D) human blood

9. In dogs, the trait for long tail is dominant (*L*), and the trait for short tail is recessive (*l*). The trait for yellow coat is dominant (*Y*), and the trait for white coat is recessive (*y*). Mating many dogs gave multiple litters totaling 305 long-tailed, yellow dogs and 95 long-tailed, white dogs. Which of the following is most likely to be the genotype of the parent dogs?
   (A) *LLYY* × *LLYY*
   (B) *LLyy* × *LLYy*
   (C) *LlYy* × *LlYy*
   (D) *LlYy* × *LLYy*

10. Muscle contractions occur when a motor neuron releases acetylcholine (ACh) into a synapse with a muscle cell. Receptors on the muscle cell membrane bind Ach, which triggers an action potential in the muscle cell. This leads to an interaction of actin and myosin filaments and the muscle cell contracts. The enzyme *acetylcholinesterase* (AChE) breaks down ACh, which removes ACh from the synapse and ends the signal. This allows the muscle cell to relax; contraction ends. The nerve gas *sarin* causes paralysis and death because it inhibits the action of AChE. Which of the following statements **best** explains the effect of sarin on muscle cells?
    (A) Paralysis occurs because ACh is quickly taken back into the motor neuron, which stops the signal, and the muscle cells are unable to contract.
    (B) Paralysis occurs because sarin blocks the receptors of the muscle cells and actin and myosin cannot interact.
    (C) Paralysis occurs because the signaling molecule remains in the synapse bound to receptors so actin and myosin continue to interact and contraction continues.
    (D) Paralysis occurs because the motor neuron cannot transmit a second signal to the muscle cell so contraction ceases.

11. In *E. coli* replication the enzyme primase is used to attach a 5 to 10 base ribonucleotide strand complementary to the parental DNA strand. The RNA strand serves as a starting point for the DNA polymerase that replicates DNA. If a mutation occurred in the primase gene, which of the following would you expect?
    (A) Replication would only occur on the leading strand.
    (B) Replication would only occur on the lagging strand.
    (C) Replication would not occur on either the leading or lagging strand.
    (D) Replication would not be affected because the enzyme primase is involved with RNA synthesis.

12. Allolactose stimulates *Escherichia coli* to produce mRNAs that code for the enzyme β-galactosidase, which breaks down lactose into glucose and galactose. Which of the following is the best description of the role of allolactose?
    (A) It inhibits DNA replication when lactose levels are high.
    (B) It inhibits translation of the mRNAs for β-galactosidase.
    (C) It is an allosteric inhibitor of β-galactosidase.
    (D) It regulates expression of the gene for β-galactosidase.

13. In a certain group of iguanas, the presence of brown skin is the result of a homozygous recessive condition in the biochemical pathway producing skin pigment. If the frequency of the genotype for this condition is 36%, which of the following is closest to the frequency of the heterozygote genotype in this population? (Assume that the population is in Hardy-Weinberg equilibrium.)
    (A) 16%
    (B) 24%
    (C) 48%
    (D) 64%

Use the pedigree below to answer questions 14 and 15.

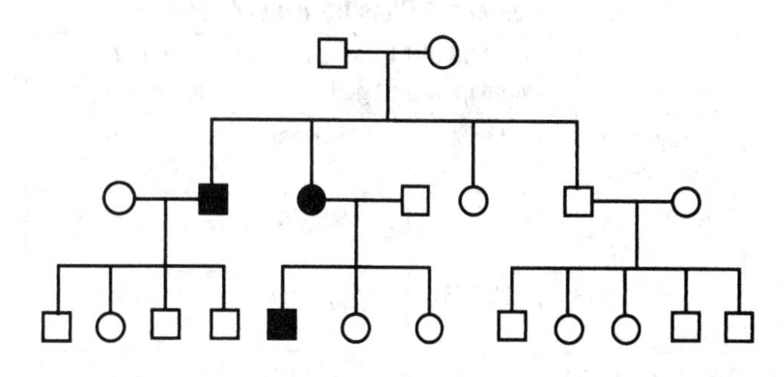

14. In the pedigree, squares represent males and circles represent females. Shaded figures represent individuals who possess a particular trait. Which of the following patterns of inheritance best explains how this trait is transmitted?
    (A) partially dominant
    (B) autosomal dominant
    (C) autosomal recessive
    (D) sex-linked recessive

15. What is the genotype of the shaded figure in the third generation if *A* is the dominant trait and *a* is the recessive trait?
    (A) *aa*
    (B) *Aa*
    (C) X^aX^a
    (D) X^ay

16. In chemiosmosis, ATP is produced as hydrogen ions flow through channels in the ATP synthase complex. This movement of hydrogen ions is accomplished by which of the following processes?
    (A) active transport
    (B) facilitated diffusion
    (C) osmosis
    (D) cotransport

Questions 17 and 18

Table 1 Fate of Goldenrod Galls from an Old Field

Gall Fate	Percent of total galls	Mean diameter of gall	Number of galls in the category
Unsuccessful (larvae died)	23.5%	17.5 mm	110
Galls parasitized (larvae died)	21%	19.9 mm	98
Successful galls (larvae do not die)	32.55%	20.9 mm	152
Gall larvae eaten by birds	21.91%	22.9 mm	107

17. If this pattern of gall fates shown in Table 1 continued over many generations, what evolutionary pattern would emerge?
    (A) Stabilizing selection, which causes the average gall size to be pushed toward an intermediate value
    (B) Directional selection, which causes the average gall size to become larger over time
    (C) Diversifying selection, which causes the average gall size to become either larger *or* smaller over time with very few galls exhibiting the intermediate size
    (D) No selection because too many of the larvae die to change the evolutionary pattern.

18. If the starting population showed a normal distribution of gall sizes as seen in Graph 1, what would the graph look like after natural selection pressures indicated in Table 1?

Graph 1

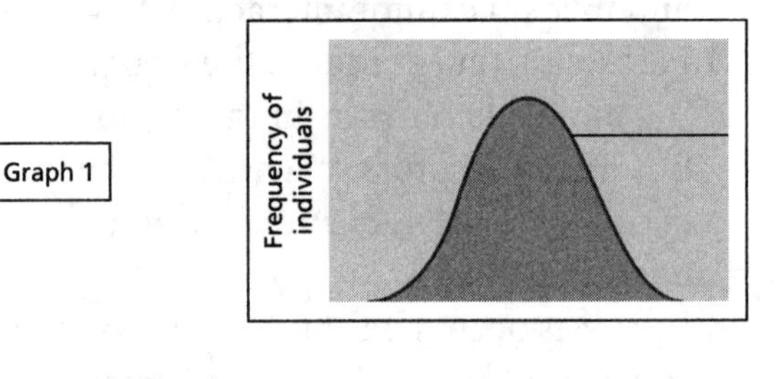

Graph A Graph B Graph C

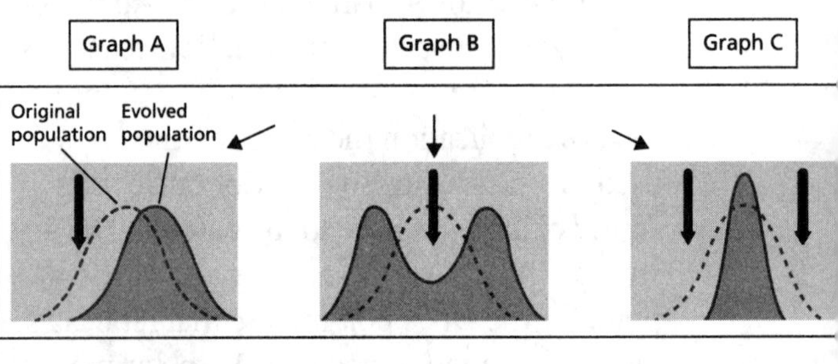

    (A) Graph A
    (B) Graph B
    (C) Graph C
    (D) No change (It would look like the original.)

19. In plant evolution there is a group of plants that close their stomata during the day to prevent water loss in hot, dry conditions. These are CAM plants. The production of an acid intermediate that "captures" CO_2 occurs in CAM plants, which enables them to produce sugars even when their stomata are closed during the day. Another group of plants, called C4 plants, overcome the loss of CO_2 under hot, dry conditions by producing a similar acid intermediate. However, the production of this intermediate has an energy cost—ATP is required to generate it.

Which of the following is a correct statement about C_3, C_4, and CAM plants?
(A) C_3 plants grow better in all conditions because their stomata are not closed during the day.
(B) C_4 plants grow better in cold, moist conditions than do C_3 plants.
(C) C_3 plants grow better in hot, arid conditions than do CAM plants.
(D) CAM plants and C_4 plants grow better in hot, arid conditions than do C_3 plants.

Use this figure for questions 20 and 21.

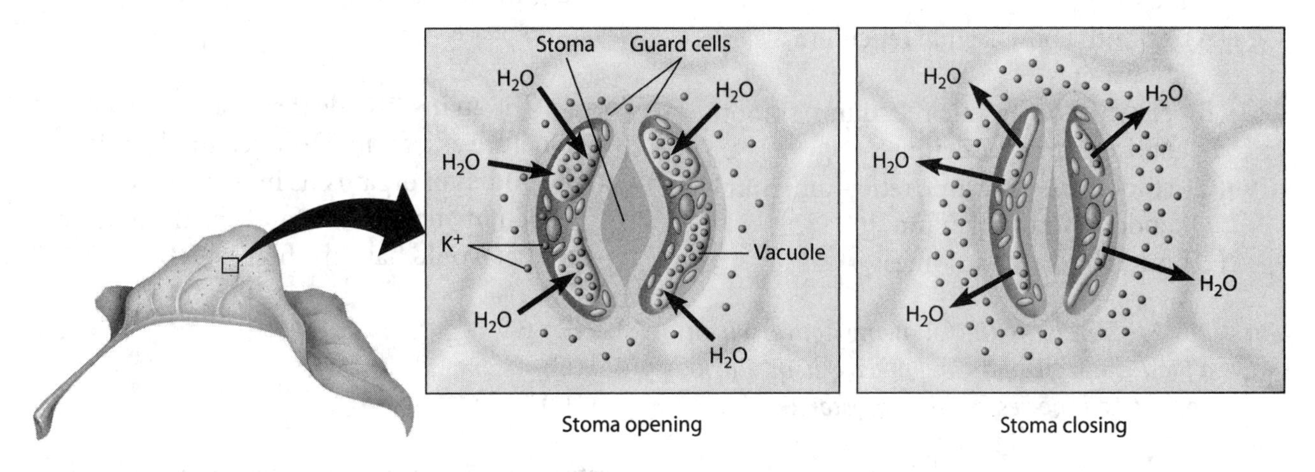

20. In the left-hand panel, K^+ and water are both moving into the guard cells, and the use of ATP is high. Why?
(A) Energy is required to move water molecules into the guard cells.
(B) Water is being lost through the stoma by transpiration.
(C) Energy is required to move oxygen into the leaf through the stoma.
(D) Energy is required to move K^+ into the guard cells against the gradient.

21. What is true of the guard cells shown in the right-hand panel of this figure?
(A) Their turgor pressure is increasing.
(B) Their water potential is lower than the surrounding cells.
(C) These cells are allowing gas exchange for photosynthesis.
(D) These cells are hypotonic to their immediate surrounding.

22. Considering the possibilities of recombination, which two genes on this chromosome are most likely to segregate together into a daughter cell?

A — 5 — W — 3 — E — 12 — G

(A) A and W
(B) W and E
(C) A and G
(D) A and E

23. Cells that produce digestive enzymes must contain a lot of_____, whereas the cells that detoxify poisons in the liver would have an abundance of_____.
(A) smooth endoplasmic reticulum, lysosomes
(B) rough endoplasmic reticulum, smooth endoplasmic reticulum
(C) smooth endoplasmic reticulum, rough endoplasmic reticulum
(D) microbodies, lysosomes

Questions 24–26 refer to the following figure, which shows a food chain in a particular ecosystem. Each letter represents a species in this ecosystem.

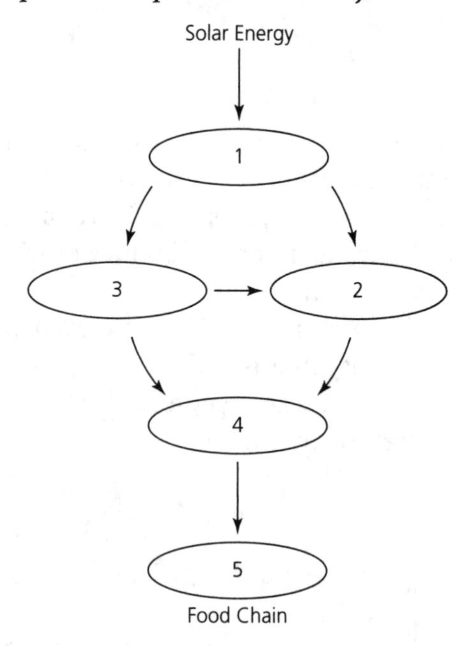

24. Which organism in the food chain obtains energy directly from both plants and animals?
(A) 1
(B) 2
(C) 3
(D) 4

25. Which of the following is a true statement about this food chain?
(A) Energy flows one way through food chains, but nutrients are recycled.
(B) Energy levels are highest in level 5.
(C) Nutrient recycling only occurs from level 5, the top of the food chain.
(D) The biomass of level 4 is greater than level 1.

26. DDT is an insecticide that accumulates in nature as a toxin. DDT accumulates in the bodies of organisms in this food chain. Which member of the food chain would you predict would have the highest level of DDT?
(A) 2
(B) 3
(C) 4
(D) 5

Questions 27–29 refer to the following gel, which was produced from four samples of a radioactively labeled strand of DNA that were cut with one type of restriction enzyme. The samples were separated by gel electrophoresis. Answer the questions based on the bands you can visualize below.

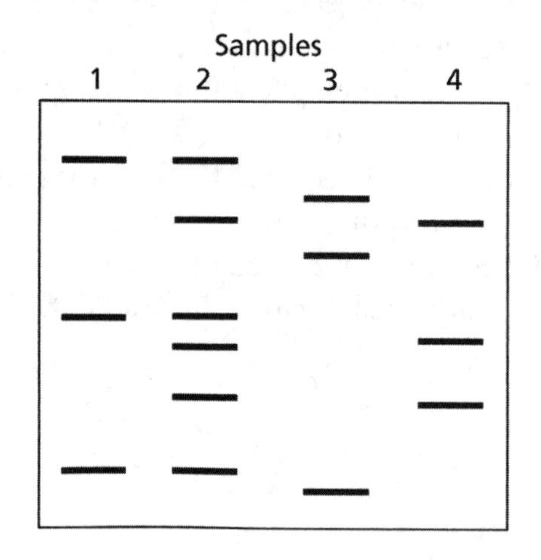

27. The DNA fragments in the gel were separated when an electric field was applied and they migrated to the positions shown on the gel. The position of the bands on the final gel are due primarily to the
    (A) concentration of the agarose gel.
    (B) amount of time the gels were run.
    (C) electrical charge of the DNA samples.
    (D) size of the fragments within the samples.

28. Which of the following is true about the DNA samples that were loaded onto the gel?
    (A) The DNA strand of sample 2 was originally the longest.
    (B) The DNA strand of sample 4 was originally the shortest.
    (C) Samples 2 and 4 are the same DNA sample.
    (D) Sample 2 was cut at more restriction sites than was sample 4.

29. Which of the following statements is supported by the data from the gel?
    (A) All samples have the same number of restriction sites.
    (B) Samples 3 and 4 have the same number of restriction sites in the same locations on the DNA strand.
    (C) Samples 1, 3, and 4 have the same number of restriction sites but in different locations on the DNA strand.
    (D) Samples 1, 3, and 4 have the same number of restriction sites in the same locations on the DNA strand.

Questions 30–32 refer to an experiment in which there is an initial setup of a U-tube with its two sides separated by a membrane that permits the passage of water and NaCl but not molecules of glucose. The U-tube is filled on one side with a solution of 0.4 M glucose and 0.5 M NaCl, and on the other, 0.8 M glucose and 0.4 M NaCl.

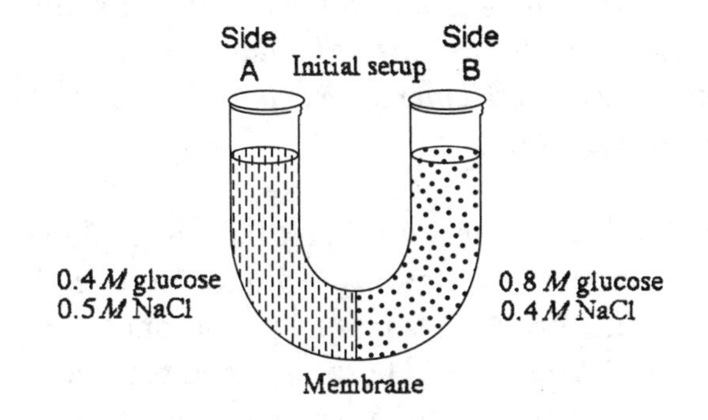

30. When this U-tube was set up, at time = 0 in the experiment, which of the following was true?
    (A) The solution on side A was hypertonic to the solution on side B.
    (B) The solution on side B was hypertonic to the solution on side A.
    (C) The two solutions were isotonic.
    (D) Active transport moved glucose from side A to side B.

31. After the experiment ran for 1 hour, the expected fluid levels in the U-tube would be
    (A) higher on side A than side B.
    (B) higher on side B than side A.
    (C) even.
    (D) equal, but the solutes would be reversed.

32. At the conclusion of the experiment, the NaCl would
    (A) show no movement.
    (B) show some movement from side A to side B.
    (C) show some movement from side B to side A.
    (D) make side A hypertonic.

Questions 33 and 34 are based on the following reading.

Recent studies have shown that the onset of puberty in American girls has decreased from an average of 12 to 13 years of age to as young as 8 to 10 years of age. Scientists who study premature puberty suggest that steroids in our food and in the environment may be contributing factors because steroids are known to cross cell membranes and bind to receptors inside cells.

33. Which of the following is the best explanation for how steroid hormones are able to pass through the plasma membrane?
    (A) Both steroid hormones and plasma membranes have receptor proteins.
    (B) Steroid hormones are polar, and the cell membrane is polar on the inside of the lipid bilayer.
    (C) Steroid hormones are nonpolar lipids, and the cell membrane is lipid based.
    (D) Steroid hormones diffuse through open channel proteins in the membrane.

34. Based on the description above, how do steroid hormones specifically alter the onset of puberty?
    (A) Steroid hormones bind to intracellular receptors and serve as transcription factors, turning on genes.
    (B) Steroid hormones bind to receptors on the cell membrane, turning on genes in the nucleus.
    (C) Steroid hormones activate G-proteins to initiate an intracellular cascade that amplifies the amount of hormone within the target cell.
    (D) Steroid hormones bind directly to DNA promoters, attract RNA polymerase, and initiate transcription.

35. A scientist wanted to determine if the metabolic rate of cancer cells is different from that of normal cells. Which of the following procedures would provide data to answer this question?
    (A) Measure the CO_2 consumed by the cell.
    (B) Measure the O_2 consumed by the cell.
    (C) Measure the amount of water consumed by the cell.
    (D) Measure the glucose produced by the cell.

36.

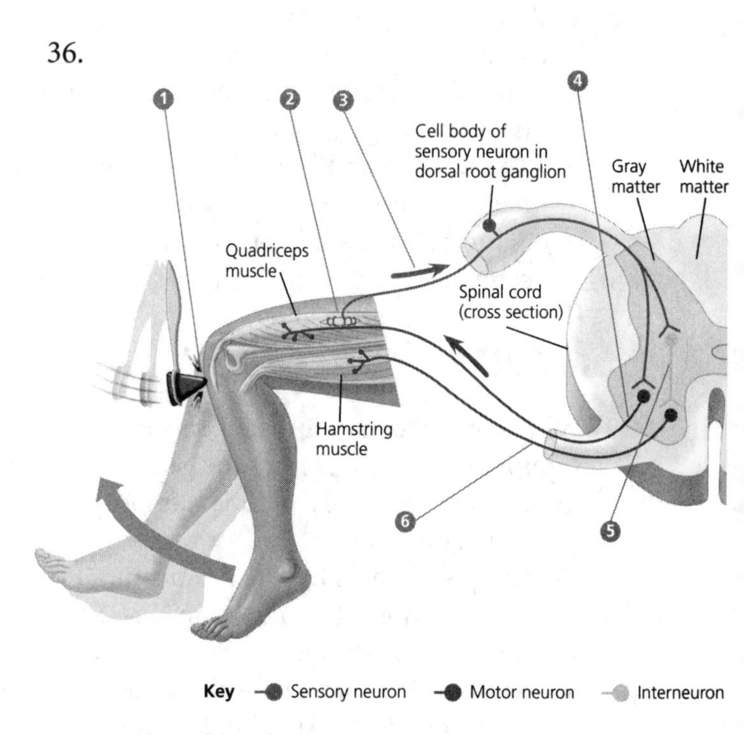

The most basic nervous system response is a reflex arc, as shown in the accompanying figure. Predict how this response would be altered if structure 6 could not function.
(A) Reception of the signal would not occur, but artificial stimulation of structure 6 would allow the response to occur.
(B) The sensory neuron would be damaged so the signal would not be transmitted to the interneuron.
(C) Damage to structure 6 would prevent contraction of the hamstring muscle.
(D) The signal would be rerouted to the brain where another pathway would be initiated.

37. One hypothesis from climate scientists is that annual mean precipitation will increase if the planet continues to become warmer. With more precipitation there is more evapotranspiration. Predict how this might affect the net primary productivity of ecosystems. (NPP = Gross PP − respiration).
   (A) Net Primary productivity would be unchanged because only temperature affects productivity.
   (B) Net Primary productivity would decrease because increased precipitation would decrease solar energy into the system.
   (C) Net primary productivity would increase because an increase in precipitation would increase productivity within the ecosystems.
   (D) Net primary productivity would be unchanged, but gross primary productivity would decrease.

38. Which of the following is an example of simple diffusion across a membrane?
   (A) the movement of H^+ across the thylakoid membrane during photosynthesis
   (B) the uptake of neurotransmitters by the postsynaptic membrane during the transmission of a nerve impulse
   (C) the movement of oxygen into the alveoli across the epithelial membrane and into the bloodstream
   (D) the exchange of sodium and potassium across a cell membrane through the sodium-potassium pump

39. Insulin is a protein that is produced by pancreatic cells and secreted into the bloodstream. Which of the following options correctly lists the order of the structures through which insulin passes from its production to its exit from the cell?
   (A) rough ER, transport vesicles, Golgi apparatus, transport vesicles, cell membrane
   (B) rough ER, lysosomes, transport vesicles, cell membrane
   (C) rough ER, Golgi apparatus, smooth ER, cell membrane
   (D) rough ER, transport vesicles, Golgi apparatus, vacuole, cell membrane

40. The figure shows a hypothetical embryo. Assume that for this embryo, a high concentration of a morphogen, called morpho, is needed to activate gene *P*; gene *Q* is active at or above medium concentrations of morpho; and gene *R* is expressed so long as any quantity of morpho is present. A different morphogen, called phogen, activates gene *S* and inactivates gene *Q* when at medium to high concentrations. If morpho and phogen are diffusing from their sites of production at the opposite ends of the embryo, which genes will be expressed in region 2 of this embryo? (Assume a gradient of morphogen concentrations in the three regions, from high at the source, to medium in the middle, and to low at the opposite end.)

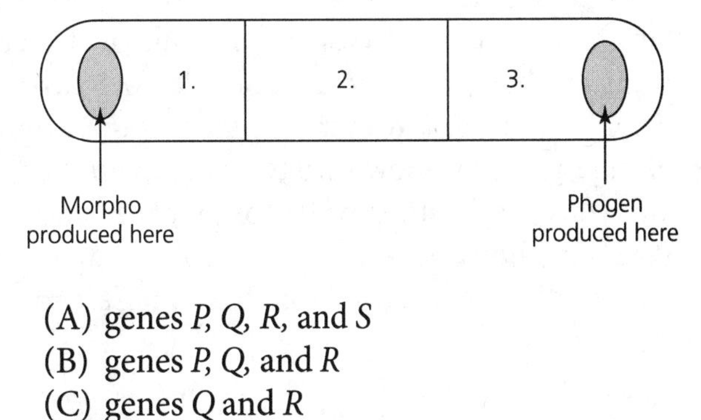

   (A) genes *P, Q, R,* and *S*
   (B) genes *P, Q,* and *R*
   (C) genes *Q* and *R*
   (D) genes *R* and *S*

41. In prokaryotic cells, genes that are involved in the same pathway are clustered in operons under the control of a single promoter. How are the coordinated transcriptions of eukaryotic genes involved in the same pathways regulated?
    (A) Each gene has its own promoter and the genes are all located on the same chromosome.
    (B) The genes are located in the same region of the chromosome, and enzymes methylate the entire region to activate transcription.
    (C) The genes respond to different transcription factors that are all activated simultaneously.
    (D) The genes may be on different chromosomes, but each responds to the same combination of control elements.

After reading the paragraphs, answer questions 42 and 43.

Scientists believe that a shift from pollination by insects to pollination by birds occurred several times over the course of angiosperm evolution. Two researchers designed an experiment to investigate how these shifts might evolve using two species of monkey flower. *Mimulus lewisii* has violet-pink flowers and is pollinated by bumblebees. *Mimulus cardinalis* has orange-red flowers and is pollinated by hummingbirds.

The researchers switched flower-color genes between the two species. As a result of the gene transfer, they produced a variation of *M. cardinalis* with dark pink flowers (instead of the original orange-red). The new variety of *M. lewisii* had orange flowers (instead of the original violet-pink). Plants of both genetically altered varieties were placed in their original habitats and observed. The genetically altered variety of *M. cardinalis* was visited by bumblebees 74 times more often than plants with the original color flowers. The genetically altered variety of *M. lewisii* was visited by hummingbirds 68 times more often than plants with the original color flowers.

42. Which of the following represents a valid conclusion based on the data?
    (A) Petal color won't contribute to speciation because pollinators will select familiar plant species regardless of petal color.
    (B) Gene mutations that affect petal color will also affect nectar production and therefore pollinator choice.
    (C) Gene mutations affecting petal color can contribute to speciation by causing a shift in pollinator species.
    (D) The flowers of the two species must be so different in their shape that they can be pollinated by only bumblebees or only hummingbirds.

43. Which of the following is a scientific justification for the statement that the evolution of easily modified flower colors that make plants attractive to animals was an important factor in the diversification of flowering plants?
    (A) Flowering plants have no way to cross-pollinate without the intervention of animals.
    (B) Changes in flower color can result in a shift of animal pollinators, which can serve as an isolating mechanism.
    (C) Bright flowers are important to attract animals that can disperse the seeds of the plant.
    (D) Successful evolution requires interactions between plants and animals.

44. An allergic reaction occurs when the immune system overreacts to an allergen such as shellfish, wasp venom, or peanuts. When the immune system recognizes the foreign allergen, it releases histamines that cause localized swelling and other allergy symptoms such as red eyes or sneezing. These normally harmless allergens can cause life-threatening reactions in some individuals. Which of the following best explains this?
(A) The initial exposure to the allergen causes the immune system to become overactive and release increasing amounts of histamine due to negative feedback mechanisms.
(B) The primary immune response results in the formation of antibodies that continue to circulate in the blood and can trigger a more aggressive response upon subsequent exposures to the allergen.
(C) Repeated exposures to the allergen lead to escalating increases in the release of histamines through an escalating positive feedback response.
(D) Circulating memory B cells are activated by exposure to the allergen and produce large quantities of histamines in response to a second exposure to the allergen.

Use the figure showing the formation of complementary DNA (cDNA) from a eukaryotic gene for questions 45 and 46.

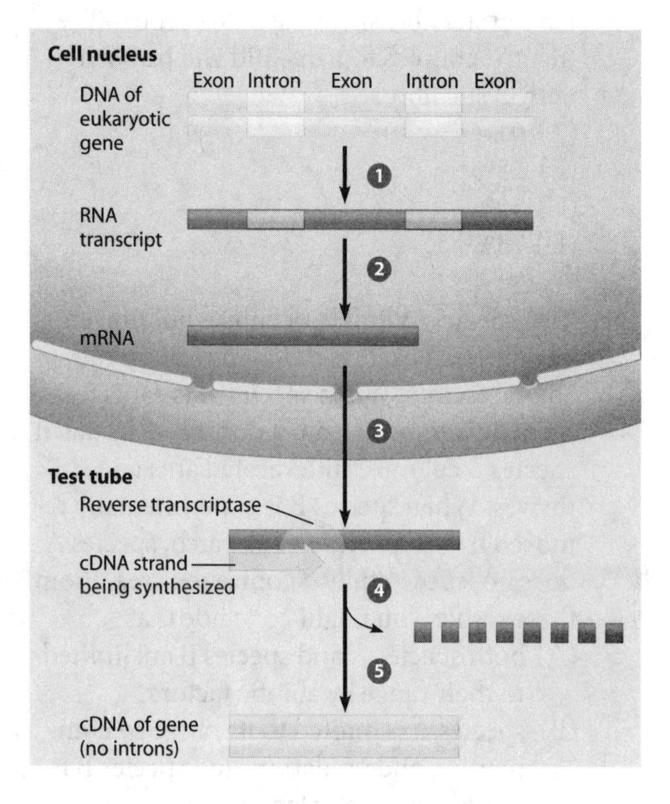

45. In the figure, how is the size of the original RNA transcript reduced in step 2?
(A) DNA polymerase removes some of the original DNA.
(B) RNA polymerase produces a second transcript.
(C) RNA acts as an enzyme in the spliceosomes.
(D) Reverse transcriptase changes the larger transcript to a smaller one.

46. What complex is responsible for the step in the figure where the single strand of DNA, formed from RNA by reverse transcriptase, is used to produce the cDNA strand?
(A) ligase
(B) RNA polymerase
(C) DNA polymerase
(D) helicase

47. Two individuals who are carriers for cystic fibrosis (a recessively inherited disorder) have three children together. None of the children have cystic fibrosis. What is the probability that the couple's fourth child will be born with cystic fibrosis?
(A) 0%
(B) 25%
(C) 50%
(D) 75%

48. Two species, A and B, occupy adjoining environmental patches that differ in several abiotic factors. When species A is experimentally removed from a portion of its patch, species B colonizes the vacated area and thrives. When species B is experimentally removed from a portion of its patch, species A does not successfully colonize the area. From these results you might conclude that
(A) both species A and species B are limited to their range by abiotic factors.
(B) species A is limited to its range by competition and predation, and species B is limited by abiotic factors.
(C) both species are limited to their range by competition although species B has broader tolerance for different conditions.
(D) species A is limited to its range by abiotic factors, and species B is limited to its range because it cannot successfully compete with species A.

49. A child is brought to the hospital with a fever of 107°F. Doctors immediately order an ice bath to lower the child's temperature. Which of the following statements offers the most logical explanation for this action?
(A) Elevated body temperature will increase reaction rates in the child's cells and overload the limited number of enzymes found in the cells of a small child.
(B) Elevated body temperatures may denature enzymes. This would interfere with the cell's abilities to catalyze various reactions.
(C) Elevated body temperatures will increase the energy of activation needed to start various chemical reactions in the body. This will interfere with the ability of enzymes to catalyze vital chemical reactions.
(D) Elevated body temperatures cause molecules to vibrate more quickly and prevent enzymes from easily attaching to reactants. This would slow vital body reactions.

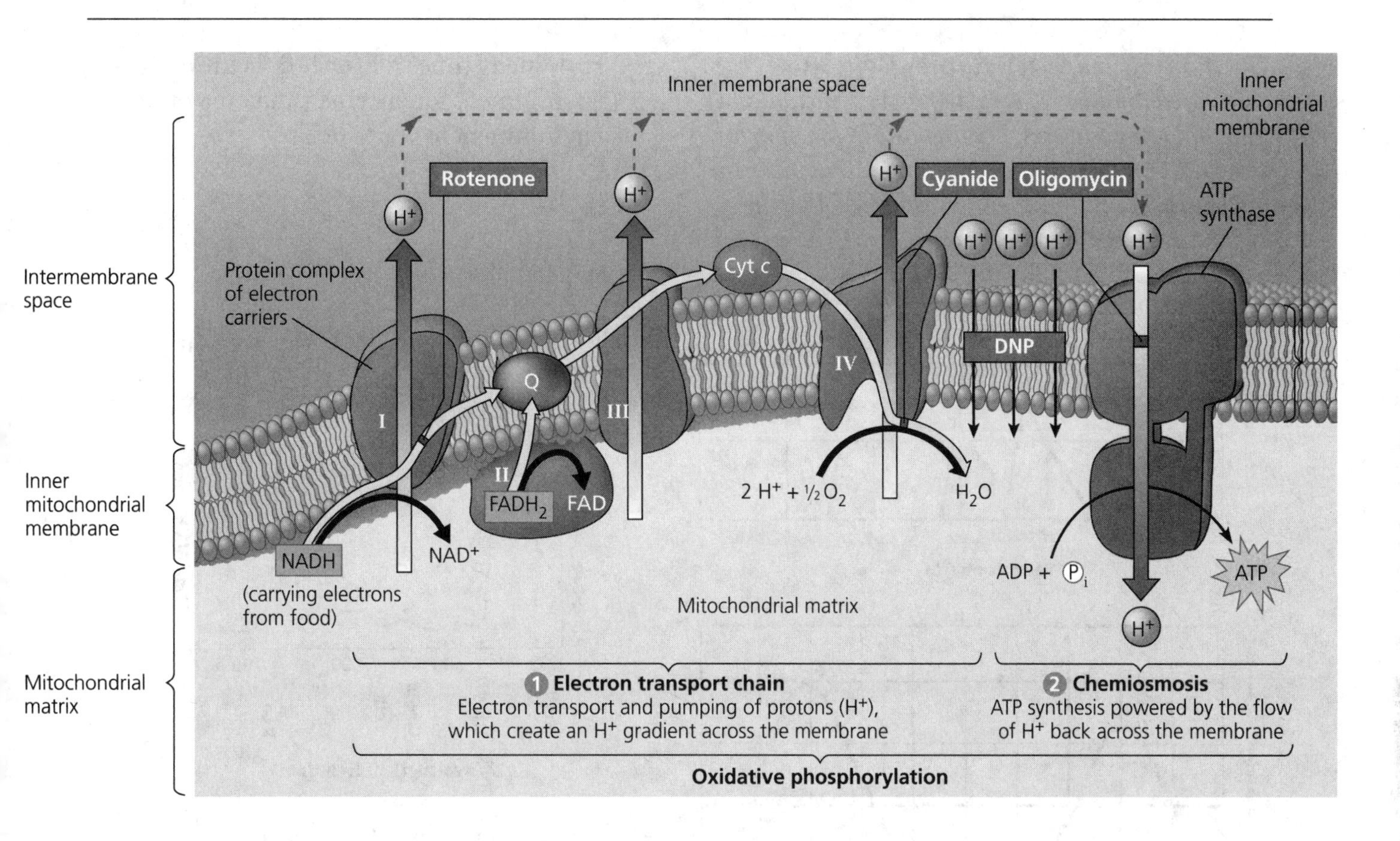

Use the figure above to answer questions 50 and 51.

50. Rotenone binds tightly with one of the electron carrier molecules in the first protein complex of the electron transport chain. It is often used as an insecticide and to kill fish. Predict the specific change to normal cellular respiration that occurs with Rotenone.
    (A) Exposed animals will no longer be able to perform anaerobic respiration.
    (B) ATP synthases are denatured and can no longer generate ATP.
    (C) An electrochemical gradient can no longer be established to drive ATP synthesis.
    (D) NADH is oxidized and its high-energy electrons are moved into the inner membrane space.

51. DNP (dinitrophenol) is a poison that makes the membrane of the mitochondrion leaky to hydrogen ions. Which of the following results from this alteration of normal cellular respiration?
    (A) The electron transport chain is no longer able to transfer electrons to oxygen and so its activity is halted.
    (B) A hydrogen ion gradient cannot be established to drive the production of ATP by chemiosmosis.
    (C) Hydrogen ions bind to the electron transport molecules and interfere with their ability to generate an electron gradient.
    (D) Abnormally large quantities of water will be produced as a result of excess hydrogen ions in the matrix, leading to dehydration and death.

Questions 52 and 53 refer to the graphs below. The rate of reaction for three enzymes was calculated at different temperatures, and the rate of reaction for two additional enzymes was calculated at different pH levels. The results are shown in the following graphs. Assume that the y axes share the same scale.

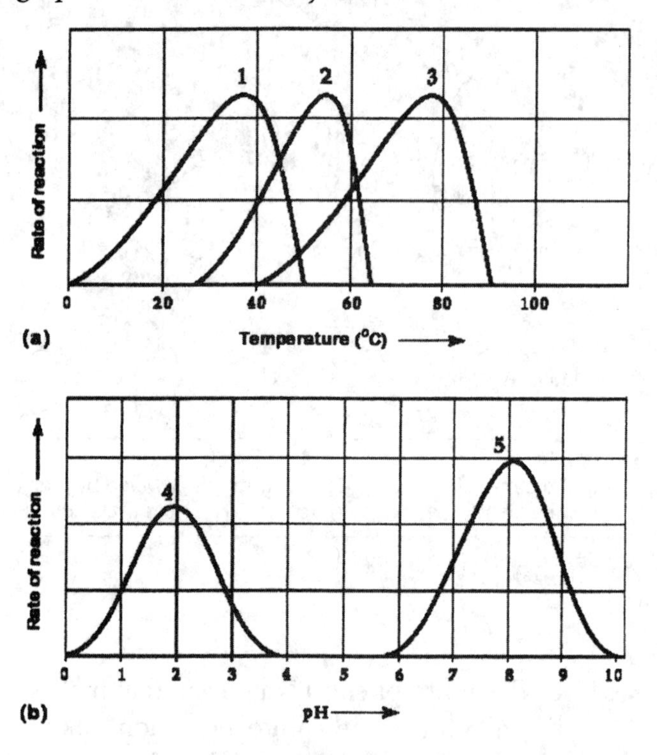

52. In enzymes 1, 2, and 3, why does the curve of the rate of reaction line for each enzyme drop sharply after the optimal temperature is reached?
    (A) The increasing thermal energy of the solution denatures the enzyme, rapidly decreasing the rate of reaction.
    (B) The substrate is completely converted at the optimal temperature, and without substrate the rate of reaction decreases rapidly.
    (C) The optimal point of conversions by the enzyme results in a decreasing number of enzyme molecules due to molecular fatigue, resulting in a rapidly decreasing rate of reaction.
    (D) The final product of the enzyme conversions is a competitive inhibitor interfering at the active site of the enzyme, resulting in a rapidly decreasing rate of reaction.

53. How many times more acidic is the optimal environment for enzyme 4 than the optimal environment for enzyme 5?
    (A) 6
    (B) 60
    (C) 10^6
    (D) 10^8

Question 54 refers to the graph shown below.

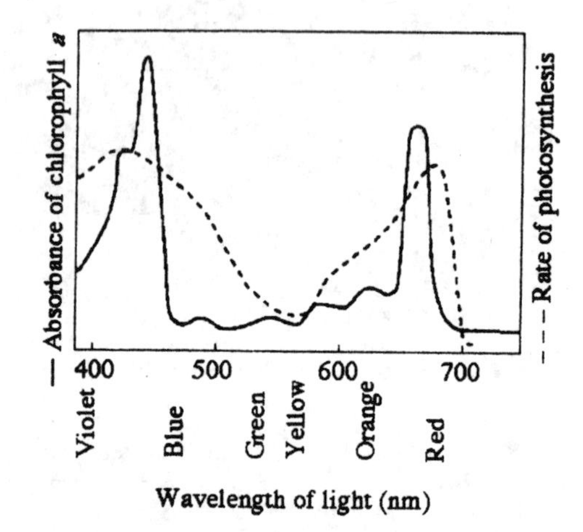

54. Which of the following is the best reason the curve for the absorbance of light by chlorophyll *a* does not perfectly match the rate of photosynthesis?
    (A) The rate of photosynthesis is always fractionally slower than the rate of absorbance by chlorophyll *a*.
    (B) Chlorophyll *a* transmits light in the green and yellow parts of the spectrum and so the rate of photosynthesis is low at those wavelengths.
    (C) There are fewer chlorophyll *a* molecules in the cell than the other molecules involved in photosynthesis, so chlorophyll *a* is the rate-limiting reagent.
    (D) Chlorophyll *a* is not the only photosynthetically important pigment in chloroplasts.

Questions 55 and 56 refer to the following table, which compares the % sequence homology of four different parts (two introns and two exons) of a gene that is found in five different eukaryotic species. Each part is numbered to indicate its distance from the promoter (for example, Intron I is the one closest to the promoter). The data reported for species A were obtained by comparing DNA from one member of species A to another member of species A.

Species	Intron I	Exon I	Intron VI	Exon V
A	100%	100%	100%	100%
B	96%	99%	82%	96%
C	88%	99%	89%	96%
D	92%	99%	92%	97%
E	86%	99%	86%	94%

55. Which of the following is the best explanation for the high degree of sequence homology observed in Exon I among these five species?
    (A) It is relatively close to the promoter, so fewer mutations have accumulated.
    (B) This exon may often be included as part of Intron I because of their proximity.
    (C) Exon I may be highly conserved because it codes for a critical portion of a polypeptide.
    (D) These species are very closely related, as indicated by the similarity of Exon I.

56. Based on the data shown in the chart, which of the following is most plausible?
    (A) Mutations are more likely to accumulate in introns compared to exons.
    (B) Species E is most closely related to species B.
    (C) A cladogram of these species would show species E as the common ancestor of the others.
    (D) With more time, the variation in Intron VI would decline due to natural selection forces.

After reading the paragraphs, answer question 57.

You're conducting an experiment to determine the effect of different wavelengths of light on the absorption of carbon dioxide as an indicator of the rate of photosynthesis in aquatic ecosystems. If the rate of photosynthesis increases, the amount of carbon dioxide in the environment will decrease and vice versa. You've added an indicator to each solution. When the carbon dioxide concentration decreases, the color of the indicator solution also changes.

Small aquatic plants are placed into three containers of water mixed with carbon dioxide and indicator solution. Container A is placed under normal sunlight, B under green light, and C under red light. The containers are observed for a 24-hour period.

57. Based on your knowledge of the process of photosynthesis, the plant in the container placed under red light would probably
    (A) absorb more CO_2 than the plant in normal light.
    (B) absorb the same amount of CO_2 as the plants under both the green light and normal sunlight.
    (C) absorb less CO_2 than the plants under green light.
    (D) absorb more CO_2 than the plants under the green light.

58. Which of the following mutations would be *most* likely to have a harmful effect on an organism?
    (A) a deletion of three nucleotides near the middle of a gene
    (B) a single nucleotide deletion in the middle of an intron
    (C) a single nucleotide deletion near the end of the coding sequence
    (D) a single nucleotide insertion downstream of, and close to, the start of the coding sequence

59. The diagram below shows a fertilized human egg dividing to form a two-celled embryo. After the initial cleavage division, chemical analysis will reveal that the cells of the embryo are not identical. These two cells will have different fates in the embryo due to differential gene expression. Which of the following statements gives the most scientifically plausible explanation?

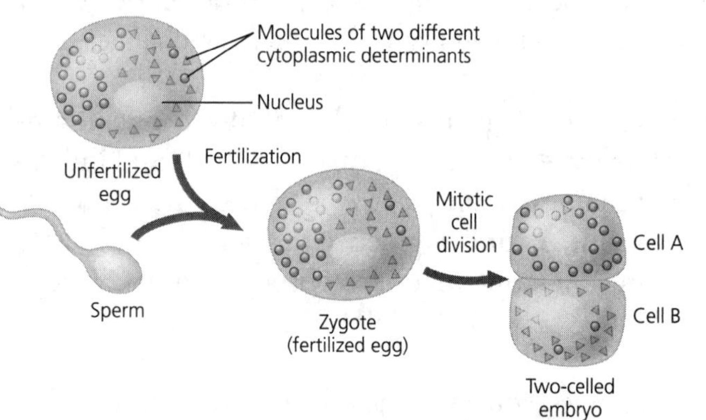

(A) Each of the daughter cells receives slightly different cytoplasmic determinants because these are not evenly distributed in the fertilized egg.
(B) As a result of cleavage, each cell has different receptor proteins expressed on the cell surface, which respond differently to hormones produced by the embryo.
(C) When the zygote divides, one cell receives cytoplasmic determinants from the sperm, while the other receives cytoplasmic determinants from the egg.
(D) The first mitotic cell division of the embryo results in independent assortment of the homologous chromosomes and therefore slightly different genes in each cell.

60. The pancreas produces insulin in response to high blood glucose levels, and insulin lowers blood glucose by stimulating its uptake into body cells. In type 2 diabetes, the pancreas produces normal insulin, but body cells are unable to efficiently uptake glucose. This disease is believed to have a genetic component, but increased age and weight gain also increase the likelihood of it occurring. Which of the following genetic mutations would be most likely to make a person more susceptible to type 2 diabetes?
(A) a mutation in a gene responsible for the production of the precursor of normal insulin
(B) a mutation in a gene that allows detection of blood glucose levels and so provides feedback regulation
(C) a mutation in a gene for a receptor protein for insulin in target cells
(D) a mutation in liver cells that prevents the breakdown of glycogen to glucose

Part B Directions: The next six questions require numeric answers. Determine the correct answer for each question and enter it in the grids provided at the end of this section.

1. Results of a *Drosophila* mating between F_1 flies resulted in 58 flies showing red eyes (a dominant trait) and 42 flies showing sepia eyes (recessive). Calculate the Chi-square value for the hypothesis that both F_1 flies were heterozygous for eye color. Give your answer to the nearest tenth.

2.

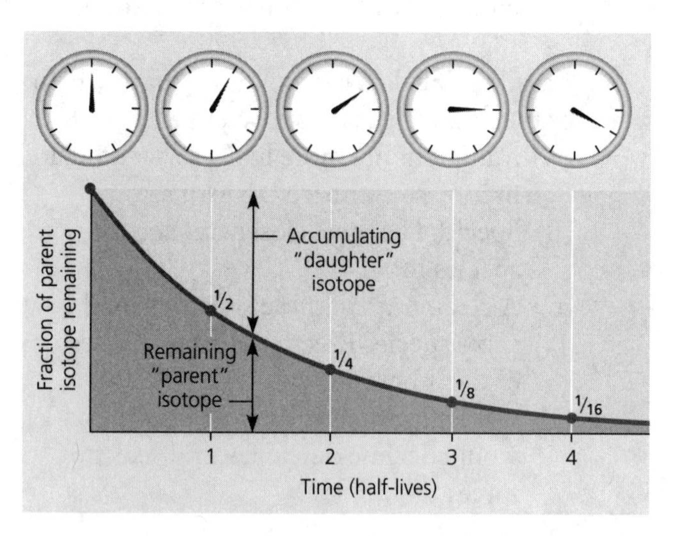

The figure provides information about radiometric dating. Note that the radioactive "parent" isotope decays to a daughter "isotope" at a characteristic rate. The age of artifacts containing wood can be dated accurately to about 75,000 years old using carbon dating. The half-life of radioactive decay of carbon-14 to carbon-12 is about 5,730 years. In a sample to be analyzed, the carbon-14 to carbon-12 ratio is 1/8. What is your estimate of the age of the artifact to the nearest year?

3. A population of deer mice on an island has a carrying capacity of 350 individuals. If the maximum rate of increase is 1.0 per individual per year and the population size is 275, determine the logistic population growth rate to the nearest mouse.

4. An enzyme in the liver removes a phosphate group from glucose so the glucose molecule can enter the bloodstream, providing energy for cellular respiration to the cells of the body. The rate of enzyme activity can be monitored by measuring the phosphate concentration over time.

   Data from the experiment:

Time (min)	Phosphate Concentration (μmol/mL)
0	0
5	10
10	90
15	180
20	270

   What is the rate of phosphate formation per minute from 5 to 15 minutes?

5. In a population of trogons (a type of bird) tail banding (B) is dominant to no tail banding (b). If 68% of the population shows tail banding, what percentage of the population, to the nearest tenth, is heterozygous for tail banding?

6. Ecologists often cannot count all the individuals in a population. In such cases, the mark-recapture method may be used to estimate population size. An ecologist wished to estimate the population of snapping turtles in an isolated farm pond. On the first trapping 19 turtles were caught and marked with a harmless mark on the top of the shell. Two weeks later a second trapping was conducted where 22 turtles were caught, of which 9 were marked. Using the formula

$$N = sn/x$$

   where N = population size estimate, s = number of turtles marked in the first trapping, n = number of turtles caught in the second trapping, and x = the number of turtles that were marked in the second trapping, what is the population estimate of snapping turtles to the nearest *hundredth*?

Grid-In Questions

These call for a numerical response.

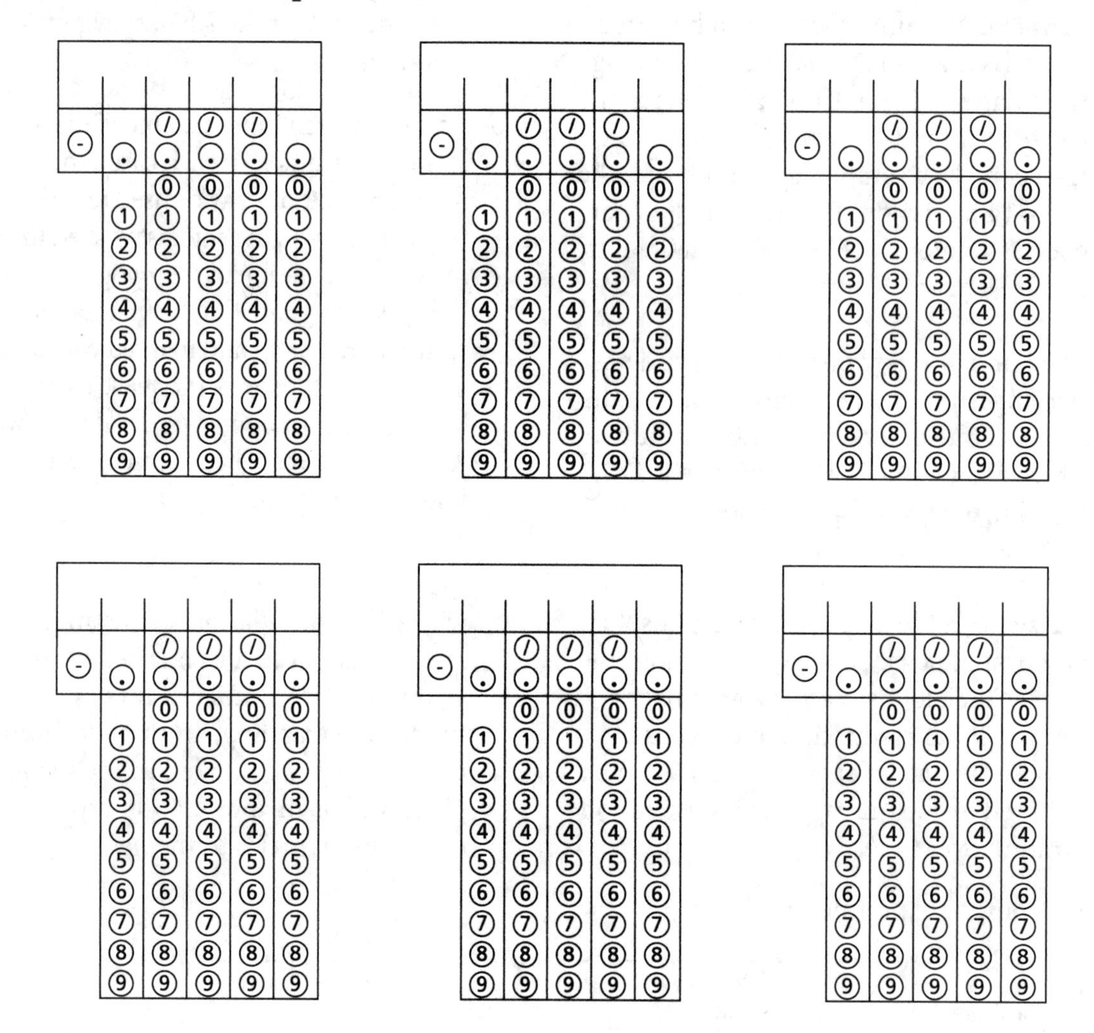

Biology
Section II

8 Questions
Planning Time—10 minutes
Writing Time—80 minutes

Directions: Questions 1 and 2 are long free-response questions that require about 22 minutes each to answer and are worth 10 points each. Questions 3–9 are short free-response questions that require about 6 minutes each to answer. Questions 3–5 are worth 4 points each and questions 6–8 are worth 3 points each.

1. The plasma membrane allows the selective passage of materials into and out of the cell.
   - **(a)** Several macromolecules are involved in the formation of the plasma membrane. **Identify three** macromolecules of a typical animal plasma membrane that are involved in the transport of materials across the plasma membrane, and **explain** the role each of these components plays in regulating the cell's internal environment.
   - **(b)** **Describe** how each of the following can enter an animal cell:
     - viral RNA
     - steroid hormones
     - water molecules
     - O_2

2. Human activities are having a significant impact on biodiversity. The diagram below represents different forest communities.

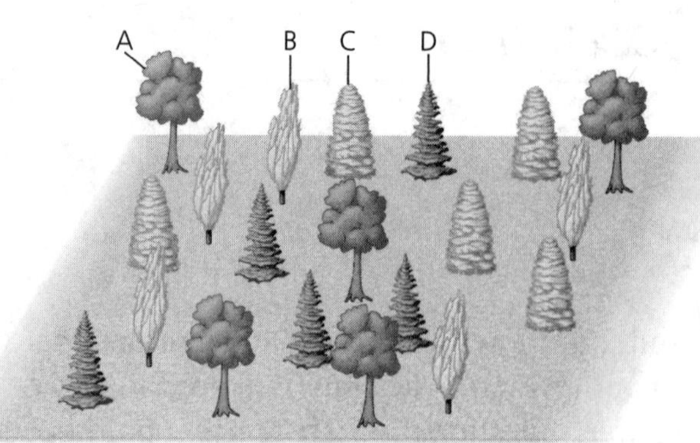

Community 1

Community 2

(a) **Describe** the factors involved in species diversity.
(b) **Calculate** the percent of *each* community represented by species A.
(c) Which of the two communities in the figure is more diverse? **Explain** how two communities that contain the same number of species can differ in species diversity.
(d) **Predict** and **justify** the impact on community stability in each forest community of a disease that affects species A.

3. Gene expression in a cell is influenced by a variety of factors. Not all genes on the eukaryotic chromosome are expressed, and, in fact, only a small fraction of the genes are transcribed into working proteins. **Describe** and **explain** two ways chromatin packaging influences gene expression.

4. Artificial selection and genetic drift can each affect the genetic makeup of a population.
   (a) Describe each, using a specific example.
   (b) Predict the effect each may have on future populations based on your example.

First mRNA base (5′ end of codon)	Second mRNA base: U	Second mRNA base: C	Second mRNA base: A	Second mRNA base: G	Third mRNA base (3′ end of codon)
U	UUU Phe	UCU Ser	UAU Tyr	UGU Cys	U
U	UUC Phe	UCC Ser	UAC Tyr	UGC Cys	C
U	UUA Leu	UCA Ser	UAA Stop	UGA Stop	A
U	UUG Leu	UCG Ser	UAG Stop	UGG Trp	G
C	CUU Leu	CCU Pro	CAU His	CGU Arg	U
C	CUC Leu	CCC Pro	CAC His	CGC Arg	C
C	CUA Leu	CCA Pro	CAA Gln	CGA Arg	A
C	CUG Leu	CCG Pro	CAG Gln	CGG Arg	G
A	AUU Ile	ACU Thr	AAU Asn	AGU Ser	U
A	AUC Ile	ACC Thr	AAC Asn	AGC Ser	C
A	AUA Ile	ACA Thr	AAA Lys	AGA Arg	A
A	AUG Met or start	ACG Thr	AAG Lys	AGG Arg	G
G	GUU Val	GCU Ala	GAU Asp	GGU Gly	U
G	GUC Val	GCC Ala	GAC Asp	GGC Gly	C
G	GUA Val	GCA Ala	GAA Glu	GGA Gly	A
G	GUG Val	GCG Ala	GAG Glu	GGG Gly	G

5. The DNA template strand of a gene contains the sequence

   3′-TACTTCAAACCGATT-5′

   (a) Draw the mRNA sequence from this strand, indicating 5′ and 3′ ends.
   (b) Predict the amino acid sequence from this gene.
   (c) What would be the effect of a mutation that substitutes a thymine for the second cytosine?
   (d) What would be the effect of a mutation that deletes the second cytosine?

6. A tray of germinating oat seedlings is placed in a window that has light shining through most of the day, and it is given adequate water and soil nutrients. After 3 days, the tray is rotated 180°.
   (a) Predict and explain what will occur.
   (b) What is the evolutionary significance of this response?

7. It is theorized that glycolysis was the first metabolic pathway for the production of ATP. Glycolysis begins the process of making ATP by breaking glucose into two molecules of pyruvate. **Justify** this claim with three pieces of evidence that support this point.

8. Sickle-cell anemia is caused by a single base-pair substitution in the amino acid sequence for the β-globin gene.
   (a) **Predict** how this change might alter the overall structure of the protein.
   (b) Explain why this mutation is common in sub-Saharan Africa.
   (c) If the frequency of the allele is 0.07 in sub-Saharan Africa, predict a biologically plausible frequency for the allele in a population of Native Americans in Alaska.

END OF EXAMINATION

Part V

Answers and Explanations

Topic 1: The Chemistry of Life

ANSWERS AND EXPLANATIONS

Level 1: Knowledge/Comprehension Questions

1. **(B) is correct.** RNA is made up of a phosphate group, a ribose sugar, and one of the following four nitrogenous bases: cytosine, guanine, uracil, and adenine. DNA is similar to RNA in many ways but different in two important ones: It contains deoxyribose instead of ribose as its sugar and it contains the base thymine instead of uracil.
2. **(C) is correct.** The answer is water, H_2O. Polar covalent bonds are those in which valence electrons are shared between atoms but unequally. (The more electronegative atom will attract the electrons more strongly, and that end of the molecule will have a slightly negative charge, whereas the less electronegative atom will attract the electron less strongly and be slightly positive.) The two atoms involved in the bond must differ significantly in electronegativity in order to form a polar covalent bond.
3. **(A) is correct.** Use this question to review the unique properties of carbon. The electronegativities of carbon and hydrogen are not different enough to form polar covalent bonds.
4. **(D) is correct.** The three terms you should keep in mind as you think of water traveling up through the xylem of a plant are transpiration (in which water evaporates from the plant's leaves); cohesion, in which the water molecules stick together due to the hydrogen bonds; and adhesion, whereby the water molecules stick to plant cell walls and resist the downward pull of gravity. Translocation is the movement of solutes in phloem.
5. **(A) is correct.** Lipids are the only one of the four major classes of biological molecules that are not polymers. They are grouped together because they are hydrophobic. Nucleic acids are polymers of nucleotide monomers, proteins are polymers of amino acid monomers, and carbohydrates are polymers of monosaccharide monomers.
6. **(B) is correct.** The linkages between the amino acids of proteins are peptide bonds. Peptide bonds are covalent bonds formed in dehydration reactions. The carboxyl group of one amino acid is joined to the amino group of an adjacent amino acid, resulting in the loss of one molecule of water.
7. **(C) is correct.** One common secondary structure of proteins is the alpha (α) helix; another is the beta (β) pleated sheet. The secondary structure of a protein refers to hydrogen bonding along the backbone (not the side chains) of the amino acid chain.
8. **(D) is correct.** Cellulose is the polysaccharide that forms the strong cell walls of plant cells. It is a polymer of glucose.
9. **(D) is correct.** Recall that hydrolysis means to use water to split a molecule, so look for a large molecule reduced to its monomers.
10. **(C) is correct.** Denaturation is the process by which proteins lose their overall structure, or conformation, as a result of changes in pH, temperature, or salt concentration. Denatured proteins have reduced biological activity.

11. (B) is correct. The ratio of carbon, hydrogen, and oxygen atoms in carbohydrates is 1:2:1. For example, glucose is $C_6H_{12}O_6$.

12. (C) is correct. To get this correct, you should first add up all the C, H, and O in three fatty acid chains plus one glycerol. This would be 51 C, 74 H, and 9 O. Then, recall that to join to molecules by dehydration synthesis, one molecule of water must be removed. Because three fatty acid chains will be attached to one glycerol, three water molecules will be removed. Subtract 6 H and 3 O to arrive at the answer.

13. (B) is correct. Proteins have many functions, which encompass most of a cell's metabolic activity.

14. (B) is correct. Phospholipids are unique macromolecules. Their hydrophilic heads and hydrophobic tails contribute to the semipermeability of cell membranes and they comprise most of the membrane. Proteins function as surface receptors and channels, and cholesterol stabilizes the fluid membrane. Carbohydrates form part of the glycocalyx.

15. (D) is correct. The negative charge comes from the electronegative oxygen of one water molecule attracted to the partial positive charge of hydrogen of another water molecule.

16. (C) is correct. The monomers in macromolecules are joined when a molecule of water is removed during dehydration, or condensation reactions. If you selected choice A, perhaps you missed that this applies to proteins as well as carbohydrates.

17. (B) is correct. Because the pH scale is logarithmic, each unit change is by a factor of 10. A drop of pH means the solution is more acidic and has 10 times more H^+ ions.

18. (C) is correct. An amino acid is composed of a central carbon, bonded to a hydrogen, with a variable (R) group, and with a carboxyl (the acid part) at one end and an amino group at the other end (the amino part). Aldehydes and ketones are found in sugars; the sulfhydryl group is found in one amino acid.

Level 2: Application/Analysis/Synthesis Questions

1. (D) is correct. Hydrogen bonds occur when a slightly positive hydrogen atom of a polar covalent bond in one molecule is attracted to a slightly negative atom of a polar covalent bond in another molecule. In living systems hydrogens are often attracted to the highly electronegative elements oxygen (as in this question with water) or nitrogen.

2. (A) is correct. Proteins fold into precise shapes because of interactions between R groups. If you chose C, you recognize that the primary structure is determined by the DNA sequence, but perhaps that this is important because it determines the pattern of folding.

3. (B) is correct. Peptide bonds occur between the carboxyl group of one amino acid and the amino group of another amino acid. All that is required in this question is for you to carefully examine the information given in the diagram. Take a look at every question, even if you may think the content area is a hard one; often the answers are easier than you might expect.

4. (D) is correct. Hydrolysis is a chemical process that splits molecules by the addition of water. Digestive enzymes work by hydrolysis. Water is removed to join the molecules; water is added to separate the molecules.

5. (D) is correct. Recall that enzymes are proteins and proteins are made of amino acids. With all molecules, *change the shape, change the function.*

6. (D) is correct. Two common environmental factors that affect the three-dimensional structure of enzymes are temperature and pH. As the structure of the enzyme is altered, the enzyme will become less effective.

7. (C) is correct. Ten glucose molecules would have a combined molecular formula of $C_{60}H_{120}O_{60}$. To form a polymer, a molecule of water would have to be removed as each glucose is added to the chain. Because 10 glucose molecules are bonded together, 9 H_2O molecules must be removed and 18 hydrogen atoms and 9 oxygen atoms. This leaves a formula of $C_{60}H_{102}O_{51}$.

8. (D) is correct. This conceptual question is based on your knowledge of the structure of DNA. Recall that the molecule is antiparallel, meaning one strand runs 5′ to 3′, whereas the opposite strand of the double helix runs 3′ to 5′. Choices B, C, and D are given with both strands running 5′ to 3′. To get the proper answer convert the second strand given in the answer to 3′ to 5′ and see which one has the proper, matching base-pair sequence. Choice A shows correct complementary base pairing for RNA, not DNA.

9. (A) is correct. B is a false statement, and although C and D are true statements, they do not justify how water acts as a solvent. Only choice A gives a true property with correct justification.

Grid-In Questions

1. 92 hydrogen atoms. To get this, remember that you must remove 8 molecules of H_2O as you join together 9 molecules by dehydration synthesis. Multiply 12 hydrogen atoms $\times$ 9 = 108, but you must remove 8 molecules of H_2O, so subtract 16 H, and you will get 92.
2. 10,000. To obtain this answer, remember that the pH scale is logarithmic, so every number of difference results in a 10X difference of H^+. Since from pH 6 to pH 2 is a change of 4 units, $10 \times 10 \times 10 \times 10 = 10{,}000$. Because there are *more* hydrogen ions, the sign is positive and does not need to be recorded in your response. If you had been asked to compare pH 2 to pH 6, there are 10,000 fewer H^+ ions in a solution with pH 6.

Free-Response Questions

(a) A phospholipid molecule contains a hydrophilic "head" (containing a glycerol molecule and a phosphate group) and two hydrophobic fatty acid tails. In cell membrane surfaces, phospholipids are arranged in a bilayer in which the hydrophilic heads are in contact with the cell's watery interior and exterior, whereas the tails are pointed away from water and toward each other in the interior of the membrane. The fatty acid chains of phospholipids can contain double bonds, which makes them unsaturated. Because of the kinks in the tails, phospholipids aren't packed together tightly, which contributes to the fluidity of the membrane. The fluidity of the cell membrane is very important in its function; the less fluid the membrane is, the more impermeable it is. There is an optimum permeability for the cell membrane at which all the substances necessary for metabolism can pass into and out of the cell.

The fluidity of cell membranes enables hydrophobic molecules such as hydrocarbons, carbon dioxide, and oxygen to dissolve in the bilayer and easily cross the membrane. However, ions and polar molecules (including water, glucose, and other sugars) cannot readily pass through because of the hydrophobic interior. Protein channels and transport proteins allow these required substances to cross membranes.

(b) If all the fatty acid chains of phospholipids were saturated, no "kinks" would occur in the phospholipids resulting in membranes that would be less fluid. At very low temperatures, this would lead to plant membranes with drastically reduced fluidity. The loss of fluidity would make the movement of essential nutrients across the membrane much less efficient, which could lead to the death of the plant.

(c) There are numerous functions of proteins in the membrane. One important function is that some proteins protrude on the extracellular side of the membrane and serve as receptors for signaling molecules. A second function is seen with proteins that extend through the interior of the bilayer and serve as channels for the passage of molecules or ions that cannot pass through the phospholipids.

(d) Receptor proteins have regions or domains that are made up of unique amino acids to bind only specific ligands. The specificity of amino acid sequences makes this possible.

Transport proteins may act as channels for molecules and ions that cannot pass through the phospholipid bilayer. Ligands that bind to transport proteins can help alter their conformation to permit the passage of molecules through them and into the cell interior.

This response shows thorough knowledge of the processes of the structure of phospholipids, cell membrane structure and components, and movement across membranes. A strong response to this item requires an understanding of topics from Units 1 and 2 of the textbook.

The student's response could have used other protein functions, including specific examples of cell-cell signaling or enzymatic function.

Topic 2: The Cell

ANSWERS AND EXPLANATIONS

Level 1: Knowledge/Comprehension Questions

1. **(B) is correct.** This question is centered around proteins for export, which would require large numbers of ribosomes, rough ER, and many Golgi bodies.
2. **(C) is correct.** Linear chromosomes are the only cell structure on this list not found in both prokaryotes and eukaryotes. Eukaryotic cells have a true

nucleus, which is surrounded by a double membrane called a nuclear envelope and contain linear chromosomes with their DNA wrapped around histone proteins. The genetic material of prokaryotes is localized in a clump in one particular region of the cell called the nucleoid region, and the main chromosome is a single molecule of circular DNA without associated histone proteins.

3. **(D) is correct.** The smooth ER proliferates in cells that detoxify poisons.

4. **(B) is correct.** The Golgi apparatus is the organelle that has a *cis* and *trans* face, and it acts as the packaging and secreting center of the cell. It consists of a series of flattened sacs of membranes called cisternae.

5. **(A) is correct.** Cellular respiration takes place in the mitochondria, forming ATP, the cell's energy currency. Mitochondria are bound by double membranes, and the proteins involved in ATP production are embedded in the inner membranes of the mitochondria. Red blood cells lack mitochondria as well as nuclei, making them well adapted to their function of oxygen transport.

6. **(C) is correct.** Lysosomes are characteristic of animal cells but not most plant cells. They are large membrane-bound structures that contain hydrolytic enzymes, and they are responsible for the breakdown of proteins, polysaccharides, fats, and nucleic acids. They function best at a low pH (around 5), so they pump hydrogen ions from the cytosol into their lumen to achieve this acidic pH.

7. **(C) is correct.** Proteins destined for export from the cell are produced from ribosomes bound to the ER (rough ER) then transported in vesicles to the Golgi apparatus for further processing, again packaged in secretory vesicles, which then migrate to the plasma membrane and release the modified protein by exocytosis.

8. **(A) is correct.** Surface proteins bind to signaling molecules (ligands) and this binding activates a signal transduction pathway, resulting in intracellular responses.

9. **(D) is correct.** The only substance listed that can passively diffuse through the cell membrane unaided by proteins is carbon dioxide. Remember that passive diffusion occurs without the cell doing any work. Other choices need the processes of facilitated diffusion and transport proteins to cross the membrane. This is true of all of the answer choices listed except for D. **This is a question a student could answer at the end of the course (we hope), but not if you are covering cell biology at the beginning of your course.**

10. **(B) is correct.** Large molecules are moved out of the cell by exocytosis. In exocytosis, vesicles that are to be exported from the cell (often coming from the Golgi apparatus) fuse with the plasma membrane, and their contents are expelled into the extracellular matrix.

11. **(A) is correct.** When a signal molecule binds to the receptor protein, the gate of the ion channel opens or closes, allowing or blocking the flow of specific ions.

12. **(A) is correct.** Because the substance is being moved against the concentration gradient, energy is required. All the other choices are modes of passive transport and require no energy.

13. (B) is correct. Choice A is a reference to a G protein-coupled receptor but is a quick pick if the question is not read carefully. The G protein is activated by the G protein-coupled receptor, which is a protein and eliminates D as a possible answer.

14. (D) is correct. Kinase enzymes are involved with ATP. Protein kinase enzymes are used to amplify the signal during the transduction phase of cell signaling by activating cell proteins with a phosphate from ATP. Protein phosphatases remove phosphates from proteins.

15. (C) is correct. Many signaling pathways involve small, nonprotein water-soluble molecules or ions called second messengers. Calcium ions and cyclic AMP are two common second messengers. The second messengers, once activated (and always found on the inside of the membrane), can initiate a phosphorylation cascade, resulting in a cellular response.

16. (D) is correct. Intracellular receptors work with signal molecules that are hydrophobic compounds and are therefore able to cross the plasma membrane. Testosterone, as indicated in the question, is a steroid hormone and thus hydrophobic. Intracellular receptors often act as transcription factors.

17. (C) is correct. Cytokinesis is the division of the cytoplasm to form two separate daughter cells.

18. (C) is correct. During cytokinesis, the cytoplasm of the cell is divided approximately equally as the cell membrane pinches off (in animal cells), forming two daughter cells; a cell plate forms in plant cells.

19. (A) is correct. In eukaryotic cells DNA is replicated during S-phase, a subphase of interphase. The important thing is to recognize that before identical cells can be produced by mitosis, the amount of DNA must be doubled by replication.

20. (B) is correct. Interphase is a busy time in the cell as growth and cell functions are occurring. Interphase is divided into Gap 1, S-phase, and G2. In G1 the cell has just completed mitosis and is growing and performing its specific cell functions. Because the cell has not passed the G1 checkpoint, no energy is being used for preparations for cell division.

21. (A) is correct. During S-phase the DNA is replicated. If S-phase was attempted and failed, mitosis would be halted and apoptosis would occur.

22. (D) is correct. In anaphase, the sister chromatids, which were lined up along the equator of the cell, begin to separate, pulled apart by the retracting microtubules. By the end of anaphase, the opposite ends of the cell contain complete and equal sets of chromosomes.

23. (D) is correct. The fluid mosaic model is a reference to a membrane with a mosaic of various proteins embedded or attached to a double layer of phospholipids.

24. (A) is correct. Because membranes are hydrophobic, hydrophobic molecules easily pass across the membrane. Remember: Hydrophilic dissolves hydrophilic; hydrophobic dissolves hydrophobic.

25. (D) is correct. The important part of this question are the three correct statements in choices A, B, and C. Think about how these big picture ideas are incorporated into the cell signaling sequence of reception, transduction, and cellular response.

Level 2: Application/Analysis/Synthesis Questions

1. (D) is correct. The radioactive tracking starts with the formation of the protein, which occurs on the ER. The protein then moves to the Golgi and out of the cell via vesicles that will fuse with the membrane. Choice C is the proper pathway, but the radioactive amino acids will not be used in the nucleus and would therefore not be tracked with this system.

2. (A) is correct. During S phase of interphase, DNA is replicated. If this did not occur and the cells divided anyway, the daughter cells would have half the genetic material found in the parent cell.

3. (A) is correct. Water moves in a hypotonic to hypertonic direction. Cells A, B, and C are all animal cells (red blood cells) and lack a cell wall. Cell D is a plant cell with a cell wall. Cell A is in a hypotonic solution, cell B is in an isotonic solution, and cell C is in a hypertonic solution. When answering this type of question, pay close attention to whether the solution or the cell is referenced as being hypertonic or hypotonic. In A the solution is hypotonic but the cell is hypertonic.

4. (C) is correct. In plant cells the relatively inelastic cell wall exerts a back pressure on the cells, called turgor pressure. In cell D the plant cell is immersed in a hypotonic solution, causing the cell to uptake water, thus creating the highest levels of turgor pressure. The animal cell A lyses ("pops") because it has only a thin, flexible membrane.

5. (C) is correct. The solution in Tank A started with more solutes, then as the water is purified its concentration of solutes increases through the process of reverse osmosis. Tank A becomes increasingly hypertonic over the course of the purification treatment.

6. (B) is correct. Tank A with the tap water is hypertonic to the purified water. Because water flows from hypotonic to hypertonic, water would move from Tank B into Tank A.

7. (C) is correct. Epinephrine is the ligand that activates the G protein-coupled receptor responsible for glycogen breakdown. Epinephrine does not enter the cell, suggesting a second messenger. Only in intact cells could the first messenger (epinephrine) be translated to a cellular response–glycogen breakdown.

8. (C) is correct. G protein-coupled receptors are activated by their specific ligand, not by a phosphorylation event.

9. (D) is correct. Substances will move down their concentration gradient until their concentration is equal on either side of a membrane. The concentration of glucose on side B is 2.0 *M*, whereas the concentration of glucose on the A side is 1.0 *M*; therefore, glucose will move from side B to side A.

10. (B) is correct. First, you must know that glucose is the smaller molecule and only focus on where it is in higher concentration. Because its highest concentration is on side B, it will move to side A.

11. (A) is correct. As glucose moves into side A, the total amount of solute on that side increases, becoming hypertonic, causing water to follow.

12. (A) is correct. Notice that the H^+ ions in the diagram are moving against their concentration gradient. Movement of solutes against their concentration gradient requires energy and, therefore, active transport.

13. **(B) is correct.** Recall that decreasing the extracellular pH (for example, going from a pH 7 to pH 6) would increase the H^+ ion concentration, providing more H^+ ions for cotransport as well as a steeper gradient into the cell.

14. **(C) is correct.** Cell A in the figure for question 3 demonstrates what happens when red blood cells are placed in a hypotonic solution. With the RBCs destroyed, the WBCs will be easier to observe.

15. **(C) is correct.** The clue is given in the stem of the question when melanin is identified as a protein for export. With that information an increase in ribosomes, rough ER, Golgi bodies, and vesicles would be expected.

16. **(B) is correct.** When the DNA is damaged beyond the ability of the cell to repair it, apoptosis pathways are initiated. Damaged cells may become cancerous, so it is better for the organism to destroy the cell than take a chance on a cancerous cell.

17. **(B) is correct.** Because both prokaryotes and eukaryotes use ATP, choice B does not provide evidence for the endosymbiotic hypothesis. Notice that the other three choices do provide evidence and would be the key points in developing an answer to essay questions about the evidence for endosymbiosis.

Grid-In Questions

1. **The correct answer is 8,000 (μm^3).**

2. **The correct answer is 8.**

3. **The correct answer is −7.84 (bars).** The formula of solute potential is −iCRT. In this case the formula is $-(1.0)(0.32)(0.0831)(295) = -7.84$. If doing these problems is difficult, go to the lab section of this book to Investigation 4, Diffusion and Osmosis, where an explanation of water potential and how to work water potential problems can be found.

Free-Response Questions

(a) Eukaryotic cells might have evolved by taking in free-living prokaryotes. One piece of evidence is that prokaryotic cells are much smaller than eukaryotic cells—they range from 100 nm to 10 μm, compared to the average size of eukaryotic cells: 10 to 100 μm. Mitochondria and chloroplasts are comparable in size to prokaryotic cells, ranging from about 1 to 10 μm.

A second piece of evidence is that mitochondria and chloroplasts both consist of a double membrane exterior, which would be a possible result of a prokaryotic cell being engulfed by another cell.

Mitochondria and chloroplasts also contain their own DNA and are capable of dividing on their own. The presence of DNA could be evidence that they were independent organisms at some time.

(b) In order to trace the path of proteins in the cell from their creation to their expulsion, we must start in the nucleus. In the nucleus, mRNA is transcribed from DNA. The mRNA travels out of the nucleus through a nuclear pore to the cytoplasm, ending up at ribosomes, some of which are associated with the endoplasmic reticulum (called rough endoplasmic reticulum because of this association). As the mRNA is translated into protein, the amino acid chain is threaded into the lumen of the ER where it undergoes folding to assume its final shape, or conformation.

Secretory proteins travel from the endoplasmic reticulum to the series of flattened membranous sacs known as the Golgi apparatus. They enter at the *cis* face and eventually bud from the *trans* face after undergoing a series of modifications to prepare them for secretion. The vesicles may then fuse with the cell membrane, and the contents are released from the cell in a process called exocytosis.

The student response in (a) might have used the presence of ribosomes in mitochondria and chloroplasts that translate genes unique to the organelles as one of the arguments for endosymbiosis. However, the question asked for three explanations, so this would be substituted for one of the others. Specific sizes of cells and organelles are not expected. Additional points would probably be awarded in (b) if the student included a discussion of signal peptides and signal recognition particles. (This is covered in Chapter 17.)

Topic 3: The Energy of Life

ANSWERS AND EXPLANATIONS

Level 1: Knowledge/Comprehension Questions

1. (D) is correct. The shape of the curve in the art shown most closely depicts an exergonic reaction. The free energy of the products is lower than that of the reactants—meaning that in the course of the reaction, energy is given off. This is characteristic of exergonic reactions. Conversely, in an endergonic reaction, energy is taken in during the course of the reaction.

2. (D) is correct. Catalysts speed up chemical reactions by providing an alternate reaction pathway that lowers the activation energy of the reaction. Less energy is required to start the reaction, so it runs more quickly.

3. (A) is correct. In allosteric regulation, the enzyme is usually composed of more than one polypeptide chain, with an allosteric site remote from the active site. When an allosteric activator binds to the allosteric site, the enzyme assumes a stable conformation with a functional active site, and the reaction can proceed. When an allosteric inhibitor binds to the allosteric site, the enzyme is stabilized in an inactive form.

4. (C) is correct. Competitive inhibitors compete for the active site of the enzyme. They are able to bind because they closely resemble the normal substrate. One way to overcome the effects of competitive inhibitors is to increase the amount of substrate so that chances are greater that a substrate molecule (rather than the competitive inhibitor) will bind.

5. (B) is correct. In feedback inhibition, the product of a metabolic pathway switches off the pathway by binding to and inhibiting an enzyme involved somewhere along the pathway.

6. (D) is correct. In noncompetitive inhibition, the inhibitor binds to a site other than the active site of the enzyme, and this causes the enzyme to change shape. The change in conformation makes the substrate unable to bind to the active site of the enzyme, and this prevents the action from taking place.

7. (C) is correct. The purpose of cellular respiration in eukaryotes is to produce energy for cellular work in the form of ATP. Respiration is an aerobic process, meaning that it requires oxygen. Answer choices A and B are incorrect because respiration involves the breakdown (not the synthesis) of carbohydrates, fats, and proteins. Choice D is wrong because oxygen is required, not produced, for cellular respiration.

8. (B) is correct. Note that this question is framed per glucose, not per single turn of the citric acid cycle. In the breakdown of glucose in the citric acid cycle, 2 ATP are produced. The citric acid cycle takes in a molecule called acetyl CoA (pyruvate is converted into acetyl CoA before it enters the citric acid cycle), and this is joined to a four-carbon molecule of oxaloacetate to form a six-carbon compound citrate that is then broken down again to produce oxaloacetate; the oxaloacetate reenters the cycle. In the course of the citric acid cycle, the following are produced by two trips through the citric acid cycle *per glucose*: 4 CO_2, 2 ATP, 6 NADH, and 2 $FADH_2$.

9. (D) is correct. In glycolysis, glucose is oxidized to two molecules of pyruvate. This is the first step in cellular respiration, showing a net production of 2 ATP and 2 NADH.

10. (A) is correct. In chemiosmosis, the hydrogen ion gradient created by the transfer of electrons in the electron transport chain provides the power to perform cell work, including synthesizing ATP from ADP.

11. (C) is correct. Fermentation is a way of harvesting chemical energy without using either oxygen or an electron transport chain. It consists of glycolysis and several reactions that serve to regenerate NAD^+. Electrons are transferred from NADH to pyruvate or its derivatives; then NAD^+ can return to glycolysis to once again accept electrons, continuing the production of small amounts of ATP. There are many types of fermentation, including alcohol fermentation (which creates ethanol as a product) and lactic acid fermentation (which creates lactate).

12. (B) is correct. The electron transport chain is a series of inner membrane-embedded molecules that are capable of being oxidized and reduced as they pass along electrons. The energy produced from the passage of these electrons down the chain is used to create an H^+ gradient across the membrane, and the flow of H^+ down the gradient and back across the membrane powers the phosphorylation reaction of ADP to from ATP. Both cellular respiration and photosynthesis make ATP, utilizing an electron transport chain.

13. (A) is correct. Groups of photosynthetic pigment molecules in the thylakoid membrane are called photosystems. The two photosystems involved in photosynthesis are photosystem I and photosystem II. Both contain chlorophyll molecules and many proteins and other organic molecules, and both have a light-harvesting complex that harnesses incoming light. Each of these photosystems contains a reaction center, where chlorophyll *a* and the primary electron acceptor are located.

14. (B) is correct. The main products of the light reactions of photosynthesis are NADPH and ATP. NADPH and ATP are used to convert CO_2 to sugar in the Calvin cycle. The enzyme rubisco combines CO_2 with ribulose bisphosphate (RuBP), and electrons from NADPH and energy from ATP to synthesize a three-carbon molecule called glyceraldehyde 3-phosphate. You should recognize that $FADH_2$ is a product of the citric acid cycle (aerobic respiration) and so choices A and C could not be correct.

15. (D) is correct. The organic product of the Calvin cycle, which may be used later to build large carbohydrates in the cell, is glyceraldehyde 3-phosphate, or G3P. This molecule is created as a result of the fixation of three molecules of CO_2, which costs the cell ATP and NADPH that were created in the light reactions of photosynthesis.

16. (B) is correct. During the Calvin cycle, chemical energy produced during the light reactions (ATP and NADPH) is used to reduce CO_2 and produce G3P (sugars). D is wrong because the electron transport chains pump protons across membranes from regions of *low* H^+ concentrations to regions of high H^+ concentrations. This proton pumping occurs in both mitochondria and chloroplasts, and the protons then diffuse (with the concentration gradient) back across the membrane through ATP synthases. This drives the synthesis of ATP. Choice C describes the light reaction, not the Calvin cycle, and so is incorrect.

17. (D) is correct. ATP synthase is located in the thylakoid membrane. Notice the direction of flow of protons through ATP synthase in photosynthesis: The protons flow from the thylakoid space to the stroma. ATP is produced in the stroma, where it will be used by the Calvin cycle.

18. (C) is correct. When water is split, three products are formed: two protons, an oxygen atom that immediately bonds with another oxygen to form O_2, and two electrons. The electrons immediately feed the P680 chlorophyll *a* in the reaction center of photosystem II. The ultimate electron donor in photosynthesis is water.

19. (A) is correct. CAM plants separate the two stages of photosynthesis temporally to reduce photorespiration. This is accomplished by fixing CO_2 at night using PEP carboxylase and storing the carbon in organic acids. During the day when CAM plants have their stomata closed to conserve water, the carbon from the organic acids is chemically released and used in the Calvin cycle. Choice C would be correct for C_4 plants but not for CAM plants.

20. (D) is correct. Each turn of the Calvin cycle involves the enzyme rubisco fixing one atom of carbon. It follows that it would take six turns to produce the six-carbon sugar glucose. Students sometimes miss this question by confusing the Calvin cycle with the citric acid cycle of cellular respiration. Read these questions carefully, being disciplined enough to carefully identify what the question is asking.

21. (C) is correct. The light reactions of photosynthesis move electrons from their low-energy state in water to a higher energy level when the electrons are donated to $NADP^+$ to make NADPH. In cellular respiration electrons pass down the electron transport chain from high to low potential energy, ultimately combining with O_2 and hydrogen ions to form water.

22. (A) is correct. The Calvin cycle occurs in the stroma.

Level 2: Application/Analysis/Synthesis Questions

1. **(D) is correct.** Plant leaves are green because they reflect and refract green light, which is not utilized in photosynthesis. Red light is used in photosynthesis, meaning the plant would absorb CO_2 for photosynthesis. Check the action and absorption spectra for photosynthesis in your text and be prepared to explain the peaks and valleys shown in the graphs.
2. **(A) is correct.** The light reactions convert solar energy to the chemical energy of ATP and NADPH, which are utilized in the Calvin cycle to reduce CO_2 to sugar.
3. **(C) is correct.** In step 3 the five-carbon α-ketoglutarate is converted to the lower energy four-carbon compound succinate. The drop in energy allows for the production of 1 ATP and the reduction of NAD^+ to NADH. No other step in the citric acid cycle accomplishes this.
4. **(A) is correct.** With the uncoupling of ATP synthase from the H^+ gradient, the inner membrane no longer has a sufficient electrochemical gradient to generate normal ATP production.
5. **(C) is correct.** In energy coupling an exergonic process (catabolic reactions) is used to drive an endergonic process (anabolic reactions).
6. **(B) is correct.** The breakdown of glucose requires many chemical reactions involving energy transfers. The second law of thermodynamics states that with each energy transfer some of the energy is lost to the system, primarily as heat. When you exercise heavily, sweat is an attempt to cool your body from the gain of heat as a result of carrying out cellular respiration.
7. **(D) is correct.** Molecular shape is directly related to molecular function (change the shape—change the function). As the pH dropped to 6, the enzyme was a more active shape. Choice A is not correct because we do not know the enzyme is at its optimum at pH 6. For example, the enzyme might work even better at pH 5.
8. **(D) is correct.** Noncompetitive inhibitors bind to the enzyme away from the active site, thereby changing the shape of the enzyme and reducing the effectiveness of the active site. Because more substrate did not increase product formation, choice C was eliminated.
9. **(B) is correct.** The process in photosynthesis that bears the closest resemblance to chemiosmosis and oxidative phosphorylation in cellular respiration is linear electron flow. In this process, energy from the transfer of electrons down the electron transport chain is used to create a hydrogen ion gradient for the making of ATP. Later, the energy stored in this ATP is used during the formation of carbohydrates in the Calvin cycle.
10. **(D) is correct.** The proteins imbedded in the inner mitochondrial membrane are proton pumps powered by the flow of electrons through the system. Cyanide stops the flow of electrons, which shuts down the pumps and stops the H^+ gradient, causing ATP synthesis to cease.
11. **(B) is correct.** ATP and NADH are obvious products of glycolysis, but most of the energy of glucose at the end of glycolysis is held in the pyruvate molecules.
12. **(D) is correct.** This question would make a good short essay question with your response centering on the three reasons cited in choice D.

All three reasons given are reflective of a pathway that evolved early in the evolution of life on earth.

13. (C) is correct. In feedback inhibition a metabolic pathway is switched off by the binding of its end product to an enzyme that functions early in the pathway, causing the pathway to be inhibited. Tryptophan is the end product of the pathway and serves as the molecule that binds to an enzyme early in the pathway, causing a change in the shape of the enzyme and inhibiting the pathway.

14. (C) is correct. Making the inner membrane leaky to H^+ ions means the mitochondria are much less efficient and must burn much more fuel to produce the ATP necessary for life. The burning of the extra fuel creates extra heat, which is used to maintain body temperature. Can you think of how surface-area-to-volume ratios are also part of this answer?

15. (A) is correct. B is not correct because glycolysis does not perform oxidative phosphorylation. C is not correct because the electron transport system does not occur in the cytoplasm. D is not correct because no CO_2 is produced in the mitochondrial membrane from NADH or ATP. Everything in A is correct, even if most of the ATP formed in the citric acid cycle will occur when the electrons from NADH and $FADH_2$ are fed into the electron transport chain. A is the best answer, even if you might feel it isn't the perfect answer. This is often the case in college-level examinations.

Grid-In Questions

1. **The answer is 90 molecules of ATP.** Each molecule of glucose that goes through fermentation will produce 2 ATP and 2 NADH. The NADH will not enter the electron transport system (there is no oxygen available) and will generate no ATP. Total for three glucose molecules is 6 ATP.

   Each molecule of glucose that goes through aerobic respiration will yield 2 ATP and 2 NADH from glycolysis; 2 NADH from pyruvate oxidation (2 molecules of pyruvate); the citric acid cycle will yield 2 ATPS, 6 NADH, and 2 $FADH_2$ (two pyruvates). Using the conversions to ATP given, each molecule of glucose will yield 32 ATP. Therefore, 3 molecules of glucose yields 96 total ATP. The difference between the two scenarios is 90 molecules.

   The point of this exercise, in part, is to drive home the huge difference the evolution of aerobic respiration made in making energy available to the processes of evolution. With this energy base, the evolution of the complex web of life was possible. A second point is that this calculation required quite a bit of time. When this occurs on your AP exam, wise students move on to the easier problems and return to ones like this once they have gathered all the easier points.
2. **The answer is − 3.9 kcal/mole. Note that you must grid in the negative sign to get credit for this one.** The coupled reaction requires + 3.4 kcal/mole but obtains − 7.3 kcal/mole from ATP which leaves −3.9 kcal/mole. Be sure you recognize that endergonic reactions have a positive delta G and that exergonic reactions have a negative delta G. This reaction is detailed in the 10th edition of *Campbell Biology*, Figure 8.10, or the *Focus* book, Figure 6.9.
3. **The answer is 18 μmol/mL/min.** This is determined by 270 μmol/mL − 180 μmol/mL/ 5 minutes = 18 μmol/mL/min.

Free-Response Questions

(a) Although plants have two photosystems, they both work in the same way. Both photosystems have two components: the light-harvesting complexes and the reaction center. The light-harvesting complex is made up of many chlorophyll and accessory pigment molecules. When one of the pigment molecules absorbs light energy in the form of photons, one of the molecule's electrons is raised to an orbital of higher potential energy. The pigment molecule is then said to be in an "excited" state. The increase in potential energy is transferred to the reaction center of the photosystem. The reaction center consists of two chlorophyll *a* molecules, which use the increased potential energy passed to them by the photosynthetic pigments to donate electrons to the primary electron acceptor. The solar-powered transfer of an electron from the reaction-center chlorophyll *a* pair to the primary electron acceptor is the first step of the light reactions. This is the conversion of light energy to chemical energy.

(b) Glycolysis is the first stage of cellular respiration and occurs in the cytoplasm. Glycolysis involves the breakdown of glucose to two pyruvate molecules. To accomplish this, 2 ATP molecules are invested, which helps to destabilize glucose, making it more reactive and allowing glucose to break into two three-carbon molecules. By the time the pathway has produced pyruvate, 4 ATP molecules have been produced along with 2 NADH molecules. This gives a net energy gain of 2 ATP and 2 NADH. Thus, one important role of glucose is to produce energy molecules for the cell to use in its life processes. The second role is to produce pyruvate, which can feed into the citric acid cycle in the mitochondria and ultimately into the electron transport chain, where most of the ATP in cellular respiration is produced.

(c) In cellular respiration water is a product of the reaction, whereas in photosynthesis water is a reactant. In cellular respiration water is formed when the electrons at the end of the electron transport chain in the inner membrane of the mitochondria combine with hydrogen ions and an atom of oxygen to form water. Oxygen is the ultimate electron acceptor in cellular respiration and when combined with protons, water is formed. In photosynthesis an enzyme splits a water molecule into two electrons, two hydrogen ions, and an oxygen atom. The electrons are supplied as needed directly to the chlorophyll molecules in the reaction center of photosystem II. In photosynthesis water is the ultimate electron donor. Additionally, the oxygen atom released immediately joins with another oxygen atom to form O_2. The oxygen on Earth comes almost entirely from the splitting of water in photosynthesis.

This student has written a particularly clear essay. Notice how carefully sequenced the responses are. Also note the clarity of each sentence and the absence of third-person pronouns. Avoiding pronouns may require a little more time and patience, but it pays off in added clarity and higher scores.

Topic 4: Mendelian Genetics

ANSWERS AND EXPLANATIONS

Level 2: Application/Analysis/Synthesis Questions

1. **(D) is correct.** The probability that the woman will have a seventh child who is a daughter is ½. Because the probability that a sperm carrying an X chromosome and the probability that a sperm carrying a Y chromosome will fertilize an egg is equal—both 50%—fertilization is considered an independent event. The outcome of independent events is unaffected by what events occurred before or will occur after. Therefore, the probability that this woman's next child will be a girl is ½. Likewise, the probability that she will have a child that is a boy is also ½.
2. **(A) is correct.** If the probability of allele *R* segregating into a gamete is ¼ and that of *S* segregating is ½, calculate the probability of two independent events occurring in a specific combination, order, or sequence by multiplying their probabilities; in this case, multiply ¼ by ½.
3. **(B) is correct.** Note that this cross involves incomplete dominance because there are three possible phenotypes in the offspring in this monohybrid cross. Let's say that the yellow coat parent is C^YC^Y and the homozygous brown coat parent is C^BC^B. Because the yellow coat parent can produce only gametes C^Y and the brown coat parent can produce only gametes C^B, the F_1 generation will all have genotype C^YC^B. Crossing two members of this generation would give you a ratio of 1 yellow coat: 2 gray coats: 1 brown coat. This means that 25% of the offspring would have brown coats, 25% would have yellow coats, and 50% would have gray coats.
4. **(A) is correct.** All of the statements about meiosis are true except A. During the first division, homologous pairs separate.
5. **(D) is correct.** Immediately rule out choices A and B because A would give you only offspring that exhibited the recessive traits—long hair and blue eyes—and B would give you all offspring that had the dominant traits—short hair and green eyes. Choice C yields a 9:3:3:1 ratio and also would give long hair, which is not seen in our results. Choice D fits the data because it can only produce short hair and yields both green and blue eyes in a ratio of 3:1.
6. **(A) is correct.** If the boy has hemophilia, then he must have inherited the recessive hemophilia gene from his mother. Sex-linked genes are usually located on the X chromosome. In order for the child to be a boy, he must have inherited a Y chromosome from his father. Because the gene causing hemophilia is located on the X chromosome, you can rule out choices B and C because males cannot pass on their X chromosomes to their sons. Therefore, you need to look for an answer choice that shows that the source of his X chromosome was a carrier of the allele—afflicted or not. This is choice A.
7. **(A) is correct.** Although genes that are on the same chromosome tend to be inherited together, the process of crossing over enables "linked" genes to sort independently. Those that are linked but located farther apart on the

chromosome will undergo crossing over more frequently than those located very close together on a chromosome simply because there are more sites between the two genes at which crossing over can take place.

8. (D) is correct. Autosomal dominant traits appear with equal frequency in both sexes, and they do not normally skip generations. These qualities are all exhibited by the trait that is illustrated in the pedigree. All three generations are affected with the trait, and both sexes are affected.

9. (A) is correct. The father either has type A, B, or O blood. The mother, who has the phenotype for type A blood, has the genotype $I^{A}i$. In order to produce a son with genotype *ii* (the genotype of people with type O blood), she would need to reproduce with a man who had genotype $I^{A}i$, genotype $I^{B}I$, or genotype *ii*. Try writing out the Punnett square if you aren't confident of this.

10. (B) is correct. If the individual with type O blood were to mate with an individual who has type AB blood, since I^{A} and I^{B} are both dominant over *i*, the genotypic ratio would be 1 $I^{A\,i}$:1 $I^{B\,i}$, and the phenotypes would be in a ratio of 1 type A blood:1 type B blood.

11. (B) is correct. The most likely reason for this 2:1 ratio in the offspring is that *Y* is lethal in homozygous form, and this caused the death of all of the *YY* individuals in the litter. The expected ratio of this cross would be 1 *YY*:2 *Yy*:1 *yy*. If you remove the *YY*, you get a ratio of 2 *Yy* (yellow mice, since the gene for yellow, *Y*, is dominant) to 1 *yy* (nonyellow mouse).

12. (D) is correct. The only process listed that does not lead directly to genetic recombination, or the recombining (scrambling) of genes in the offspring, is gene linkage. If genes are linked, they are located on the same chromosome and are more likely to segregate together into the same cell, reducing genetic recombination.

13. (C) is correct. Recall that a dihybrid cross between two heterozygotes produces a 9 double dominants : 3 dominant, recessives : 3 recessive dominants : 1 double recessive offspring ratio. The question is asking for one of the heterozygotes (which would be one of the 3s in the above ratio). To come up with the answer, multiply $\frac{3}{16}$ by 144 to get 27; the closest answer choice to 27 is 28.

14. (B) is correct. Fertilization restores the diploid number in a sexually reproducing organism. The two major events in the life cycle of sexually reproducing organisms are meiosis and fertilization.

15. (D) is correct. Homologous chromosomes are a key conceptual point in Mendelian genetics. Choices A, B, and C all occur during the first meiotic division, in which homologous pairs synapse, crossing over occurs, and then they separate. Also note that the cells at the end of the first meiotic division are haploid! If you missed this question, review more thoroughly the concept of homologous chromosomes.

16. (B) is correct. The number of possible gametes can be determined by using the formula $2n$, where n equals the haploid number. With a diploid number of 6, thus a haploid number of 3, the answer is $2 \times 2 \times 2 = 8$.

17. (A) is correct. Recall that in G_1 the chromosomes have not replicated. After S phase of interphase the chromosomes have replicated, which doubles the amount of DNA. When the homologs separate in meiosis I the amount of DNA is reduced by one-half and back to the original amount.

18. (A) is correct. In meiosis II the most important event is the separation of sister chromatids. Mitosis involves only the separation of sister chromatids; homologs do not form in mitosis.

19. (D) is correct. Begin by drawing a straight line to represent a chromosome, and then place genes A and B 8 units apart. Note that A and C are 28 units apart with B and C 20 units apart. This means the genes are found in order A, B, C. Next, consider the map distances between A and D and B and D. Because map distance represents the frequency of crossing over, it will only work if D precedes A on the chromosome. The correct gene sequence must be D-A-B-C.

20. (B) is correct. Because males have only one X chromosome they have only a single allele for genes found on this sex chromosome. A single defective allele can therefore be the source of the X-linked condition.

21. (D) is correct. Given that the diploid ($2n$) number is 4 and some of the cells in the diagram have an (n) or haploid number with two chromosomes, this could be meiosis. However, the improper separating of a chromosome in meiosis, termed nondisjunction, would yield cells with differing chromosome number. Specifically, the diagram shows sister chromatids not separating, thus leaving one cell $n + 1$ and the other cell $n - 1$. This is not normal meiosis but an example of a nondisjunction event.

22. (B) is correct. Type O blood is recessive. This means that the woman could be heterozygous for blood type A ($I^{A}i$) and the father heterozygous for type B blood ($I^{B}i$). If each parent donates his or her recessive allele, the baby would be *ii*, type O blood.

23. (C) is correct. It appears that the mother has inherited the color-blind gene from her father and passed it to her son. This is a typical inheritance pattern seen in sex-linked genes.

24. (C) is correct. The question and correct answer describe independent assortment as it occurs in meiosis. Independent assortment increases variability, or as phrased in the question, increases the possible combinations of characteristics.

Additional Sample Problems

1. Albino (b) is a recessive trait; black (B) is dominant. First cross: parents $BB \times bb$; gametes B and b; offspring all Bb (black coat). Second cross: parents $Bb \times bb$; gametes $1/2\ B$ and $1/2\ b$ (heterozygous parent) and b; offspring ½ Bb and ½ bb.

2. Recognize that this is a heterozygous, dihybrid cross (GgIi × GgIi) and all heterozygous dihybrid crosses yield the same results: 9 double dominants: 3 dominant, recessives: 3 recessive dominants: 1 double recessive. For this specific cross, 9 green, inflated: 3 green, constricted: 3 yellow, inflated: 1 yellow, constricted. You could do this problem with a Punnett square, but it would take a lot of time. Know the common monohybrid and dihybrid crosses and the expected ratios.

3. Parental cross is $AAC^{R}C^{R} \times aaC^{W}C^{W}$. F_1 genotype is $AaC^{R}C^{W}$, phenotype is all axial-pink. F_2 phenotypes are 3 axial-red : 6 axial-pink : 3 axial-white : 1 terminal-red : 2 terminal-pink : 1 terminal-white.

Grid-In Questions

1. **The answer is 12 (fg).** As with mitotic cell division, the DNA content in meiosis is replicated during the synthesis portion of interphase, bringing the DNA total to 48 fg, which is equally distributed to the four haploid cells.
2. **The answer is 3⁄8.** This question is most quickly answered by calculating probabilities. To do this, consider the two traits of the dihybrid cross individually. For example, both plants are heterozygous for tall, so ¾ of the offspring will be tall to ¼ dwarf. (Tt × Tt = 1TT:2Tt:1tt, or three tall to one dwarf) The second trait crosses Rr × rr, resulting in a ratio of ½ round to ½ wrinkled. Multiplying the two independent events together, ¾ Tall × ½ wrinkled = 3⁄8 Tall **and** wrinkled. The problem could be solved with a Punnett Square, but that would be quite time consuming on a test where every minute is important.
3. **The answer is 17 (%).** All the information to solve the problem is given, but you must recognize that the phenotypes Grey body/vestigial wings and Black body/normal wings are recombinants (phenotypes that differ from the true breeding parents), whereas Grey body/normal wings and Black body/vestigial wings are parental types. The data indicate 195 total recombinants divided by 1149 total flies times 100 equals 17%. The genes are 17 map units apart.

   Answer from page 100: The karyotype in Figure 4.1 is that of a male. Note the unpaired X and Y in the bottom right corner.

Topic 5: Molecular Genetics

ANSWERS AND EXPLANATIONS

Level 1: Knowledge/Comprehension Questions

1. **(B) is correct.** Viruses have *either* a DNA genome or an RNA genome surrounded by a protein coat. The answer is not A because retroviruses have an RNA genome. HIV is a retrovirus.
2. **(D) is correct.** Restriction enzymes can be used to cut DNA at specific locations and this enables researchers to perform recombinant DNA techniques. When specific restriction enzymes are added to the DNA, they produce cuts in the sugar-phosphate backbone and create "sticky ends," which can bind to DNA fragments from a different source to produce recombinant DNA. DNA ligase is then added to seal the strands together permanently.
3. **(C) is correct.** Transposons are also called transposable genetic elements, and they are pieces of DNA that can move from location to location in a chromosome—or a genome. Transposons are also called "jumping genes," and most of them are capable of moving to many different target sites in the genome. They are important in generating genetic diversity.
4. **(D) is correct.** One of the two important ways that the cell has of controlling gene expression is through DNA methylation. In DNA methylation, methyl groups are attached to certain DNA bases after DNA is synthesized. This appears to be responsible for the long-term inactivation of genes.

5. **(B) is correct.** *Be sure you don't confuse the vocabulary! Many students reverse the terms* transcription *and* translation. The process by which genetic information flows from mRNA to protein is called translation. Translation occurs in the cytoplasm of the cell, at the ribosomes. A molecule of mRNA is moved through the ribosome, and codons are translated into amino acids one at a time. Transfer RNAs add their associated amino acids onto a growing polypeptide as its anticodon pairs with a codon on the mRNA and then departs from the ribosome to bind more free amino acids.

6. **(A) is correct.** In transcription, RNA is synthesized using the genetic information encoded by DNA. Transcription occurs in the nucleus of the cell. The double-stranded DNA helix unwinds, allowing enzymes and proteins to synthesize a new complementary single-stranded RNA molecule from the template strand of DNA.

7. **(B) is correct.** In histone acetylation, acetyl groups are attached to certain amino acids of histones. Acetylation makes the histones change shape so they are less tightly bound to DNA, and this allows the proteins involved in transcription to move in and begin the process. Therefore, acetylation is one way for the cell to initiate transcription and to control the expression of its genes.

8. **(B) is correct.** In the lysogenic cycle, the phage genome becomes incorporated into the host cell's DNA without destroying the host cell. There, it lurks, much like a Trojan horse, waiting to receive the chemical signals that would trigger it to enter the lytic cycle. In the lytic cycle, the phage hydrolyzes the host cell's DNA and uses the cell's machinery to produce phage proteins and to replicate its genome. The phage proteins are then assembled in the cell until the host cell lyses (breaks open), and the new phages are released to infect other cells.

9. **(C) is correct.** In genetic engineering (the manipulation of genes for practical purposes), DNA ligase is an enzyme that is used to seal the strands of newly recombinant DNA (DNA that is spliced together from two different sources) by catalyzing the formation of phosphodiester bonds.

10. **(C) is correct.** The addition of a poly-A tail after transcription is one example of post-transcriptional modifications that the mRNA undergoes. This poly-A tail inhibits the degradation of the newly synthesized mRNA strand and is thought also to help ribosomes attach to it. Another important modification that mRNA undergoes is the addition of a 5′ cap. The 5′ cap helps protect mRNA from degradation and also acts as the point of attachment for the ribosomes, just prior to translation.

11. **(D) is correct.** RNA polymerase is the most prominent enzyme involved in the transcription of DNA to make mRNA. It is responsible for binding to the promoter sequence on the template DNA, prying the two DNA strands apart, and hooking the RNA nucleotides together as they base-pair along the DNA template. RNA polymerases add nucleotides to the 3′ end of the growing chain until a terminator sequence is reached—it transcribes entire transcription units.

12. **(A) is correct.** tRNA, or transfer RNA, interprets the genetic message coded in mRNA. It transfers amino acids taken from the cytoplasmic pool to a ribosome, which adds the specific amino acid brought to it by tRNA to the end

of a growing polypeptide chain. Each type of tRNA binds to a specific amino acid at one end; its other end contains an anticodon, which base-pairs with a complementary codon on the mRNA strand.

13. (B) is correct. mRNA, also known as messenger RNA, is a type of RNA that is synthesized from DNA and attaches to ribosomes in the cytoplasm to specify the primary structure of a protein. Because mRNA is the product of transcription, which occurs in the nucleus of eukaryotic cells, it must travel out of the nucleus and into the cytoplasm in order to participate in translation.

14. (B) is correct. In the modification of mRNA that occurs after transcription, a process called RNA splicing occurs. In this process, noncoding regions of nucleic acid that are situated between coding regions are cut out. These noncoding regions are called introns. The remaining regions are called exons, and these are spliced together to form the final mRNA product. When you think of exons, think expressed—because they are actually translated into proteins, whereas introns are not.

Level 2: Application/Analysis/Synthesis Questions

1. (D) is correct. The process of genetic engineering is, in part, possible because of the universal nature of DNA and RNA as carriers of genetic information. Once the spider gene or genes that were responsible for coding for the silk proteins were isolated and then inserted into a bacterial plasmid (which would serve as the vector), the cloning vector would be taken up by the goat's cells, and the goat's cells' transcription/translation machinery would begin the process of producing the spider protein, following the universal codon chart.

2. (B) is correct. DNA ligase is the enzyme that joins DNA fragments; it is used by cells to join Okazaki fragments as well as in biotechnology to generate engineered plasmids. You are expected to know the following four enzymes involved in replication: DNA polymerase, ligase, helicase, and topoisomerase.

3. (D) is correct. PCR, the polymerase chain reaction, is a technique by which any piece of DNA can be copied many times without the use of cells. With each turn of the cycle, the amount of DNA is multiplied by 2. After only 30 cycles, which can occur in just a few hours, over 1 billion molecules match the target sequence.

4. (C) is correct. The fragments of DNA separated out from one another along the gel once the electric field was applied because they differ in size. For DNA in gel electrophoresis, how far a molecule travels through a gel (while the current is applied) is inversely proportional to its size. The larger a fragment is, the more slowly it will migrate.

5. (D) is correct. The restriction enzyme used to cut the DNA that was placed into the first well of the gel must have cut the DNA at eight sites, because it produced nine DNA fragments. In linear DNA the number of fragments produced is always one more than the number of restriction sites cut. (How would this answer be different if the DNA was a circular plasmid?)

6. (C) is correct. You should recall that a T4 phage injects its DNA, which is used to assemble new protein coats and viral DNA. The protein coat of the infecting phages is left outside the bacterium and is not involved in forming new viruses.

7. (B) is correct. Reverse transcriptase is an enzyme that is required to produce DNA from the RNA of a retrovirus.

8. (C) is correct. The introduction of a "stop" codon will result in an abbreviated polypeptide chain and is therefore likely to do more harm to the final protein than choice B, a frameshift mutation occurring near the terminal end of the gene.

9. (B) is correct. The amount of adenine in DNA will equal the amount of thymine; guanine will equal cytosine, which leads to the mathematical equality $A + G = C + T$. Try this: If a certain species has 15% adenine in its DNA, what would be the percentage of guanine? (It would be 35%. $A + T = 30\%$, so $G + C = 70\%$, and $G = 35\%$.)

10. (B) is correct. In gel electrophoresis the smallest fragments travel farthest; the largest fragments are closest to the well. Fragment b is shortest, followed by a and c. We would expect a gel with c closest to the well, then a, with b at the far end.

11. (A) is correct. Choice A is correct because the signals that control gene expressions and promoter regions are different between prokaryotes and eukaryotes. Several of the other possible answers have errors that would be instructive to note. Choice B is false because the genetic code is nearly universal; choice C is false because the genetic code includes start and stop codons and is universal; choice D is false because ribosomes are not governed by the length of the gene.

12. (A) is correct. Genes that are active are producing mRNAs. The mRNAs of the active genes are used to make cDNA, which will bind to the single-stranded DNA on the microarray. The DNA on the chip serves as a probe and will bind to a complementary sequence of DNA, so cDNA must be made from the mRNA of the tissue. In the light, the chlorophyll genes are on, producing mRNAs, which are used to produce cDNA and washed across the chip.

13. (B) is correct. The number of genes in the human genome has been revised sharply downward since the Human Genome Project. Less than a decade ago the number of human genes was at 100,000; now it is about 20,000! How can only 20,000 genes be enough for a human? The answer is alternative gene splicing yielding multiple proteins per gene.

14. (D) is correct. This is an example of a multigene family. The alpha-globin gene family is on chromosome 16, whereas the beta-globin gene family is on chromosome 11. The genes arose by duplication and translocation.

15. (D) is correct. In HIV (the virus that causes AIDS) the envelope glycoprotein enables the virus to bind to specific receptors on certain white blood cells. After binding, the virus fuses with the cell's plasma membrane, entering the cell. The mutated gene *CCR5* prevents binding and subsequent entering of the cell by HIV.

16. (D) is correct. In HIV infections, recall that the infectious agent is a retrovirus with RNA as the genetic material. In the cytoplasm of the host cell the viral RNA is converted to DNA, then incorporated as a provirus into the host cell's DNA. The provirus can direct the production of new viral particles.

17. (C) is correct. Bacteria do not perform mitosis, meiosis, or sexual reproduction. However, the processes of transformation, transduction, and conjugation are processes that bacteria use to increase their genetic diversity. Be sure to review all three processes.

18. (B) is correct. These profiles were done using short tandem repeats as genetic markers. All 13 sites must match because this shows the highest degree of certainty between the pilot's DNA and the DNA from the three families. In addition, because the DNA for comparison comes from a probable twin of the pilot, it is expected that the bands would be identical for the twins of the same family.

19. (D) is correct. Family 3 matches in all 13 sites. This is a good time to review the principles of gel electrophoresis, making sure you understand which poles on the gel are positive and negative, how the bands are arranged by size on the gel, and the role of restriction enzymes in preparing the DNA.

20. (A) is correct. Introns, not exons, are removed before mRNA leaves the nucleus. The rest of the choices in this question are true and worthy of your perusal and understanding. RNA editing is a fundamental concept in molecular genetics.

21. (C) is correct. The polymerase enzymes, both DNA and RNA, can only add nucleotides to the 3′ end of a growing strand. This is the underlying reason for the leading strand, where DNA polymerase is adding a continuous, unbroken new strand of nucleotides and the lagging strand where DNA polymerase is moving away from the unwinding replication fork, forming new nucleotides in Okazaki fragments. Knowing all the enzymes of replication is not required for the exam, but understanding how DNA polymerase works and the formation of leading and lagging strands is fundamental and worth your time investment.

22. (C) is correct. Species 4 has the greatest number of amino acid differences (two) when compared to the other species. Note the assumptions made in the question. Mutation rates for all species for all genes are not equal over time and answering questions about relatedness between species would require much more data. The assumptions made the question one that could be answered in the limited time of an exam.

Grid-In Questions

1. **The answer is 6.** On questions like this check to see if the DNA being cut is circular or linear. Draw a linear strip of DNA, then mark the cut sites, and count the fragments.
2. **The answer is 26%.** If guanine is 24% then thymine is also 24%, totaling 48% of the DNA. That leaves 52% of the DNA to be split evenly between guanine and cytosine. These calculations are based on Chargaff's rules and are the same rules that Franklin, Watson, and Crick used in solving the structure of DNA.

Free-Response Questions

(a) Nondisjunction occurs when the chromosomes fail to separate properly in meiosis and one cell ends up with two copies of a chromosome, whereas the other gets no copies. This results in a condition called aneuploidy. If a gamete received an extra copy of chromosome #21 and was successfully fertilized, the child would exhibit Trisomy 21, which is characterized by mental retardation and reduced stature.

A base-pair deletion can occur when a DNA copying error results in the loss of a single nucleotide, and a frameshift means that the triplets being read will all be different. If this occurs within an intron there will be no phenotypic effect because introns are removed prior to translation.

A base-pair substitution results when one nucleotide, such as thymine, is substituted for another, such as guanine. This type of error occurs to cause sickle-cell disease. When a different amino acid is placed into the polypeptide, it can affect the secondary or tertiary structure of the protein so that it does not fold into the correct conformation. Change the shape, change the function. If it is an enzyme or other protein, it may no longer function as it did.

(b) Colorblindness is more common in males than females because males are hemizygous for these traits. The trait is found on the X chromosome, and because males have only one copy of the X chromosome, they will show the phenotype for whatever allele they receive. Females, on the other hand, have two copies of the X chromosome and, if one X chromosome has an allele for colorblindness and the other X chromosome carries an allele for normal vision, the female will have normal vision. She is a "carrier" for colorblindness because she can pass it on to her sons.

(c) When fertilization occurs and a sperm carrying a Y chromosome penetrates the egg first, a male zygote with one X and one Y chromosome is produced. If a sperm carrying an X chromosome penetrates the egg first, a female zygote with XX is produced. Although it seems as though the female zygote would have twice the cell product as the male, due to its double dose of genes located on the two X chromosomes, this is not the case. The reason for this is that, in every cell of the female human body, one of the X chromosomes is inactivated. The mechanisms for this are not fully understood, but the X chromosome that is inactivated condenses into a structure called a Barr body, which then associates with the nuclear envelope. This is called X-inactivation, and the condensed DNA is heterochromatin. As a Barr body, most of the X chromosome's genes are not expressed. As a result of this, females are a mosaic consisting of cells with the X chromosome from their mother activated and other cells with the X chromosome from their father activated in about a 50:50 ratio. This is also the reason sex-linked disorders are usually not expressed in females. Although one of the X chromosomes may be incapable of producing a crucial gene product, this mosaic effect ensures that the other half of the somatic cells produces sufficient amounts of the protein in question.

Note that the response for part (c) covers more than the question asks. The graded response does not actually begin until the fourth sentence. Also, the last two sentences go beyond the question. Because your time is limited, always focus on JUST the question.

Topic 6: Mechanisms of Evolution

ANSWERS AND EXPLANATIONS

Level 1: Knowledge/Comprehension Questions

1. **(B) is correct.** When two populations of the same species are prevented from breeding by a geographic barrier such as a mountain range or other physical obstruction, the two populations are said to be geographically isolated.
2. **(C) is correct.** The concept of punctuated equilibrium was put forth recently by Niles Eldredge and Stephen Jay Gould. It explains periods of apparently little change (stasis) "punctuated" by sudden changes observed in the fossil record. Punctuated equilibrium is about the pace of evolution, not about differences in how evolution occurs.
3. **(B) is correct.** Because selection favors the average-sized wings and reduces the frequency of two extremes, this is an example of stabilizing selection. This would be a good time to also review directional and disruptive selection, shown in Figure 6.1 of this guide.
4. **(D) is correct.** The founder effect occurs when a few individuals from a population colonize a new, isolated habitat. The smaller the number of individuals who start this new population, the more limited will be the starting gene pool for the population—and the less the new gene pool will resemble that of the parent population.
5. **(B) is correct.** Homology is the result of descent from a common ancestor. It can be described as the underlying structural or molecular similarities (even in structures that are no longer used for the same function) that exist in organisms as a result of common ancestry. An example would be the wing of a bat, which is homologous to the flipper of a whale, because both have the same bones.
6. **(A) is correct.** Bottleneck effects are often the result of a natural disaster such as a flood, drought, fire, or anything that destroys most members of a population. The gene pool of the surviving members of the population may not resemble the gene pool of that of the parent population—some genes will be overrepresented, and some will be underrepresented. Bottlenecking reduces the genetic variability in a population because of the loss of alleles.
7. **(C) is correct.** An example of analogous structures would be a bat wing and a butterfly wing—both useful for flight but not anatomically similar.
8. **(A) is correct.** Prezygotic barriers are those that prevent or hinder the mating of two species, or they prevent fertilization even if two species mate. All the answers are examples of prezygotic barriers except hybrid breakdown. It is an example of a postzygotic barrier (postzygotic barriers are those that prevent hybrid zygotes from developing into viable adults) in which the second generation of offspring from two species is either weak or sterile.
9. **(C) is correct.** In convergent evolution, species from different evolutionary branches appear alike as a result of undergoing evolution in very similar ecological roles and environments. Similarity between species that have undergone convergent evolution is known as analogy, and structures they share are

analogous (not homologous) structures. (Remember, similar problem, similar solution. However, convergence does *not* indicate a common ancestor.) Homologous structures are those that are shared in two species as a result of those species with a common ancestor.

10. (C) is correct. It is important to recall that the early atmosphere did not have oxygen and that the evolution of linear electron flow in photosynthesis by cyanobacteria resulted in the accumulation of oxygen in the atmosphere.

11. (C) is correct. The smallest unit capable of evolution is the population. Individuals cannot undergo evolution because they exist for only one generation, and evolution is the changing and refinement of a group's gene pool to best fit the group's environment.

Level 2: Application/Analysis/Synthesis Questions

1. (A) is correct. Darwin's observations can be summarized as follows: Organisms produce more offspring than the environment can support. Further, the offspring vary in their heritable characteristics. Those individuals well suited to their environment leave more offspring than other individuals (differential reproductive success). Over time, favorable traits accumulate in the population.

2. (D) is correct. If the frequency of the recessive allele is 0.09, then we know that the frequency of the other allele is 1 − 0.9, which is equal to 0.91. If your answer was B, you assigned 0.09 as *q2* rather than *q*. Remember, the allelic frequency is *q*, the genotypic frequency is *q2*. If the genotypic frequency was 0.09, B would have been the correct answer. In every H-W problem note if the information given is about allelic frequency (*q*) or genotypic frequency (*q2*).

3. (C) is correct. Did you select A? You were too hasty and forgot a very important concept: The 64% of the population that show the dominant trait includes heterozygotes, so you cannot simply take the square root of 0.64. You must subtract 0.64 from 1.00 to obtain $0.36 = q^2$. Taking the square root of q^2 yields $q = 0.6$, so $p = 0.4$.

4. (C) is correct. This scenario mirrors the human disorder of sickle-cell disease and is an example of heterozygote advantage. In sickle-cell disease two dominant alleles for normal hemoglobin leave the individual susceptible to malaria, but two recessive alleles cause sickle-cell disease. The heterozygote does not have sickle-cell disease and tends to have milder cases of malaria. Hence, the heterozygote is the advantageous genotype. The mouse population appears to be undergoing a similar situation.

5. (B) is correct. This question describes the two major categories of speciation, and you must analyze the data to select the most plausible mechanism. Although choice A could account for 30 different species, it would not explain why the species seem to occur in related pairs. Only choice B does this.

6. (D) is correct. One mechanism that can lead to sympatric speciation is the formation of autopolyploids through nondisjunction in meiosis. These plants have $4n$ chromosomes, instead of the normal $2n$ number, and they are unable to breed with members of the parent population—although they are still able to breed with other tetraploids.

7. (A) is correct. For this question recall that the endosymbiotic hypothesis proposes that mitochondria and plastids were engulfed by other cells and have their own genomes. It is most likely that over time some of their genes were incorporated into the host cell's genome. (*Hint:* Review the endosymbiotic hypothesis.)

8. **(C) is correct.** The key to this selection is the phrase "natural selection can only edit existing variations." The other answers deal mostly with misconceptions. Evolution does not produce perfect organisms and it does not operate by a preconceived plan. Read and note the misconceptions in the wrong answers in this question.
9. **(C) is correct.** Mammals share branch point 3 with birds and branch point 2 with amphibians. The most recent common ancestor branch point is 3; thus, mammals and birds are more closely related than mammals and amphibians.
10. **(D) is correct.** The lizard and ostrich share the most recent branch point, number 4.
11. **(D) is correct.** Of the groups listed, hawks and other birds share the most recent common ancestor with snakes and lizards, branch point 4. Do not be confused by which group of animal is closest on the figure. Mammals are shown next to snakes and lizards, but they share branch point 3, a more distant branch point than 4.
12. **(B) is correct.** Directional selection is a form of natural selection where individuals at one end of the phenotypic range (darker fur in this question) survive or reproduce more successfully than do other individuals. Use this question to review directional, stabilizing, and disruptive selection.
13. **(C) is correct.** Polyploidy is a common form of sympatric speciation in plants. Many plant species originated from accidents of cell division that resulted in extra *sets* of chromosomes being inherited. Note that sympatric speciation takes place without geographic isolation. Of the choices offered only *T. turgidum* is a polyploid. You may have noticed that *T. aestivum* is also a polyploid but was not offered as an answer.
14. **(B) is correct.** Allopatric speciation is the formation of new species in populations that are geographically isolated. The stem of the question gives us information indicating that *H. erectus* became isolated on an island, out of genetic contact with the rest of its species.
15. **(D) is correct.** Reproductive barriers impede members of two closely related species from interbreeding and producing viable, fertile offspring. Barriers are divided into prezygotic and postzygotic barriers and have been used as the basis for essay questions in the past.

Grid-In Questions

1. **The answer is 2.6 (billion years).** The rock layer shows one-fourth the current ratio, which indicates that two half-lives have passed (½ × ½). 1.3 billion × 2 = 2.6 billion years ago.
2. **The answer is 45.2%.** The 12% that show no banding represent q^2, because no banding is the recessive genotype. This means q equals the square root of 0.12 or 0.346; therefore, p equals 0.654. The heterozygote is represented as $2pq$, so 2(0.654) (0.346) = .452. Change to a percent by multiplying by 100 to yield 45.2%. The exam allows for different rounding of figures, so a range of answers around this one will be acceptable.

3. **The answer is 7.8.** Note that the dominant allele (p) is given, meaning the frequency of the recessive allele (q) can be found by $1.0 - 0.72 = 0.28$. The genotype for the recessive phenotype is q^2 or $(0.28)^2$, which equals 0.078. This number times 100 yields the percent that will have red shells.

Free-Response Questions

(a) Miller demonstrated the formation of organic molecules by using an apparatus that mimicked the conditions thought to have existed on the early Earth. Using electricity sparks to mimic the energy from lightning, and an atmosphere containing a mixture of hydrogen gas, methane, ammonia, and water vapor, Miller generated a variety of organic molecules that are common in organisms. This was a demonstration that abiotic synthesis of organic molecules was possible.

(b) RNA has two important features that make it likely to have been the earliest genetic material. RNA can function as an enzyme, known as a ribozyme. Some ribozymes can make complementary copies of short pieces of RNA so RNA can be self-replicating. This combination of catalytic and self-replicating ability makes it likely it was the earliest genetic material.

(c) Because the earliest prokaryotes evolved in an atmosphere without oxygen, it is likely that most species were anaerobic and the development of oxygen in the atmosphere may have resulted in a mass extinction of species that were not able to tolerate some level of oxygen. This would have opened many niches and made it a time of adaptive radiation for organisms capable of dealing with oxygen in the atmosphere.

Note that this question has three parts, and these are clearly labeled on the response. The student chose to describe the Miller experiment but might also have discussed reactions on solid reactive surfaces. The formation of vesicles would also have been appropriate in this discussion. Essay questions on evolution are common occurrences on the AP Biology exam. When answering essay questions on evolution, avoid clichés like "survival of the fittest" because the phrase does not convey any specific information about evolution. Be descriptive in your discussions of evolution and remember to define key terms as they are used.

Topic 7: The Evolutionary History of Life

ANSWERS AND EXPLANATIONS

Level 1: Knowledge/Comprehension Questions

1. **(B) is correct.** Animals are heterotrophic—they are not capable of fixing carbon and must obtain it from other organisms. They belong to the Eukarya and are multicellular. The fungi are also eukaryotic, multicellular heterotrophs but have a cell wall made of chitin.

2. (B) is correct. The three domains into which all the living organisms are placed by systematists are Bacteria, Archaea, and Eukarya. Domains are one taxonomic level above kingdoms. Prokaryotes make up Archaea and Bacteria, whereas eukaryotes make up Eukarya.

3. (C) is correct. In mutualistic symbiosis, both organisms benefit from the association; in commensalistic symbiosis, one organism benefits while the other is neither helped nor harmed; and in parasitic symbiosis, one organism benefits while the other is harmed.

4. (B) is correct. Mosses are bryophytes and so lack vascular tissue, ferns have vascular tissue and reproduce with spores, and gymnosperms have pollen and seeds. Angiosperms, and the evolution of flowering plants, were the last to appear.

5. (A) is correct. The amniote egg is composed of extraembryonic membranes that function in gas exchange, waste storage, and the delivery of nutrients to the embryo. The amniote egg was a key evolutionary adaptation for terrestrial life, allowing the embryo to have its own "pond" in which to develop.

6. (D) is correct. Each node or branch point represents a common ancestor of the two branches that come from the node. Lizards share a more recent common ancestor with humans than salamanders and are therefore more closely related to humans. The "not supported" part of the question should be a red flag to be very careful with the question. Questions most often ask for the supported, so this one can be confusing. Read carefully!

7. (D) is correct. Mitosis and meiosis do **not** occur in bacteria. In plants and animals, however, meiosis *will* introduce variation through independent assortment and crossing over. You should be able to explain the three ways in which bacteria receive new genes (transduction, transformation, and conjugation) as well as know that mutation is the ultimate source of genetic variation.

8. (A) is correct. This is the definition of a shared derived character. An example that may help you to remember is that hair is a shared derived characteristic for all mammals. All mammals have hair, and no organism that is not a mammal has hair.

Level 2: Application/Analysis/Synthesis Questions

1. (A) is correct. This question hinges on knowing the difference between shared ancestral characters and shared derived characteristics. Shared derived characters are evolutionary novelties unique to a specific clade, whereas shared ancestral characters originate in the ancestor of the taxon. Having four limbs is not unique to birds and mammals but is an ancestral trait shared by other clades.

2. (C) is correct. A, B, and D are all cats. Outgroups are used as points of comparison to help determine shared ancestral and shared derived characters.

3. (C) is correct. Transformation is the uptake of external DNA by a cell. Transformation is one of the ways in which bacteria are able to increase genetic diversity. When bacteria take up DNA from a different species, transformation results in horizontal gene transfer.

4. (C) is correct. Scientists can use mtDNA sequences for faster evolving gene sequences and rRNA sequences for distantly related groups.

5. (C) is correct. *Archaeopteryx* had teeth, flat sternum, and claws—all reptile characteristics that the fossil displayed.

6. (C) is correct. This question is concerned with the ecological roles of fungi. Fungi play an important role as decomposers as well as being involved in many symbiotic relationships with plants, such as mutualistic mycorrhizal associations.

7. (A) is correct. Moving onto land was an evolutionary challenge for both plants and animals. The scales of reptiles are waterproof and protect the organism from desiccation while the amniotic egg protects the embryo from drying while on land. In plants the cuticle protects the plant body from desiccation while the seed protects the embryo from drying while on land. Note the parallels.

8. (D) is correct. Remember that a clade includes all the descendants. All of the other choices included are only part of the descendants.

9. (C) is correct. In many phylogenetic trees the x axis is given in relative time rather than the specific millions of years ago as in this figure. In this case specific data can be derived concerning time from the x axis.

10. (C) is correct. In both cases homologous genes are exchanged. Many evolutionary biologists think that the enzymes used in bacteria are the evolutionary forerunners of the recombination enzymes used in eukaryotic crossing over.

Grid-In Questions

1. The answer is 15 million years ago. The acceptable answer will have a range around this number, perhaps accepting any number between 14 and 16 millions of years ago.

Free-Response Question

Three methods or types of evidence that scientists use to classify organisms and study their degree of evolutionary relatedness are fossil evidence, homologies, and molecular evidence.

1. Fossil evidence can be used to determine the relative ages of different groups because deeper strata contain older species. For example, there are no terrestrial species in the most ancient fossil layers, an indication that emergence of life onto land was a later evolutionary development. Fossils can also provide evidence for the appearance of key features, such as feathers or emergence of the tetrapod body plan. Feathers coincided with the emergence of birds.

2. Homologous structures can be used to determine evolutionary relationships. Species that were derived from the same ancestor should have similarities, called homologies. Here, scientists would look for homologous structures—structures that are similar in different species—and use these to tie organisms together. For example, the bones of a bat wing are the same as those of a lion's front paw—radius, ulna, humerus, and so on. This provides evidence of relatedness.

3. Biochemical similarities can indicate relatedness. Today, DNA analysis is widely used. Because DNA is heritable, related species should share common genes, and the more recently the species branched off from a common ancestor, the more similar their DNA should be. Studying the DNA of organisms makes

it possible for systematists to determine the degree of evolutionary difference between two species that are nearly identical in appearance, and it also allows systematists to judge the relatedness of two species that they might not guess would be related at all, based on external appearance. It is a much more precise and quantitative method for appraising evolutionary relatedness. For example, in 2013 DNA evidence confirmed that there was a new carnivore species, the Olinguito. This is a raccoon relative and thought to be part of the species of Olingos. Guess what—they could never get the two to mate!

This response clearly describes three methods for studying the evolutionary relatedness of species and provided an example of each method's use. However, the introductory paragraph is nice fluff—save your time! The three methods are also numbered, making it hard for the Reader to miss the information. Always check your final response to make sure it addresses the original question.

Topic 8: Plant Form and Function

ANSWERS AND EXPLANATIONS

Level 1: Knowledge/Comprehension Questions

1. (D) is correct. When the light source on a plant is uneven, the plant will grow toward the light source. In some plants this is the result of auxin moving from the apex down to the cells that are less exposed to light and causing them to elongate faster than the cells on the side that is illuminated.

2. (C) is correct. The driving force behind the movement of sap in xylem (in the direction from the roots to the leaves) is the transpiration of water through the stomata on the leaves. The mechanism responsible for movement up through the xylem is the transpiration-cohesion-tension mechanism, and it occurs through bulk flow, in which fluid moves because of a pressure difference at opposite ends of a tube. The pressure is created by transpiration from the leaves, and contributing to the movement of water and minerals up the plant are gradients of water potential from cell to cell within the plant.

3. (D) is correct. In the transpiration-cohesion-tension mechanism, water is lost through transpiration from the leaves of the plant due to the lower water potential of the air. The cohesion of water due to hydrogen bonding plus the adhesion of water to the plant cell walls by hydrogen bonding enables the water to form a water column. Water is drawn up through the xylem as water evaporates from the leaves, each evaporating water molecule pulling on the one beneath it through the attraction of hydrogen bonds.

4. (A) is correct. The production of ATP in plant cells occurs in the mitochondria, as it does in the cells of animals, but it also occurs in the chloroplasts during photosynthesis. The light reactions of photosynthesis convert solar energy to the chemical energy of ATP and NADPH, and these light reactions take place in the chloroplasts in the mesophyll cells of the plant leaf.

5. (D) is correct. All of the factors listed aid in the uptake of water and minerals by the roots of a plant except the last choice, gravity. The large surface area of leaves plays an important role in transpiration.

6. (A) is correct. This question requires you to know that the sperm of angiosperms do not swim! This is the adaptive value of pollen: It allows fertilization without water. Sperm nuclei enter a pollen tube that grows down the style, discharging the two sperm nuclei for double fertilization.

7. The ovule contains the egg and, when fertilized, will become the seed. The ovule is found within the ovary. The ovary may contain one ovule and, therefore, one seed, or many ovules, like a watermelon. A watermelon is a single swollen ovary!

Level 2: Application/Analysis/Synthesis Questions

1. (B) is correct. Short-day plants require a period of continuous darkness longer than a critical period in order to flower. These plants flower in early spring or fall. Short-day plants are actually long-night plants; that is, what the plant measures is the length of the night.

2. (B) is correct. Long-day plants require a night period that is shorter than a critical period. The critical length for flowering is 9 hours, but the question asks which choice prevents flowering. Only one choice has a 24-hour period with more than 9 consecutive hours of darkness, thus preventing flowering.

3. (A) is correct. Understanding signal transduction pathways is an important concept for this unit and in biology. Keep its role in allowing plants (and other organisms) to adapt to their environment in your mind as you study. A hormone or environmental stimulus activates the receptor protein in a signal transduction pathway, not a relay molecule.

4. (C) is correct. This experiment by Went centered on phototropism—the growth of a shoot in response to light. Light causes auxins to move to the dark side of the stem, where the increase in auxins results in cell elongation. Cell elongation causes the dark side of the stem to expand more rapidly, causing the plant to grow toward the light. Consider this question in the context of the environment causing changes in the behavior of an organism.

5. (B) is correct. Because Went applied only the plant tip to the agar, he must have used knowledge from prior experiments (by Darwin) that demonstrated that only the tip influenced movement of the plant toward light. This question tests your ability to design an experiment.

6. (C) is correct. This is the type of question you can expect to see on the exam. Choice A is wrong because it is not a conclusion (they are a control) about the treatment. Choice B in incorrect because you see that plant A does not bend. When you compare what happens with plants A, B, and C, you will see that elongation has occurred in each case. Therefore, the chemical must be necessary for cell elongation.

7. (A) is correct. In plants, phloem is responsible for carrying sugar from a source (leaves for example) to sinks (other locations that are consuming energy). This process is called translocation. As sugar is added to the phloem, water follows and the solutes are moved by bulk flow (pressure flow) toward the sink, where sugar is unloaded and water is lost. A sugar sink is any tissue that is a net consumer or depository of sugar.

8. (B) is correct. Signal transduction pathways are often started by a hormone or other small molecule, but they can also be initiated by light. This is possible because the receptor is a protein (a phytochrome) that is activated by light, changes shape, and starts the transduction process. Not surprisingly, a number of pathways in plants are started with light as the stimulus.

9. (B) is correct. This is the transduction step, which enables the plant to amplify the signal received in reception.

10. (D) is correct. Activation of the cellular receptor protein (phytochrome) leads to the opening of the calcium ion channels and a temporary but marked increase in calcium in the cytosol. The high calcium levels activate protein kinase 2, which leads to the activation of transcription factor 2.

Free-Response Question

(a) It was originally thought that plant flowering depended on the length of the daylight, but then scientists concluded that it is actually night length that determines when a plant will flower. Plants monitor the length of the night by the molecular switching of two forms of the phytochrome pigment. Phytochromes respond to a shift in red **(r)** light to far-red (fr) light. This is a photoreversible response, and the threshold to trigger flowering is called the critical dark period. The accumulation of specific phytochrome isomers combined with the biological clock of plants allows for the specific monitoring of the length of the night. Long-day plants and short-day plants actually monitor the length of the night, not the day. Short-day plants must have a period of darkness longer than a specific critical period to flower, so you get chrysanthemums to flower by keeping them in a greenhouse where shades are pulled to block light to simulate a long night. The total amount of light must not exceed the critical period, and this will induce flowering.

(b) The response described is phototropism. Early experiments determined that only the tip of the coleoptiles responds to light by producing a chemical, now known to be auxin, whose concentration increases on the dark side of the stem. Auxin causes elongation (not cell division) of the cells on the dark side, and this will bend the seedling toward the light source.

(c) The first response is photoperiodism. Because in many plants, flowering is related to day length, this serves as an adaptation to ensure that flowering resources are not committed until growing conditions are most favorable for survival of flowers, fruits, and seeds. For example, short-day plants will not set flower buds unless they receive a critical period of darkness, and any interruption in this will keep them from flowering. These plants will, therefore, in nature, flower in the earliest days of spring, whereas long-day plants will not flower until the longer days of summer. In each case, the time of flowering will enhance reproductive success. The second response is phototropism. This response ensures that plants will maximize their exposure to light, which increases their ability to do photosynthesis.

Topic 9: Animal Form and Function

ANSWERS AND EXPLANATIONS

Level 1: Knowledge/Comprehension Questions

1. (C) is correct. The only condition listed that is necessary in all organisms that breathe is the presence of moist membranes. The movement of O_2 and CO_2 across the membranes between the environment and the respiratory surface occurs by diffusion. Respiratory surfaces are generally thin and because living animal cells must be wet in order to maintain their plasma membranes, these respiratory surfaces must be moist.

2. (D) is correct. You will want to review events at the synapse that trigger the release of neurotransmitter from the presynaptic cell. Vesicles filled with neurotransmitter migrate to the plasma membrane, fuse with it, and neurotransmitter is released by exocytosis into the synapse, where it will bind receptors on the next neuron or cell in the pathway and generate a response.

3. (D) is correct. Epinephrine is a hormone that is secreted by the adrenal glands—specifically the adrenal medulla. It functions in raising the blood glucose level, increasing metabolic activities, and constricting blood vessels; all of this prepares the animal for the fight-or-flight response that is elicited in the body during stressful times.

4. (A) is correct. Erythrocytes are red blood cells, and they transport oxygen around the body. They are the most numerous blood cells and are small and disk-shaped. In mammals, erythrocytes have no nuclei. Instead, they contain millions of molecules of hemoglobin, which is the iron-containing protein that transports oxygen. One molecule of hemoglobin can bind four oxygen molecules. Red blood cells are good examples of structure following function.

5. (D) is correct. You need to know that depolarization can occur at a minimal level and that no impulse is transmitted. When the threshold level of depolarization occurs, the nerve impulse or action potential is transmitted. This is sometimes referred to as an "all-or-none" response.

6. (A) is correct. Salivary amylase is an enzyme that hydrolyzes starch, a glucose polymer found in plants, and glycogen, a glucose polymer found in animals. After hydrolysis, smaller polysaccharides and maltose remain.

7. (B) is correct. Pepsin begins the hydrolysis of proteins by breaking peptide bonds between amino acids, thereby cleaving proteins into smaller polypeptides. The digestion of proteins continues in the small intestine.

8. (D) is correct. The role of the sinoatrial (SA) node, or pacemaker, is to control the rate and timing of the contraction of heart muscles. It generates electric impulses that spread rapidly through the walls of the atria, making them contract in unison.

9. (C) is correct. The lymphatic system collects fluid and some proteins lost during regular circulation and returns them to the blood. This system is composed of a network of lymph vessels throughout the body with lymph nodes, which are the sites at which lymph is filtered and viruses and bacteria are encountered by the immune system.

10. (D) is correct. Carbon dioxide is most commonly transported in the blood in the form of bicarbonate. Carbon dioxide reacts with water in the presence of the enzyme carbonic anhydrase to form carbonic acid, which disassociates to a hydrogen ion and a bicarbonate ion (HCO_3^-). Less commonly, carbon dioxide is transported by hemoglobin, or transported in solution in the blood.

11. (B) is correct. All of the answers listed—except phagocytes—are examples of barrier defenses. Phagocytosis is one of the body's cellular innate defenses against infectious agents; it is the process by which invading organisms are ingested and destroyed by white blood cells.

12. (B) is correct. When a lymphocyte is activated by an antigen, it is stimulated to divide and differentiate, and it forms two clones. One clone is of effector cells that combat the antigen. One clone is of memory cells that stay in circulation, recognize the antigen if it infects the body in the future, and launch an attack against it.

13. (D) is correct. The sliding-filament model of muscle contraction states that the thin and thick filaments do not shrink during muscle contraction. Instead, the filaments slide past each other so that the degree of their overlap increases; this sliding is based on the interactions of actin and myosin molecules that make up the filaments.

14. (B) is correct. Three successive stages of development follow fertilization. The first is cleavage, which is rapid cell division that produces a mass of new cells that share the cytoplasm of the original cell. The second stage is gastrulation, and the third is organogenesis.

15. (A) is correct. The reflex arc is the simplest type of nerve circuit (automatic response). A sensory neuron receives information from a receptor; it is passed to interneurons in the spinal cord and then to a motor neuron, which signals an effector cell (such as a muscle fiber) to respond to the stimulus. Only choice A has these in the correct order.

16. (C) is correct. Neurotransmitters are secreted by the synaptic vesicles and act as signaling molecules. When they bind with receptors on the postsynaptic plasma membrane, the nerve impulse is transmitted from one neuron to the next neuron or another cell. A single postsynaptic neuron can receive signals from many neurons that secrete different neurotransmitters.

17. (C) is correct. Reabsorption occurs when materials that are in the filtrate are taken back into the blood. The filtrate is found in the nephron tubules.

18. (D) is correct. You will want to review the role of each of these important cell types. Helper T cells are targeted by HIV infections, which is why AIDS patients have compromised immune systems. When helper T cells are compromised, both arms of the adaptive immune system (B cells and T cells) malfunction.

19. (C) is correct. Gastrulation is an important event in embryonic development because it establishes the three germ or tissue layers. Choice D describes the process of neurulation, which follows gastrulation in chordates.

20. (D) is correct. You should review the primary structures seen in this figure and know the function of each region. The cerebellum coordinates motor activities.

21. **(C) is correct.** This is the medulla, a part of the brainstem that controls many involuntary activities.

22. **(C) is correct.** The autonomic nervous system is also referred to as the involuntary nervous system and has two components, the sympathetic branch and the parasympathetic branch. Nerves of the sympathetic nervous system activate organ systems to deal with stress, while the parasympathetic nervous system maintains the steady, relaxed situation.

23. **(C) is correct.** This figure depicts clonal selection, a central idea in immunology. The active plasma cell is the plasma cell producing antibodies. Use this figure to review clonal selection. This would be a good time to label antigen, plasma cells, memory cells, and antibody on the figure.

Level 2: Application/Analysis/Synthesis Questions

1. **(B) is correct.** Maintenance of glucose homeostasis by insulin and glucagon is the classic system for explaining negative feedback. The two hormones are antagonistic with insulin triggering the uptake of glucose by the cells, thus lowering blood glucose levels, and glucagon promoting the release of glucose into the blood from liver glycogen, raising blood glucose levels. Sensors monitor glucose levels while the two antagonistic hormones are used to maintain steady glucose levels in the blood. It is important to be able to explain both negative and positive feedback systems.

2. **(B) is correct.** Glucagon raises blood glucose levels, as noted in question 1.

3. **(C) is correct.** How the environment affects energy budgets and behavior and how free energy flows through organisms are the key points. In this case, if the organism is increasing in mass the inflow of energy must be greater than the outflow.

4. **(C) is correct.** B cells are part of the humoral immune response, which occurs in the blood and lymph. In the humoral response, antibodies help neutralize or eliminate toxins and pathogens in the blood and lymph. Cytotoxic T cells are part of the cell-mediated immune response, which destroys infected host cells. Always pick the *best* answer; choice B is close, but C is better.

5. **(D) is correct.** Lymphocytes (B cells and T cells) have receptors for a single antigen. Antigens are *anti*body *gen*erating and generally stimulate more than one lymphocyte, but each lymphocyte is specific for the one antigen it is genetically equipped to recognize. Also note the true statements in this question to help review basic information about antibodies, antigens, and epitopes.

6. **(C) is correct.** This question is at the center of understanding active immunity, vaccinations, and how B cells protect the body by producing antibodies and memory cells.

7. **(B) is correct.** The secondary immune response is a heightened response to the same antigen that generated a primary immune response. Figure 9.3 on page 235 has been on numerous AP exams, where it clearly shows primary and secondary immune responses.

8. **(B) is correct.** Membranes are hydrophobic, so only hydrophobic molecules can pass directly through them. Hydrophilic molecules remain on the outside, unless transported across the membrane. The signal molecule in Figure A does not enter the cell; therefore, it must be hydrophilic and not lipid soluble.

9. (B) is correct. When a receptor protein is embedded in the plasma membrane, it is positioned to bond with a hydrophilic ligand (signal molecule).

10. (D) is correct. This diagram uses a steroid hormone (testosterone and estradiol are examples) that is lipid soluble. The hormone functions in cell-to-cell communication by causing the production of transcription factors that in turn activate genes.

11. (A) is correct. Structure C is the myelin sheath that functions as insulation that surrounds the axon, increasing the rate of impulse transmission.

12. (D) is correct. Altruism is thought to occur in populations because helping other close relatives increases the chances that they will survive to pass on genes that are shared between them and the altruistic member.

13. (C) is correct. The insecticide blocks the enzyme acetylcholinesterase, which normally removes acetylcholine from the synapse after the signal is received. Without this removal the neurotransmitter acetylcholine remains in the synapse, causing constant muscle contraction. Knowing specific neurotransmitters is not required, but understanding how neurotransmitters work is required.

14. (B) is correct. Because the neurotransmitter acetylcholine stimulates muscle contraction in both roaches and people, it is logical to conclude that the mechanism of stimulating skeletal muscle contraction must be similar in humans and roaches.

15. (C) is correct. If depolarization shifts the membrane potential enough, the result is a change in membrane potential called an action potential. Action potentials occur fully or not at all; they represent an all-or-none response to stimuli.

16. (A) is correct. The action potential depolarizes the presynaptic membrane, which opens voltage-gated calcium channels. The key to this question is the emphasis on the word *next.*

17. (B) is correct. Immediately behind depolarization caused by Na^+ inflow is a zone of repolarization caused by K^+ outflow. In the repolarized zone, the sodium channels remain inactivated. This prevents an action potential from moving backward. The interval between the onset on an action potential and end of the refractory period is only 1–2 milliseconds.

18. (D) is correct. The graph shows a clear peak at the 7-mm mark along the *x* axis. This question is based on the idea of optimal foraging, the analyzing of feeding behavior as a compromise between feeding cost and feeding benefits. Understanding how organisms use energy, how energy use affects evolution, and how organisms interact with their environment are all themes to guide your review.

19. (B) is correct. Learned behavior occurs when behavior is modified based on specific experiences. If the salmon returned to Cold Bay, then they learned the new magnetic bearings during the experiment.

20. (A) is correct. Imprinting occurs at a specific stage in life and results in a long-lasting behavioral response. Imprinting has a sensitive or critical period when the learning must occur. The pictures of young geese following Konrad Lorenz or whooping cranes following an ultralight plane are both examples of imprinting in action as the young animals imprint during their sensitive period. In the case of salmon, it is believed they imprint on chemical cues in the water of their home stream.

21. (A) is correct. Behavior is best understood as a consequence of natural selection. Many important behaviors to survival are innate, so this statement is not supported. An organism's reproductive success depends *in part* on how the behavior is performed. Part of the behavior has a genetic basis, but the entire behavior does not have to be determined by genes.

22. (D) is correct. The traditional example of a negative feedback system is the thermostat example. In the body, one very prominent example of negative feedback is the regulation of our body temperature at about 37°C. A section of the brain is responsible for keeping track of the temperature of the blood, and if the blood is too warm, for example, it tells the sweat glands to increase production. Remember, more gets you less!

23. (D) is correct. Both HCl acid and pepsin can convert pepsinogen to its active form, pepsin. As more pepsin is converted, it can convert even more pepsinogen, all of which would digest the stomach wall. Note that this is an example of positive feedback. (More gets you more!)

24. (C) is correct. Homeostasis is the ability of many animals to regulate their internal environment. In case C, the body fails to maintain body temperature within narrow limits. All the other choices describe events that are triggered when the body is outside normal homeostatic bounds with regard to a specific factor.

Grid-In Questions

1. **The answer is 302.4 L.** One way to solve this is the number of beats per minute times 60 minutes per hour ($72 \times 60 = 4{,}320$ heartbeats per hour), then multiply that by 70 mL per beat ($4{,}320 \times 70 = 302{,}400$) and convert to liters by dividing by 1000 to give 302.4 L/hour!
2. **The answer is 52.5 mm Hg.** The peak has a partial pressure of oxygen ($510 \times 0.21 = 107.1$ mm Hg) of 107.1 mm Hg. Subtracting that from the sea level value of 159.6 mm Hg, the final answer is 52.5 mm Hg, or about one-third the amount of oxygen at sea level. Did you remember to find the *difference?* Be sure to read the questions carefully.

Free-Response Question

(a) Innate behaviors are behaviors that are not learned but genetically programmed. An example of an innate behavior would be when birds begin migration based on day length cues. This behavior enhances survival because the changes in day length correlate to seasonal changes in food supplies and place the birds in habitats with more food for survival and reproduction.

(b) An example of a learned behavior would be the species-specific song of a bird. It is learned during a critical period by imprinting. If a bird is fostered by another species, it may learn the song of the foster species. Because the song is important to attract mates, failure to learn the species-specific song limits reproductive success.

(c) Cooperative behavior is seen in colonial insects such as bees. A reproductive female produces all the eggs, nonreproductive females tend the larvae and obtain food for the group, and the males mate with the queen. The needs of the entire population are met through this cooperative behavior. As an example, the worker bees will fan their wings to cool the hive, and this cooperative behavior ensures survival of the entire group.

(d) Pheromones are one type of chemical signal. Termites will deposit a pheromone trail that other blind members of the colony will follow to locate a food supply. This behavior is adaptive because it enables the entire colony access to a resource. One worker may stumble upon the resource and be able to deposit this chemical trail that leads others to the source.

This question could have been answered with many different examples; what is important to note is that the student not only provided a description of the behavior in a particular species, but carefully explained how the behavior increased survival. Be careful to always follow directions! Many students would answer this question with great examples, but then forget to relate this behavior to natural selection.

Topic 10: Ecology

ANSWERS AND EXPLANATIONS

Level 1: Knowledge/Comprehension Questions

1. (C) is correct. The ozone layer is located in the stratosphere and surrounds Earth. It is composed of O_3, and it absorbs UV radiation, reducing the level of UV radiation reaching the organisms in the biosphere. Researchers have been observing the thinning of the ozone layer since about 1975. One reason for the destruction of the ozone layer has been attributed to the widespread use of chlorofluorocarbons.

2. (B) is correct. The carrying capacity of a population is defined as the maximum population size a particular environment can support at a particular time with no degradation of the habitat. It is fixed at certain times, but it varies over the course of time with the amount of resources that exist in an environment.

3. (C) is correct. Tundra is characterized by having permafrost (which is a permanently frozen layer of soil), very cold temperatures, and high winds. These factors prevent tall plants from growing in the tundra. Tundra generally does not receive much rainfall throughout the year, and what rain does fall cannot soak into the soil because of the permafrost.

4. (B) is correct. Tropical forests generally have thick canopies that prevent much sun from filtering through. This means that in breaks in the canopy, other plants grow quickly to compete for sunlight. Tropical forests are home to epiphytes, and rainfall is frequent.

5. **(A) is correct.** Temperate grasslands are characterized by having thick grass, seasonal drought, occasional fires, and large grazing animals. Their soil is generally rich with nutrients, making them good areas for agriculture. Most of the temperate grassland in the United States is used today for agriculture.

6. **(D) is correct.** Deserts experience very little rainfall, so they are home to many plants and animals that have adaptations for storing and saving water. Deserts are marked by drastic temperature fluctuation; they can be very hot in the day but freezing at night. Many desert plants rely on CAM photosynthesis.

7. **(D) is correct.** A bacterial colony growing where it has limitless nutrients, and other ideal conditions, will experience what is called exponential growth. In exponential growth, all members are free to reproduce at their physiological capacity.

8. **(C) is correct.** All the others are +/− interactions. This notation indicates one species benefits (+), whereas the other species is harmed (−). Choice C does not fit this notation because both the honeybee and the flower benefit, an example of mutualism (+/+).

9. **(D) is correct.** Figure 10.5 shows the effect of the removal of a keystone species from a tide pool. In this case the removal of *Pisaster* allowed for the unrestrained growth of mussels, which eventually took over the rock faces of the tide pool and eliminated most other invertebrates.

10. **(D) is correct.** An early fall frost is a *density-independent factor*—it occurs without regard to the density of the population. *Density-dependent factors* increasingly slow population growth as density increases.

11. **(A) is correct.** An increase in the carrying capacity does not directly relate to an increase in human population size; it could allow better nutrition and health for humans. All of the other answers are directly contributing to increased human population growth. Death rates have fallen, but birth rates are much slower to change.

12. **(B) is correct.** Type I survivorship curve is the pattern described in the question. Type II curves show an equal chance of death throughout the life span, whereas Type III shows heavy mortality in early stages of the life cycle.

13. **(C) is correct.** The greenhouse effect is the process by which carbon dioxide and water vapor in the atmosphere intercept reflected infrared radiation from the sun and re-reflect it to Earth. Global warming is the process by which the amount of carbon dioxide in the atmosphere is increasing because of humans' combustion of fossil fuels, leading to higher temperatures on Earth.

14. **(C) is correct.** A species' ecological niche is defined as the sum of its use of the abiotic and biotic factors in an environment. For instance, a particular bird's niche refers to many things, including the food it consumes, where it builds its nest, the time of day it is active, and what climate it lives in.

15. **(C) is correct.** The dominant species in a community has the greatest biomass, or sum weight of all of the members of a population. Dominant species are also hypothesized to be the most competitive in exploiting the resources in its ecosystem.

16. **(B) is correct.** Secondary succession refers to a situation in which a community has been cleared by a disturbance of some kind, but the soil is left intact. The area will begin to return to its original state through the process of plants invading the area and recolonizing.

Level 2: Application/Analysis/Synthesis Questions

1. **(C) is correct.** The answer gives the sequence of events leading to the environmental degradation of the ecosystem. Had the stem of this question been the basis for an essay question, you might have been asked to explain how the alteration of an abiotic factor impacted the ecosystem. Choice C would have been the backbone of your response.
2. **(C) is correct.** Nitrogen and phosphorus are often limiting factors in both aquatic and terrestrial ecosystems because they are often in short supply for producers. This is why our lawn fertilizers are rich in nitrogen and phosphorus. When producers are limited, this limits energy flow throughout the ecosystem. Note the impact of an abiotic factor on populations in the ecosystem.
3. **(C) is correct.** The carrying capacity is the number of individuals that the environment can sustain. This is shown by a decrease of growth rate, shown by leveling off on a graph. When the graph line is horizontal, the rate of growth is zero. Population growth reaches that point in this graph very close to 1940.
4. **(A) is correct.** When $K = N$, then the population estimate and the carrying capacity are equal. This causes the right side of the equation to equal zero, as indicated by the flat line in 1945.
5. **(B) is correct.** This is an example of a biotic factor impacting an ecosystem. Introduced species may reduce species diversity by increasing competition for resources and sometimes replacing native species that cannot compete as well. This is a good time for you to review the components of species diversity, and reasons introduced species are often invasive.
6. **(B) is correct.** Option B is the best of the answers suggested. In practical terms, finding a predator that only eats snakeheads is problematic at best.
7. **(C) is correct.** As with any population, if the number of individuals entering the reproductive years is much greater than the number of postreproductive individuals, the population will expand. The large increase in the number of individuals entering the childbearing years in Nigeria foretells rapid population growth.
8. **(B) is correct.** The expanding population will place increased demands on resources. Although this question and question 7 deal with human populations, the underlying concepts of population ecology apply. The demands of a population that is above carrying capacity can result in the degradation of the ecosystem.
9. **(C) is correct.** Almost all organisms use solar energy stored in organic molecules produced by green plants to power life processes. Autotrophs convert solar energy to a form useful to both autotrophs and heterotrophs. At each successive trophic level, less energy is available because so much is converted to a form not useful to the organisms. Energy cannot be recycled.
10. **(B) is correct.** The graph shows an increase in mortality as the density of kelp perch increases. Thus, an increase in density leads to a decrease in population size—more gets you less—an example of negative feedback. The key step in answering the question is found in the first part of sentence two.

Grid-In Questions

1. **The answer is 32.** Using a starting population of 20 bobcats, multiply that number by the birth rate of 0.48 to yield the increase (9.6), then get the number of deaths ($20 \times 0.21 = 4.2$). The overall gain is $9.6 - 4.2$ or $+ 5.4$ bobcats. Add this to the beginning population of 20 to get the estimate for year one, 25.4 bobcats. Follow the same procedure for the second year using 25.4 bobcats as the starting population. The question specifies to the nearest whole number, so round to the nearest tenth with all numbers until the final answer; only then should you round to the nearest whole number. Rounding every number may put your answer outside the acceptable range.
2. **The answer is 461.** $N = (142)(133)/41 = 460.6$, rounded to 461.

Free-Response Questions

(a) The hawk, mouse, and plant in this particular ecosystem are related by the passage of energy. Together they comprise a food chain—the mouse is a primary consumer, and it consumes the plant, which is a primary producer (the plant is an autotroph—capable of trapping the energy of the sun and converting it into chemical energy in the form of carbohydrates). The hawk then is a predator of the mouse—and a secondary consumer. Secondary consumers eat herbivores. Energy is lost at each trophic level.

(b) One example in which the biotic factors of the biosphere impact the abiotic factors is seen in the case of global warming. We rely on the greenhouse effect (in which atmospheric carbon dioxide acts as an insulator, trapping infrared radiation from the sun and re-reflecting it) to help maintain the hospitable temperature of Earth. Yet, due to the burning of fossil fuels—beginning during the Industrial Revolution—the concentration of carbon dioxide in the atmosphere has increased significantly, and this has led to an increase in global temperatures.

The thinning of the ozone layer is another way in which humans (a biotic factor of the biosphere) impact abiotic processes. Organisms are protected from ultraviolet radiation from the sun by a protective layer of ozone that surrounds Earth. Decreased ozone levels are expected to increase human skin cancer rates as well as increase DNA damage in a wide variety of organisms. The ozone layer has been degraded by humans' use of chlorofluorocarbons (CFCs), which are chemicals used in refrigeration and other industrial processes. Many countries have stopped using these chemicals, but chlorine molecules already in the atmosphere continue to have an effect on ozone.

This response clearly marks part (a) and (b) of the response, which is a good technique to use to be sure the Reader follows your answer. In part (b), the student limited himself to only two examples as directed and gave a clear explanation of each.

BIG IDEA 1: Investigation 1, Artificial Selection

1. **The correct answer is 8.1.** The *mean is determined by adding the height of all the plants (80.7) and dividing by the number of plants (10) = 8.07, then rounding to the nearest tenth.*
2. **The correct answer is 8.8.** The mode is the most common value in the data set.
3. **The correct answer is 5.8(%).** This is determined by taking the change in height from Generation 1 to Generation 2 and dividing it by the mean height for Generation 1. (0.47/8.47) This gives 0.0582; convert to a percentage by multiplying by 100, and note that your answer is to be given to the nearest tenth. If you used the rounded figures (8.1 and 8.5) your answer would be 4.9%. Any answer between these two figures (4.9% and 5.8%) would be considered correct on the exam.
4. **The correct answer is (C).** This is the only choice that selects for height.

BIG IDEA 1: Investigation 2, Mathematical Modeling: Hardy-Weinberg

1. (C) is correct. The question tells you that $p = 0.9$ and $q = 0.1$. From this, you can calculate the heterozygotes: $2pq = 2\ (0.9)\ (0.1) = 0.18$. If you selected E as your response, you may have confused the allele frequency with genotypic frequency. This problem gives you the allele frequency of *a*, which is 10%.

2. (B) is correct. The conditions described all contribute to genetic equilibrium, where it would be expected for initial gene frequencies to remain constant generation after generation. If you chose E, remember that genetic equilibrium does not mean that the frequency of A = the frequency of a.

3. (D) is correct.

$$q^2 = 0.09, \text{ so } q = 0.3$$
$$p = 1 - q, \text{ so } p = 1 - 0.3 = 0.7$$
$$AA = q^2 = 0.49$$

BIG IDEA 1: Investigation 3, Comparing DNA Sequences to Understand Evolutionary Relationships with BLAST

1. (C) is correct. This is a good time to recall convergent evolution, the independent evolution of similar features (similar problem – similar solution), as seen in comparing a shark to a dolphin. They look similar but are distantly related. Organisms with similar gene sequences are demonstrating a recent shared ancestry.

2. (D) is correct. The nodes on a phylogenetic tree can rotate like a mobile and not change the relationship between species. Choices a, b, and c all do this, but not choice d. In choice d the taxons C and D have changed positions, making their relationships to each other and taxa E and F different.

3. (B) is correct. This answer provides maximum parsimony, the simplest explanation that is consistent with the facts.

4. (B) is correct. Understanding this answer involves the idea of a molecular clock: the number of nucleotide substitutions in related genes is proportional to the time that has elapsed since the genes branched from their common ancestor. The greater the number of changes, the greater the time that has passed since the two species shared a common ancestor.

BIG IDEA 2: Investigation 4, Diffusion and Osmosis

1. (B) is correct. The water potential of distilled water in a container open to the environment is 0. The water potential of the beet core is −0.2. (Since water potential = solute potential (-0.4) + pressure potential (0.2), water potential of the beet core = −0.2.)

2. (A) is correct. The water potential for both the distilled water and the beet core in beaker a is 0. In choice C the beet core in (b) has a pressure of 0.2 whereas the beet core in (a) is 0.4, so the beet in (b) would be less turgid.

3. The solute potential is –8.55 bars. $\psi_S = -iCRT$. $\psi_S = -(1)(0.35M)$ (0.0831 liter bar/mole K)(273 + 21) . Check your gridded response—did you remember to use the minus sign?

-	8	.	5	5	
		/	/	/	
●	.	●	.	.	.
		0	0	0	0
	1	1	1	1	1
	2	2	2	2	2
	3	3	3	3	3
	4	4	4	4	4
	5	5	● 5	● 5	5
	6	6	6	6	6
	7	7	7	7	7
	● 8	8	8	8	8
	9	9	9	9	9

BIG IDEA 2: Investigation 5, Photosynthesis

1. **(C) is correct.** DPIP is reduced in this experiment because it receives high-energy electrons from chlorophyll. With the disruption of the thylakoid membranes (remember the blender) $NADP^+$ is no longer positioned to capture electrons. DPIP replaces $NADP^+$ as the electron acceptor, turning from blue to colorless as it does so.
2. **(B) is correct.** When DPIP becomes colorless very rapidly either there are too many chloroplasts or the concentration of DPIP is so dilute that the quantity can be reduced rapidly.
3. **(C) is correct.** A flat line indicates that no DPIP is being reduced. This must mean the chloroplasts are not functioning. In the DPIP lab boiled chloroplasts are used as a comparison to normal chloroplasts. The boiled chloroplasts give a flat line when graphed.
4. **(B) is correct.** You should recall the equation for photosynthesis. Without a source of CO_2, photosynthesis cannot proceed.

BIG IDEA 2: Investigation 6, Cellular Respiration

1. **(B) is correct.** To calculate this, it is easiest to find the change in *y* at 10 minutes (0.4 mL − 0 mL = 0.4 mL) and divide by the change in *x* (10 minutes − 0 minutes = 10 minutes). 0.4 mL/10 minutes = 0.04 mL/min. (As the data form a straight line, you get the same answer if you use 20 minutes.)
2. **(A) is correct.** Study Figure 6.2 carefully to see that at 10 minutes the 22°C germinating corn consumed 0.8 mL of oxygen, whereas the 12°C germinating corn consumed 0.04 mL of oxygen.
3. **(D) is correct.** This is the only statement that is supported by information provided on the graph in Figure 6.2.
4. **(D) is correct.** As carbon dioxide is released, it is removed from the air in the vial. Because oxygen is being consumed during cellular respiration, the total gas volume in the vial decreases. This causes pressure to decrease inside the vial, and water begins to enter the pipette.

BIG IDEA 3: Investigation 7, Cell Division: Mitosis and Meiosis

<u>Null Hypothesis:</u> There is no significant difference between the observed number of cells in interphase and mitosis in the control group compared to the experimental group.

	Observed (*o*)	**Expected (*e*)**	$(o\text{-}e)^2/e$
Interphase cells	186	220	5.25
Mitosis cells	64	30	5.25
Total	250	250	10.50
			$X^2 = \mathbf{10.50}$

Degrees of Freedom = 1   ***p* value** = 0.05   **critical value** = 3.84

Accept or reject the null hypothesis? Reject the null hypothesis.

Explanation: The Chi-square value of 10.50 is greater than the critical value of 3.84. This means there was a significant difference in the number of cells in mitosis between the control group and the experimental group.

1. **(D) is correct.** There are two nuclear divisions in meiosis, and only one in mitosis. Crossing over occurs only in meiosis; it does not occur at all in mitosis. Replication occurs only once in preparation for both mitosis and meiosis. The daughter cells of mitosis are identical to the parent cell, but in meiosis the daughter cells have only one of each homologous chromosome pair. Synapsis occurs only in prophase I of meiosis.
2. **(D) is correct.** Remember that if crossing over does not occur, the arrangement of spores will be 4 light and 4 dark. All other combinations are the result of crossing over.
3. **(B) is correct.** Map distance = number of crossovers divided by total number of asci × 100 divided by 2.

BIG IDEA 3: Investigation 8, Biotechnology: Bacterial Transformation

1. **(B) is correct.** Because not every cell takes up the plasmid with resistance to kanamycin, the result will be scattered colonies of bacteria.
2. **(D) is correct.** Plate IV has kanamycin but the bacteria cells were not exposed to the kanamycin plasmid, and so there was no transformation. The antibiotic killed all the *E. coli* cells.

3. **(D) is correct.** Plate II has colonies of transformed cells, so these should be used to verify transformation. If these cells are spread on agar with kanamycin, all cells will be able to grow and a lawn will result.

4. **(C) is correct.** Because not every cell takes up the plasmid with resistance to kanamycin, the result will be scattered colonies of bacteria when the cells are spread on LB agar with kanamycin.

BIG IDEA 3: Investigation 9, Biotechnology: Restriction Enzyme Analysis of DNA

1. **(D) is correct.** There are approximately 3,500 base pairs in the circled fragment. To determine this, measure the migration distance of the circled fragment. It is 22 mm. Locate 22 mm on the graph, and take a straightedge up to the point of intersection with the curve. Read the number of bp from the y axis (left side of graph).

2. **(B) is correct.** Small fragments migrate farther than large fragments. All DNA is negatively charged and so moves toward the positive electrode.

3. **(C) is correct.** Because *Bam*HI has three restriction sites on the plasmid, there will be three fragments measuring 6 kb, 12 kb, and 2 kb. Does lane III have the expected banding pattern?

4. **(C) is correct.** There are four restriction sites for both enzymes, so there will be four fragments. Does the banding pattern match the fragment sizes?

BIG IDEA 4: Investigation 10, Energy Dynamics

1. **(C) is correct.** This is the point where the line crosses the x axis and where respiration therefore exceeds photosynthesis. Imagine a dark, rainy day where algae in a pond are not able to carry out photosynthesis but are still performing cellular respiration. On these days cellular respiration may exceed photosynthesis.

2. **(A) is correct.** Net primary productivity is determined by subtracting the energy lost by cellular respiration from gross primary productivity, or NPP = GPP − Cellular respiration.

3. **(C) is correct.** Primary productivity is directly correlated with photosynthesis, so having more sunny days as well as more frost-free days will allow greater productivity. Although Ecosystem B gets more rain, light and growing days are most significant.

BIG IDEA 4: Investigation 11, Transpiration

- **1. (D) is correct.** When K^+ leave the guard cells, water follows, the guard cells become flaccid, and the stomata close. As a result, transpiration doesn't take place.
- **2. (D) is correct.** Water moves from a region of high water potential to a region of low water potential. The region of lowest water potential would be at the farthest end of the transpiration pathway.

BIG IDEA 4: Investigation 12, Fruit Fly Behavior

- **1. (C) is correct.** The amount of fertilizer is the variable factor, and the use of two similar plants in the same location receiving equal amounts of water implies the concept of a control. However, with such a small sample size (two plants) and no repetition, this is a flawed experiment.
- **2. (D) is correct.** However, this experiment has no control so any of the other choices are reasonable conclusions from the data. In order to reach a reliable single conclusion, you must design a new controlled experiment.
- **3. (B) is correct.** Much of the early movement is simply exploration, so data analysis should not begin until the pill bugs have acclimated. Choice C may seem reasonable, but it is not as good as B because it looks only at a single point in time.
- **4. (A) is correct.** It is the only statement that gives both the conditions and a measurable predicted result.

BIG IDEA 4: Investigation 13, Enzyme Activity

- **1. (C) is correct.** The rate of the reaction decreases when the number of substrate molecules has been reduced by the enzyme. There are fewer and fewer substrate molecules available, reducing the number of enzyme-substrate collisions. Above the maximum substrate concentration, the rate will not be increased by adding more substrate; the enzyme is already working as fast as it can. An enzyme can catalyze a certain number of reactions per second, and if there is not sufficient substrate present for it to work at its maximum velocity, the rate will decrease. Therefore, to keep the enzyme working at its maximum, you must add more substrate.
- **2. (B) is correct.** H_2SO_4 lowers the pH so that the globular shape of the protein is altered. The active site is distorted to the point that the enzyme no longer functions.

Statistical Analysis: Chi-Square Analysis of Data

1. **(D) is correct.** There is no evidence for any type of linkage because both males and females show the traits in approximately equal proportions, and eye color and wings appear to sort independently. If the parents were homozygous for these traits, the offspring would not show different phenotypes from both parents.
2. **(C) is correct.** This is obtained by using the formula x^2 the sum of $(o - e)^2/e$.

	o	*e*	*o – e*	$(o - e)^2/e$
Red eyes/normal wings	98	90	8	0.7111
Red eyes/no wings	22	30	8	2.1333
Sepia eyes/normal wings	26	30	4	0.5333
Sepia eyes/no wings	14	10	4	1.6
				$x^2 = 4.977$

3. **(A) is correct.** The expected results are based on obtaining a 9:3:3:1 ratio from two heterozygous parents. There are three degrees of freedom. Because the Chi-square value is below 7.82, the results support the null hypothesis. Be sure you understand why A is correct and D is incorrect!

Sample Test

ANSWERS AND EXPLANATIONS

Part A

1. (D) is correct. The growth rate is for 10 days, not the total number of days on the x axis. Read the questions carefully! The density at 10 days is 900 paramecia for a rate of 90 paramecia/mL/day. The carrying capacity of a population is defined as the maximum population size that a certain environment can support without being degraded. You should note that the population increases in number until it reaches about 900 members and at that point it stabilizes. This is carrying capacity.

2. (A) is correct. When $K = N$, then the population estimate and the carrying capacity are equal. This causes the right side of the equation to equal zero as indicated by the flat line at carrying capacity. The equation calculates change over time. When the population is at carrying capacity the change in population is zero.

3. (A) is correct. This is a simple probability question. In order to calculate the chance that two or more independent events will occur together in a specific combination, you can use the multiplication rule. Take the probability that gene P will segregate into a gamete (1/4), and multiply it by the probability that gene Q will segregate into a gamete (1/4). Then multiply that by the probability that gene R will segregate into a gamete to get $1/4 \times 1/4 \times 1/4 = 1/64$.

4. (A) is correct. Recall from Investigation 4 that water moves from a region where water potential is high to a region where water potential is low. Keep in mind that the numbers are negative, so water will move from -0.15 MPa to -0.23MPa, or from the root tissue to the sucrose solution.

5. (D) is correct. This cross is an example of incomplete dominance because each genotype has its own phenotype. If you consider that the homozygous long-haired deer is *HH*, and the homozygous short-haired deer is *hh*, then crossing them would give all offspring with the genotype *Hh* and medium-length hair. Crossing the heterozygotes would give you offspring in the ratio of 1HH:2Hh:1hh. This means that 25% of the offspring would have long hair (*HH*), 25% of them would have short hair (*hh*), and 50% of them would have medium-length hair (*Hh*).

6. (D) is correct. Choice A indicates incorrectly that chloroplasts function in cellular respiration, whereas choice B indicates incorrectly that all eukaryotes have a cell wall and choice C incorrectly states that all prokaryotes have a double membrane. Only choice D, which states that mitochondria, chloroplasts, and prokaryotes have similar DNA and chromosomes, is correct and supports the idea of symbiosis.

7. (C) is correct. Each of the four surrounding water molecules will form hydrogen bonds with three other water molecules, yielding 16 water molecules plus the one in the middle of the complex for a total of 17.

8. (D) is correct. In any aqueous solution at 25° C the product of H^+ and OH^- concentrations is constant at 10^{-14}. pH is defined as the −log of H^+ concentration. Note that the exponents are negative. The pH with the highest number will have the highest OH− concentration and the pH with the lowest number will have the highest H+ concentration. A pH of 2 has a H^+ concentration of 10^{-2} and a OH^- concentration of 10^{-12}; a pH of 12 has a H^+ concentration of 10^{-12} and a OH− concentration of 10^{-2}. Watch the negative exponents.

9. (D) is correct. Think of this problem as two monohybrid crosses. With tail length, all offspring show the dominant long tail. Choices A, B, and D satisfy this requirement. The coat color gene shows a 3 yellow to 1 white ratio. Choices C and D satisfy this requirement. Only choice D satisfies both requirements.

10. (C) is correct. Because sarin inhibits the action of acetylcholinesterase (AChE), the enzyme that removes the signaling molecule (acetylcholine) from the synapse, the muscle fiber does not relax. This continuous signal results in a continued contraction and, therefore, paralysis.

11. (C) is correct. Both the leading and the lagging strand require the primer before DNA polymerase can begin replication. DNA polymerase cannot start replication without a nucleotide—even a ribonucleotide—already in place. The RNA primer is later removed and replaced with DNA.

12. (D) is correct. Allolactose is an inducer molecule. Specifically, it regulates gene expression in the following manner: When it is present in the cell, it binds to the repressor protein on the operator site of the *lac* operon. The operator molecule prevents expression of the operon genes. The binding of allolactose causes the repressor protein to change shape and fall from the operator site. The operon can now be transcribed, allowing mRNA to be formed. The lac operon is an example of an inducible operon.

13. (C) is correct. Because the frequency of the recessive homozygote is 36%, we know that $q^2 = 0.36$, so $q = 0.6$ and $p = 0.4$ (because $p + q = 1.0$). The heterozygote condition, represented by $2pq$ in the Hardy-Weinberg equation, would be 2 (0.4)(0.6) in our problem, or 0.48. If the answer is converted to a percent by multiplying by 100, the result is a final answer of 48%. If this answer is puzzling, review the *Evolution of Populations* chapter in Topic 6, paying particular attention to the information that is boxed.

14. (C) is correct. This is an autosomal recessive trait. In the first generation neither parent shows the trait, but they have children who do. Also note that if the trait were sex linked the female with the trait in the second generation would have a father with the trait.

15. (A) is correct. This is an autosomal recessive trait, which means the trait is not on the X chromosome, eliminating choices C and D. Choice B indicates a dominant autosomal, not a recessive autosomal, as in the given answer, A.

16. (B) is correct. Facilitated diffusion is the transport of certain substances across a membrane with their concentration gradient, aided by transport proteins that span the membrane. These proteins are specialized for the solute that they transport. In the case of chemiosmosis in cellular respiration, an H^+ gradient is established across the inner mitochondrial membrane by the electron transport chain. A similar mechanism established a hydrogen ion gradient in photosynthesis. In both cases, hydrogen ions move through channels in the ATP synthase complex. This is an excellent time to review types of passive and active transport as well as the production of ATP by chemiosmosis. Both areas are fundamental to biology and often show up in multiple choice and essay questions.

17. (A) is correct. The pattern of selection is an indication of the natural selection pressures on the population. In this population, goldenrod galls that are too small are unsuccessful or parasitized, whereas those that are too large are eaten by birds. The most successful galls are intermediate in size, forming a pattern of stabilizing selection.

18. (C) is correct. The graphs show patterns of directional, diversifying, and stabilizing evolution. With both extremes selected against, the pattern for goldenrod galls is stabilizing.

19. (D) is correct. C_4 and CAM plants both grow better than do C_3 plants under conditions of increased median air temperature and decreased relative humidity. Both C_4 and CAM plants use an alternative method of carbon fixation that enables them to fix carbon into an acid intermediate for later deposit into the Calvin cycle. In cooler conditions, the advantage goes to the C_3 plant.

20. (D) is correct. The guard cells are hypertonic due to the accumulation of K^+, a process of active transport (note from the figure that the K^+ is moving against the gradient) that requires ATP. Stomata open when guard cells actively accumulate K^+ from neighboring epidermal cells, causing the guard cells to become hypertonic. This causes water to enter the guard cells, as shown in the figure.

21. (D) is correct. Hypotonic means having less solute and, therefore, more water than the comparison tissue. Because water is moving out of the cells in this figure, they are hypotonic to the surrounding cells. Their turgor pressure is decreasing, and they are becoming flaccid. Although the AP Biology framework does not have many specific details about plants, this question is an example of taking knowledge from an area that is required, osmosis, and applying it in one of many possible illustrative examples. Be prepared for this and do not quit on a question that seems to be about an area you might not have studied.

22. (B) is correct. In a recombination event W and E will be most likely to stay together because they are only 3 map units apart, closer than any other two genes. The closer the genes, the more likely they will stay together in a crossover event; the further apart genes are, the more likely they will be separated by a crossover event. This idea is the basis for making gene maps based on crossover percentages.

23. (B) is correct. Digestive enzymes, like all enzymes, are proteins that are produced on rough ER and exported from the cell. Detoxifying poisons in humans is a function of smooth ER where the detoxifying enzymes are embedded in the smooth ER membranes in the liver.

24. (B) is correct. Only species 2 is an omnivore, obtaining energy from both the producer (species 1) and primary consumer (species 3).

25. (A) is correct. The statement in choice A is a fundamental tenet of ecology: energy flows one way through ecosystems, but nutrients are recycled.

26. (D) is correct. Because DDT is not broken down by normal biological means, it accumulates in ever higher concentrations as the poison moves through the food chain. This is referred to as biological magnification.

27. **(D) is correct.** The DNA fragments migrated along the gel at rates according to their size—the smaller DNA fragments migrated more quickly through the dense gel and can be found near the bottom of the gel, whereas the larger fragments migrated more slowly and can be found closer to the top.

28. **(D) is correct.** Sample 2 must have been cut at more restriction sites than was sample 4 because more DNA fragments of different sizes were produced. This is shown by the greater number of bands on the gel in the lane of sample 2. The length of the original DNA molecule cannot be determined by the gel because multiple fragments of the same length (for example, three different fragments each 1,000 bp) will only form a single band.

29. **(C) is correct.** Samples 1, 3, and 4 have the same number of restriction sites because they form the same number of bands. Because the bands are in different locations the restriction sites must also be in different locations, forming bands of different lengths.

30. **(B) is correct.** The solution on side B is hypertonic to the solution on side A. It is more concentrated (total molarity of 1.2) than the solution on side A (total molarity of 0.9) at the time this experiment began.

31. **(B) is correct.** Because side B is hypertonic to side A, water will cross the membrane, increasing the level of fluid on side B.

32. **(B) is correct.** Each solute moves down its own concentration gradient. NaCl is more concentrated on side A than side B and the membrane is permeable to Na^+ and Cl^-. This means we can expect Na^+ and Cl^- ions to move down their concentration gradients from side A to side B.

33. **(C) is correct.** This question requires you to know that steroid hormones are lipids and so can easily diffuse across the plasma membrane phospholipid bilayer. They are bound by intracellular receptors, not by cell-surface receptors.

34. **(A) is correct.** Because steroid hormones are hydrophobic and can cross the plasma membrane, they bond with intracellular receptors forming transcription factors that help to stimulate the transcription of the gene into mRNA. Choices B and C both involve plasma membrane receptors and choice D does not include the binding of the hormone to an intracellular receptor.

35. **(B) is correct.** The cell consumes O_2 as it receives electrons from the electron transport chain and forms water during the process of oxidative phosphorylation. The cell makes ATP through oxidative phosphorylation, so measuring the rate of consumption of O_2 by the cell is a good way to determine its metabolic rate.

36. **(C) is correct.** This question requires you to know the essential parts of a reflex arc: receptors, sensory neuron, interneuron, motor neuron, and effector. Structure 6 is a motor neuron, and it stimulates the muscle to contract.

37. **(C) is correct.** With other parameters such as light and temperature in the ecosystem staying more or less constant, an increase in precipitation will increase photosynthesis and, therefore, net primary productivity.

38. **(C) is correct.** The movement of oxygen from the clusters of alveoli at the tips of the bronchioles in the lung, across the epithelial walls, and into the bloodstream is an example of passive diffusion. All of the choices except C involve transport mechanisms more complicated than simple diffusion.

Transport is a favorite topic on the AP exam. Be sure to review transport before the exam.

39. (A) is correct. Choice A is an outline of the organelles involved in producing proteins for export from the cell. Integrating cell organelles into one function, like protein export, is a good area for essay questions, so be sure to review this topic until you can write a complete essay about protein export from cells, starting with DNA in the nucleus.

40. (D) is correct. The morpho compound will turn on genes *Q* and *R* in region 2 (gene *P* requires a high concentration, which is found only in region 1). Phogen will turn on *S* and turn off *Q* at medium to high concentrations. This leaves genes *R* and *S* on in region 2.

41. (D) is correct. Eukaryotes are able to coordinately control genes in the same pathway because the genes all have the same combination of control elements. Copies of activators that recognize the control elements bind to them, promoting simultaneous transcription.

42. (C) is correct. The change in petal color causes a change in pollinator, which can serve as an evolutionary isolating mechanism, setting the stage for the evolution of new species.

43. (B) is correct. The coevolution of insects and flowers offers many strategies for successful pollination. With the change of petal colors also comes the potential for reproductive isolation as new animals may be attracted to the new flower.

44. (C) is correct. Choice A centers around a negative feedback system that would turn down the response, not up. Choice B is not correct because antibodies are not directly tied to histamine releases and choice D is incorrect because memory B cells produce antibodies on activation, not histamines.

45. (C) is correct. The removal of introns requires spliceosomes. Spliceosomes are complexes of snRNA and proteins that interact with certain sites along an intron, excising the intron and joining together the two exons that flanked the intron.

46. (C) is correct. Although the figure may seem complicated, ultimately the question is asking what complex converts single-stranded DNA to double-stranded DNA. In the list of possible answers, only the enzyme DNA polymerase can copy DNA to DNA.

47. (B) is correct. The carrier parents would each have the genotype *Aa*. This means their children would have a 25% chance of inheriting both of the recessive alleles. The fact that their first three children do not have cystic fibrosis does not affect the probability of the fourth, unrelated event. Chance has no memory.

48. (D) is correct. When species B is removed species A cannot colonize the area, indicating an abiotic factor could be preventing the colonization. When species A is removed species B colonizes the area, indicating it cannot compete with species B or it would be growing and competing in the same plot as B.

49. (B) is correct. An ice bath or other procedure to lower body temperature helps to prevent enzymes and other proteins from denaturing. Although many proteins in the body could be replaced without harm, damage done to brain enzymes and proteins may be permanent.

50. (C) is correct. Rotenone disrupts the electron transport chain, reducing its efficiency and preventing an effective H^+ gradient from forming. Without the H^+ gradient ATP synthase cannot convert enough ADP to ATP to keep the organism alive.

51. (B) is correct. DNP disrupts the integrity of the membrane, allowing H^+ to leak back across into the matrix. Without the H^+ gradient, ATP synthase cannot produce adequate quantities of ATP.

52. (A) is correct. Temperature above the optimal causes thermal agitation of the enzyme, resulting in disruption of hydrogen bonds, ionic bonds, and other weak interactions that normally stabilize the structure of the enzyme. With these disruptions the protein enzyme molecule denatures.

53. (C) is correct. The pH scale is logarithmic, so the difference between scale numbers is a power of 10. Enzyme 4 has an optimal pH of 2 and enzyme 5 has an optimal pH of 8. The difference is 10^6 or 1,000,000 times more acidic for enzyme 4!

54. (D) is correct. The reason the action spectrum for photosynthesis doesn't match the absorption spectrum for chlorophyll *a* is because chlorophyll *a* is not the only photosynthetically important pigment in the chloroplast. Two other photosynthetically important pigments in plants are chlorophyll *b* and the carotenoids. Choice B is a correct statement but does not explain the discrepancy between that absorption spectrum of chlorophyll and the action spectrum of photosynthesis.

55. The answer is (C). The high level of nucleotide matching in Exon 1 indicates that few mutations have accumulated in this exon over time. This usually means that any change from the original sequence has a negative selection impact on the individual, removing the nucleotide mutation from the gene pool. This would occur when the exon plays a specific, critical role in the functioning of the polypeptide.

56. The answer is (A). Mutations often accumulate at a higher level in introns because the majority of the nucleotides in introns do not play critical roles in the expression of the gene. If random changes in the nucleotide sequence in introns have no effect on natural selection, the changes in nucleotides accumulate over time.

57. (D) is correct. Plant leaves are green because they reflect and refract green light, which is not utilized in photosynthesis. Red light is used in photosynthesis, meaning the plant would absorb CO_2 for photosynthesis. Choice A requires you to know that normal light has all the wavelengths of the visible spectrum and, therefore, even more available energy that red light alone. The plant in normal light would use even more CO_2! Check the action spectrum for photosynthesis in your text (or look back at question 54) and be prepared to explain the peaks and valleys shown in the graph.

58. (D) is correct. A nucleotide insertion downstream and close to the start of the coding sequence creates a frameshift mutation, causing the regrouping of codons and a completely new list of amino acids, leading to a nonfunctional protein. If the amino acid sequence is changed after the insertion, a *missense mutation* occurs. If the regrouping leads to the formation of a stop codon, a *nonsense mutation* occurs, which terminates the forming protein. As you study different types of mutations, give emphasis to the *effect* of a mutation, and be able to predict and justify which type in a particular location would cause the greatest change in nucleotide sequence.

59. (A) is correct. The cytoplasmic determinants are produced by the mother and unevenly distributed across the egg before fertilization. After fertilization the nuclei of the embryo are exposed to different sets of cytoplasmic determinants and, as a result, express different genes.

60. (C) is correct. In type 2 diabetes, the receptors for insulin no longer function well, so body cells cannot take in as much glucose, which raises blood glucose levels. A careful reading of the stem will allow you to reject choices A and B.

Part B

1. The correct answer is between 15.4 and 15.5. (Depending on when you round—the College Board scoring machine is set to accept a range of answers in situations like this.) To get the expected values, if the parents were both heterozygous, you would predict a ratio of 3 red-eyed to 1 sepia-eyed. If there were 100 offspring, that would be 75 and 25.

Observed (o)	**Expected (e)**	$(o - e)$	$(o - e)^2$	$(o - e)^2$
58 red-eyed	75	17	289	289/75 = 3.9
42 sepia-eyed	25	17	289	289/25 = 11.6
				$x^2 = 15.4$

2. The answer is 17,190. From the graph it is shown that radioactive decay to 1/8 requires three half-lives. With each half-life given at 5,730 years, a total of three half-lives would be 17,190 years.

3. The answer is 59. Use the formula: $dN/dt = r_{max}N(K - N/K)$. Substituting the values given, (1.0) (275) (350 − 275)/350 = 58.9 deer mice. Especially when working on grid-ins, keep the formula sheet in mind as a possible resource in finding the answer.

4. The answer is 17 μmol/mL/min. This is determined by 180 μmol/mL − 10 μmol/mL/ 10 minutes = 17 μmol/mL/min.

5. The answer is 49.0%. The 68% that show banding represent $P^2 + 2pq$, which is not an expression we can work with. However, it also means 32% have no banding and that equals q^2, so q equals the square root of 0.32 or 0.57; therefore, p equals 0.43. The heterozygote is represented as $2pq$, so 2(0.43) (0.57) = .4902, changing to a percent by multiplying by 100 and rounding to the nearest tenth yields 49.0%. *In this example, an answer of 49 would be counted wrong because it is not given to the precision specified in the question.* Read carefully!

6. The answer is 46.44. N = (19) (22)/9 = 46.44.

Grid-In Questions

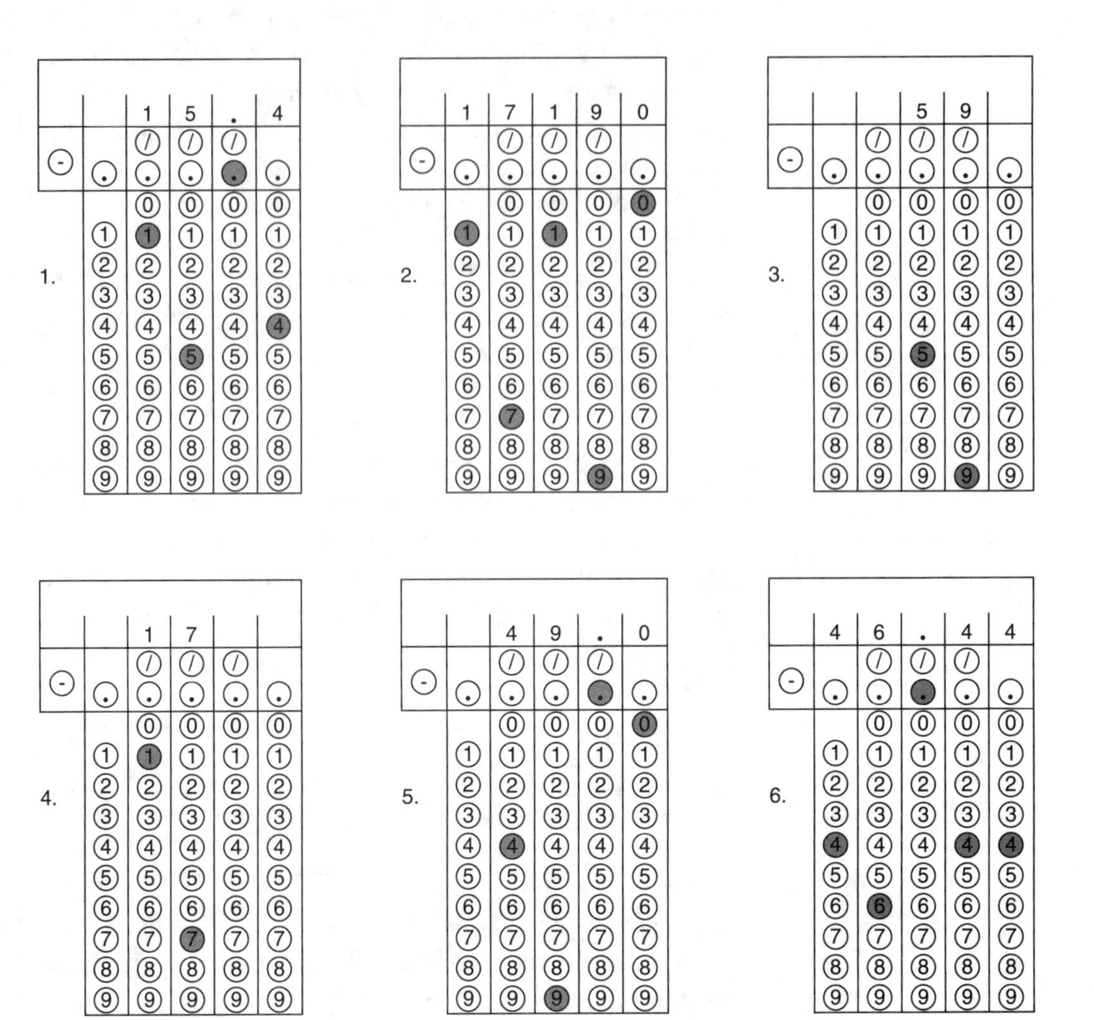

Free-Response Questions

> *Each question includes a sample answer, our evaluation of the answer with test-taking hints, and a rubric for you to use in evaluating your own response. We strongly suggest you write a practice response to each question, and grade it using the rubric.*

Question 1 Sample Response

1. (a) The cell membrane of a typical animal cell is composed of three main components that are involved in transport across the membrane: phospholipids, which are two fatty acids joined to two glycerol hydroxyl groups and a phosphate group connected to the third glycerol hydroxyl group; proteins, both integral (embedded in the cell membrane) and peripheral (associated with the outside of the membrane); and cholesterol a hydrophobic steroid embedded in the hydrophobic portion of the phospholipids. Carbohydrates are also an important macromolecule found on cell membranes, but they are not involved with cell transport.

Phospholipids form a hydrophobic foundation for the rest of the molecules in the cell membrane. Some very small nonpolar molecules, like O_2 and CO_2, can pass through the lipid membrane unaided, but hydrophilic and large molecules either cannot pass through the phospholipids, or require special channels.

Proteins are involved in providing channels for the passive transport of water and certain other solutes across the membrane. Proteins can also act as pumps that use ATP energy to actively transport substances against their concentration gradient.

Cholesterol is also found in animal (but not plant) plasma membranes where it functions as a fluidity buffer. At high temperatures, such as human body temperature (37°C), cholesterol makes the membrane less fluid. At lower temperatures cholesterol prevents packing of phospholipids, helping to maintain fluidity.

(b) Glycoproteins found on viral envelopes bind to specific receptor molecules on the surface of a host cell. This promotes entry of the capsid and viral genome into the cytoplasm where cellular enzymes digest the capsid and release the genetic material. This type of entry is a form of receptor-mediated endocytosis.

Lipid-soluble hormones such as steroids diffuse through the cell membrane to bind to a specific receptor protein. Often, the protein receptor/hormone combination acts as a transcription factor, turning on specific genes.

Water enters the cell through a process called facilitated diffusion. This means that water crosses the cell membrane down its concentration gradient, but with the help of specific transport proteins termed "aquaporins." Transport proteins are specific for the molecules they assist across the membrane, but they do not require the input of energy.

Oxygen molecules diffuse across the membrane because they are very small and hydrophobic.

The student clearly addresses all parts of this question but digresses in the first paragraph to describe phospholipids and types of proteins in the membrane. This is outside the scope of the question, and you should avoid doing this because there is not enough time to add extraneous details and answer every question well. This student would have done just as well if the entire first paragraph were omitted! Remember to always read the questions carefully, answer all parts of the question, and organize your answer so it is easy for the person grading your response to follow your answer.

Question 1 Rubric

(a) **Identify three** components of a typical animal plasma membrane, and **explain** the role each of these components plays in regulating the cell's internal environment. **(6 points maximum)**

Identify (1 pt for each macromolecule)	**Explain** the role in regulating the cells internal environment. (1 pt for each explanation.)
Phospholipids	Create a hydrophobic barrier which prevents movement of hydrophilic molecules into the cell.
Proteins	May span the membrane and provide a channel across the membrane for a hydrophilic solute.
Cholesterol	"Fluidity buffer," resists changes in membrane fluidity caused by changes in the temperature. Proper fluidity allows for working proteins and diffusion across the membrane.

(b) **Describe** how each of the following can enter an animal cell. **(4 points maximum)**

Object entering the cell	**Describe how object enters the cell. (1 point each)**
Viral RNA	An animal RNA virus enters the cell by fusing its envelope or outer membrane with the plasma membrane.
Steroid hormone	Steroid hormones are hydrophobic and can diffuse through the plasma membrane.
Water molecules	The polar water molecule diffuses into the cell through aquaporins, protein channels imbedded in the plasma membrane. (facilitated diffusion)
O_2	Diffuses through the plasma membrane as O_2 is hydrophobic.

Question 2 Sample Response

2. (a) Species richness, the number of species in the community, and relative abundance, the proportions of the community represented by the various species, both contribute to species diversity.

(b) Community 1 has a distribution of species A at 5 trees / 20 total trees = 25%; Community 2 has a distribution of species A at 16 trees / 20 total trees = 80%.

(c) Community 1 is considered more diverse because although species diversity is equal between the two communities, the relative abundance is more equitable in Community 1. Compared to a community with a very high proportion of one species, a community with a more even relative abundance, like Community 1, has greater species diversity.

(d) If the blight damaged species A, the structure of Community 2 would be most seriously impacted. This is because Community 2 has a much higher relative abundance of species A (species A makes up 80% of the individual trees in Community 2 compared to 25% in Community 1). Loss of species A from Community 2 would result in large light gaps, more erosion, and greater loss of nutrients from the community. Communities with greater species diversity are more stable. Community 1 is more diverse and more stable; thus it is better able to recover from environmental insults and repel invasive species.

Question 2 Rubric

(a) **Describe** the factors involved in species diversity. **(2 points maximum)**

- **1 point** for **species richness** or definition: number of different species in the community.
- **1 point** for **relative abundance** or definition: proportion each species represents of all individuals in the community.

(b) **Calculate** the percent of *each* community represented by species A. **(2 points maximum)**

- **1 point** for the calculation in Community 1: 5 trees of species A/ 20 total trees = **25%.**
- **1 point** for the calculation in Community 2: 16 trees of species A/ 20 total trees = **80%.**

(c) Which of the two communities in the figure is more diverse? **Explain** how two communities that contain the same number of species can differ in species diversity. **(2 points maximum)**

- **1 point** for stating that Community 1 is more diverse.
- **1 point** for explaining that Community 1 has greater **relative abundance** than Community 2 because it has a more equal distribution of individuals of each species.

(d) **Predict** and **justify** the impact on community stability in each forest community of a disease that affects species A. **(4 points maximum)**

- **1 prediction point** for stating that Community 1 will have less damage to its community structure than Community 2.
- **1 justification point** for noting the higher relative abundance of species A in Community 2 than in Community 1 (80% of species A in Community 2 compared to 25% of species A in Community 1).

- **1 justification point** for naming a specific environmental impact that would be more severe in Community 2. Examples include but are not limited to: increased light penetration, increased erosion, increased susceptibility to invasive species, and increased loss of nutrients.
- **1 justification point** for the overall conceptual statement that more diverse communities are more stable communities; thus Community 1 is more diverse and more stable.

Question 3 Sample Response

3. In the euchromatin form DNA is more accessible to proteins involved in transcription, like RNA polymerase. In chromatin packing, when the genetic material is in heterochromatin form, it is highly condensed and proteins involved in transcription do not have access to the DNA. In DNA methylation, methyl groups are attached to specific regions of DNA immediately after it is synthesized. In some cases, this is thought to be responsible for these genes' long-term inactivation. In histone acetylation, acetyl groups are attached to certain amino acids of the tails of histone proteins. When the histones are acetylated, their shape alters so that they are less tightly bound to DNA; this enables the proteins involved in transcription greater access to the gene. When histones are deacetylated, DNA transcription is greatly reduced.

This response would earn all possible points, but it would have been easier for the Reader to find them if the student wrote, "One way chromatin packaging affects . . ." and then a second paragraph clearly describing a second way. Also, the student gives three ways packaging can affect gene expression. Remember that only the first two will be graded! Don't waste time that can be used for another answer.

Question 3 Rubric

Gene expression in a cell is influenced by a variety of factors. Not all genes on the eukaryotic chromosome are expressed, and, in fact, only a small fraction of the genes are transcribed into working proteins. **Describe** and **explain** two ways chromatin packaging influences gene expression. (**4 points)**

Description: 1 point each, 2 maximum	**Explanation:** 1 point each, 2 maximum
Highly packaged DNA (heterochromatin) compared to unpackaged DNA (euchromatin)	Heterochromatin does not allow access for transcription proteins; euchromatin does
Histone acetylation	Unfolds DNA, which allows access to transcription proteins
DNA methylation	Reduces transcription of DNA, inactivating genes

Question 4 Sample Response

4. Artificial selection occurs when humans choose a trait for selective breeding. An example would be to use artificial selection to develop the largest collie dog possible. In future populations only the largest of collies would be allowed to breed out of each litter, eventually forming a larger and larger collie breed. In this example the alleles involved in large size would become increasingly common.

Genetic drift occurs when chance events cause changes in allelic frequencies and this is most common when populations are small. Bottlenecking events can cause genetic drift. For example, cheetahs have probably undergone two bottleneck events in their species history. Cheetah populations were decimated during the last ice age and again in the early 20th century when cheetahs were hunted almost to extinction. Once hunting was stopped, the few remaining cheetahs have given rise to all the cheetahs currently in Africa. Today's populations of cheetahs have very little genetic diversity due to the bottleneck events and, consequently, suffer fertility issues. Future populations can also be expected to suffer problems such as low fertility due to limited genetic diversity.

All the rubric points are included in this essay, but as a reminder, it makes it less likely a Reader will miss an important part of your answer if you divide it into sections. For this question, each example, description, and prediction could be separate paragraphs, or you could clearly mark sections "(a)" and "(b)."

Question 4 Rubric (4 points)

Artificial selection and genetic drift can each affect the genetic makeup of a population.

(a) **Describe** each, using a specific example. **(2 points maximum)**

(b) Predict the effect each may have on future populations, based on your example.

	Description Must be linked to an example	**Example** (1 point each)	**Prediction (1 point each)** Must explain in reference to the example
Genetic drift	Chance events cause changes in allelic frequencies.	Biologically reasonable example of founder effect or bottleneck effect.	Random changes in allelic frequency may result in loss alleles or different allelic frequencies in new or small population compared to the parent population.
Artificial selection	Human-directed selective breeding for desirable traits.	Biologically reasonable example of artificial selection (not natural selection).	The allele or alleles selected for will become more common in the gene pool because only organisms with the desired traits are allowed to breed.

Question 5 Sample Response

5. Template sequence from problem: 3′-TAC TTC AAA CCG ATT-5′.

(a) The mRNA sequence would be 5'-AUG AAG UUU GGC UAA-3'.

(b) The amino acid sequence would be met-lys-phe-gly-stop.

(c) The substitution would cause the second codon to become AAA, but because both AAG and AAA code for the same amino acid, lysine, there would be no change to the polypeptide chain.

(d) Deletion of the second cytosine would be a frameshift mutation, and all codons after the first one would be different, resulting in many changes to the original amino acid sequence.

The student clearly addressed each part of the question. This is a typical 4-point question with a relatively narrow focus. Although the student could have figured out each of the changed mRNAs and amino acids in part (d), for example, it was not called for in the question. Limit your answers to what is being specifically asked in order to have time to answer every question on the test!

Question 5 Rubric (4 points)

The DNA template strand of a gene contains the sequence

3′-TAC TTC AAA CCG ATT-5′.

(a) Draw the mRNA sequence, indicating 5′ and 3′ ends.

- **(1 point)** Correct mRNA sequence: 5'-AUG AAG UUU GGC UAA-3'

(b) Predict the amino acid sequence from this gene.

- **(1 point)** Correct prediction: met-lys-phe-gly-stop

(c) What would be the effect of a mutation that substitutes a thymine for the second cytosine?

- **(1 point)** The second codon would be AAA. Because it also codes for lysine, there would be no effect in the amino acid sequence.

(d) What would be the effect of a mutation that deletes the second cytosine? **(1 point maximum)**

- Only the first amino acid in the chain would be unchanged.
  and/OR
- This would be a frameshift mutation.

Question 6 Sample Response

6. (a) If the seedlings are rotated 180° away from the window, over several days they will change the direction in which they are growing and all lean back toward the window. This is because a plant responds to light by an asymmetrical distribution of auxin, which causes the cells on the darker side of the plant to elongate more than the cells on the brighter side of the plant. The auxin hormone activates enzymes in the cell wall that pump hydrogen ions (protons) into the cell wall, causing a drop in pH. The lowering pH activates other enzymes in the cell wall to break cross-links between cellulose fibrils, which weaken the cell wall. This weakening of the cell wall allows turgor pressure from the central vacuole to stretch the cell wall, making the cell larger and in the process turning the plant toward the light. Growth of a plant toward a light source is known as positive phototropism.

(b) This response enables a plant to gain maximum exposure to light, which is the energy source for photosynthesis.

The student gives a very detailed explanation of why the seedlings move toward light. Because this is only a 3-point question, this amount of detail exceeds what would be expected. You will see this reflected in the rubric.

Question 6 Rubric (3 points)

A tray of germinating oat seedlings is placed in a window that has light shining through most of the day, and it is given adequate water and soil nutrients. After 3 days, the tray is rotated 180°.

(a) Predict and explain what will occur. **(2 points maximum)**

- Prediction that the seedlings will all bend toward the light
- Explanation that a hormone (auxin) is produced in response to uneven light. This hormone causes cells to elongate on the dark side of the seedlings, causing them to bend toward light.

(b) What is the evolutionary significance of this response? **(1 point maximum)**

- Maximum light exposures results from this bending, which increases the amount of photosynthesis/maximizes use of energy.

Question 7 Sample Response

7. Three pieces of evidence that glycolysis was the first metabolic pathway for the production of ATP include:

1. Long ago, Earth's atmosphere contained almost no oxygen, and only relatively recently have the current atmospheric levels of gases come to be what they are. Glycolysis does not require oxygen, so it is possible that prokaryotes (which evolved before eukaryotes) used glycolysis for making ATP at a time when no free oxygen was present in the early atmosphere.

2. The second piece of evidence is that glycolysis is a very common method for making ATP conserved across organisms in the three domains: Bacteria, Archaea, and Eukarya. This commonality implies that it originated very early in the evolution of metabolic pathways.
3. The final reason has to do with the site of glycolysis—that is, it takes place in the cytosol, and not in an organelle. Prokaryotic cells, which evolved first, are much simpler than eukaryotic cells, and they contain no membrane-bound organelles (not even a nucleus). Therefore, if glycolysis were to take place in an early prokaryotic cell, it would have to evolve in the cytosol—for instance, it would have to evolve such that it did not rely on a specialized membrane in order to function.

This response gives the required three pieces of evidence and good reasoning to support them.

Question 7 Rubric (3 points)

It is theorized that glycolysis was the first metabolic pathway for the production of ATP. Glycolysis begins the process of making ATP by breaking glucose into two molecules of pyruvate. **Justify** this claim with three pieces of evidence that support this point. (**3 points**)

- Glycolysis does not require oxygen. Oxygen was not present in the early atmosphere.
- Glycolysis occurs in across all domains/enzymes of the pathway are widely conserved.
- Glycolysis occurs in cytosol/does not require membrane-bound organelles.

Question 8 Sample Response

8. (a) A substitution mutation can result in a different amino acid being incorporated into the protein. Because of R-group interactions, the protein may fold differently and therefore function differently.

(b) Malaria kills many people in sub-Saharan Africa, but individuals who are heterozygous for sickle-cell disease are less likely to die of it than people without the mutant form of hemoglobin. Because these people are able to live and therefore reproduce, the sickle-cell allele will continue to be passed on.

(c) The frequency of the allele would be lower in a population in Alaska because malaria is not found there. There would be no selective advantage to having the mutation, and because it has a negative effect in individuals who receive two copies of the allele, the allele would become less common over time. The value of the sickle-cell allele will decline over time, dropping from 0.07 to 0.06, to 0.05 and down toward 0.01.

The student addresses each part of the question. In part (c) the student would not have received credit without the last sentence. If you reread the question, it asks for a frequency, so a wise student will give a number, not simply say "lower."

Question 8 Rubric (3 points)

Sickle-cell disease is caused by a single base-pair substitution in the amino acid sequence for the β-globin gene.

(a) **Predict** how this change might alter the overall structure of the protein. (**1 point maximum**)

- changes interactions between amino acids
- changes the secondary structure or tertiary due to alteration in folding
- changes the quaternary structure of hemoglobin

(b) Explain why this mutation is common in sub-Saharan Africa. (**1 point maximum**)

- Individuals who are heterozygous for the sickle-cell allele are less likely to die of malaria (a common disease in sub-Saharan Africa) and so pass on the mutant allele.

(c) If the frequency of the allele is 0.07 in sub-Saharan Africa, predict a biologically plausible frequency for the allele in a population of Native Americans in Alaska. (**1 point maximum**)

- Any value less than 0.07. (Because the allele would not offer an advantage to individuals in a region free of malaria, the frequency of the allele would decrease.)

Index

B

D

E

F

G

H

I

K

L

M

N

O

P

T

U

V

W

X

Y

Z